SPECTROMETRIC IDENTIFICATION OF ORGANIC COMPOUNDS

SPECTROMETRIC IDENTIFICATION OF ORGANIC COMPOUNDS

THIRD EDITION

Robert M. Silverstein
SUNY College of Environmental Science and Forestry

G. Clayton Bassler
Hills Brothers Coffee, Inc.

Terence C. Morrill
Rochester Institute of Technology

John Wiley & Sons, Inc.
New York   London   Sydney   Toronto

Copyright © 1963, 1967, and 1974, by John Wiley & Sons, Inc.

All rights reserved. Published simultaneously in Canada.

No part of this book may be reproduced by any means, nor
transmitted, nor translated into a machine language without the written permission of the publisher.

Library of Congress Cataloging in Publication Data:

Silverstein, Robert Milton, 1916–
    Spectrometric Identification of Organic Compounds.

    Includes bibliographies.
    1. Spectrum analysis.   2. Chemistry, Organic.
I. Bassler, G. Clayton, joint author.   II. Morrill,
Terence C., joint author.   III. Title.
QD272.S6S55   1974   547'.34'6   74-3008
ISBN A 471-79178-4

Printed in the United States of America

10 9 8

PREFACE

The third edition was prepared to keep pace with the rapid changes within the field of spectrometry. This book assumes no prior knowledge of spectrometry and is primarily an introduction to spectral interpretation for organic structure determination. However, considerable reference material makes the book a useful first resource for the practicing organic chemist.

The major changes in Chapter 4, "Nuclear Magnetic Resonance Spectrometry," have clarified some of the basic concepts. Chemical shift equivalence and magnetic equivalence on the basis of symmetry have been explained more thoroughly. Many of the newer developments, such as shift reagents, are succinctly described.

Modest changes—inevitably resulting in some expansion—make the other chapters more readable for the student. We have presented an interpreted IR spectrum for each class of compounds as a useful pedagogical device that relieves the tedium of a stream of data.

As in the second edition, Chapter 6 consists of sets of spectra with interpretation, and Chapter 7 consists of sets of spectra without interpretation but with a Beilstein reference. The old Chapter 8, dealing with sets of spectra without interpretation or reference, has been omitted for two reasons. First, we saved space to compensate for the expanded treatment in the descriptive chapters. Second, any such set of Problems available to students has a very short useful life.

In place of this Problem set, an Instructor's Supplement will be made available only to teachers.

This will contain interpretations of the sets of spectra in Chapter 7 and a new set of problems with interpretations, which teachers can use for assignments and exams.

In the first edition we referred to "methods now available and used in research but *not systematically taught* in universities. . . ." The concept behind the first edition was that organic structures could be elucidated from the complementary information available from mass, infrared, NMR, and UV spectrometry. Since that time, several excellent books have appeared, organized similarly to this text, and brief treatments of the four areas of spectrometry are a part of almost every elementary organic textbook. In the first edition we stated: "In one form or another, such material [i.e., spectrometric interpretation] should soon become part of the training of every organic chemist." We repeated this sentence in the second edition. We are now constrained to change "should soon" to "has." One of us (T.C.M.) has a unique perspective. He can look back over the past decade to his first contact with spectrometry in a graduate course at San Jose State College in 1962. The instructors were R. M. Silverstein and G. C. Bassler, whose lecture notes were the basis for the first edition of this text.

We are grateful to the various sources that provided us with illustrations. We are especially indebted to Sadtler Research Laboratories and Aldrich Chemical Company, that have provided us with many very useful NMR and infrared spectra. The helpful comments and contributions of Dr.

Robert T. LaLonde (SUNY College of Environmental Science and Forestry), Dr. Robert E. Gilman (Rochester Institute of Technology) and Dr. Robert Silberman (State University of New York, College at Cortland) are deeply appreciated. We also express our appreciation to our wives who were "author-widows" during the course of this revision.

R. M. Silverstein
T. C. Morrill

PREFACE
TO SECOND EDITION

We undertake a complete rewriting of the first edition for the following reasons: (1) The numerous deficiencies of the first edition became painfully apparent on further teaching and on using the book for reference in the laboratory. (2) The spectacular recent advances in mass and NMR spectrometry made Chapters 2 and 4 obsolete. (3) We are frankly delighted by the reception accorded our initial effort.

This book is still a teaching manual at the introductory level. However, it also appears to be useful to the practicing chemist as a quick reference source.

In this second edition we have expanded the coverage of each of the four areas of spectrometry, added much new reference material, and added a number of new problems.

The extensive rewriting was made possible by a generous grant from Consolidated Electrodynamics Corporation.

We are indebted to our colleagues at Stanford Research Institute: Mr. W. R. Anderson, Jr. who ran many of the NMR spectra and is responsible for much of the Appendix to the NMR chapter; Miss J. S. Whittick and Mr. F. M. Church who ran many of the mass spectra; and Dr. R. F. Muraca, Director of Analysis and Instrumentation.

The mass spectrometry chapter was read by Dr. W. H. McFadden of the Western Regional Laboratories USDA, the IR chapter by Dr. J. P. Collman of the University of North Carolina, the NMR chapter by Dr. L. J. Durham of Stanford University and the UV chapter by Dr. R. H. Eastman of Stanford University. Their many helpful comments are deeply appreciated.

We gratefully acknowledge the following sources for permission to use the material indicated.

Consolidated Electrodynamics Corporation: Chapter 2, Fig. 1.

J. H. Beynon, *Mass Spectrometry and its Application to Organic Chemistry,* Elsevier, Amsterdam, 1960: Chapter 2, Appendix A.

Sadtler Research Laboratories, 3314–20 Spring Garden St., Philadelphia, Pa.: The Infrared Spectra Used as Illustrations in the Discussion of Characteristic Group Frequencies.

Varian Associates: Chapter 4, Figs. 19, 20, 21, 22, 23, and 26.

E. D. Becker, *J. Chem. Ed.,* 42, 591 (1965): Chapter 4, Fig. 27, with permission of the *Journal of Chemical Education.*

F. A. L. Anet, *Can. J. Chem.,* 39 2262 (1961): Chapter 4, Fig. 28b, with permission of the *Canadian Journal of Chemistry.*

Perkin-Elmer Corporation: Chapter 3, Fig. 2; Chapter 5, Fig. 5.

A. E. Gillam and E. S. Stern, *An Introduction to Electronic Absorption Spectroscopy in Organic Chemistry,* Edward Arnold, London, 2nd ed., 1957: Chapter 5, Table III.

A. I. Scott, *Interpretation of the Ultraviolet Spectra of Natural Products*, Pergamon Press Ltd., Oxford, England, 1964: Chapter 5, Tables VI and XI.

R. P. Bauman, *Absorption Spectroscopy*, John Wiley and Sons, New York, 1962, and R. A. Friedel and Milton Orchin, *Ultraviolet Spectra of Aromatic Compounds*, John Wiley and Sons, New York, 1951: Chapter 5, Fig. 7.

R. A. Friedel and Milton Orchin, *Ultraviolet Spectra of Aromatic Compounds*, John Wiley and Sons, New York, 1951: Chapter 5, Fig. 8.

As are most authors, we are acutely aware of the burden carried by our wives during manuscript preparation.

Menlo Park, California
January 1967

R. M. Silverstein
G. C. Bassler

PREFACE
TO FIRST EDITION

During the past several years, we have been engaged in isolating small amounts of organic compounds from complex mixtures and identifying these compounds spectrometrically.

At the suggestion of Dr. A. J. Castro of San Jose State College, we developed a one unit course entitled "Spectrometric Identification of Organic Compounds," and presented it to a class of graduate students and industrial chemists during the 1962 spring semester. This book has evolved largely from the material gathered for the course and bears the same title as the course.

We should first like to acknowledge the financial support we received from two sources: The Perkin-Elmer Corporation and Stanford Research Institute.

A large debt of gratitude is owed to our colleagues at Stanford Research Institute. We have taken advantage of the generosity of too many of them to list them individually, but we should like to thank Dr. S. A. Fuqua, in particular, for many helpful discussions of NMR spectrometry. We wish to acknowledge also the cooperation at the management level, of Dr. C. M. Himel, chairman of the Organic Research Department, and Dr. D. M. Coulson, chairman of the Analytical Research Department.

Varian Associates contributed the time and talents of its NMR Applications Laboratory. We are indebted to Mr. N. S. Bhacca, Mr. L. F. Johnson, and Dr. J. N. Shoolery for the NMR spectra and for their generous help with points of interpretation.

The invitation to teach at San Jose State College was extended by Dr. Bert. M. Morris, head of the Department of Chemistry, who kindly arranged the administrative details.

The bulk of the manuscript was read by Dr. R. H. Eastman of Stanford University whose comments were most helpful and are deeply appreciated.

Finally, we want to thank our wives. As a test of a wife's patience, there are few things to compare with an author in the throes of composition. Our wives not only endured, they also encouraged, assisted, and inspired.

Menlo Park, California
April 1963

R. M. Silverstein
G. C. Bassler

CONTENTS

1 Introduction *1*

2 Mass Spectrometry *5*

    I.   Introduction *5*
   II.   Instrumentation *6*
  III.   The Mass Spectrum *11*
  IV.   Determination of a Molecular Formula *12*
    V.   Recognition of a Molecular Ion
          (Parent Peak) *14*
  VI.   Use of the Molecular Formula *16*
 VII.   Fragmentation *16*
VIII.   Rearrangements *18*
  IX.   Derivatives *18*
    X.   Mass Spectra of Some Chemical
          Classes *19*

3 Infrared Spectrometry *73*

    I.   Introduction *73*
   II.   Theory *74*
  III.   Instrumentation *78*
  IV.   Sample Handling *80*
    V.   Interpretation of Spectra *82*
  VI.   Characteristic Group Frequencies of
          Organic Molecules *83*

4 Magnetic Resonance Spectrometry *159*

    I.   Introduction and Theory *159*
   II.   Apparatus and Sample Handling *162*

  III.   Chemical Shift *164*
  IV.   Simple Spin-Spin Coupling *169*
    V.   Protons on Heteroatoms *174*
  VI.   Coupling of Protons to Other Nuclei *179*
 VII.   More Complex Spin-Spin Coupling *181*
VIII.   Virtual Coupling and Deceptively Simple
          Spectra *187*
  IX.   Effects of an Asymmetric Center *189*
    X.   Vicinal and Geminal Coupling in Rigid
          Systems *190*
  XI.   Long-Range Coupling *191*
 XII.   Spin-Spin Decoupling *192*
XIII.   Shift Reagents *194*
XIV.   Carbon 13 NMR *196*

5 Ultraviolet Spectrometry *231*

    I.   Introduction *231*
   II.   Theory *232*
  III.   Instrumentation *236*
  IV.   Sample Handling *238*
    V.   Characteristic Absorption of Organic
          Compounds *238*

6 Sets of Spectra Translated into Compounds *259*

7 Sets of Spectra with Beilstein References *315*

Index *335*

introduction

Our sole purpose in writing this book is to teach the organic chemist how to identify organic compounds from the complementary information afforded by four spectra: mass, infrared, nuclear magnetic resonance, and ultraviolet. Essentially, the molecule in question is subjected to four energy probes, and the molecule's responses are recorded as spectra.

The small amounts of pure compounds that can be isolated from complex mixtures by gas chromatography or thin-layer chromatography present a challenge to the chemist concerned with identification and structure elucidation of organic compounds. These techniques have two characteristics: they are rapid, and they are most effective in milligram and microgram quantities. There is neither enough time nor enough material to accommodate the classical manipulations involving sodium fusion, boiling point, refractive index, solubility tests, functional group tests, derivative preparation, mixture melting point, combustion analysis, molecular weight, and degradation with similar manipulations of the degradation products.

Our goal in this book is a rather modest level of sophistication and expertise in each of the four areas of spectrometry. Even this level will permit solution of a gratifying number of identification problems with *no history and no other chemical or physical data*. Of course, in practice other information is usually available: the sample source, details of isolation, a synthesis sequence, or information on analogous material. Often complex molecules can be identified because partial structures are known and specific questions can be formulated; the process is more confirmation than identification. In practice, however, difficulties arise in physical handling of minute amounts of compound: trapping, elution from adsorbents, solvent removal, prevention of contamination, and decomposition of unstable compounds. Water, air, stopcock greases, solvent impurities, and plasticizers have frustrated many investigations. The quality of spectra obtained in practice is usually inferior to that presented here.

For pedagogical reasons, we deal only with pure organic compounds. *Pure* in this context is a relative term, and all we can say is: the purer, the better. Probably the ultimate criterion of purity (for a sufficiently volatile compound) is gas chromatographic homogeneity on each of polar and nonpolar substrate capillary columns. Various forms of liquid-phase chromatography (adsorption and liquid-liquid columns, paper, thin layer) are applicable to relatively nonvolatile compounds. The spectra presented in this book were obtained on samples that were purified by recrystallization to constant melting point, or by gas chromatography.

In many cases, identification can be made on a fraction of a milligram, or even on several micrograms, of sample. Identification on the milligram scale is routine. Of course, not all molecules yield so easily. Chemical manipulations may be necessary. But the information obtained from the four spectra will permit intelligent selection of chemical treatment, and the energy probe methodology can be applied to the resulting products.

There are limitations to the methodology we espouse. A mass spectrum is dependent on a degree of volatility and thermal stability. However, mass spectra have been obtained on many high molecular weight compounds—steroids, terpenoids, peptides, polysaccharides, and alkaloids—by inserting the sample very close to the ionizing beam. Solubility is a limiting factor in nuclear magnetic resonance spectrometry. However, the availability of many deuterated solvents and the development of microsample tubes, CAT units, and FT NMR (Chapter 4) make it possible to obtain adequate spectra in dilute solutions.

The problem of cost of necessary instrumentation is raised, and answered by pointing to the amazing evolution of commercial instruments. The time saved, the smaller samples required, and the information made available far overbalance the cost. Identifications are frequently made on the basis of several hours of a student's, technician's, or analyst's time. Under a classical regime, much larger samples and several days or even weeks of a skilled analyst's time would probably be necessary. Infrared and ultraviolet spectrometers have been developed beyond the stage of reliable instruments in the hands of a trained technician. They are now cheap, rugged, and simple enough to be used as a bench tool by the organic chemist. Certain types of simple nuclear magnetic resonance spectrometers are now available to even the more modest institutions; these are essentially as easy to run as infrared instruments. More sophisticated magnetic resonance spectrometers, especially those with computer facets, require skilled technicians.

Almost from its inception, the utility of nuclear magnetic resonance spectrometry to the organic chemist has been evident. Mass spectrometry, however, has had a somewhat different history. Developed by the physicist and utilized extensively by the petroleum chemist, it has been ignored almost completely by the organic chemist concerned with identification and structure determination. Mass spectrometers have become available to a wide range of institutions. Most instruments, however, are expensive and complex and require considerable skill in operation and maintenance. Inexpensive, rugged instruments are now beginning to appear on the market.

We spend very little time on instrumentation per se for three reasons: it is not requisite to our goal; we are not qualified; and excellent treatises are available. The four chapters on spectrometry are designed to give the analyst an appreciation of the potentialities of each technique as applied to the identification of organic compounds. The rest of the book consists of selected spectra. These mass, infrared, nuclear magnetic resonance, and ultraviolet spectra are presented as sets, each set representing a compound. These are translated, as exercises, into the chemical structure (Chapter 6) or are identified only by a Beilstein reference (Chapter 7). Aside from practical applications, the considerations involved in translating spectra into organic compounds lead to an appreciation of modern concepts of structural organic chemistry.

If we have been judicious in our selection of spectra, they should serve as useful reference material for teachers and for chemists in industry. In one form or another such material has become part of the training of every organic chemist.

SPECTROMETRY NOMENCLATURE

To make the reader aware of the more common conventions used in papers on spectrometry, the following definitions are given. These are taken from the December, 1972 issue of Analytical Chemistry; one should consult succeeding December issues of this journal to become aware of any changes in these conventions.

Absorbance, A. (Not optical density, absorbancy, or extinction.) Logarithm to the base 10 of the reciprocal of the transmittance $A = \log_{10}(1/T)$.

Absorptivity, a. (Not k.) (Not absorbancy index, specific extinction, or extinction coefficient.) Absorbance divided by the product of the concentration of the substance and the sample path length,

$$a = \frac{A}{bc}$$

Absorptivity, Molar, ϵ. (Not molar absorbancy index, molar extinction coefficient or molar absorption coefficient.) Product of the absorptivity, a, and the molecular weight of the substance.

Angstrom, Å. Unit of length equal to 1/6438.4696 of wavelength of red line of Cd. For practical purposes, it is considered equal to 10^{-8} cm.

Beer's Law. (Representing Beer-Lambert law.) Absorptivity of a substance is a constant with respect to changes in concentration.

Concentration, c. Quantity of the substance contained in a unit quantity of sample. (In absorption spectrometry it is usually expressed in grams per liter.)

Frequency. Number of cycles per unit time.

Infrared. The region of the electromagnetic spectrum extending from approximately 0.78 to 300 micrometers.

Micrometer, μm. Unit of length equal to 10^{-6} meter. (Do not use micron.)

Nanometer, nm. Unit of length to 10^{-9} meter. (Do not use millimicron.)

Sample Path Length, b. (Not l or d.) Internal cell or sample length, usually given in centimeters.

Spectrograph. Instrument with an entrance slit and dispersing device that uses photography to obtain a record of spectral range. The radiant power passing through the optical system is integrated over time, and the quantity recorded is a function of radiant energy.

Spectrometer, Optical. Instrument with an entrance slit, a dispersing device, and with one or more exit slits, with which measurements are made at selected wavelengths within the spectral range, or by scanning over the range. The quantity detected is a function of radiant power.

Spectrometry. Branch of physical science treating the measurement of spectra.

Spectrophotometer. Spectrometer with associated equipment, so that it furnishes the ratio, or a function of the ratio, of the radiant power of two beams as a function of spectral wavelength. These two beams may be separated in time, space, or both.

Transmittance, T. (Not transmittancy or transmission.) The ratio of the radiant power transmitted by a sample to the radiant power incident on the sample.

Ultraviolet. The region of the electromagnetic spectrum from approximately 10 to 380 nm. The term without further qualification usually refers to the region from 200 to 380 nm.

Visible. Pertaining to radiant energy in the electromagnetic spectral range visible to the human eye (approximately 380 to 780 nm).

Wavelength. (One word.) The distance, measured along the line of propagation, between two points that are in phase on adjacent waves; units Å, μm, and nm.

Wavenumber. (One word.) Number of waves per unit length. The usual unit of wavenumber is the reciprocal centimeter, cm^{-1}. In terms of this unit, the wavenumber is the reciprocal of the wavelength when the latter is in centimeters in vacuo.

REFERENCES

The following books deal with the utilization of several aspects of spectrometry for the organic chemist. Several collections of problems are also included. Additional references are cited in each chapter.

1. Ault, A., *Problems in Organic Structure Determination*, McGraw-Hill, New York, 1967.
2. Baker, A. J., and T. Cairns, *Spectroscopic Techniques in Organic Chemistry*, Heyden, London, 1965.
3. Bentley, K. W., Ed., *Elucidation of Structure by Physical and Chemical Methods*, vol. 4 of *Techniques of Organic Chemistry*, John Wiley, New York, 1972.
4. Brame, Jr., E. G., Ed., *Applied Spectroscopy Reviews*, Marcel Dekker, New York, 1969.
5. Brand, J. C. D., and G. Eglinton, *Applications of Spectroscopy to Organic Chemistry*, Oldbourne Press, London, 1965.
6. Brittain, E., W. O. George, and C. H. J. Wells, *Introduction to Molecular Spectroscopy Theory and Experiments*, Academic Press, New York, 1970.
7. Browning, D. R., Ed., *Spectroscopy*, McGraw-Hill, Maidenhead, England, 1970.
8. Cairns, T., Ed., *Spectroscopy in Education*, Heyden, London: T. Cairns, *Spectroscopic Problems in Organic Chemistry*, 1964, vol. 1; A. J. Baker and T. Cairns, *Spectroscopic Techniques in Organic Chemistry*, 1965, vol. 2; H. C. Hill, *Introduction to Mass Spectrometry*, 1966, vol. 3; A. J. Baker, G. Eglinton, P. J. Preston, and T. Cairns, *More Spectroscopic Problems in Organic Chemistry*, 1967, vol. 4.
9. Dyer, John R., *Applications of Absorption Spectroscopy of Organic Compounds*, Prentice-Hall, Englewood Cliffs, N.J., 1965; *Organic Spectral Problems*, Prentice-Hall, Englewood Cliffs, N.J., 1972.
10. Freeman, S. K., Ed., *Interpretive Spectroscopy*, Reinhold, New York, 1965.
11. Laszlo, P., and P. Stang, *Organic Spectroscopy*, Harper and Row, New York, 1971.
12. Mathieson, D. W., *Interpretation of Organic Spectra*, Academic Press, New York, 1965.

13. Nachod, F. C., and W. D. Phillips, Eds., *Determination of Organic Structures by Physical Methods*, vol. 2, Academic Press, New York, 1962.

14. Nachod, F. C., and J. J. Zuckerman, *Determination of Organic Structures by Physical Methods*, vols. 3, 4, Academic Press, New York, 1971.

15. Pasto, D., and C. Johnson, *Organic Structure Determination*, Prentice-Hall, Englewood Cliffs, N.J., 1969.

16. Philip, J. P., *Spectra-Structure Correlation*, Academic Press, New York, 1964.

17. Scheinmann, F., Ed., *An Introduction to Spectroscopic Methods for the Identification of Organic Compounds*, 1970, vol. 1; *Nuclear Magnetic Resonance and Infrared Spectroscopy*, Pergamon Press, Oxford.

18. Schwarz, J. C. P., Ed., *Physical Methods in Organic Chemistry*, Oliver and Boyd, Edinburgh, 1964.

19. Shapiro, R. H., *Spectral Exercises in Structural Determination of Organic Compounds*, Rinehart and Winston, New York, 1969.

20. Trost, B., *Problems in Spectroscopy*, W. A. Benjamin, New York, 1967.

21. Whiffen, D. H., and I. Fleming, *Spectroscopy*, John Wiley, New York, 1966.

22. Williams, D. H., and I. Fleming, *Spectroscopic Problems in Organic Chemistry*, McGraw-Hill, New York, 1967.

23. Winstead, M. B., *Organic Chemistry Structural Problems*, Sadtler Research Laboratories, Philadelphia, 1968.

two

I. INTRODUCTION

A mass spectrometer bombards a substance under investigation with an electron beam and quantitatively records the result as a spectrum of positive ion fragments. This record is a mass spectrum. Separation of the positive ion fragments is on the basis of mass (strictly, mass/charge, but the majority of ions are singly charged). How this is accomplished will be sketched in just sufficient detail to impart some appreciation to the organic chemist for the potentialities of mass spectrometry as applied to compound identification.

A number of good general texts are available for the broad field of mass spectrometry; these include the books by Biemann,[1] Beynon,[3] the more recent work of Hamming and Foster,[10f] and the extensive volumes by Budzikiewicz, Djerassi, and Williams.[2,20] Brief introductions to the field are available in the paperback books by McLafferty[10c] and Shrader.[10d] Pertinent journals are *Organic Mass Spectrometry, the Journal of Mass Spectrometry and Ion Physics,* and the *Archives of Mass Spectral Data.* Compilations of spectra are available;[11] references to data include Appendix I (organized by compound class) of Hamming and Foster,[10f] the Atlas of Mass Spectral Data[10m] and the Aldermaston collection.[12c] Indexes to mass spectral information include the ASTM index,[12a,b] the Eight-Peak Index,[21] the chemical compound index of the journal *Org. Mass Spec.* and the Grenoble compilation.[22]

Mass spectrometers are still characterized by high cost and the need for highly skilled technicians for operation and maintenance. Mass spectral data, however, are routinely reported in professional journals. In addition, mass spectral discussions have become commonplace in organic chemistry textbooks and problem manuals for structure determination.

II. INSTRUMENTATION

Recent developments in the instrumentation are described in the books by Shrader[10f] and by Hamming and Foster.[10d] Mass spectrometers for structure elucidation can be classified according to the method of separating the charged particles:

A. Magnetic Field Deflection (Direction Focusing)
   1. Magnetic field only (unit resolution)
   2. Double focusing (electrostatic field before magnetic field, high resolution)
B. Time of Flight
C. Quadrupole

The minimum instrumental requirement for the organic chemist is the ability to record the molecular weight of the compound under examination to the nearest whole number. Thus the recording should show a peak at, say, mass 400, which is distinguishable from a peak at mass 399, and at mass 401. In order to select possible molecular formulas by measuring isotope peak intensities (see Section IV), adjacent peaks must be quite cleanly separated. Arbitrarily, a valley between two adjacent peaks should not be more than about 10% of the height of the larger peak. This latter degree of resolution is termed "unit" resolution and can be obtained up to about mass 500 on several single-focusing (magnetic-focusing) instruments.

To determine the resolution of an instrument, consider two adjacent peaks of approximately equal intensity. These peaks should be chosen so that the height of the valley between the peaks is about 10% of the intensity of the peaks. The resolution (R) is

$$R = \frac{M}{\Delta M}$$

where M is the higher mass number of the two peaks and ΔM is the difference between the two mass numbers.

There are two important categories of magnetic deflection mass spectrometers: low resolution and high resolution. Low-resolution instruments can be defined arbitrarily as the instruments that separate unit masses up to m/e 2000 [$R = 2000/(2000-1999) = 2000$]. Unit mass (or low-resolution) spectra are obtained from these instruments. An instrument is generally considered high resolution if it can separate two ions differing in mass by at least one part in ten thousand to fifteen thousand ($R = 10,000-15,000$). An instrument with 10,000 resolution can separate an ion of mass 500.00 from one of mass 499.95 ($R = 500/0.05 = 10,000$). This important class of mass spectrometers can measure the mass of an ion with sufficient accuracy to determine its atomic composition. High-resolution mass spectrometry will be discussed only briefly, since the instruments are not accessible to most laboratories.

A schematic diagram of a typical 180° single-focusing mass spectrometer is shown in Figure 1. There are five component parts.

1. SAMPLE HANDLING SYSTEM. This consists of a device for introducing the sample, a micromanometer for determining the amount of sample introduced, a device (molecular leak) for metering the sample to the ionization chamber, and a pumping system. Introduction of gases is usually a simple matter of transfer from a gas bulb into the metering volume, thence to the ionizing chamber. Liquids are introduced with various break-off devices, by touching a micropipette to a sintered glass disc or an orifice under mercury or gallium, or simply by hypodermic needle injection through a silicone rubber dam or serum cap. A bulb containing the sample may be pumped out under dry ice, then warmed to vaporize the sample into the inlet system. Heated inlet systems are used for less volatile liquids and for solids. Insertion of the sample directly into the ionization chamber further extends the limitations imposed by lack of volatility and of thermal stability. Reproducible breakdown patterns have been obtained on high molecular weight terpenoids, steroids, polysaccharides, peptides, and alkaloids. Even with these special techniques, a compound must be stable at a temperature at which its vapor pressure is of the order of 10^{-7} to 10^{-6} Torr. For routine work, using a molecular leak, a vapor pressure of about 10^{-1} to 10^{-3} Torr. is desired. Sample sizes for liquids and solids range from several milligrams to less than a microgram, depending on the method of introduction and the detector.

2. IONIZATION AND ACCELERATING CHAMBERS. The gas stream from the molecular leak enters the ionization chamber (operated at a pressure of about 10^{-6} to 10^{-5} Torr) in which it is bombarded at right angles by an electron beam emitted from a hot filament. The positive ions, produced by interaction with the electron beam, are forced through the first accelerating slit by a small electrostatic field between the repellers and the first accelerating slit. A strong electrostatic field between the first and second accelerating slits accelerates the ions to their final velocities. Additional focusing of the ion beam is provided between the accelerating slits. To obtain a spectrum, either the magnetic field applied to the analyzer tube (Figure 1) or the accelerating voltage between the first and second ion slits is varied. Thus, the ions are successively focused at the collector slit as a function of mass (strictly mass/charge). In modern instruments, a scan from mass 12 to mass 500 may be performed in 1 to 4 minutes. However, scan speeds of a few seconds have been used to obtain spectra of gas chromatography fractions (see below).

3. ANALYZER TUBE AND MAGNET. The analyzer tube is an evacuated (10^{-7} to 10^{-8} Torr), curved (in Figure 1, a 180° curve), metal tube through which the ion beam passes from ion source to collector. The magnetic pole pieces (electromagnets are usually used for the larger

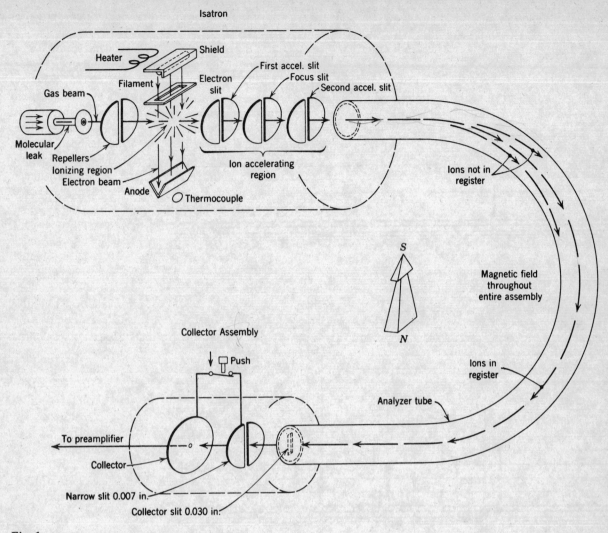

Fig. 1.
Schematic diagram of CEC model 21-103 Mass Spectrometer, a single-focusing, 180° sector mass analyzer. The magnetic field is perpendicular to the page.

instruments) are mounted perpendicular to the plane of the diagram (Figure 1). The main requirement is a uniform, stable magnetic field.

4. ION COLLECTOR AND AMPLIFIER. A typical ion collector consists of one or more collimating slits and a Faraday cylinder; the ion beam impinges axially into the collector, and the signal is amplified by a vacuum-tube electrometer or an electron multiplier.

5. RECORDER. A widely used recorder employs five separate galvanometers that record simultaneously on photographic paper (ultraviolet recording paper, which does not require wet development, is also common). Figure 2a presents a spectrum traced by a five-element galvonometer system at sensitivity levels decreasing from top to bottom in the ratios of 1:3:10:30:100. Peak heights from the base

line are read on the most sensitive trace remaining on scale and are multiplied by the appropriate sensitivity factor. Peak heights are proportional to the number of ions of each mass. The recorder trace can be presented as a table or a graph (Figures 2b and 2c). Throughout this book, and in most published work, a peak is reported as a value of mass divided by charge (m/e). Since we are usually dealing with singly charged ions (i.e., $e = 1$), such a value is the mass (m) of the ion corresponding to that peak. That multiple ionization does occur is indicated by the peaks at half-mass units in the tracing; these represent odd-numbered masses that carry a double charge. A broad, weak, "metastable" peak (see below) can be seen between m/e 90 and 91 (Figure 2a).

Assignment of mass to the peaks of the recorder tracing

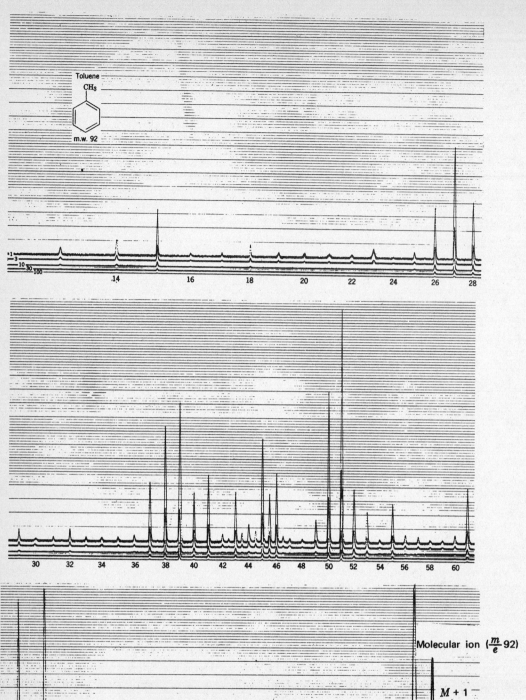

Fig. 2a.

Mass Spectrum traced by a five-element galvanometer. Note the metastable peak at m* = (91)² /(92) = ca. 90.

Toluene
CH$_3$

m.w. 92

Isotope Abundances

m/e	% of Base Peak		m/e	% of M
38	4.4		92 (M)	100
39	5.3		93 (M + 1)	7.23
45	3.9		94 (M + 2)	0.29
50	6.3			
51	9.1			
62	4.1			
63	8.6			
65	11			
91	100	(Base)		
92	68	(Parent or Molecular Ion Peak)*		
93	4.9	(M + 1)		
94	0.21	(M + 2)		

*These terms are interchangeable; most of the time we have used "molecular ion" herein in accordance with the trend in the literature.

Fig. 2*b*.
Tabular presentation of Fig. 2a.

can be a problem at the high mass end of the scan. The common practice is to start at the low mass end of the scan, which can be accurately set, and count the peaks to the last recorded peak. This is generally feasible because the most sensitive galvanometer will record a slight ion current at each mass unit. Sometimes the peaks at the high end of the spectrum may be widely spaced, and the background trace indistinct; in this case, a calibration compound may be added to the sample. Some instruments are equipped with automatic mass markers, but the usual experience is that these are not reliable where they are most needed—at the high mass end of the spectrum. A mass digitizer that prints out the mass number and relative intensities is a valuable auxiliary piece of equipment. However, a slight maladjustment may result in loss or gain of a full mass unit in a scan; this can be disastrous.

It is well to check the digitizer print-out against the galvanometer trace.

A mass analyzer using four electric poles (a "quadrupole") and no magnetic field has been developed[10d] (Figure 3). Ions entering from the top travel with constant velocity in the direction parallel to the poles (z direction), but acquire oscillations in the x and y directions. This is accomplished by application of both a dc voltage and a radio frequency voltage to the poles. There is a "stable oscillation" that allows an ion to pass from one end of the quadrupole to the other without striking the poles; this oscillation is dependent on the mass-to-charge ratio of an ion. Therefore ions of only a single m/e value will traverse the entire length of the analyzer. All other ions will have unstable oscillations and will strike the poles. Mass scanning is carried out by varying each of the dc and radio frequencies while keeping their ratios constant. Resolutions as high as $R = 10,000$ have been achieved with this type of analyzer.

The introduction of an electrostatic field ahead of the magnetic field (double focusing) permits high resolution so that the mass of a particle can be obtained to three or four decimal places.[1,3-7,10c,10d] Figures 4*a* and 4*b* show examples of such double-focusing instrument schematics. A positive ion in an electric field experiences a force in the direction of the field; the path of an ion moving through the field is

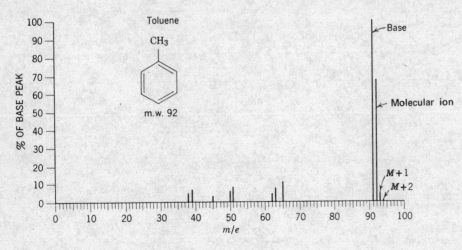

Fig. 2*c*.
Graphical presentation of Fig. 2a.

Fig. 3.
Quadrupole arrangement.

magnetic deflection analyzers. Proper choice of the angle of deflection in the electric field also results in directional focusing of an ion beam. The Mattauch-Herzog double-focusing mass analyzer incorporates 31°50' electrostatic and 90° magnetic analyzers (Figure 4a). The multimass ion beam is uniform in energy as it leaves the electrostatic analyzer, and the magnetic field produces the desired mass dispersion while focusing each unimass ion beam to a different point. Another common double-focusing arrangement was developed by Nier and Johnson (Figure 4b). The

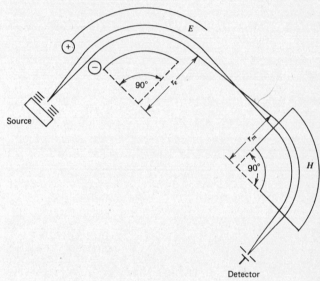

Fig. 4b.
Nier-Johnson double-focusing mass analyzer. E = electric field. H = magnetic field.

thus curved. In a radial electric field (always perpendicular to the direction of flight of the ions) the radius of curvature, r_e, of the ion path is dependent on the energy of the ion and the strength of the electric field. The electric field is an energy analyzer, instead of a mass analyzer, and serves to limit the energy spread of the ion beam before it enters the magnetic field. The energy spread is one of the major factors in decreasing the resolution of the simple

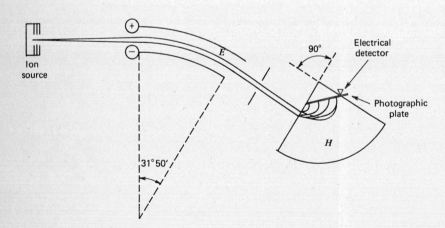

Fig. 4a.
Mattauch-Herzog double-focusing mass analyzer. E = electric field. H = magnetic field.

electric and magnetic sectors are both 90°, and all ions are focused at the same point in the detector.

Resolutions on the order of 40,000 are routinely obtainable with the high-resolution commercial instruments available (either geometry) when relatively monoenergetic ions are produced; both geometries have been used extensively for accurate measurement of the masses of ions in organic structure determinations. However, the high cost and complexity make this instrument available to relatively few laboratories.

Several firms offer a gas chromatographic instrument coupled to a mass spectrometer through an interface that enriches the concentration of the sample in the carrier gas (Figure 5).[10d,10f] Scan times are rapid enough so that several mass spectra can be obtained during the elution of a single peak from the gas chromatographic unit. Several efficient enrichment devices have been developed.[10d,10f]

III. THE MASS SPECTRUM

Mass spectra are routinely obtained at an electron beam energy of 70 electron volts. The simplest event that occurs is the removal of a single electron from the molecule in the gas phase by an electron of the electron beam to form a molecular (parent)* ion (M). This is a radical cation. For example, for methanol,

$$CH_3OH + e \rightarrow CH_3OH^{\ddagger}\,(m/e\ 32) + 2e^-$$

In accordance with the convention adopted in the first edition, the parent radical ion is depicted with a $\ddagger$ symbol. When the charge can be localized on one particular atom, the charge is shown on that atom. The dot represents the

* P (parent) was used in the 1st and 2nd editions. See footnote to Fig. 2b, p. 9.

odd electron. Many of these molecular (parent) ions disintegrate in 10^{-10} to 10^{-3} second to give, in the simplest case, a positively charged fragment and a radical. A number of fragment ions are thus formed, and each of these can cleave to yield smaller fragments. Again, illustrating with methanol:

$$CH_3OH^{\ddagger} \rightarrow CH_2OH^+\,(m/e\ 31) + H\cdot$$

$$CH_3OH^{\ddagger} \rightarrow CH_3^+\,(m/e\ 15) + \cdot OH$$

$$CH_2OH^+ \rightarrow CHO^+\,(m/e\ 29) + H_2$$

If some of the molecular (parent) ions remain intact long enough (about 10^{-6} seconds) to reach the detector, we see a parent peak. It is important to recognize the parent peak because this gives the molecular weight of the compound. This molecular weight is the molecular weight to the nearest whole number, and not merely the approximation obtained by all other molecular weight determinations familiar to the organic chemist.

A mass spectrum is a presentation of the masses of the positively charged fragments (including the parent ion) versus their relative concentrations. The most intense peak in the spectrum, called the base peak, is assigned a value of 100%, and the intensities (height × sensitivity factor) of the other peaks, including the parent peak, are reported as percentages of the base peak. Of course, the molecular ion peak or parent peak may sometimes be the base peak. In Figure 2a, the parent peak is m/e 92, and the base peak is m/e 91.

A tabular or graphic presentation of a spectrum may be used. A graph has the advantage of presenting patterns that, with experience, can be quickly recognized. However, a graph must be drawn so that there is no difficulty in distinguishing mass units. Mistaking a peak at, say, m/e 79 for m/e 80 can result in total confusion. Unfortunately, many of the graphs presented in the journals are illegible. Computer programs are available for printing out such line graphs. The grid that we use permits ready recognition of individual mass units as well as patterns. Except for isotope

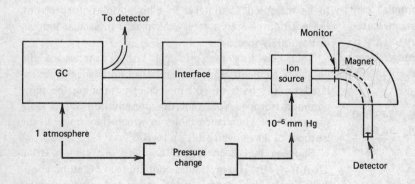

Fig. 5.
Gas Chromatograph-mass spectrometer combination.

peaks and parent peaks (molecular ion), peaks of less than 3% intensity are not shown unless they have special significance.

The parent peak is the peak of highest mass number except for the isotope peaks. These isotope peaks are present because a certain number of molecules contain heavier isotopes than the common isotopes. We shall show how the intensities of the isotope peaks relative to the parent peak can lead to the determination of a molecular formula. In a separate table accompanying the mass spectral graphs in the problems in Chapters 6 and 7, the parent peak is given an intensity of 100%, and the isotope peak intensities are given relative to the parent peak intensity (Figure 2b).

If an ion (m_1) fragments after acceleration but before entering the magnetic field, it will have been accelerated as mass m_1, but dispersed in the magnetic field as m_2. The resulting ion current will be recorded as a low-intensity, broad peak at apparent mass m^*. The numerical value of m^* is given by

$$m^* = \frac{(m_2)^2}{m_1}$$

The peak caused by the ion current corresponding to mass m^* is called a *metastable peak*. Measurement of the mass of the metastable peak affords information that m_2 is derived directly from m_1 by loss of a neutral fragment. An example of a metastable peak is the broad peak at about m/e 90 in Figure 2a. The m_1 peak is the molecular ion (m/e 92) and the m_2 peak is at m/e 91. Although this technique is useful in interpretations of spectra, it will not be used in this book. Its use is discussed in Reference 1, pp. 153-157, and in Reference 3, pp. 251-262.

IV. DETERMINATION OF A MOLECULAR FORMULA

A unique molecular formula (or fragment formula) can often be derived from a sufficiently accurate mass measurement alone (high-resolution mass spectrometry). This is possible because the atomic masses are not integers (see Table IA). For example, we can distinguish at a nominal mass of 28 among CO, N_2, CH_2N, and C_2H_4:

^{12}C	12.0000	$^{14}N_2$	28.0062	^{12}C	12.0000	$^{12}C_2$	24.0000
^{16}O	15.9949			$^1H_2$	2.0156	$^1H_4$	4.0312
	27.9949			^{14}N	14.0031		28.0312
					28.0187		

Thus the mass observed for the molecular ion of CO is the sum of the exact masses of the most abundant isotope of carbon and of oxygen. This differs from a molecular

Table IA Exact Masses of Isotopes

Element	Atomic Weight	Nuclide	Mass
Hydrogen	1.00797	^1H	1.00783
		$D(^2H)$	2.01410
Carbon	12.01115	^{12}C	12.00000 (std)
		^{13}C	13.00336
Nitrogen	14.0067	^{14}N	14.0031
		^{15}N	15.0001
Oxygen	15.9994	^{16}O	15.9949
		^{17}O	16.9991
		^{18}O	17.9992
Fluorine	18.9984	^{19}F	18.9984
Silicon	28.086	^{28}Si	27.9769
		^{29}Si	28.9765
		^{30}Si	29.9738
Phosphorus	30.974	^{31}P	30.9738
Sulfur	32.064	^{32}S	31.9721
		^{33}S	32.9715
		^{34}S	33.9679
Chlorine	35.453	^{35}Cl	34.9689
		^{37}Cl	36.9659
Bromine	79.909	^{79}Br	78.9183
		^{81}Br	80.9163
Iodine	126.904	^{127}I	126.9045

weight of CO based on atomic weights that are the average of weights of all natural isotopes of an element (e.g., C = 12.01, O = 15.999). It is a tedious task to find a molecular formula by arithmetic trial and error from the output of a high-resolution mass spectrometer. Tables, algorithms, and computer programs have been assembled for this purpose.[10d,13-18] We will use unit mass resolution results and intensities of isotope peaks to arrive at possible molecular formulas, since most laboratories have limited access to high-resolution spectrometers. Determination of a possible molecular formula from the isotope peak intensities is limited to the cases in which the parent peak is relatively intense so that the isotope peaks are large enough to be measured accurately. In a high-resolution spectrum, the position of even a very weak parent molecular ion can be accurately measured.

Table IB lists the principal stable isotopes of the common elements and their relative abundance as calculated on the basis of 100 molecules containing the most common isotope. Note that this presentation differs from many isotope abundance tables in which the sum of all the isotopes of an element adds up to 100%.

Suppose that a compound contains one carbon atom. Then for every 100 molecules containing a ^{12}C atom, about 1.08 "molecules" contain a ^{13}C atom, and these molecules will produce an $M + 1$ peak of about 1.08% the intensity of the parent peak; the ^2H atoms present will make an

Table IB

Elements						
Carbon	^{12}C	100	^{13}C	1.08		
Hydrogen	^{1}H	100	^{2}H	0.016		
Nitrogen	^{14}N	100	^{15}N	0.38		
Oxygen	^{16}O	100	^{17}O	0.04	^{18}O	0.20
Fluorine	^{19}F	100				
Silicon	^{28}Si	100	^{29}Si	5.10	^{30}Si	3.35
Phosphorus	^{31}P	100				
Sulfur	^{32}S	100	^{33}S	0.78	^{34}S	4.40
Chlorine	^{35}Cl	100	^{37}Cl	32.5		
Bromine	^{79}Br	100	^{81}Br	98.0		
Iodine	^{127}I	100				

additional very small contribution to the $M + 1$ peak. If a compound contains one sulfur atom, the $M + 2$ peak will be about 4.4% of the parent peak. The $M + 1$ and $M + 2$ peaks are so designated in Figure 2.

Selection of likely molecular formulas appropriate to particular mass and isotope abundance measurements is greatly facilitated by the table constructed by Beynon.* This table, which has recently been extended[13] to mass 500, can be used for high resolution mass spectrometry. Beyond about mass 250, use of the isotopes to determine a molecular formula loses its effectiveness. A reduced version of Beynon's table is presented as Appendix A. Its use will become more evident as we work through the spectra in this book. In practice, the measured isotope peaks are usually slightly higher than the calculated contributions because of incomplete resolution, bimolecular collisions (see below), or a contribution from the coincident peak of an impurity. The table is limited to compounds containing C, H, O, and N. The presence of S, Cl, or Br is usually readily apparent because of a large isotope contribution to $M + 2$. We shall see that the number of chlorine and bromine atoms can be determined. Iodine, fluorine, and phosphorus are monoisotopic. Their presence can usually be deduced from a suspiciously small $M + 1$ peak relative to the molecular weight, from the fragmentation pattern, from the other spectra with which we are concerned, or from the history of the compound.

If only C, H, N, O, F, P, I are present, the approximate expected % $(M + 1)$ and % $(M + 2)$ data can be calculated by use of the following formulas:

$$\% \, (M + 1) = 100 \left\{ \frac{[(M + 1)]}{[M]} \right\} \simeq$$

$$\simeq 1.1 \times \text{number of C atoms}$$
$$+ 0.36 \times \text{number of N atoms*}$$

$$\%(M + 2) = 100 \left\{ \frac{[(M + 2)]}{[M]} \right\}$$

$$\simeq \frac{(1.1 \times \text{number of C atoms})^2}{200}$$

$$+ 0.20 \times \text{number of O atoms*}$$

The use of such equations is almost completely limited to cases in which one has a preconceived notion about the molecular formula for the compound of interest. In this book, however, we shall be concerned almost solely with problems in which there is no prior information regarding the molecular formula.

It is difficult to overemphasize the importance of locating the parent peak. It will be stressed again that this gives an exact numerical molecular weight. Even in cases in which the parent peak is very small (and therefore an accurate determination of $M + 1$ and $M + 2$ is impossible), only a little extra information can often lead to identification. This information may be available from the source and history of the sample, from the fragmentation pattern, and from other spectra. Let us work through the selection of a molecular formula from the isotope abundance data obtained on an organic compound. We are given the following information:

m/e	%
150 (M)	100
151 (M + 1)	10.2
152 (M + 2)	0.88

The parent peak is mass 150; thus we have the molecular weight. The parent + 2 $(M + 2)$ peak obviously does not allow for the presence of sulfur or halogen atoms. We look in Appendix A under mass 150. Our $M + 1$ peak is 10.2% of the parent peak. We list the formulas whose calculated isotope contribution to the $M + 1$ peak falls—to be arbitrary—between 9.0 and 11.0; we also list the calculated $M + 2$ values:

Formula	$M + 1$	$M + 2$
$C_7H_{10}N_4$	9.25	0.38
$C_8H_8NO_2$	9.23	0.78
$C_8H_{10}N_2O$	9.61	0.61
$C_8H_{12}N_3$	9.98	0.45
$C_9H_{10}O_2$	9.96	0.84
$C_9H_{12}NO$	10.34	0.68
$C_9H_{14}N_2$	10.71	0.52

---

* Beynon's first table[3] is based on O = 16.000000. His expanded table[13] is based on the currently accepted standard, C = 12.000000.

* [M + 1] is the *intensity* of the M + 1 peak, and [M] is the intensity of the M peak.

On the basis of the "nitrogen rule" (see Section V), we immediately eliminate three of these formulas because they contain an odd number of nitrogen atoms. Our $M + 2$ peak is 0.88% of the parent. This best fits $C_9H_{10}O_2$. However, $C_8H_{10}N_2O$ cannot be ruled out without additional evidence. Note that mass 150 is the sum of the masses of the common isotopes of these molecular formulas; the isotope masses used are whole numbers (12 for carbon, 14 for nitrogen, etc.).

When elements other than C, H, O, and N are present, their kind and number must be determined (see the discussion under the appropriate chemical class) and their mass subtracted from the molecular weight. The composition of the remainder of the molecule is then determined from Appendix A.

Quite often the organic chemist must identify a by-product or an unexpected main product from a reaction. Under these circumstances, even rather complex molecules can be handled expeditiously. Since the compound has a history, an intelligent guess can be made as to what elements are present. The following example[19] may be instructive. In the course of polymerizing the fluorinated silicon-containing monomer, a crystalline sublimate was obtained as a

by-product. The parent peak of the mass spectrum of the sublimate was m/e 434, and the base peak, m/e 419 (i.e., $M - 15$). The $M + 1$ peak was 31.2% of the parent peak, and the $M + 2$ peak, 10.4%. On the assumption that the compound contained six fluorine and two silicon atoms, the molecular formula $C_{19}H_{20}F_6OSi_2$ was written. The intensities of the $M + 1$ and $M + 2$ peaks for this formula were calculated as follows:

Contributor	$M + 1$	$M + 2$	Source
$C_{19}H_{20}$ (mass 248)	20.85	2.06	Appendix A*
O	0.04	0.20	Table IB
F_6	.....	....	Table IB
Si_2	10.20†	6.70†	Table IB
Sum:	31.09	8.96	

*The C, H, and O content could have been obtained directly from Beynon's extended tables.[13] An alternative method for obtaining C, H, and O contributions is the use of the equations given earlier, thus:

$\%(M + 1) = (1.1)(19) + (0.36)(0) = 20.9\%$ of parent

$\%(M + 2) = [(1.1)(19)]^2/200 + 0.20(1) = 2.38\%$ of parent

†2 × (contribution to $M + 1$ or $M + 2$ per Si atom).

The good agreement between the calculated and the determined values provided strong support for the molecular formula written. This information combined with the fragmentation pattern and the infrared and NMR spectra led to the following structure.

The foregoing is a good example of both the possibilities and limitations of spectrometric identification without any chemical manipulation on the compound. It is doubtful that even an experienced mass spectrometrist would have derived a molecular formula without some indication that silicon and fluorine were present. Confirmation is certainly easier than diagnosis.

V. RECOGNITION OF THE MOLECULAR ION (PARENT) PEAK

There are two situations in which identification of the molecular ion peak may be difficult.

1. The molecular ion does not appear or is very weak. The obvious remedy in most cases is to run the spectrum at maximum sensitivity (and accept the resulting loss in resolution) and to use a larger sample. (Sometimes a large sample exaggerates the $M + 1$ peak. See below.) Still the molecular ion may not be evident, and other sources of information may be useful. The type of compound may be known, and the parent mass may be deduced from the breakdown pattern. For example, alcohols usually give a very weak parent molecular ion peak, but often show a pronounced peak resulting from loss of water ($M - 18$). A molecular weight determination, which can be done on the mass spectrometer or by any of the usual "wet lab" methods, or a combustion analysis together with consideration of the fragmentation pattern may help us to arrive at the parent mass. Preparation of a suitable derivative is another device that has been used in a limited way and will probably come to be used more extensively (see Section VIII).

2. The molecular ion is present but is one of several peaks which may be as prominent or even more prominent. In this situation, the first question is that of purity. If the compound can be assumed to be pure, the usual problem is to distinguish the parent peak from a more

prominent $M - 1$ peak. One good test is to reduce the energy of the bombarding electron beam to near the appearance potential. This will reduce the intensities of all peaks, but will increase the intensity of the molecular ion relative to other peaks, including fragmentation peaks (but not parent peaks) of impurities. Another test frequently used is to increase the size of the sample, or increase the time the sample spends in the ionization chamber by decreasing the ion repeller voltage (Reference 1, pp. 55-57). In either case, the net effect is to increase the opportunity for bimolecular collisions to occur in the ion chamber. The most common result of a bimolecular collision of a parent radical ion containing a heteroatom (O, N, or S) is a contribution to the $M + 1$ peak (that is, the net effect is the transfer of a hydrogen atom from a neutral molecule to the molecular ion).

$$RCH_2 - \overset{\overset{+}{\cdot}}{\underset{\cdot\cdot}{O}} - CH_2 R + RCH_2 - \overset{\cdot\cdot}{\underset{\cdot\cdot}{O}} - CH_2 R \cdot \rightarrow$$
$$(M^{\overset{+}{\cdot}})$$

$$RCH_2 - \overset{\overset{+}{\cdot}}{\underset{\cdot\cdot}{O}} - CH_2 R + R\overset{\cdot}{C}H - \overset{\cdot\cdot}{\underset{\cdot\cdot}{O}} - CH_2 R$$
$$H (M + 1)^{+}$$

Thus an increase in peak size relative to other peaks, as sample size is increased or the repeller voltage is decreased, designates that peak as the $M + 1$ peak and affords an indirect identification of the molecular ion. Of course, the dependence of the $M + 1$ peak on sample size must be kept in mind when this peak is used to establish a molecular formula of a compound containing a heteroatom.

Many peaks can be ruled out as possible molecular ions simply on grounds of reasonable structure requirements. The "nitrogen rule" is often helpful in this regard. It states that a molecule of even-numbered molecular weight must contain no nitrogen or an even number of nitrogen atoms; an odd-numbered molecular weight requires an odd number of nitrogen atoms. This rule holds for all compounds containing carbon, hydrogen, oxygen, nitrogen, sulfur, and the halogens, as well as many of the less usual atoms such as phosphorus, boron, silicon, arsenic, and the alkaline earths. A useful corollary states that fragmentation at a single bond gives an odd-numbered ion fragment from an even-numbered molecular ion, and an even-numbered ion fragment from an odd-numbered molecular ion; for this corollary to hold, the ion fragment must contain all of the nitrogen (if any) of the molecular ion. Consideration of the breakdown pattern coupled with other information will also assist in identifying molecular ions. It should be kept in mind that Appendix A contains fragments and trivial formulas as well as molecular formulas.

The presence of appreciable amounts of impurities that give rise to prominent peaks near the molecular ion can be troublesome. Here again, the expediency of reducing the energy of the electron beam will cause a relative increase in the intensity of the parent peak (and also of the parent peak of an impurity). The fragmentation pattern will often furnish clues. Another useful technique for detection of impurities is microeffusiometry. A fixed volume of sample is allowed to flow through a molecular leak, and the logarithms of the intensities of the peaks in question are plotted as a function of time. All peaks belonging to the same molecule will give lines of the same slope. Those due to other components of different molecular weight will give lines of different slopes.

The intensity of the molecular ion peak depends on the stability of the parent ion. The most stable molecular ions are those of purely aromatic systems. If substituents are present that have favorable modes of cleavage, the parent peak will be less intense, and the fragment peaks relatively more intense. In general, aromatics, conjugated olefins, saturated ring compounds, certain sulfur-containing compounds, and short, straight-chain hydrocarbons will give a prominent parent peak. The molecular ion will usually be recognizable in straight chain ketones, esters, acids, aldehydes, amides, ethers and halides. The molecular ion is frequently not detectable in aliphatic alcohols, amines, nitrites, nitrates, nitro compounds, nitriles, and in highly branched compounds.

At this time, the most promising technique for locating the molecular ion peak for compounds that give no peak (or a very weak one) is chemical ionization. The sample is introduced near atmospheric pressure with a large excess of methane. The methane is ionized (in the usual way) to the primary ions: CH_4^{+}, CH_3^{+}, etc. These react with the excess methane to give secondary ions:

$$CH_4^{+} + CH_4 \rightarrow CH_5^{+} \text{ and } CH_3 \cdot$$

$$CH_3^{+} + CH_4 \rightarrow C_2H_5^{+} \text{ and } H_2$$

The secondary ions react with the sample (RH):

$$CH_5^{+} + RH \rightarrow RH_2^{+} \text{ and } CH_4$$

$$C_2H_5^{+} + RH \rightarrow RH_2^{+} \text{ and } C_2H_4$$

These $M + 1$ ions (*quasi-molecular ions*) are often prominent. They can fragment (loss of H_2) to give prominent $M - 1$ ions.

The presence of an $M - 15$ peak (loss of CH_3) or an $M - 18$ peak (loss of H_2O) or an $M - 31$ peak (loss of OCH_3 from methyl esters), etc., is taken as confirmation of a molecular ion peak. An $M - 1$ peak is common, and occasionally an $M - 2$ peak (loss of H_2 by either fragmentation or thermolysis), or even a rare $M - 3$ peak (from alcohols) is reasonable. But peaks in the range of $M - 3$ to

$M - 14$ indicate that contaminants may be present or that the presumed molecular ion peak is actually a fragment ion peak. Losses of fragments of masses 19 to 25 are also unlikely (except for loss of $F = 19$ or $HF = 20$ from fluorinated compounds). Loss of 16 (O), 17 (OH), or 18 (H_2O) are likely only if an oxygen atom is in the molecule.

VI. USE OF THE MOLECULAR FORMULA

If an organic chemist had to choose a single item of information above all others that are usually available to him from spectra or from chemical manipulations, he would certainly choose the molecular formula.

In addition to the kinds and numbers of atoms, the molecular formula gives the index of hydrogen deficiency. The index of hydrogen deficiency is the number of *pairs* of hydrogen atoms that must be removed from the "saturated" formula (e.g., C_nH_{2n+2} for alkanes) to produce the molecular formula of the compound of interest. The index of hydrogen deficiency has also been called the number of "sites (or degrees) of unsaturation"; this description is unsatisfactory since hydrogen deficiency can be due to cyclic structure features as well as multiple bonds. The index is thus the sum of the number of rings, the number of double bonds, and twice the number of triple bonds.

The concept of the index of hydrogen deficiency may extend to oxygen, sulfur, nitrogen, halogen, and other compounds by the following:

For the generalized molecular formula $I_yII_nIII_zIV_x$ (e.g., $C_xH_yN_zO_n$) where

I can be H,F,Cl,Br,I,D, etc. (i.e., any monovalent atom).
II can be O,S or any other bivalent atom.
III can be N,P, or any other trivalent atom.
IV can be C,Si, or any other tetravalent atom.

The index of hydrogen deficiency $= x - y/2 + z/2 + 1$.

As an example, consider the molecular formula $C_{13}H_9N_2O_4BrS$. The index of hydrogen deficiency would be $13 - 10/2 + 2/2 + 1 = 10$ and a consistent structure would be:

(Index of hydrogen deficiency = 4 per benzene ring and 1 per NO_2 group.)

For simple molecular formulas, one can arrive at the index by comparison of the formula of interest with the molecular formula of the corresponding saturated compound.

The above formula can be applied to fragment ions as well as the molecular ion. When it is applied to even electron (all electrons paired) ions, the result is always an odd multiple of 0.5. As an example, consider $C_7H_5O^+$ with an index of 5.5. A reasonable structure is

since five and a half pairs of hydrogens would be necessary to obtain the corresponding saturated formula $C_7H_{16}O$ ($C_nH_{2n+2}O$). Odd-electron fragment ions will always give integer values of the index.

Terpenes often present a choice between a double bond and a ring structure. This question can readily be resolved on a microgram scale by catalytically hydrogenating the compound and rerunning the mass spectrum. If no other easily reducible groups are present, the increase in the mass of the molecular peak is a measure of the number of double bonds; the other "unsaturated sites" must be rings.

Such simple considerations give the chemist very ready information about structure. As another example, a compound containing a single oxygen atom may quickly be determined to be an ether or a carbonyl compound simply by counting "unsaturated sites."

VII. FRAGMENTATION

As a first impression, fragmenting a molecule with a huge excess of energy would seem a brute-force approach to molecular structure. The rationalizations used to correlate spectral patterns with structure, however, can only be described as elegant. The insight of such pioneers as McLafferty, Beynon, Stenhagen, Ryhage, and Myerson have led to a number of rational mechanisms for fragmentation. These have been masterfully summarized and elaborated by Biemann.[1] Generally, the tendency has been to represent the molecular ion with a delocalized charge. Djerassi's approach[2] has been to localize the positive charge on either a π bond (except in conjugated systems), or on a heteroatom. Whether or not this concept is totally rigorous, it is at the least a pedagogic *tour de force*. We shall use such locally charged molecular ions in this book.

In the first edition of this book, we discussed mechanisms in terms of single-electron shifts, which seems appropriate for unimolecular processes in the gas phase;

these shifts were shown by the same curved arrow generally used by organic chemists to show two-electron shifts. The "fishhook" symbolism used by Djerassi et al.[2,20] is an apt device and will be used here. A single-barbed fishhook ($\frown$) designates the shift of a single electron. Cleavage of a bond requires the movement of two electrons. However, to prevent clutter, only one of a pair of fishhooks will be drawn. This practice can be illustrated for cleavage of a C–C bond next to a heteroatom:

$$CH_3-CH_2 \overset{\frown}{-} \overset{+}{\underset{\cdot\cdot}{O}}-R \equiv CH_3-\overset{\frown}{CH_2}-\overset{+}{\underset{\cdot\cdot}{O}}-R$$

The probability of cleavage of a particular bond is related to the bond strength, to the possibility of low-energy transitions, and to the stability of the fragments both charged and uncharged formed in the fragmentation process. Our knowledge of pyrolytic cleavages can be used, to some extent, to predict likely modes of cleavage of the parent ion. Because of the extremely low vapor pressure in the mass spectrometer, there are very few fragment collisions; we are dealing largely with unimolecular decompositions. This assumption, backed by a file of reference spectra, is the basis for the vast amount of information available from the fragmentation pattern of a molecule. Whereas conventional organic chemistry deals with reactions initiated by chemical reagents or by thermal or actinic energy, mass spectrometry is concerned with the consequences suffered by an organic molecule struck by an ionizing electronic beam at a vapor pressure of about 10^{-5} mm Hg. A number of general rules for predicting prominent peaks in a spectrum can be written and rationalized, using standard concepts of physical organic chemistry.

1. The relative height of the molecular ion peak is greatest for the straight-chain compound and decreases as the degree of branching increases.
2. The relative height of the molecular ion peak usually decreases with increasing molecular weight in a homologous series. Fatty esters appear to be an exception.
3. Cleavage is favored at branched carbon atoms; the more branched, the more likely is cleavage. This is a consequence of the increased stability of a tertiary carbonium ion over a secondary, which in turn is more stable than a primary.

$$[R-\overset{|}{\underset{|}{C}}-]^{+} \cdot \rightarrow R\cdot + \cdot\overset{|}{\underset{|}{C}}-$$

Cation stability order:

$$\underset{+}{C}H_3 < R'\underset{+}{C}H_2 < R'_2\underset{+}{C}H < R'_3\underset{+}{C}$$

Generally, the largest substituent at a branch is eliminated most readily as a radical, presumably because a long-chain radical can achieve some stability by delocalization of the lone electron.

4. Double bonds, cyclic structures, and especially aromatic (or heteroaromatic) rings stabilize the molecular ion, and thus increase the probability of its appearance.
5. Double bonds favor allylic cleavage and give the resonance-stabilized allylic carbonium ion.

$$CH_2 \overset{+}{:}CH\overset{\frown}{-}CH_2-R \xrightarrow{-R\cdot} \overset{+}{C}H_2-\overset{\frown}{C}H=CH_2$$
$$\updownarrow$$
$$CH_2=CH-\overset{+}{\underset{\frown}{C}}H_2$$

6. Saturated rings tend to lose side chains at the α-bond. This is merely a special case of branching (Rule 3). The positive charge tends to stay with the ring fragment. Unsaturated rings can undergo a retro-Diels-Alder reaction:

7. In alkyl-substituted aromatic compounds, cleavage is very probable at the bond beta to the ring, giving the resonance-stabilized benzyl ion or, more likely, the tropylium ion directly:

8. C–C bonds next to a heteroatom are frequently cleaved, leaving the charge on the fragment containing the heteroatom whose nonbonding electrons provide resonance stabilization.

$$CH_3 - \overbrace{CH_2} - \overset{+}{\underset{\cdot}{Y}} - R \xrightarrow{-CH_3} CH_2 \overset{+}{=} \overset{}{Y} - R$$

$$+ CH_2 \overset{\ddagger}{\underset{}{Y}} - R$$

$$Y = O, N, \text{ or } S$$

$$R - \underset{\underset{+}{\overset{||}{O}:}}{C} - CH_2 R' \xrightarrow{-R\cdot} \underset{\underset{+}{\overset{|||}{O}:}}{C} - CH_2 R' \leftrightarrow \overset{+}{C} - CH_2 R'$$

9. Cleavage is often associated with elimination of small stable neutral molecules such as carbon monoxide, olefins, water, ammonia, hydrogen sulfide, hydrogen cyanide, mercaptans, ketene, or alcohols.

VIII. REARRANGEMENTS

Rearrangement ions are fragments whose origin cannot be described by simple cleavage of bonds in the parent ion, but are a result of intramolecular atomic rearrangement during fragmentation. Rearrangements involving migration of hydrogen atoms in molecules that contain a heteroatom are especially common. One important example is the so-called McLafferty rearrangement:

$$\underset{CH_2}{\overset{:O^+}{Y-C}} \overset{H}{\underset{CH_2}{CR_2}} \xrightarrow[\substack{Y = H, R, OH, \\ OR, NR_2}]{-R_2C=CH_2} \underset{CH_2}{\overset{\overset{\cdot}{O}^+ \cdots H}{Y-C}} \leftrightarrow$$

$$\underset{CH_2\cdot}{\overset{\overset{\cdot}{O}^+ \cdots H}{Y-C}} \longleftrightarrow \underset{CH_2\cdot}{\overset{\overset{\cdot}{O} \cdots H}{Y-C+}}$$

To undergo a McLafferty rearrangement, a molecule must possess: the appropriately located heteroatoms (e.g., O), π-system (usually a double bond) and abstractable hydrogen (γ to the C = O system).

Such rearrangements often account for prominent characteristic peaks, and are consequently very useful for our purpose. They can frequently be rationalized on the basis of low-energy transitions and increased stability of the products. Rearrangements resulting in elimination of a stable neutral molecule are common (e.g., the olefin product in the McLafferty rearrangement) and will be encountered in the discussion of mass spectra of chemical classes.

Rearrangement peaks can be recognized by considering the mass (m/e) number for fragment ions and for their corresponding molecular ions. A simple (no rearrangement) cleavage of an even-numbered molecular ion gives an odd-numbered fragment ion and simple cleavage of an odd-numbered molecular ion gives an even-numbered fragment. Observation of a fragment ion mass one unit different from that expected for a fragment resulting from simple cleavage (e.g., an even-numbered fragment mass from an even-numbered molecular ion mass) indicates rearrangement of hydrogen has accompanied fragmentation. Rearrangement peaks may be recognized by considering the corollary to the "nitrogen rule" (Section V). Thus an even-numbered peak derived from an even-numbered molecular ion is a result of two cleavages, which may involve a rearrangement.

"Random" rearrangements of hydrocarbons were noted by the early mass spectrometrists in the petroleum industry. For example:

$$\left[\underset{CH_3}{\overset{CH_3}{CH_3 - C - CH_3}} \right]^{\ddagger} \rightarrow [CH_3CH_2]^+$$

These rearrangements defy straightforward explanations based on ground-state theory.

IX. DERIVATIVES

If a compound has low volatility or if the parent mass cannot be determined, it may be possible to prepare a suitable derivative. The derivative selected should provide enhanced volatility, a predictable mode of cleavage, a simplified fragmentation pattern, or increased stability of the parent ion.

Compounds containing several polar groups may have very low volatility, e.g., sugars, amino acids, and dibasic carboxylic acids. Acetylation of hydroxyl and amino groups, and methylation of free acids are obvious and effective choices to increase volatility and give characteristic peaks. Perhaps less immediately obvious is the use of trimethylsilyl derivatives of hydroxyl, amino, sulfhydryl, and carboxylic acid groups. Trimethylsilyl derivatives of sugars and of amino acids are volatile enough to pass through gas chromatographic columns. The molecular ion peak of trimethylsilyl derivatives may not always be

present, but the $M-15$ peak due to cleavage of one of the Si–CH$_3$ bonds is always prominent.

Reduction of ketones to hydrocarbons has been used to elucidate the carbon skeleton of the ketone molecule. Polypeptides have been reduced with LiAlH$_4$ to give polyamino alcohols that were volatile and gave predictable fragmentation patterns. Methylation and trifluoroacylation of tri- and tetrapeptides have been used to obtain mass spectra.

X. MASS SPECTRA OF SOME CHEMICAL CLASSES

Mass spectra of a number of chemical classes are briefly described in this section in terms of the most useful generalizations for identification. For more details, the references cited (in particular, the thorough treatment by Budzikiewicz, Djerassi and Williams [2,20]) should be consulted. The references are selective rather than comprehensive. A table of frequently encountered fragment ions is given in Appendix B. A table of fragments (uncharged) that are commonly eliminated is presented in Appendix C. More exhaustive listings of common fragment ions have been compiled.[10f,29]

HYDROCARBONS

Saturated Hydrocarbons[11]

Most of the work in mass spectrometry has been done on hydrocarbons of interest to the petroleum industry. Rules 1 to 3 (p. 17) apply quite generally; rearrangement peaks, though common, are not usually intense (random rearrangements), and numerous reference spectra are available.

The molecular ion peak (M) of a straight-chain, saturated hydrocarbon is always present, though of low intensity for long-chain compounds. The fragmentation pattern is characterized by clusters of peaks, and the corresponding peaks of each cluster are 14 (CH$_2$) mass units apart. The largest peak in each cluster represents a C_nH_{2n+1} fragment; this is accompanied by C_nH_{2n} and C_nH_{2n-1} fragments. The most intense fragments are at C_3 and C_4, and the fragment intensities decrease in a smooth curve down to M–C$_2$H$_5$; the M–CH$_3$ peak is characteristically very weak or missing. Compounds containing more than 8 carbon atoms show fairly similar spectra; identification then depends on the molecular ion peak.

Spectra of branched saturated hydrocarbons are grossly similar to those of straight-chain compounds, but the smooth curve of decreasing intensities is broken by pre-

ferred fragmentation at each branch. Thus, compare Figures 6a and 6b.

In Figure 6b, the peaks at m/e 169 and 85 represent cleavage on either side of the branch with charge retention on the substituted carbon atom. Subtraction of the molecular weight from the sum of these fragments accounts for the fragment –CH–CH$_3$. Note also the absence of the C$_{11}$ fragment which cannot form by a single cleavage. Finally, the presence of a distinct $M-15$ peak also indicates a methyl branch. The fragment resulting from cleavage at a branch tends to lose a single hydrogen atom so that the resulting C_nH_{2n} peak is prominent and sometimes more intense than the corresponding C_nH_{2n+1} peak. Random rearrangements are common, and the use of reference compounds for final identification is good practice.

A saturated ring in a hydrocarbon increases the relative intensity of the molecular ion peak, and favors cleavage at the bond connecting the ring to the rest of the molecule (Rule 6). Fragmentation of the ring is usually characterized by loss of two carbon atoms as C$_2$H$_4$ (28) and C$_2$H$_5$ (29). This tendency to lose even numbered fragments such as C$_2$H$_4$ gives a spectrum that contains a greater proportion of even-numbered mass ions than the spectrum of an acyclic hydrocarbon. As in branched hydrocarbons, C–C cleavage is accompanied by loss of a hydrogen atom. The characteristic peaks are therefore in the C_nH_{2n-1} and C_nH_{2n-2} series.

Olefins[11]

The molecular ion peak of olefins, especially polyolefins, is usually distinct. Location of the double bond in acyclic olefins is difficult because of its facile migration in the fragments. In cyclic (especially polycyclic) olefins, location of the double bond is frequently evident as a result of a strong tendency for allylic cleavage without much double bond migration (Rule 5). Conjugation with a carbonyl group also fixes the position of the double bond. As with saturated hydrocarbons, acyclic olefins are characterized by clusters of peaks at intervals of 14 units. In these clusters the C_nH_{2n-1} and C_nH_{2n} peaks are more intense than the C_nH_{2n+1} peaks.

Cyclic olefins usually show a distinct molecular ion peak. A unique mode of cleavage is a type of homolytic retro-Diels-Alder reaction as shown by limonene.

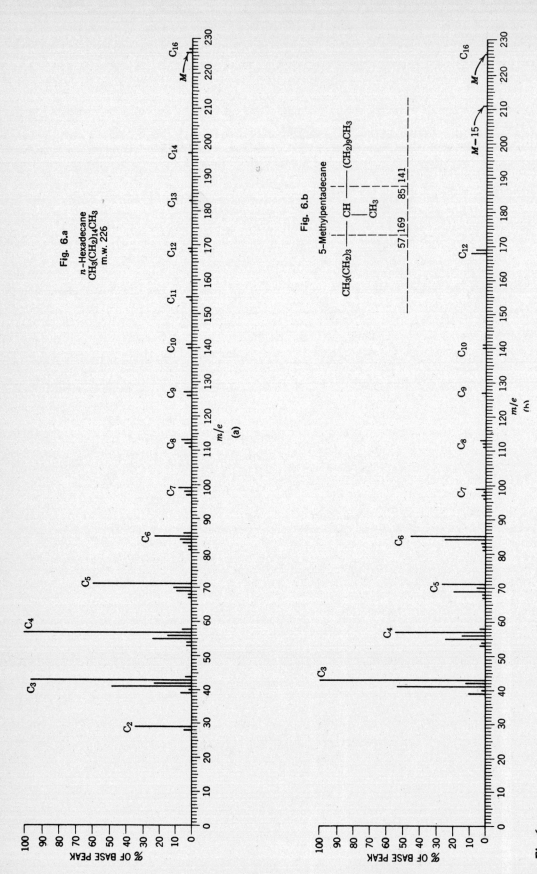

Fig. 6.a

n–Hexadecane
$CH_3(CH_2)_{14}CH_3$
m.w. 226

Fig. 6.b

5–Methylpentadecane

$CH_3(CH_2)_3 \quad CH \quad (CH_2)_9CH_3$
$\quad\quad\quad\quad\quad | $
$\quad\quad\quad\quad\quad CH_3$

$57 | 169 \quad\quad 85 | 141$

Fig. 6.
Isomeric C^{16} hydrocarbons.

Aralkyl Hydrocarbons

An aromatic ring in a molecule stabilizes the molecular ion peak (Rule 4), which is usually sufficiently large that accurate measurements can be made on the $M + 1$ and $M + 2$ peaks.

A prominent peak (often the base peak) at m/e 91 ($C_6H_5CH_2{}^+$) is indicative of an alkyl substituted benzene ring. Branching at the α-carbon leads to masses higher than 91 by increments of 14, the largest substituent being eliminated most readily (Rule 3). The mere presence of a peak at mass 91, however, does not preclude branching at the α-carbon because this highly stabilized fragment may result from rearrangements. A distinct and sometimes prominent $M - 1$ peak results from similar benzylic cleavage of a C–H bond.

It has been shown that in most cases the ion of mass 91 is a tropylium rather than a benzylic cation. This explains the ready loss of a methyl group from xylenes although toluene does not easily lose a methyl group. The incipient molecular ion rearranges to the parent tropylium radical ion, which then cleaves to the simple tropylium ion (C_7H_7+):

$$m/e\ 91$$

The frequently observed peak at m/e 65 results from elimination of a neutral acetylene molecule from the tropylium ion:

$$m/e\ 65$$

Hydrogen rearrangement with elimination of a neutral olefin molecule accounts for the peak at m/e 92 observed when the alkyl group is longer than C_2.

$$m/e\ 92$$

A characteristic cluster of ions due to α-cleavage and hydrogen rearrangements of monalkylbenzenes appears at m/e 77 ($C_6H_5{}^+$), 78 ($C_6H_6{}^+$), and 79 ($C_6H_7{}^+$).

Alkylated polyphenyls and alkylated polycyclic aro-matic hydrocarbons exhibit the same β-cleavage as alkylbenzene compounds.

HYDROXY COMPOUNDS

Alcohols

The molecular ion peak of a primary or secondary alcohol is quite small and for a tertiary alcohol is undetectable. The molecular ion of n-pentanol is extremely weak compared with its near homologs. The expedients mentioned above may be used to obtain the molecular weight.

Cleavage of the C–C bond next to the oxygen atom is of general occurrence (Rule 8). Thus, primary alcohols show a prominent peak due to $CH_2{=}\overset{+}{\ddot{O}}H$ (m/e 31). Secondary and tertiary alcohols cleave analogously to give a prominent peak due to $\begin{smallmatrix}R\\ \\H\end{smallmatrix}C{=}\overset{+}{\ddot{O}}H$ (m/e 45, 59, 73, etc.), and $\begin{smallmatrix}R\\ \\R'\end{smallmatrix}C{=}\overset{+}{\ddot{O}}H$ (m/e 59, 73, 87 etc.), respectively. The largest substituent is expelled most readily (Rule 3).

$$(R'' > R' \text{ or } R) = \text{alkyl}$$

When R and/or $R' = H$, an $M - 1$ peak can usually be seen.

Primary alcohols, in addition to the principal C–C cleavage next to the oxygen atom, show a homologous series of peaks of progressively decreasing intensity at m/e 45, 59, 73, ..., resulting from cleavage at C–C bonds successively removed from the oxygen atom. In long-chain ($> C_6$) alcohols, the fragmentation becomes dominated by the hydrocarbon pattern; in fact, the spectrum resembles that of the corresponding olefin.

The spectrum in the vicinity of the very weak or missing molecular ion peak of a primary alcohol is sometimes complicated by weak $M - 2$ ($R{-}\overset{\cdot}{C}H{-}\overset{\cdot}{\ddot{O}}$) and $M - 3$ ($R{-}C{\equiv}\overset{+}{O}$) peaks.

A distinct and sometimes prominent peak can usually be found at $M - 18$ from loss of water. This peak is most noticeable in spectra of primary alcohols. This elimination by electron impact has been rationalized as follows:

This pathway is consistent with the loss of the OH and γ-hydrogen ($n = 1$) or δ-hydrogen ($n = 2$); the ring structure is not proven by the observations and is merely one possible structure for the product radical cation. The $M - 18$ peak is frequently exaggerated by thermal decomposition of higher alcohols on hot inlet surfaces. Elimination of water, together with elimination of an olefin from primary alcohols, accounts for the presence of a peak at $M - (\text{olefin} + H_2O)$, i.e., a peak at $M - 46$, $M - 74$, $M - 102, \ldots$

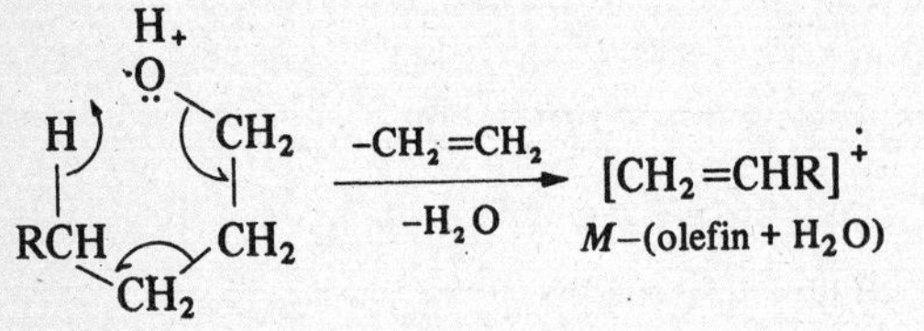

The olefinic ion then decomposes by successive eliminations of ethylene.

Alcohols containing branched methyl groups (e.g., terpene alcohols) frequently show a fairly strong peak at $M - 33$ resulting from loss of CH_3 and H_2O.

Cyclic alcohols undergo fragmentation by complicated pathways;[2] e.g., cyclohexanol ($M = m/e\ 100$) forms $C_6H_{11}O^+$ by simple loss of the α-hydrogen, loses H_2O to form $C_6H_{10}+$ (which appears to have more than one possible bridged bicyclic structure), and forms C_3H_5O (m/e 57) by a complex ring cleavage pathway.

A peak at m/e 31 (see above) is quite diagnostic for a primary alcohol provided it is more intense than peaks at m/e 45, 59, 73.... However, the first-formed ion of a secondary alcohol can decompose further to give a moderately intense m/e 31 ion.

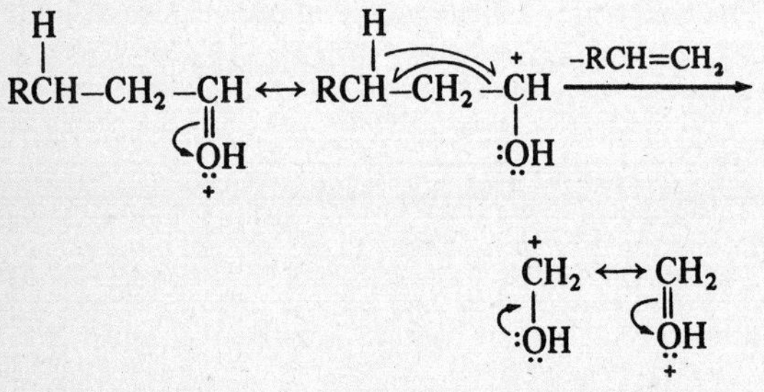

Figure 7 gives the characteristic spectra of isomeric primary, secondary, and tertiary C_5 alcohols.

Benzyl alcohols and their substituted homologs and analogs constitute a distinct class. Generally the parent peak is strong. A moderate benzylic peak ($M - OH$) is present as expected from cleavage beta to the ring. A complicated sequence leads to prominent $M - 1$, $M - 2$, and $M - 3$ peaks. Benzyl alcohol itself fragments to give sequentially the $M - 1$ ion, the C_6H_7 ion by loss of CO, and the C_6H_5 ion by loss of H_2:

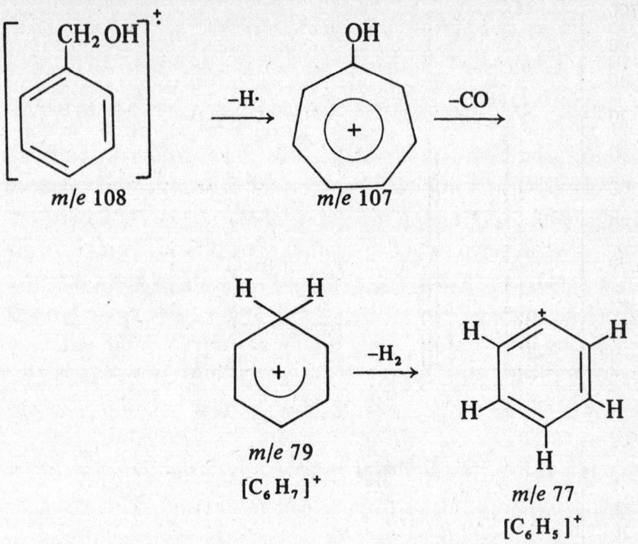

Loss of H_2O to give a distinct $M - 18$ peak is a common feature, especially pronounced and mechanistically straightforward in some ortho substituted benzyl alcohols.

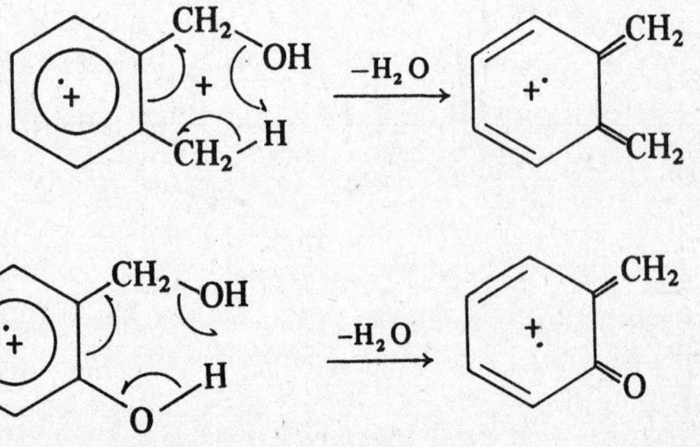

The aromatic cluster at m/e 77, 78, and 79 resulting from complex degradation is prominent here also.

Phenols

A conspicuous molecular ion peak facilitates identification of phenols. In phenol itself, the molecular ion peak is the base peak and the $M - 1$ peak is small. In cresols, the $M - 1$ peak is larger than the parent peak as a result of a facile benzylic C–H cleavage. A rearrangement peak at m/e 77 and peaks resulting from loss of CO ($M - 28$) and CHO ($M - 29$) are usually found in phenols:

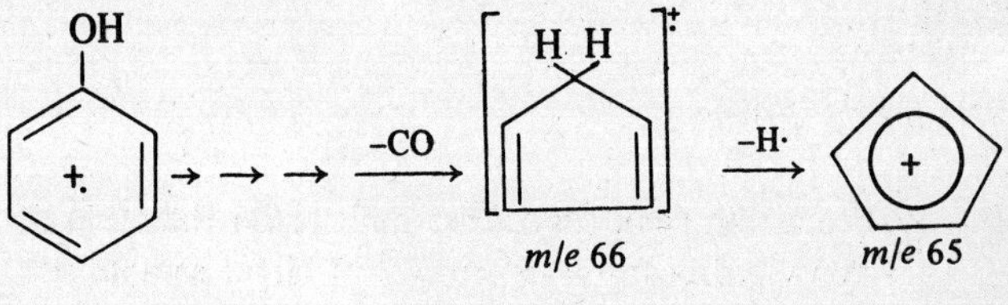

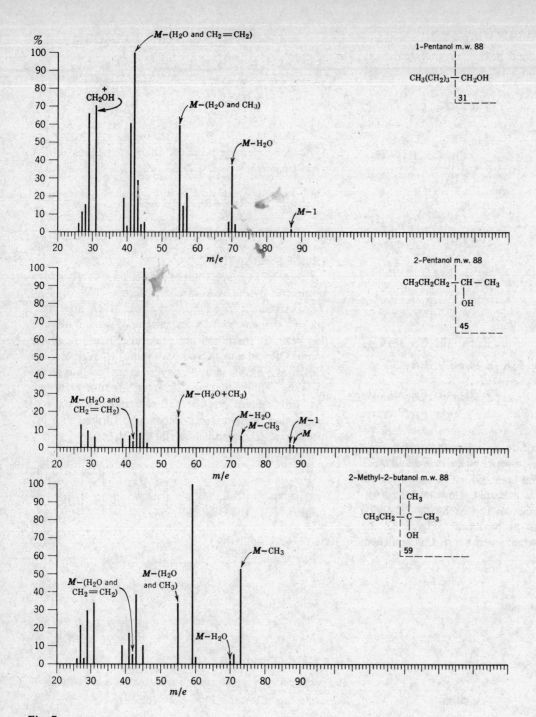

Fig. 7.
Isomeric pentanols.

ETHERS

Aliphatic Ethers

The molecular ion peak (two mass units larger than that of an analogous hydrocarbon) is small, but larger sample size will usually make the molecular ion peak or the $M + 1$ peak (result of H · transfer during ion-molecule collision, see part 2, Section V) obvious.

The presence of an oxygen atom can be deduced from strong peaks at m/e 31, 45, 59, 73, ... These peaks represent the RO^+ and $ROCH_2^+$ fragments.

Fragmentation occurs in two principal ways:

a. Cleavage of the C–C bond next to the oxygen atom (α-β bond, Rule 8, Section VII)

$$RCH_2-CH_2-\overset{CH}{\underset{CH_3}{|}}-\overset{+}{O}-CH_2-CH_3 \xrightarrow{-RCH_2CH_2\cdot}$$

In Fig. 8
R = H

$$CH\overset{+}{=}\overset{|}{\underset{CH_3}{O}}-CH_2-CH_3 \updownarrow$$

$$\overset{+}{CH}-\overset{..}{\underset{CH_3}{O}}-CH_2-CH_3 \quad m/e\ 73$$

$$RCH_2-CH_2-\overset{CH}{\underset{CH_3}{|}}-\overset{+}{O}-CH_2-CH_3 \xrightarrow{-CH_3\cdot}$$

$$RCH_2-CH_2\overset{CH}{\underset{CH_3}{|}}-\overset{+}{O}=CH_2$$

In Fig. 8, R=H, m/e 87 $\updownarrow$

$$RCH_2-CH_2\overset{CH}{\underset{CH_3}{|}}-\overset{+}{O}-CH_2$$

One or the other of these oxygen-containing ions may account for the base peak. In the case shown, the first cleavage, i.e., at the branched carbon atom to lose the larger fragment, is preferred. However, the first-formed fragment decomposes further by the following process, often to give the base peak (Figure 8); the decomposition is important when the α–C is branched.

$$\overset{CH=\overset{+}{O}-CH_2}{\underset{CH_3\ H-CH_2}{|}} \xrightarrow{-CH_2=CH_2} \overset{CH=\overset{+}{O}H}{\underset{CH_3}{|}}$$

$$\downarrow$$

$$\overset{+}{CH}-\overset{..}{O}H \\ \underset{CH_3}{|}$$

m/e 45

b. C–O bond cleavage with the charge remaining on the alkyl fragment.

$$R\overset{+}{\overset{..}{O}}-R' \xrightarrow{-\cdot\overset{..}{O}R'} \overset{+}{R}$$

$$R-\overset{+}{\overset{..}{O}}-R' \xrightarrow{-R\overset{..}{O}\cdot} \overset{+}{R'}$$

(Rule 8 Section VII)

As expected, the spectrum of long chain ethers becomes dominated by the hydrocarbon pattern.

Acetals are a special class of ethers. Their mass spectra are characterized by an extremely weak molecular ion peak, by the prominent peaks at M minus R, and M minus OR and a weak peak at M minus H. Each of these cleavages is mediated by an oxygen atom and thus facile. As usual, elimination of the largest group is preferred. As with aliphatic ethers, the first-formed oxygen-containing fragments can decompose further with hydrogen rearrangement and olefin elimination.

$$\left[\overset{H}{\underset{OR}{R\text{--}C\text{--}OR}}\right]^{\dot{+}} \rightarrow \left[\overset{}{\underset{OR}{R\text{--}C\text{--}OR}}\right]^+ + \left[\overset{H}{\underset{}{R\text{--}C\text{--}OR}}\right]^+ + \left[\overset{C\text{--}OR}{\underset{OR}{}}\right]^+$$

Ketals behave similarly.

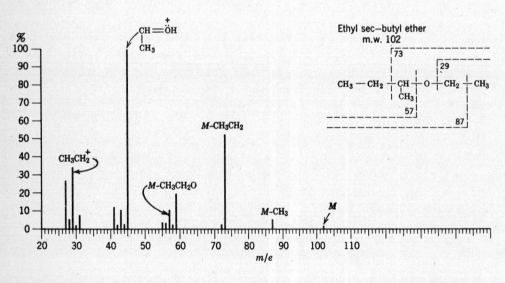

Fig. 8.
Ethyl sec-butyl ether.

Aromatic Ethers

The molecular ion peak of aromatic ethers is prominent. Primary cleavage occurs at the bond beta to the ring, and the first-formed ion can decompose further. Thus anisole, m.w. 108, gives ions of m/e 93 and 65.

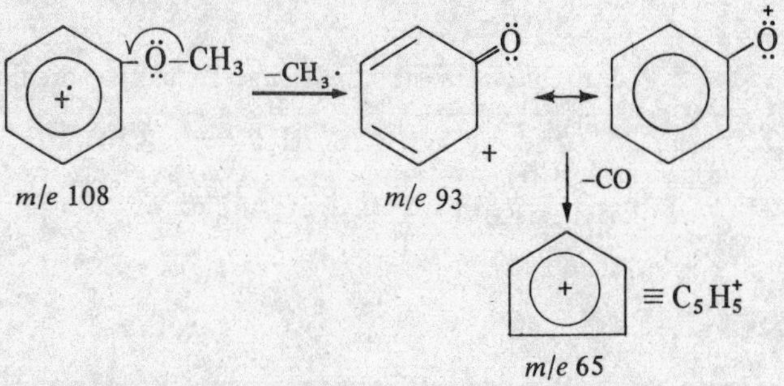

m/e 108 m/e 93 $\equiv C_5H_5^+$ m/e 65

The characteristic aromatic peaks at m/e 78 and 77 may arise from anisole as follows.

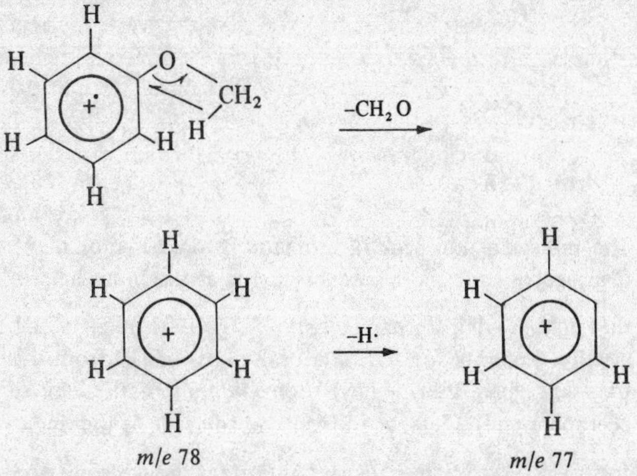

m/e 78 m/e 77

When the alkyl portion of an aromatic alkyl ether is C_2 or larger, cleavage beta to the ring is accompanied by hydrogen rearrangement as noted above for alkyl-benzenes. Clearly, cleavage is mediated by the ring rather than by the oxygen atom; C–C cleavage next to the oxygen atom is insignificant.

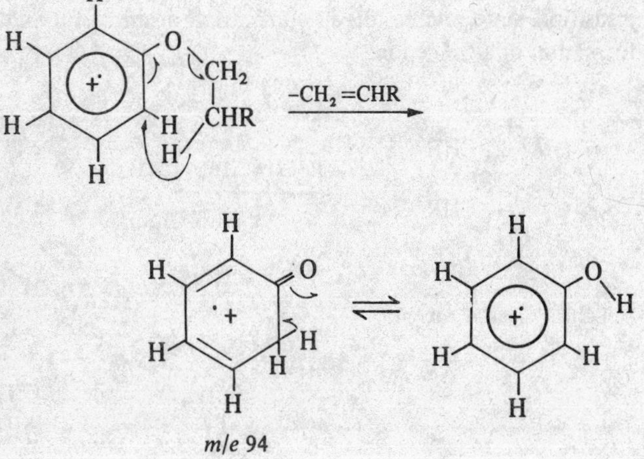

m/e 94

Diphenyl ethers show peaks at $M - H$, $M - CO$, and $M - CHO$ by complex rearrangements.

KETONES

Aliphatic Ketones

The molecular ion peak of ketones is usually quite pronounced. Major fragmentation peaks of aliphatic ketones result from cleavage at the C–C bonds adjacent (α) to the oxygen atom, the charge remaining with the oxygenated fragment. Thus, as with alcohols and ethers, cleavage is again at the C–C bond next to the oxygen atom.

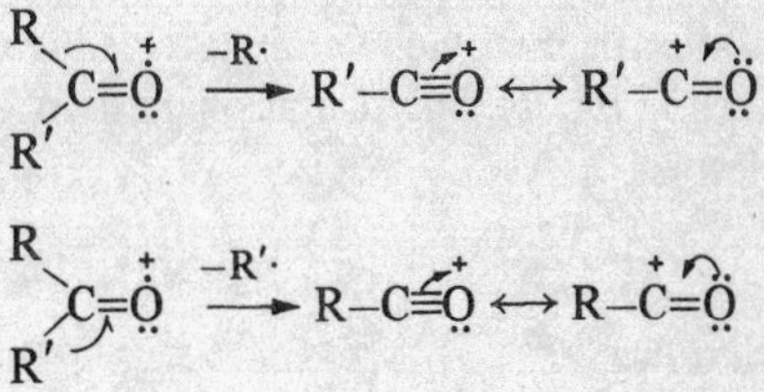

This cleavage gives rise to a peak at m/e 43 or 57 or 71.... The base peak very often results from loss of the larger alkyl group.

When one of the alkyl chains attached to the C=O group is C_3 or larger, cleavage of the C–C bond once removed ($\alpha\beta$ bond) from the C=O group occurs with hydrogen rearrangement to give a major peak at m/e 58 or 72 or 86.... (McLafferty rearrangement):

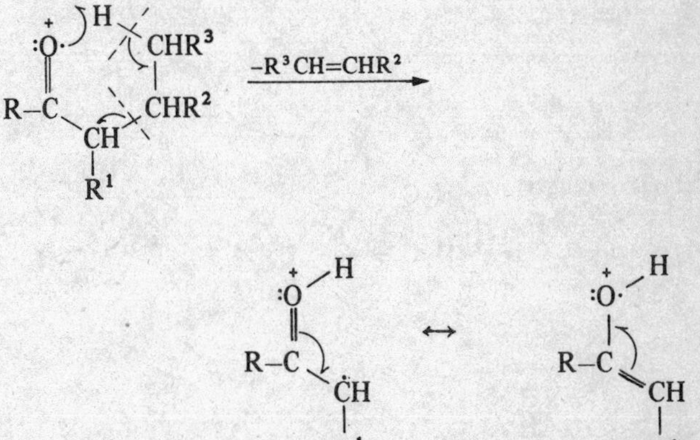

Simple cleavage of the $\alpha\beta$ bond, which does not occur to any extent, would give an ion of low stability with two adjacent positive centers $R\overset{\delta +}{-}\overset{+}{C}-CH_2$. When R is C_3 or

$$\underset{\underset{\delta -}{\overset{\|}{O}}}{}$$

longer, the first-formed ion can cleave again with hydrogen rearrangement:

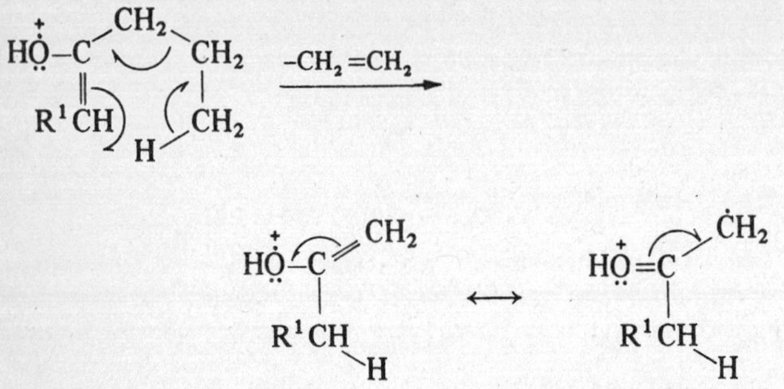

$-CH_2=CH_2 \longrightarrow$

Note that in long-chain ketones the hydrocarbon peaks are indistinguishable (without the aid of high-resolution techniques) from the acyl peaks. The multiple cleavage modes in ketones sometimes make difficult the determination of the carbon chain configuration. Reduction of the carbonyl group to a methylene group yields the corresponding hydrocarbon whose fragmentation pattern leads to the carbon skeleton.

Cyclic Ketones[60-62]

The molecular ion peak in cyclic ketones is prominent. As with aliphatic ketones, the primary cleavage of cyclic ketones is adjacent to the C=O group, but the ion thus formed must undergo further cleavage in order to produce a fragment. The base peak in the spectrum of cyclopentanone and of cyclohexanone is m/e 55. The mechanism is similar in both cases: hydrogen rearrangement from a primary radical to a conjugated secondary radical followed by formation of the resonance-stable ion, m/e 55.

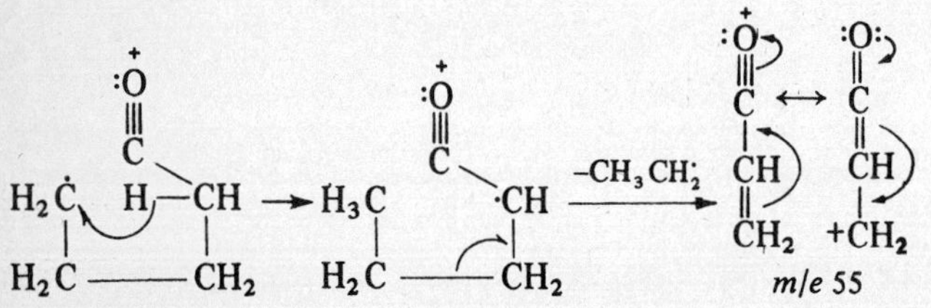

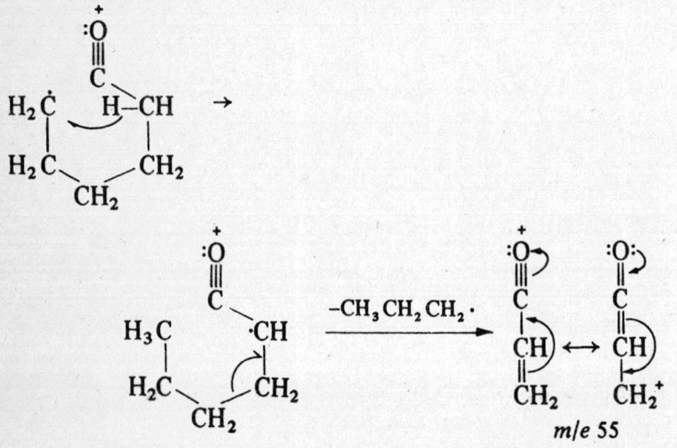

$-CH_3CH_2 \cdot$ m/e 55

$\rightarrow$

$-CH_3CH_2CH_2 \cdot$ m/e 55

The other distinctive peaks at m/e 83 and 42 in the spectrum of cyclohexanone have been rationalized as follows.

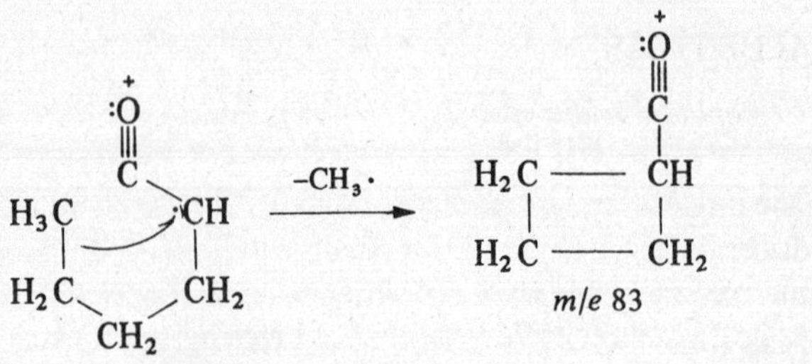

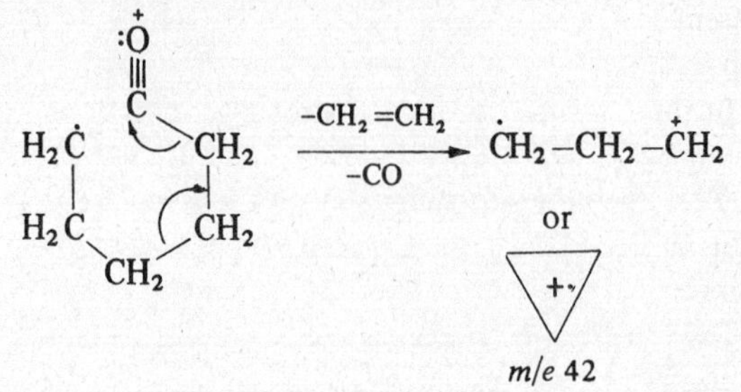

$-CH_3 \cdot$ m/e 83

$-CH_2=CH_2$
$-CO$ $\dot{C}H_2-CH_2-\overset{+}{C}H_2$

or

m/e 42

Aromatic Ketones

The molecular ion peak of aromatic ketones is prominent. Cleavage of aryl alkyl ketones occurs at the bond beta to the ring, leaving a characteristic $Ar\overset{+}{C}\equiv O \colon$ fragment which usually accounts for the base peak. Loss of CO from this fragment gives the "phenyl" ion (m/e 77 in the case of acetophenone). Cleavage of the bond adjacent to the ring to form a $R\overset{+}{C}\equiv O$ fragment is less important though somewhat enhanced by electron-withdrawing groups (and diminished by electron-donating groups) in the *para* position of the Ar group.

When the alkyl chain is C_3 or longer, cleavage of the C–C bond once-removed from the C=O group occurs with hydrogen rearrangement. This is the same cleavage noted for aliphatic ketones that proceeds through a cyclic transition state and results in elimination of an olefin and formation of a stable ion.

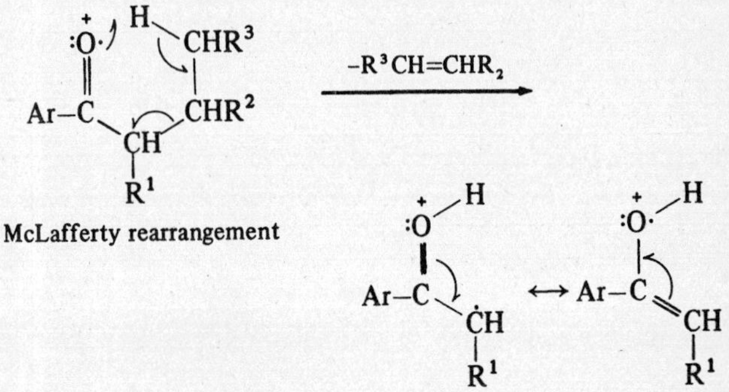

$-R^3CH=CHR_2 \longrightarrow$

McLafferty rearrangement

Unsymmetrical diarylketones cleave to give each of the $ArC\equiv\overset{+}{\underset{..}{O}}$ ions expected.

ALDEHYDES

Aliphatic Aldehydes

The molecular ion peak of aliphatic aldehydes is usually discernible. Cleavage of the C–H and C–C bonds next to the oxygen atom results in an $M - 1$ peak and in an $M - R$ peak (m/e 29, CHO+). The $M - 1$ peak is a good diagnostic peak even for long-chain aldehydes, but the m/e 29 peak present in C_4 and higher aldehydes is due to the hydrocarbon $C_2H_5^+$ ion.

In the C_4 and higher aldehydes, McLafferty cleavage of the $\alpha\beta$ C–C bond occurs to give a major peak at m/e 44, 58, or 72, . . . , depending on the α-substituents. This is the resonance-stabilized ion formed through the cyclic transition state as shown above for aliphatic ketones (R=H):

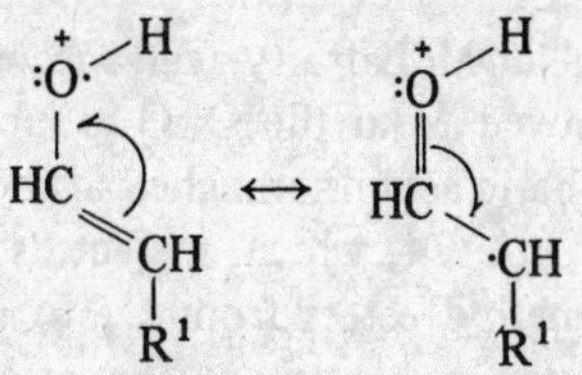

In straight chain aldehydes, the other unique, diagnostic peaks are at $M - 18$ (loss of water), $M - 28$ (loss of ethylene), $M - 43$ (loss of CH_2=CH–O ·), and $M - 44$ (loss of CH_2=CH–OH). The rearrangements leading to these peaks have been rationalized (see Reference 2). As the chain lengthens, the hydrocarbon pattern (m/e 29, 43, 57, 71, . . .) becomes dominant. These features are evident in the spectrum of nonanal (Figure 9).

Aromatic Aldehydes

Aromatic aldehydes are characterized by a large molecular ion peak and by an $M - 1$ peak ($Ar-\overset{+}{C}\equiv\overset{..}{O}$) that is always large and may be larger than the molecular ion. The $M - 1$ ion $\phi-C\equiv\overset{..}{\overset{+}{O}}$ eliminates CO to give the phenyl ion (m/e 77), which in turn eliminates HC≡CH to give the $C_4H_3^+$ ion (m/e 51).

CARBOXYLIC ACIDS

Aliphatic Acids

The molecular ion peak of a straight-chain monocarboxylic acid is weak but usually discernible. The most characteristic

(sometimes the base) peak is m/e 60 due to the McLafferty rearrangement. Branching at the α-carbon increases this fragment by the mass of the substituent.

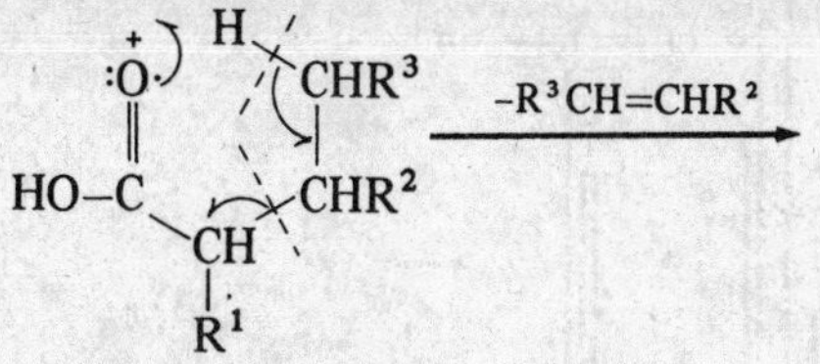

McLafferty rearrangement

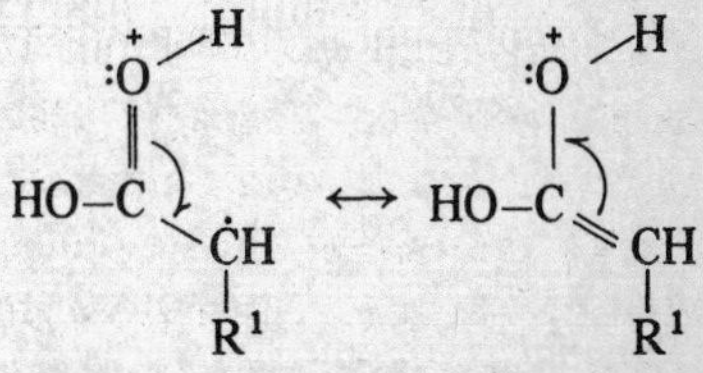

In short-chain acids, peaks at $M - OH$ and $M - COOH$ are prominent; these represent cleavage of bonds next to C=O. In long-chain acids, the spectrum consists of two series of peaks resulting from cleavage at each C–C bond with retention of charge either on the oxygen-containing fragment (m/e 45, 59, 73, 87, . . .) or on the alkyl fragment (m/e 29, 43, 57, 71, 85, . . .). As previously discussed, the hydrocarbon pattern also shows peaks at m/e 27, 28; 41, 42; 55, 56; 69, 70; . . . . In summary, besides the McLafferty rearrangement peak, the spectrum of a long-chain acid resembles the series of "hydrocarbon" clusters at interval of 14 mass units. In each cluster, however, is a prominent peak at $C_nH_{2n-1}O_2$. Caproic acid (m.w. 116), for example, cleaves as follows:

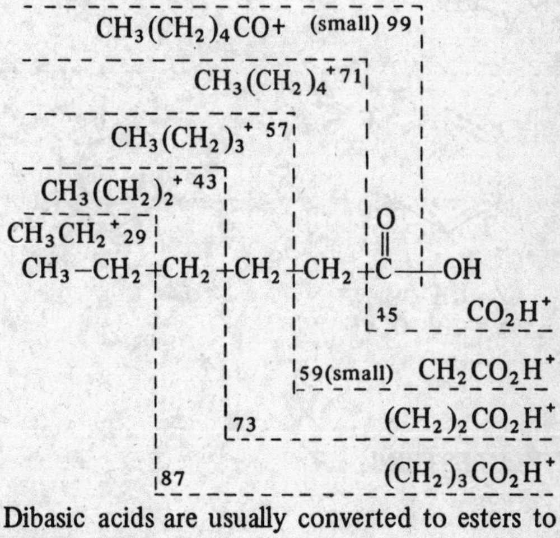

Dibasic acids are usually converted to esters to increase volatility.

Aromatic Acids

The molecular ion peak of aromatic acids is large. The other prominent peaks are formed by loss of OH ($M - 17$) and of

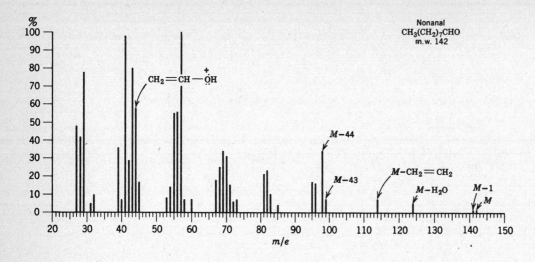

Fig. 9.
Nonanal.

COOH ($M - 45$). Loss of H_2O ($M - 18$) is noted if a hydrogen-bearing ortho group is available. This is one example of the general "ortho effect" noted when the substituents can form a 6-membered transition state to facilitate loss of a neutral molecule of H_2O, ROH or NH_3.

$$Z = OH, OR, NH_2$$
$$Y = CH_2, O, NH$$

CARBOXYLIC ESTERS

Aliphatic Esters

The molecular ion peak of a methyl ester of a straight-chain aliphatic acid is usually distinct. Even waxes usually show a discernible parent peak. The parent peak is weak in the range m/e 130 to about 200, but becomes somewhat more intense beyond this range. The most characteristic peak is due to the familiar McLafferty rearrangement and cleavage one bond removed from the C=O group. Thus a methyl ester of an aliphatic acid unbranched at the α-carbon gives a strong peak at m/e 74, which, in fact, is the base peak in straight-chain methyl esters from C_6 to C_{26}. The alcohol moiety and/or the α-substituent can often be deduced by the location of the peak resulting from this cleavage.

McLafferty rearrangement

Four ions can result from bond cleavage next to C=O.

$$\left[\begin{matrix} O \\ \| \\ R \! + \! C \! - \! OR' \end{matrix}\right]^+ \rightarrow R^+ \text{ and } \left[\begin{matrix} O \\ \| \\ C \! - \! OR' \end{matrix}\right]^+$$

$$\left[\begin{matrix} O \\ \| \\ R \! - \! C \! + \! OR' \end{matrix}\right]^+ \rightarrow R \! - \! C \! \equiv \! \overset{+}{O} \text{ and } [OR']^+$$

The ion R^+ is prominent in the short chain esters, but diminishes rapidly with increasing chain length and is barely perceptible in methyl caproate. The ion $R-C\equiv\overset{+}{O}$ gives an excellent diagnostic peak for esters. In methyl esters it occurs at $M - 31$. It is the base peak in methyl acetate, and is still 4% of the base peak in the C_{26} methyl ester. The ions

$[OR']^+$ and $\begin{bmatrix} O \\ \parallel \\ COR' \end{bmatrix}^+$ are usually of little importance. The latter is discernible when $R' = CH_3$ (see m/e 59 peak of Figure 10).

Consider, first, esters in which the acid portion is the predominant portion of the molecule. The fragmentation pattern for methyl esters of straight-chain acids can be described in the same terms used for the pattern of the free acid. Cleavage at each C–C bond gives an alkyl ion (m/e 29, 43, 57, ...) and an oxygen-containing ion, $C_nH_{2n-1}O_2{}^+$ (59, 73, 87, ...). Thus there are hydrocarbon clusters at intervals of 14 mass units; in each cluster is a prominent peak at $C_nH_{2n-1}O_2$. The peak (m/e 87) representing the ion $[CH_2CH_2COOCH_3]^+$ is always more intense than its homologs, but the reason is not immediately obvious. However, it seems clear that the $C_nH_{2n-1}O_2$ ions do not all arise from simple cleavage.

The spectrum of methyl caprylate is presented as Figure 10. This spectrum illustrates one difficulty previously mentioned (Section V, part 2) in using the $M + 1$ peak to arrive at a molecular formula. The measured value for the $M + 1$ peak is 12.9%. The calculated value (Appendix A) is 10.0%. The measured value is high due to an ion-molecule reaction because a relatively large sample was used in order to see the weak parent peak. The utility of the 5-galvanometer recorder is implied by this spectrum; thus the $M + 1$ peak can be measured accurately even though its intensity is only 0.11% of that of the base peak. The accuracy of the $M + 2$ peak measurement is marginal.

Now let us consider esters in which the alcohol portion is the predominant portion of the molecule. Esters of fatty alcohols (except methyl esters) eliminate a molecule of acid in the same manner that alcohols eliminate water. A scheme similar to that described earlier for alcohols, involving a single hydrogen transfer to the alcohol oxygen of the ester, can be written. An alternative mechanism involves a hydride transfer to the carbonyl oxygen (McLafferty rearrangement):

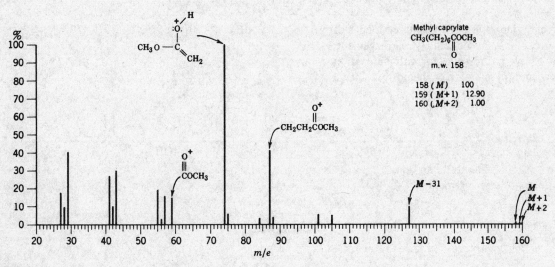

The preceding loss of acetic acid is so facile in steroidal acetates that they frequently show no detectable molecular ion peak. Steroidal systems also seem unusual in that they often display significant molecular ions as alcohols, even when the corresponding acetates do not.[6]

Esters of long-chain alcohols show a diagnostic peak at m/e 61, 75, or 89, ... from elimination of the alkyl moiety and transfer of *two* hydrogen atoms to the fragment containing the oxygen atoms.

Methyl caprylate
$CH_3(CH_2)_6COCH_3$
$\quad\quad\quad\quad\;\; O$
m.w. 158

158 (M)	100
159 ($M+1$)	12.90
160 ($M+2$)	1.00

Fig. 10.
Methyl caprylate.

Esters of dibasic acids $ROC(CH_2)_nCOR$, in general, give
$$ROC(CH_2)_nCOR$$
recognizable parent peaks. Intense peaks are found at $(ROC(CH_2)_nC)^+$ and at $(ROC(CH_2)_n)^+$.

Benzyl and Phenyl Esters

Benzyl acetate (also furfuryl acetate and other similar acetates) and phenyl acetate eliminate the neutral molecule ketene, frequently to form the base peak.

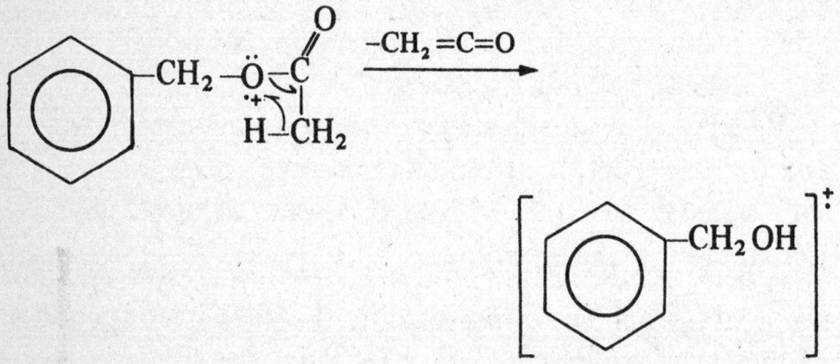

$$m/e\ 108$$

Of course, the m/e 43 peak ($CH_3C\equiv\overset{+}{O}$) and m/e 91 (C_7H_7+) peaks are prominent for benzyl acetate.

Esters of Aromatic Acids

The molecular ion peak of methyl esters of aromatic acids is prominent. As the size of the alcohol moiety increases, the intensity of the parent peak decreases rapidly to practically zero at C_5. The base peak results from elimination of $\cdot OR$, and elimination of $\cdot COOR$ accounts for another prominent peak. In methyl esters, these peaks are at $M - 31$, and $M - 59$, respectively.

As the alkyl moiety increases in length, three modes of cleavage become important: (a) McLafferty rearrangement, (b) rearrangement of two hydrogen atoms with elimination of an allylic radical, and (c) retention of the positive charge by the alkyl group.

(a)

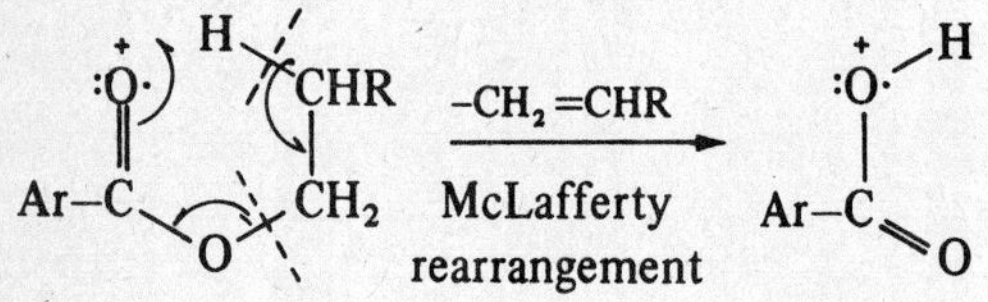

(b)

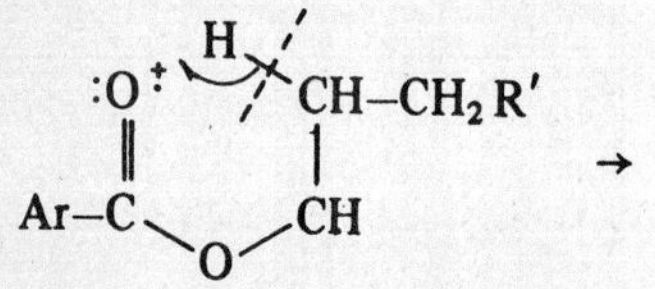

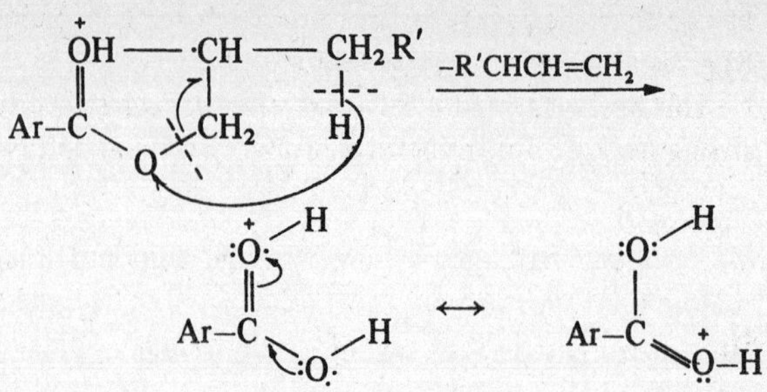

(c) 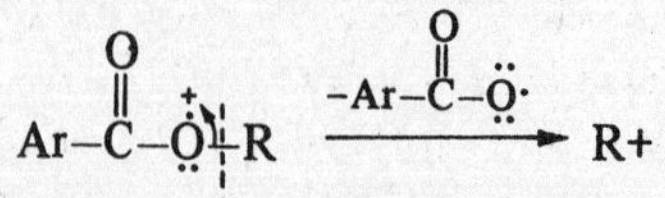 $Ar-\overset{O}{\underset{}{C}}-\overset{+}{\underset{}{O}}\overset{\cdot}{-}R \xrightarrow{-Ar-C-\overset{\cdot\cdot}{O}\cdot} R+$

Appropriately, ortho-substituted benzoates eliminate ROH through the general "ortho" effect described above under aromatic acids. Thus the base peak in the spectrum of methyl salicylate is m/e 120; this ion eliminates carbon monoxide to give a strong peak at m/e 92.

A strong characteristic peak at mass 149 is found in the spectra of all esters of phthalic acid, starting with the diethyl ester. This peak is not significant in the dimethyl or methyl ethyl ester of phthalic acid, nor in esters of isophthalic or terephthalic acids, all of which give the expected peaks at $M - R$, $M - 2R$, $M - COOR$, and $M - 2COOR$. Since long-chain phthalate esters are widely used as plasticizers and in oil diffusion pumps (often, in the diffusion pump of the mass spectrometer inlet), a strong peak at m/e 149 may indicate contamination. The m/e 149 fragment is probably formed by two ester cleavages involving rearrangement of two hydrogen atoms and one hydrogen atom, followed by elimination of H_2O:

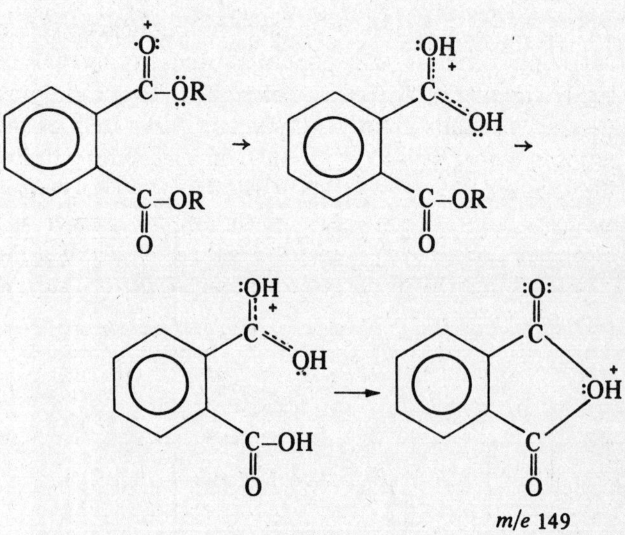

$$m/e\ 149$$

LACTONES

The molecular ion peak of 5-membered ring lactones is distinct, but is weaker when an alkyl substituent is present

at C_4. Facile cleavage of the side chain at C_4 (Rules 3 and 8) gives a strong peak at M minus alkyl.

The base peak (m/e 56) of γ-valerolactone and the same strong peak of butyrolactone probably arise as follows.

$$\xrightarrow{-CH_3CH=O}$$

m/e 56

Labeling experiments indicate that some of the m/e 56 peak in γ-valerolactone arises from the $C_4H_8^+$ ion. The other intense peaks in γ-valerolactone are at m/e 27 ($C_2H_3^+$), 28 ($C_2H_4^+$), 29 ($C_2H_5^+$), 41 ($C_3H_5^+$), and 43 ($C_3H_7^+$), and 85 ($C_4H_5O_2^+$, loss of the methyl group). In butyrolactone, there are strong peaks at m/e 27, 28, 29, 41, and 42 ($C_3H_6^+$).

AMINES

Aliphatic Amines

The molecular ion peak of an aliphatic monoamine is an odd number but is usually quite weak, and, in long-chain or highly branched amines, undetectable. The base peak frequently results from C–C cleavage next ($\alpha\beta$) to the nitrogen atom (Rule 8); for primary amines unbranched at the α-atom, this is m/e 30 ($CH_2NH_2^+$). This cleavage accounts for the base peak in all primary amines and secondary and tertiary amines that are not branched at the α-carbon. Cleavage of the largest branch at the α-C atom is preferred.

$(R^2 > R^1 \text{ or } R) = \text{alkyl}$

When R and/or R^1 = H, an $M - 1$ peak is usually visible. This is the same type of cleavage noted above for alcohols. The effect is more pronounced in amines because of the better resonance stabilization of the ion fragment by the less electronegative N atom compared with the O atom.

Primary straight-chain amines show a homologous series

of peaks of progressively decreasing intensity (the cleavage at the ϵ bond is slightly more intense than at the neighboring bonds) at m/e 30, 44, 58, . . . resulting from cleavage at C–C bonds successively removed from the nitrogen atom with retention of the charge on the N-containing fragment. These peaks are accompanied by the hydrocarbon pattern of C_nH_{2n+1}, C_nH_{2n}, and C_nH_{2n-1} ions. Thus we note characteristic clusters at intervals of 14 mass units, each cluster containing a peak due to a $C_nH_{2n+2}N$ ion. Because of the very facile cleavage to form the base peak, the fragmentation pattern in the high mass region becomes extremely weak.

Cyclic fragments apparently occur during the fragmentation of longer chain amines:

$n = 3,4$ m/e 72,86

A peak at m/e 30 is good though not conclusive evidence for a straight-chain primary amine. Further decomposition of the first-formed ion from a secondary or tertiary amine leads to a peak at m/e 30, 44, 58, or 72, ... This is a process similar to that described for aliphatic alcohols and ethers above, and similarly is enhanced by branching at one of the α-carbon atoms:

$R'' = CH_3$, m/e 44, more intense
$R'' = H$, m/e 30, less in$^+$ense

Cleavage of amino acid esters occurs at both C–C bonds (a, b, below) next to the nitrogen atom, loss of the carbalkoxy group being preferred (a). The aliphatic amine fragment decomposes further to give a peak at m/e 30.

m/e 30

CYCLIC AMINES

In contrast to acyclic amines, the molecular ion peaks of cyclic amines are usually intense; e.g., the parent peak of pyrrolidine is strong. Primary cleavage at the bonds next to the N atom leads either to loss of an α-H atom to give a strong $M-1$ peak, or to opening of the ring; the latter event is followed by elimination of ethylene to give $\cdot CH_2 \overset{+}{N}H=CH_2$ (m/e 43, base peak), thence by loss of a hydrogen atom to give $CH_2=\overset{+}{N}=CH_2$ (m/e 42). N-methyl pyrrolidine also gives a $C_2H_4N^+$ (m/e 42) peak, apparently by more than one pathway.

Piperidine likewise shows a strong molecular ion and $M-1$ (base) peak. Ring opening followed by several available sequences leads to characteristic peaks at m/e 70, 57, 56, 44, 43, 42, 30, 29, and 28. Substituents cleave at the ring junction (Rule 6).

Aromatic Amines

The molecular ion peak (odd number) of an aromatic monoamine is intense. Loss of one of the amino H atoms of aniline gives a moderately intense $M-1$ peak; loss of a neutral molecule of HCN followed by loss of a hydrogen atom gives prominent peaks at m/e 66 and 65, respectively.

It was noted above that cleavage of alkylaryl ethers occurs with rearrangement involving cleavage of the ArO—R bond; i.e., cleavage was controlled by the ring rather than by the oxygen atom. In the case of alkylaryl amines, cleavage of the C—C bond next to the nitrogen atom is dominant; i.e., the heteroatom controls cleavage:

m/e 106

ALIPHATIC AMIDES

The molecular ion peak of straight-chain monoamides is usually discernible. The dominant modes of cleavage depend on the length of the acyl moiety, and on the lengths and number of the alkyl groups attached to the nitrogen atom.

The base peak in all straight-chain primary amides higher than propionamide results from the familiar McLafferty rearrangement:

m/e 59

Branching at the α-carbon, (CH_3, etc.) gives a homologous peak at m/e 73 or 87, . . .

Primary amides give a strong peak at m/e 44 from cleavage of the R—$CONH_2$ bond: ($O\equiv C-\overset{..}{N}H_2 \leftrightarrow \overset{..}{O}=C=\overset{+}{N}H_2$); this is the base peak in C_1-C_3 primary amides and in iso-butyramide. A moderate peak at m/e 86 results from $\gamma\delta$ C—C cleavage, possibly accompanied by cyclization:

m/e 86

Secondary and tertiary amides with an available hydrogen on the γ-carbon of the acyl moiety and methyl groups on the N atom show the dominant peak resulting from the McLafferty rearrangement. When the N-alkyl groups are C_2 or longer and the acyl moiety is shorter than C_3, another mode of cleavage predominates. This is cleavage of the N-alkyl group beta to the N atom, and cleavage of the carbonyl C—N bond with rearrangement of an α-H atom of the acyl moiety:

m/e 30

ALIPHATIC NITRILES

The molecular ion peaks of aliphatic nitriles (except for acetonitrile and propionitrile) are weak or absent, but the $M+1$ peak can usually be located by its behavior on

increasing inlet pressure or decreasing repeller voltage (Section V, part 2). A weak but diagnostically useful $M - 1$ peak is formed by loss of an α-hydrogen to form the stable

ion: $RCH-C\equiv\overset{+}{N} \longleftrightarrow RCH=C=\overset{+}{N}$.

The base peak of straight-chain nitriles between C_4 and C_9 is m/e 41. This peak is the ion resulting from hydrogen rearrangement in a 6-membered transition state:

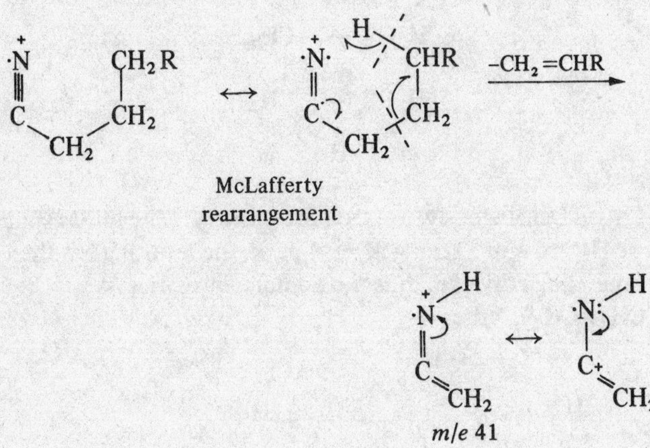

McLafferty
rearrangement

m/e 41

However, this peak lacks diagnostic value because of the presence of the C_3H_5 (m/e 41) in all molecules containing a hydrocarbon chain.

A peak at m/e 97 is characteristic and intense (sometimes the base peak) in straight-chain nitriles C_8 and higher. The following mechanism has been depicted:

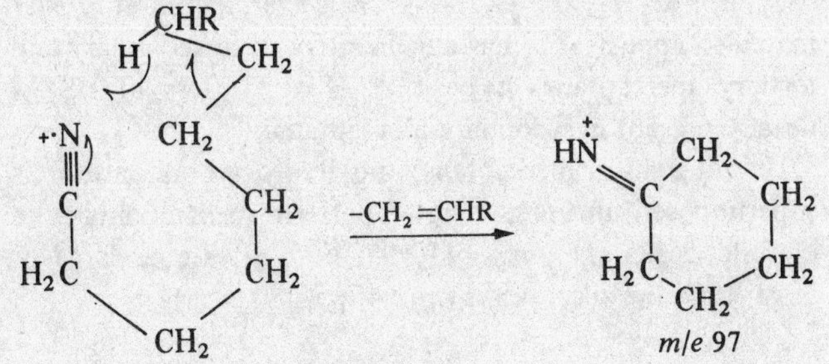

m/e 97

Simple cleavage at each C–C bond (except the one next to the N atom) gives a characteristic series of homologous peaks of even mass number down the entire length of the chain (m/e 40, 54, 68, 82, . . .) due to the $(CH_2)_n C\equiv N^+$ ions. Accompanying these peaks are the usual peaks of the hydrocarbon pattern.

NITRO COMPOUNDS

Aliphatic Nitro Compounds

The molecular ion peak (odd number) of an aliphatic mononitro compound is weak or absent (except in the lower homologs). The main peaks are attributable to the hydrocarbon fragments up to $M - NO_2$. Presence of a nitro group is indicated by an appreciable peak at m/e 30 (NO^+) and a smaller peak at mass 46 (NO_2^+).

Aromatic Nitro Compounds

The molecular ion peak of aromatic nitro compounds (odd number for one N atom) is strong. Prominent peaks result from elimination of an NO_2 radical ($M - 46$, the base peak in nitrobenzene), and of a neutral NO molecule with rearrangement to form the phenoxy cation ($M - 30$); both are good diagnostic peaks. Loss of $HC\equiv CH$ from the $M - 46$ ion accounts for a strong peak at $M - 72$; loss of CO from the $M - 30$ ion gives a peak at $M - 58$. A diagnostic peak at m/e 30 results from the NO^+ ion.

The isomeric o-, m-, and p-nitroanilines each give a strong molecular ion (even number). They all give prominent peaks resulting from two sequences:

Aside from differences in intensities, the three isomers give very similar spectra. The m- and p-compounds give a small peak at m/e 122 from loss of an O atom, whereas the o-compound eliminates $\cdot\ddot{O}H$ as follows to give a small peak at m/e 121.

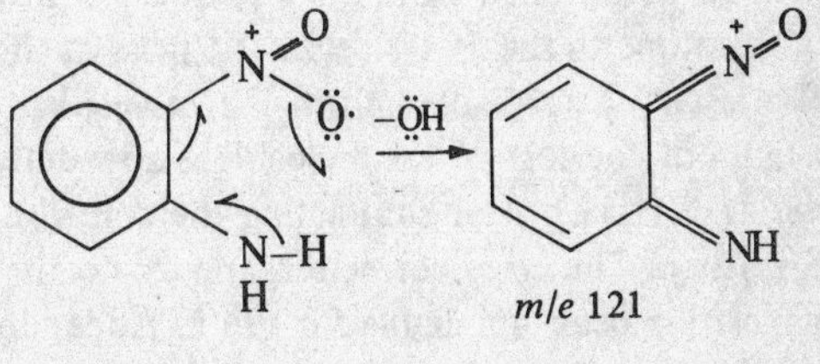

m/e 121

ALIPHATIC NITRITES

The molecular ion peak (odd number) of aliphatic nitrites (1 N present) is weak or absent. The peak at m/e 30 (NO^+) is always large and is often the base peak. There is a

large peak at m/e 60 ($CH_2=\overset{+}{O}NO$) in all nitrites unbranched at the α-carbon; this represents cleavage of the C–C bond next to the ONO group. An α-branch can be identified by a peak at m/e 74, 88, or 102, . . . Absence of a large peak at m/e 46 permits differentiation from nitro compounds. Hydrocarbon peaks are prominent, and their distribution

and intensities describe the configuration of the carbon chain.

ALIPHATIC NITRATES

The molecular ion peak (odd number) of aliphatic nitrates (1 N present) is weak or absent. A prominent (frequently the base) peak is formed by cleavage of the C–C bond next to the ONO_2 group with loss of the heaviest alkyl group attached to the α-carbon.

$$R-\overset{+}{\underset{R'}{CH}}-\overset{+}{O}-NO_2 \xrightarrow{-R\cdot} \overset{+}{\underset{R'}{CH}}=\overset{+}{O}-NO_2$$

$$R > R'$$

The NO_2^+ peak at m/e 46 is also prominent. As in the case of aliphatic nitrites, the hydrocarbon fragment ions are distinct.

SULFUR COMPOUNDS

The contribution of the ^{34}S isotope to the $M + 2$ peak, and often to a fragment + 2 peak, affords ready recognition of sulfur-containing compounds. A homologous series of sulfur-containing fragments is four mass units higher than the hydrocarbon fragment series. The number of sulfur atoms can be determined from the size of the contribution of the ^{34}S isotope to the $M + 2$ peak. The mass of the sulfur atom(s) present is subtracted from the molecular weight. The formula of the rest of the molecule is now determined from the $M + 1$ peak after subtracting the contribution of the ^{33}S isotope. The large correction applied to the $M + 2$ peak makes this peak unreliable for use in Appendix A. In Compound 3 in Chapter 6, for example, the molecular weight is 206, and the molecule contains two sulfur atoms. The formula for the rest of the molecule is therefore found under mass 142, i.e., 206 minus (2 × 32). The corrected $M + 1$ peak is 12.5 minus (2 × 0.78) which gives 10.9.

Aliphatic Mercaptans

The molecular ion peak of aliphatic mercaptans, except for higher tertiary mercaptans, is usually strong enough so that the $M + 2$ peak can be accurately measured. In general, the cleavage modes resemble those of alcohols. Cleavage of the C–C bond (αβ bond) next to the SH group gives the

characteristic ion $CH_2=\overset{+}{\underset{..}{S}}H \leftrightarrow \overset{+}{C}H_2-\overset{..}{\underset{..}{S}}H$ (m/e 47). Sulfur is poorer than nitrogen, but better than oxygen, at stabilizing such a fragment. Cleavage at the βγ bond gives a peak at m/e 61 of about one-half the intensity of the m/e 47 peak. Cleavage at the γδ bond gives a small peak at m/e 75, and cleavage at the δε bond gives a peak at m/e 89 that is more intense than the peak at m/e 75; presumably the m/e 89 ion is stabilized by cyclization:

$$
\begin{array}{ccc}
H_2C & \!\!\!\!—\!\!\!\! & CH_2 \\
| & & | \\
H_2C & & CH_2 \\
& \underset{\underset{H}{\overset{+}{S}}}{} &
\end{array}
$$

Again analogously to alcohols, primary mercaptans split out H_2S to give a strong $M - 34$ peak, the resulting ion then eliminating ethylene; thus the homologous series $M - H_2S - (CH_2=CH_2)_n$ arises:

$$
\begin{array}{c}
\underset{\underset{CH_2}{\overset{..}{\underset{|}{S}}}}{\overset{H}{\overset{|}{\underset{|}{S}}}}\;\overset{H}{\underset{CHR}{\overset{|}{}}} \\
\end{array}
\xrightarrow[-C_2H_4]{-H_2S} [CH_2=CHR]^{\ddagger} \xrightarrow{-(C_2H_4)_n}
$$

Secondary and tertiary mercaptans cleave at the α-C with loss of the largest group to give a prominent peak $M - CH_3$, $M - C_2H_5$, $M - C_3H_7$, ... However, a peak at m/e 47 may also appear as a rearrangement peak in secondary and tertiary mercaptans. A peak at $M - 33$ (loss of HS) is usually present in secondary mercaptans.

In long-chain mercaptans, the hydrocarbon pattern is superimposed on the mercaptan pattern. As in alcohols, the olefinic peaks (i.e., m/e 41, 55, 69, ...) are as large or larger than the alkyl peaks (m/e 43, 57, 71, ...).

Aliphatic Sulfides

The molecular ion peak of aliphatic sulfides is usually intense enough so that the $M + 2$ peak can be accurately measured. The cleavage modes generally resemble those of ethers. Cleavage of one or the other of the αβ C–C bonds occurs, with loss of the largest group being favored. These first-formed ions decompose further with hydrogen transfer and elimination of an olefin. The steps for aliphatic ethers also occur for sulfides; the end result is the ion $RCH=\overset{+}{\underset{..}{S}}H$ (see p. 35 for an example).

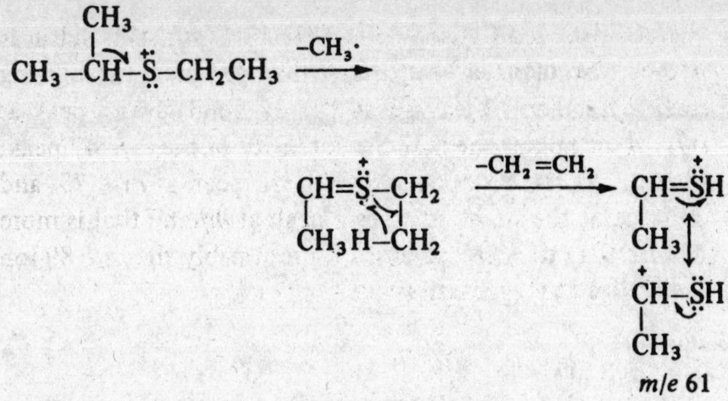

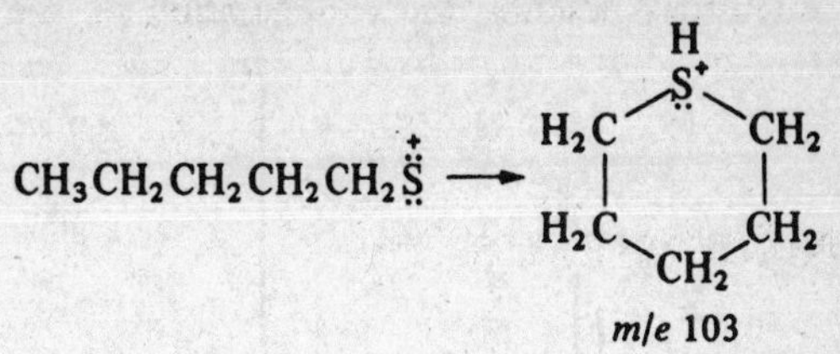

$$CH_3-CH-\overset{..}{\underset{..}{S}}-CH_2CH_3 \xrightarrow{\ -CH_3\cdot\ } \qquad CH_3CH_2CH_2CH_2CH_2\overset{+}{\underset{..}{S}} \longrightarrow$$

(top-left, above arrow: CH_3)

$$\underset{CH_3H-CH_2}{CH=\overset{+}{\underset{..}{S}}-CH_2} \xrightarrow{\ -CH_2=CH_2\ } \underset{CH_3}{CH=\overset{+}{S}H} \Big\downarrow$$

$$\underset{CH_3}{\overset{+}{CH-\underset{..}{S}H}}$$

m/e 61

(top right)

$$\underset{H_2C\qquad\ \ CH_2}{\overset{\displaystyle H\atop \displaystyle \overset{+}{S}}{\ }}\ \underset{CH_2}{\overset{H_2C\qquad CH_2}{\ }}$$

m/e 103

For a sulfide unbranched at either α-C, this ion is $\overset{+}{CH_2}=\underset{..}{S}H$ (*m/e* 47), and its intensity may lead to confusion with the same ion derived from a mercaptan. However, the absence of $M - H_2S$ or $M - SH$ peaks in sulfide spectra makes the distinction.

A moderate to strong peak at *m/e* 61 is present (see alkyl sulfide cleavage above) in the spectrum of all except tertiary sulfides. When an α-methyl substituent is present, *m/e* 61 is the ion $CH_3CH=\overset{+}{\underset{..}{S}}H$ resulting from the double cleavage described above. Methyl primary sulfides cleave at the αβ bond to give the *m/e* 61 ion, $CH_3-\overset{+}{S}=CH_2$.

However, a strong *m/e* 61 peak in the spectrum of a completely straight-chain sulfide calls for a different explanation. The following rationalization is offered.

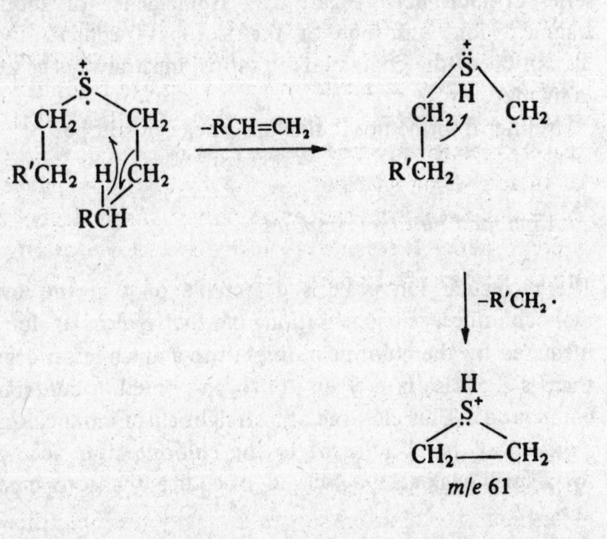

$$\xrightarrow{\ -RCH=CH_2\ }$$

(left structure, center: $\overset{..}{\underset{..}{S}}$ ring with CH_2, CH_2, $R'CH_2$, H, CH_2, RCH)

(right structure: $\overset{+}{S}H$ ring with CH_2, CH_2, $R'CH_2$)

$$\Big\downarrow\ -R'CH_2\cdot$$

$$\underset{CH_2\rule{6mm}{0.4pt}CH_2}{\overset{\displaystyle H\atop \displaystyle \overset{+}{\underset{..}{S}}}{\ }}$$

m/e 61

Sulfides give a characteristic ion by cleavage of the C—S bond with retention of charge on sulfur. The resulting $R\overset{+}{\underset{..}{S}}$ ion gives a peak at *m/e* 32 + CH_3, 32 + C_2H_5, 32 + C_3H_7, ... The ion of *m/e* 103 seems especially favored possibly because of formation of a rearranged cyclic ion:

These features are illustrated by the spectrum of di-*n*-amyl sulfide (Figure 11).

As with long-chain ethers, the hydrocarbon pattern may dominate the spectrum of long-chain sulfides; the C_nH_{2n} peaks seem especially prominent. In branched chain sulfides, cleavage at the branch may reduce the relative intensity of the characteristic sulfide peaks.

Aliphatic Disulfides[11]

The molecular ion peak, at least up to C_{10} disulfides, is strong (Chapter 6, Compound No. 3).

A major peak results from cleavage of one of the C—S bonds with retention of the charge on the alkyl fragment. Another major peak results from the same cleavage with shift of a hydrogen atom to form the RSSH fragment which retains the charge. Other peaks apparently result from cleavage between the sulfur atoms without rearrangement, and with rearrangement of one or two hydrogen atoms to give, respectively, $R\overset{+}{S}$, $R\overset{+}{S}-1$, and $R\overset{+}{S}-2$.

HALOGEN COMPOUNDS

A compound that contains one chlorine atom will have an $M + 2$ peak approximately one-third the intensity of the molecular ion peak because of the presence of molecular ions containing the ^{37}Cl isotope. A compound that contains one bromine atom will have an $M + 2$ peak almost equal in intensity to the molecular ion because of the presence of molecular ions containing the ^{81}Br isotope. A compound that contains two chlorines, or two bromines, or one chlorine and one bromine, will show a distinct $M + 4$ peak, in addition to the $M + 2$ peak, because of the presence of molecular ions containing two atoms of the heavy isotope. In general, the number of chlorine and/or bromine atoms in a molecule can be ascertained by the number of alternate peaks beyond the parent peak. Thus, three Cl atoms in a molecule will give peaks at $M + 2$, $M + 4$, and $M + 6$; in polychloro compounds, the peak of highest mass may be so weak as to escape notice.

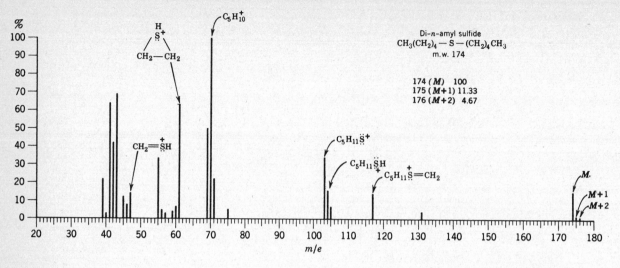

Fig. 11.
Di-n-amyl sulfide.

The relative abundances of the peaks (molecular ion, $M + 2$, $M + 4$, and so forth, have been calculated by Beynon[3] for compounds containing chlorine and bromine (atoms other than chlorine and bromine were ignored). A portion of these results is presented here, somewhat modified, as Table II. We can now tell what combination of chlorine and bromine atoms is present. It should be noted that Table II presents the isotope contributions in terms of percent of the parent peak.

The $M + 1$ peak is still useful for arriving at a molecular formula by use of Appendix A, after we substract the masses of the appropriate number of chlorine and bromine atoms.

Unfortunately, the application of isotope contributions,

though generally useful for aromatic halogen compounds is limited by the weak parent peak of many aliphatic halogen compounds of more than about six carbon atoms for a straight chain, or fewer for a branched chain. However, the halogen-containing fragments are recognizable by the ratio of the fragment + 2 peaks to fragment peaks in monochlorides or monobromides. In polychloro-or-bromo-compounds, these fragment + isotope peaks form a distinctive series of multiplets (Figure 12). Coincidence of another fragment ion with one of the isotope fragments, with disruption of the characteristic ratios, must always be kept in mind.

Neither fluorine nor iodine has a heavier isotope.

Aliphatic Chlorine Compounds

The molecular ion peak is detectable only in the lower monochlorides. Fragmentation of the molecular ion is mediated by the chlorine atom, but to a much lesser degree than is the case in oxygen-, nitrogen-, or sulfur-containing compounds. Thus cleavage of a straight-chain monochloride at the C—C bond adjacent to the chlorine atom accounts for a small peak at m/e 49 (and, of course, the isotope peak at m/e 51).

Table II   Intensities of Isotope Peaks (Relative to the Molecular Ion) for Combinations of Bromine and Chlorine

Halogen Present	% $M + 2$	% $M + 4$	% $M + 6$	% $M + 8$	% $M + 10$	% $M + 12$
Br	97.7					
Br$_2$	195.0	95.5				
Br$_3$	293.0	286.0	93.4			
Cl	32.6					
Cl$_2$	65.3	10.6				
Cl$_3$	97.8	31.9	3.47			
Cl$_4$	131.0	63.9	14.0	1.15		
Cl$_5$	163.0	106.0	34.7	5.66	0.37	
Cl$_6$	196.0	161.0	69.4	17.0	2.23	0.11
BrCl	130.0	31.9				
Br$_2$Cl	228.0	159.0	31.2			
Cl$_2$Br	163.0	74.4	10.4			

$$R\text{—}CH_2\text{—}\overset{+}{\underset{..}{C}}l: \xrightarrow{-R\cdot} CH_2\overset{+}{=}\overset{..}{C}l: \longleftrightarrow \overset{+}{C}H_2\text{—}\overset{..}{\underset{..}{C}}l:$$

Cleavage of the C—Cl bond leads to a small Cl$^+$ peak and to a R$^+$ peak which is prominent in the lower chlorides, but quite small when the chain is longer than about C$_5$.

Straight-chain chlorides longer than C$_6$ give C$_3$H$_6\overset{+}{C}$l,

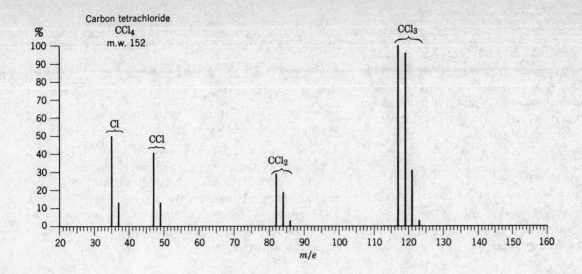

Fig. 12.
Carbon tetrachloride.

$C_4H_8\overset{+}{Cl}$, and $C_5H_{10}\overset{+}{Cl}$ ions. Of these, the $C_4H_8\overset{+}{Cl}$ ion forms the most intense (sometimes the base) peak; a five-membered cyclic structure may explain its stability.

Loss of HCl occurs, possibly by 1,2-elimination, to give a peak (weak or moderate) at $M - 36$.

In general, the spectrum of an aliphatic monochloride is dominated by the hydrocarbon pattern to a greater extent than that of a corresponding alcohol, amine, or mercaptan.

Aliphatic Bromides

The remarks under aliphatic chlorides apply quite generally to the corresponding bromides.

Aliphatic Iodides

Aliphatic iodides give the strongest molecular ion peak of the aliphatic halides. Since iodine is monoisotopic, there is no distinctive isotope peak. The presence of an iodine atom can sometimes be deduced from isotope peaks that are suspiciously low in relation to the molecular weight, and

from several distinctive peaks; in polyiodo compounds, the large interval between major peaks is characteristic.

Iodides cleave much as do chlorides and bromides, but the $C_4H_8\overset{+}{I}$ ion is not as evident as the corresponding chloride and bromide.

Aliphatic Fluorides

Aliphatic fluorides give the weakest molecular ion peak of the aliphatic halides. Fluorine is monoisotopic, and its detection in polyfluoro compounds depends on suspiciously small isotopic peaks relative to the molecular ion, on the intervals between peaks, and on characteristic peaks. Of these, the most characteristic is m/e 69 due to the ion $CF_3{}^+$, which is the base peak in all perfluorocarbons. Prominent peaks are noted at m/e 119, 169, 219...; these are increments of CF_2. The stable ions $C_3F_5{}^+$ and $C_4F_7{}^+$ give large peaks at m/e 131 and 181. The $M - F$ peak is frequently visible in perfluorinated compounds. In monofluorides, cleavage of the $\alpha\beta$ C–C bond is less important than in the other monohalides, but cleavage of a C–H bond on the α–C is more important. This reversal is a consequence of the high electro-negativity of the F atom, and is rationalized by placing the positive charge on the α-carbon. The secondary carbonium ion thus depicted by loss of a hydrogen atom is more stable than the primary carbonium ion resulting from loss of an alkyl radical.

$$[R-CH_2-F]^{\cdot+} \xrightarrow[\text{$-R\cdot$}]{\text{$-H\cdot$}} \begin{array}{l} R-\overset{+}{CH}-F \\ \overset{+}{CH_2}-F \end{array}$$

Benzyl Halides

The molecular ion peak of benzyl halides is usually detectable. The benzyl (or tropylium) ion from loss of the halide (Section VII, Rule 8) is favored even over β-bond cleavage of an alkyl substituent. A substituted phenyl ion (α-bond cleavage) is prominent when the ring is polysubstituted.

Aromatic Halides

The molecular ion peak of an aryl halide is readily apparent. The $M - X$ peak is large for all compounds in which X is attached directly to the ring.

HETEROAROMATIC COMPOUNDS

The molecular ion peak of heteroaromatics and alkylated heteroaromatics is intense. Cleavage of the bond beta to the ring, as in alkyl benzenes, is the general rule; in pyridine, the position of substitution determines the ease of cleavage of the beta bond (see below).

Localizing the charge of the molecular ion on the heteroatom, rather than in the ring π structure, provides a satisfactory rationale for the observed mode of cleavage. The present treatment follows that used by Djerassi.[2]

The five-membered ring heteroaromatics (furan, thiophene, and pyrrole) show very similar ring cleavage patterns. The first step in each case is cleavage of the carbon-heteroatom bond

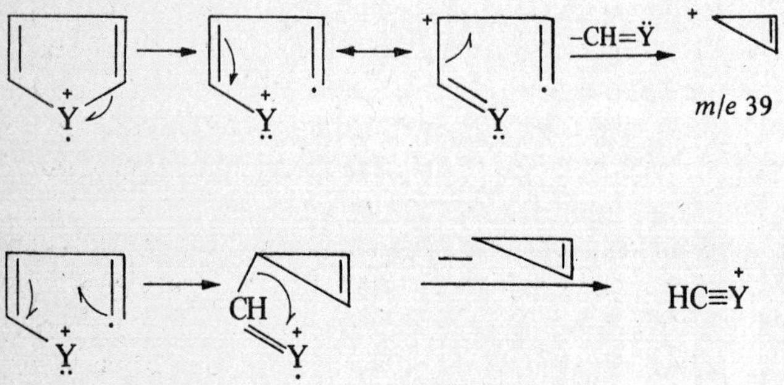

$$m/e \ 39$$

Y = O, S, NH

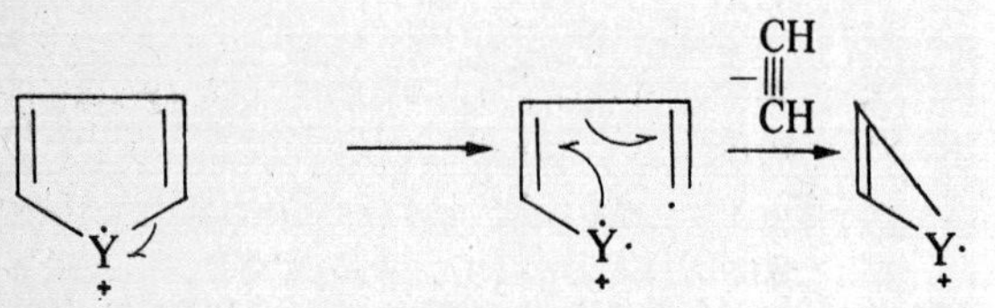

Y = S, NH

Thus, in furan, there are two principal peaks: $C_3H_3^+$ (m/e 39) and $HC{\equiv}\overset{+}{O}$ (m/e 29). In thiophene, there are three: $C_3H_3^+$ (m/e 39), $HC{\equiv}\overset{+}{S}$ (m/e 45), and $C_2H_2\overset{.+}{S}$ (m/e 58). And in pyrrole, there are three: $C_3H_3^+$ (m/e 39), $HC{\equiv}\overset{+.}{NH}$ (m/e 28) and $C_2H_2\overset{+.}{NH}$ (m/e 41). Pyrrole also eliminates a neutral molecule of HCN to give an intense peak at m/e 40.

The base peak in 2,5-dimethylfuran is m/e 43 ($CH_3C{\equiv}\overset{+}{O}$).

Cleavage of the β C–C bond in alkyl pyridines depends on the position of the ring substitution, being more pronounced when the alkyl group is in the 3-position. An alkyl group of more than three carbon atoms in the 2-position can undergo rearrangement of a hydrogen atom to the ring nitrogen.

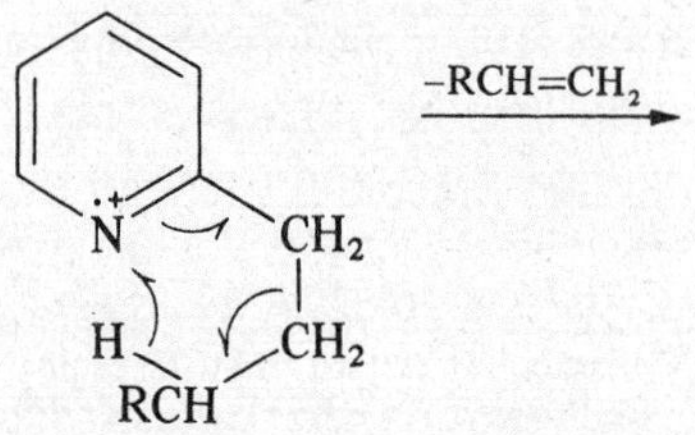

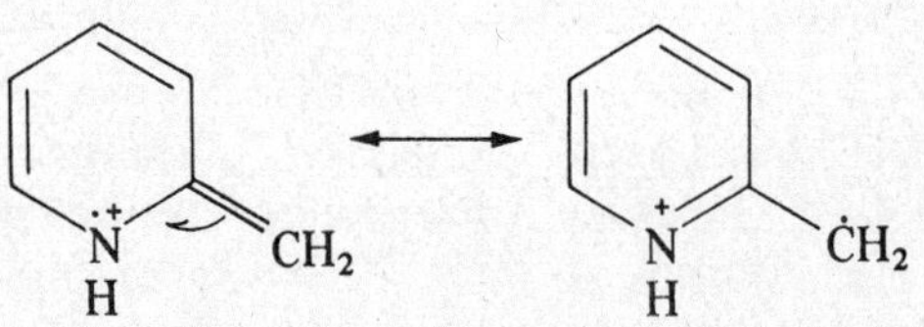

A similar cleavage is found in pyrazines since all ring substituents are necessarily ortho to one of the nitrogen atoms.

OTHER CLASSES

The following classes of organic compounds are discussed in Biemann's book (B),[1] or in Djerassi's books (D[2a], DI[20a], DII[20b]).

Alkaloids	D, Chapter 5; DI; DII, Chapter 17; B, p. 305
Amino Acids & Peptides	DII, Chapter 26; B, Chapter 7
Antibiotics	DII, p. 172
Carbohydrates	DII, Chapter 27
Cyanides & Isothiocyanates	D, Chapter 11
Estrogens	DII, Chapter 19
Glycerides	B, p. 255
Sapogenins	DII, Chapter 22

Silicones	B, p. 172
Steroids	B, Chapter 9; DII, Chapter 17, 20, 21, 22
Terpenes	B, p. 334; DII, Chapter 23, 24
Thioketals	D, Chapter 7
Tropone & Tropolones	D, Chapter 18

REFERENCES

1. Biemann, K., *Mass Spectrometry, Applications to Organic Chemistry,* McGraw-Hill, New York, 1962.
2. Budzikiewicz, H., C. Djerassi, and D. H. Williams, *Mass Spectrometry of Organic Compounds,* Holden-Day, San Francisco, 1967. This is the most complete general reference written for the organic chemist interested in employing mass spectrometry in structure determination (Reference D above).
3a. Beynon, J. H., *Mass Spectrometry and Its Application to Organic Chemistry,* Elsevier, Amsterdam, 1960.
3b. Beynon, J. H., R. A. Saunders, and A. E. Williams, *The Mass Spectra of Organic Molecules,* American Elsevier, New York, 1968.
4. McLafferty, F. W., Ed., *Mass Spectrometry of Organic Ions,* Academic Press, New York, 1963.
5. McDowell, C. A. Ed., *Mass Spectrometry,* McGraw-Hill, New York, 1963.
6. Elliott, R. M. Ed., *Advances in Mass Spectrometry,* Vol. 2, Pergamon, London. 1963.
7. Reed, R. I., *Ion Production by Electron Impact,* Academic Press, London, 1962.
8. McLafferty, F. W., Mass Spectrometry, Chapt. 2, Vol. II, in *Determination of Organic Structure by Physical Methods,* F. C. Nachod and W. D. Phillips, Eds., Academic Press, New York, 1962.
9. Waldron, J. D., Ed., *Advances in Mass Spectrometry,* Pergamon Press, London, 1959.
10. Kiser, R. W., *Introduction to Mass Spectroscopy and its Applications,* Prentice-Hall, Englewood Cliffs, N.J., 1965, Chapter 5.
10a. Bondarovich, H. A. and S. K. Freeman, Mass Spectrometry, Chapt. 4 in *Interpretive Spectroscopy,* S. K. Freeman Ed., Reinhold, New York, 1965.
10b. Reed, R. I., *Application of Mass Spectrometry to Organic Chemistry,* Academic Press, New York, 1966.
10c. McLafferty, F. W., *Interpretation of Mass Spectra,* W. A. Benjamin, New York, 1966; 2nd Ed., 1973.
10d. Shrader, S. R., *Introduction to Mass Spectrometry,* Allyn and Bacon, Boston, 1971.
10e. Pasto, D. J., and C. R. Johnson, *Organic Structure Determination,* Prentice-Hall, Englewood Cliffs, N.J., 1969.
10f. Hamming, M., and N. Foster, *Interpretation of Mass Spectra of Organic Compounds,* Academic Press, New York, 1972.
10g. Williams, D. H., Senior Reporter. *Mass Spectrometry.* Volume 1 (a review of the literature published between June 1968 and June 1970), The Chemical Society, London, 1971.
10h. Johnstone, R. A. W., *Mass Spectrometry for Organic Chemists,* Cambridge University Press, Cambridge, England, 1972.
10i. McLafferty, F. W., and J. Penzelik, *Index and Bibliography of Mass Spectrometry,* 1963-1965, Interscience, New York, 1967.
10j. McLafferty, F. W., Mass Spectrometry, Chap. 2, Vol. II, in *Determination of Organic Structure by Physical Methods,* F. C. Nachod and W. D. Phillips, Eds., Academic Press, New York, 1962.
10k. Reed, R. I., *Application of Mass Spectrometry to Organic Chemistry,* Academic Press, New York, 1966.
10l. Mead, W. L., Ed., *Advances in Mass Spectrometry,* Elsevier, Amsterdam, 1966.
10m. Stenhagen, E., S. Abrahamson, and F. McLafferty, Eds., *Atlas of Mass Spectral Data*, Interscience-Wiley, New York, 1969;
    Vol. 1:Molecular Weight: 16. 313 to 142.0089
    Vol. 2:Molecular Weight: 142.0185 to 213.2456
    Vol. 3:Molecular Weight: 213.8629 to 702.7981
11. "Catalog of Selected Mass Spectral Data" (1947 to date), American Petroleum Institute Research Project 44 and Thermodynamics Research Center (formerly MCA Research Project), Texas A & M University, College Station, Texas. Dr. Bruno Zwolinski, Director.
12a. ASTM (1963), "Index of Mass Spectral Data," STP-356, 244 pp., American Society for Testing and Materials, Philadelphia, Pennsylvania.
12b. ASTM (1969), "Index of Mass Spectral Data," AMD 11, 632 pp., American Society for Testing and Materials, 1916 Race Street, Philadelphia, Pennsylvania. Includes the Göteborg University mass spectral data collection.
12c. Ridley, R. G., and A. Quayle, Mass Spectrometry Data Centre of the Atomic Weapons Research Establishment (AWRE) Aldermaston, England (about 8200 spectra on magnetic tape). The organization functions in cooperation with ASTM as a worldwide central collection for mass spectra.
13a. Beynon, J. H., and A. E. Williams, *Mass and Abundance Tables for Use in Mass Spectrometry,* Elsevier, Amsterdam, 1963.

13b. Cornu, A., and R. Massot, *Compilation of Mass Spectral Data*, Heyden, London, 1966. First and second supplements issued.

14. Lederberg, J., *Computation of Molecular Formulas for Mass Spectrometry*, Holden-Day, San Francisco, 1964.

15. Lederberg, J., *Tables and an Algorithm for Calculating Functional Groups of Organic Molecules in High Resolution Mass Spectrometry*, NASA Sci. Techn. Aerosp. Rep., N64-21426 (1964).

16. Kendrich, E., "A Mass Scale Based on $CH_2 = 14.0000$ for High-Resolution Mass Spectrometry of Organic Compounds," *Anal. Chem.*, 35, 2146 (1963).

17. Biemann, K., P. Bommer, and D. M. Desiderio, Tetrahedron Letters, 1725 (1964). Biemann, K., W. McMurray, Tetrahedron Letters, 647 (1965). Biemann, K., *J. Pure Appl. Chem.*, 9. 95 (1964).

18. Tunnicliff, D. D., P. A. Wadsworth, and D. O. Schissler, Mass and Abundance Tables, Shell Development Co., Emeryville, Calif.

19. Fuqua, S. A. and R. M. Silverstein, *J. Org. Chem.*, 29, 395 (1964).

20. Budzikiewicz, H., C. Djerassi, and D. H. Williams, *Structure Elucidation of Natural Products by Mass Spectrometry*, Holden-Day, San Francisco, 1964.
    Vol. I, (Reference DI above)
    Vol. II, (Reference DII above)

21. "Eight Peak Index of Mass Spectra" (1970), British Information Services, 845 Third Ave., New York. A compilation of essential mass spectral data indexed from over 17,000 mass spectra contributed by many mass spectrometrists from all over the world.

22. Compilation of Mass Spectral Data, Centre d'Etudes Nucleaires de Grenoble, France. This compilation is an index to some 6000 spectra at the Centre d'Etudes Nucleaires de Grenoble, France.

23. Price, D., and J. D. Williams, Eds., "Dynamic Mass Spectrometry," Vol. I, Heyden, London, 1969.

24. Waller, G. R., Ed., "Biochemical Applications of Mass Spectrometry," Wiley, New York, 1972.

25. Wanless, G. G., *Spectrometry of Fuels* (R. A. Friedel, Ed.), pp. 15-28, Plenum, New York, 1970. Washburn, H. W., in *Physical Methods in Chemical Analysis* (W. G. Berl, Ed.), Vol. i, pp. 587-637. Academic Press, New York, 1950.

26. White, F. A., *Mass Spectrometry in Science and Technology*, Wiley, New York, 1968.

27. Benz, W., *Massenspektrometric Organischer Verbindungen*, Akad. Verlag. Frankfurt a.M., 1968.

28. Weygand, F., et al., *Angew. Chem.* Int. Ed., 2, 485 (1963).

29. McLafferty, F. W., *Mass Spectral Correlations*, Advances in Chemistry Series No. 40., American Chemical Society, Washington, D. C., 1963.

MASSES AND ISOTOPIC ABUNDANCE RATIOS FOR VARIOUS COMBINATIONS OF CARBON, HYDROGEN, NITROGEN AND OXYGEN*

	M+1	M+2
12		
C	1.08	
13		
CH	1.10	
14		
N	0.38	
CH_2	1.11	
15		
NH	0.40	
CH_3	1.13	
16		
O	0.04	0.20
NH_2	0.41	
CH_4	1.15	
17		
OH	0.06	0.20
NH_3	0.43	
CH_5	1.16	
18		
H_2O	0.07	0.20
NH_4	0.45	
19		
H_3O	0.09	0.20
24		
C_2	2.16	0.01
25		
C_2H	2.18	0.01
26		
CN	1.46	
C_2H_2	2.19	0.01
27		
CHN	1.48	
C_2H_3	2.21	0.01
28		
N_2	0.76	
CO	1.12	0.2

	M+1	M+2
CH_2N	1.49	
C_2H_4	2.23	0.01
29		
N_2H	0.78	
CHO	1.14	0.20
CH_3N	1.51	
C_2H_5	2.24	0.01
30		
NO	0.42	0.20
N_2H_2	0.79	
CH_2O	1.15	0.20
CH_4N	1.53	0.01
C_2H_6	2.26	0.01
31		
NOH	0.44	0.20
N_2H_3	0.81	
CH_3O	1.17	0.20
CH_5N	1.54	
32		
O_2	0.08	0.40
NOH_2	0.45	0.20
N_2H_4	0.83	
CH_4O	1.18	0.20
33		
NOH_3	0.47	0.20
N_2H_5	0.84	
CH_5O	1.12	
34		
N_2H_6	0.86	
36		
C_3	3.24	0.04
37		
C_3H	3.26	0.04
38		
C_2N	2.54	0.02
C_3H_2	3.27	0.04
39		
C_2HN	2.56	0.02
C_3H_3	3.29	0.04

	M+1	M+2
40		
CN_2	1.84	0.01
C_2O	2.20	0.21
C_2H_2N	2.58	0.02
C_3H_4	3.31	0.04
41		
CHN_2	1.86	
C_2HO	2.22	0.21
C_2H_3N	2.59	0.02
C_3H_5	3.32	0.04
42		
CNO	1.50	0.21
CH_2N_2	1.88	0.01
C_2H_2O	2.23	0.21
C_2H_4N	2.61	0.02
C_3H_6	3.34	0.04
43		
CHNO	1.52	0.21
CH_3N_2	1.89	0.01
C_2H_3O	2.25	0.21
C_2H_5N	2.62	0.02
C_3H_7	3.35	0.04
44		
N_2O	0.80	0.20
CO_2	1.16	0.40
CH_2NO	1.53	0.21
CH_4N_2	1.91	0.01
C_2H_4O	2.26	0.21
C_2H_6N	2.64	0.02
C_3H_8	3.37	0.04
45		
HN_2O	0.82	0.20
CHO_2	1.18	0.40
CH_3NO	1.55	0.21
CH_5N_2	1.92	0.01
C_2H_5O	2.28	0.21
C_2H_7N	2.66	0.02
46		
NO_2	0.46	0.40
N_2H_2O	0.83	0.20
CH_2O_2	1.19	0.40
CH_4NO	1.57	0.21
CH_6N_2	1.94	0.01

	M+1	M+2
C_2H_6O	2.30	0.22
C_2H_8N	2.66	0.02
47		
CH_3O_2	1.21	0.40
CH_5NO	1.58	0.21
CH_7N_2	1.96	0.01
C_2H_7O	2.31	0.22
48		
CH_4O_2	1.22	0.40
C_4	4.32	0.07
49		
CH_5O_2	1.24	0.40
C_4H	4.34	0.07
50		
C_4H_2	4.34	0.07
51		
C_4H_3	4.37	0.07
52		
C_2N_2	2.92	0.03
C_3H_2N	3.66	0.05
C_4H_4	4.39	0.07
53		
C_2HN_2	2.94	0.03
C_3HO	3.30	0.24
C_3H_3N	3.67	0.05
C_4H_5	4.40	0.07
54		
C_2NO	2.58	0.22
$C_2H_2N_2$	2.96	0.03
C_3H_2O	3.31	0.24
C_3H_4N	3.69	0.05
C_4H_6	4.42	0.07
55		
C_2HNO	2.60	0.22
$C_2H_3N_2$	2.97	0.03
C_3H_3O	3.33	0.24
C_3H_5N	3.70	0.05
C_4H_7	4.43	0.08

*Adapted with permission from J. H. Beynon, *Mass Spectrometry and Its Application to Organic Chemistry*, Elsevier, Amsterdam, 1960. The values given form a self-consistent set and can be used regardless of the mass standard.

	M + 1	M + 2
56		
CH_2N_3	2.26	0.02
C_2O_2	2.24	0.41
C_2H_2NO	2.61	0.22
$C_2H_4N_2$	2.99	0.03
C_3H_4O	3.35	0.24
C_3H_6N	3.72	0.05
C_4H_8	4.45	0.08
57		
CHN_2O	1.90	0.21
CH_3N_3	2.27	0.02
C_2HO_2	2.26	0.41
C_2H_3NO	2.63	0.22
$C_2H_5N_2$	3.00	0.03
C_3H_5O	3.36	0.24
C_3H_7N	3.74	0.05
C_4H_9	4.47	0.08
58		
CNO_2	1.54	0.41
CH_2N_2O	1.92	0.21
CH_4N_3	2.29	0.02
$C_2H_2O_2$	2.27	0.42
C_2H_4NO	2.65	0.22
$C_2H_6N_2$	3.02	0.03
C_3H_6O	3.38	0.24
C_3H_8N	3.75	0.05
C_4H_{10}	4.48	0.08
59		
$CHNO_2$	1.56	0.41
CH_3N_2O	1.93	0.21
CH_5N_3	2.31	0.02
$C_2H_3O_2$	2.29	0.42
C_2H_5NO	2.66	0.22
$C_2H_7N_2$	3.04	0.03
C_3H_7O	3.39	0.24
C_3H_9N	3.77	0.05
60		
CH_2NO_2	1.57	0.41
CH_4N_2O	1.95	0.21
CH_6N_3	2.32	0.02
$C_2H_4O_2$	2.30	0.04
C_2H_6NO	2.68	0.22
$C_2H_8N_2$	3.05	0.03
C_3H_8O	3.41	0.24
61		
CHO_3	1.21	0.60
CH_3NO_2	1.59	0.41
CH_5N_2O	1.96	0.21
CH_7N_3	2.34	0.02
$C_2H_5O_2$	2.32	0.42
C_2H_7NO	2.69	0.22
C_3H_9O	3.43	0.24
C_5H	5.42	0.12
62		
CH_2O_3	1.23	0.60
CH_4NO_2	1.60	0.41
CH_6N_2O	1.98	0.21
CH_8N_3	2.35	0.02

	M + 1	M + 2
$C_2H_6O_2$	2.34	0.42
C_5H_2	5.44	0.12
63		
CH_3O_3	1.25	0.60
CH_5NO_2	1.62	0.41
C_4HN	4.72	0.09
C_5H_3	5.45	0.12
64		
CH_4O_3	1.26	0.60
C_4H_2N	4.74	0.09
C_5H_4	5.47	0:12
65		
C_3HN_2	4.02	0.06
C_4HO	4.38	0.27
C_4H_3N	4.75	0.09
C_5H_5	5.48	0.12
66		
$C_3H_2N_2$	4.04	0.06
C_4H_2O	4.39	0.27
C_4H_4N	4.77	0.09
C_5H_6	5.50	0.12
67		
C_2HN_3	3.32	0.04
C_3HNO	3.68	0.25
$C_3H_3N_2$	4.05	0.06
C_4H_3O	4.41	0.27
C_4H_5N	4.78	0.09
C_5H_7	5.52	0.12
68		
$C_2H_2N_3$	3.34	0.04
C_3O_2	3.32	0.44
C_3H_2NO	3.69	0.25
$C_3H_4N_2$	4.07	0.06
C_4H_4O	4.43	0.28
C_4H_6N	4.80	0.09
C_5H_8	5.53	0.12
69		
CHN_4	2.62	0.03
C_2HN_2O	2.98	0.23
$C_2H_3N_3$	3.35	0.04
C_3HO_2	3.34	0.44
C_3H_3NO	3.71	0.25
$C_3H_5N_2$	4.09	0.06
C_4H_5O	4.44	0.28
C_4H_7N	4.82	0.09
C_5H_9	5.55	0.12
70		
CH_2N_4	2.64	0.03
C_2NO_2	2.62	0.42
$C_2H_2N_2O$	3.00	0.23
$C_2H_4N_3$	3.37	0.04
$C_3H_2O_2$	3.35	0.44
C_3H_4NO	3.73	0.25
$C_3H_6N_2$	4.10	0.07
C_4H_6O	4.46	0.28
C_4H_8N	4.83	0.09

	M + 1	M + 2
C_5H_{10}	5.56	0.13
71		
CHN_3O	2.28	0.22
CH_3N_4	2.65	0.03
C_2HNO_2	2.64	0.42
$C_2H_3N_2O$	3.01	0.23
$C_2H_5N_3$	3.39	0.04
$C_3H_3O_2$	3.37	0.44
C_3H_5NO	3.74	0.25
$C_3H_7N_2$	4.12	0.07
C_4H_7O	4.47	0.28
C_4H_9N	4.85	0.09
C_5H_{11}	5.58	0.13
72		
CH_2N_3O	2.30	0.22
CH_4N_4	2.67	0.03
$C_2H_2NO_2$	2.65	0.42
$C_2H_4N_2O$	3.03	0.23
$C_2H_6N_3$	3.40	0.44
$C_3H_4O_2$	3.38	0.44
C_3H_6NO	3.76	0.25
$C_3H_8N_2$	4.13	0.07
C_4H_8O	4.49	0.28
$C_4H_{10}N$	4.86	0.09
C_5H_{12}	5.60	0.13
73		
CHN_2O_2	1.94	0.41
CH_3N_3O	2.31	0.22
CH_5N_4	2.69	0.03
C_2HO_3	2.30	0.62
$C_2H_3NO_2$	2.67	0.42
$C_2H_5N_2O$	3.04	0.23
$C_2H_7N_3$	3.42	0.04
$C_3H_5O_2$	3.40	0.44
C_3H_7NO	3.77	0.25
$C_3H_9N_2$	4.15	0.07
C_4H_9O	4.51	0.28
$C_4H_{11}N$	4.88	0.10
C_6H	6.50	0.18
74		
$CH_2N_2O_2$	1.95	0.41
CH_4N_3O	2.33	0.22
CH_6N_4	2.70	0.03
$C_2H_2O_3$	2.31	0.62
$C_2H_4NO_2$	2.69	0.42
$C_2H_6N_2O$	3.06	0.23
$C_2H_8N_3$	3.43	0.05
$C_3H_6O_2$	3.42	0.44
C_3H_8NO	3.79	0.25
$C_3H_{10}N_2$	4.17	0.07
$C_4H_{10}O$	4.52	0.28
C_6H_2	6.52	0.18
75		
$CHNO_3$	1.60	0.61
$CH_3N_2O_2$	1.97	0.41
CH_5N_3O	2.34	0.22
CH_7N_4	2.72	0.03
$C_2H_3O_3$	2.33	0.62
$C_2H_5NO_2$	2.70	0.43

	M + 1	M + 2
$C_2H_7N_2O$	3.08	0.23
$C_2H_9N_3$	3.45	0.05
$C_3H_7O_2$	3.43	0.44
C_3H_9NO	3.81	0.25
C_5HN	5.80	0.14
C_6H_3	6.53	0.18
76		
CH_2NO_3	1.61	0.61
$CH_4N_2O_2$	1.99	0.41
CH_6N_3O	2.36	0.22
CH_8N_4	2.73	0.03
$C_2H_4O_3$	2.34	0.62
$C_2H_6NO_2$	2.72	0.43
$C_2H_8N_2O$	3.09	0.24
$C_3H_8O_2$	3.45	0.44
C_5H_2N	5.82	0.14
C_6H_4	6.55	0.18
77		
CHO_4	1.25	0.80
CH_3NO_3	1.63	0.61
$CH_5N_2O_2$	2.00	0.41
CH_7N_3O	2.38	0.22
$C_2H_5O_3$	2.39	0.62
$C_2H_7NO_2$	2.73	0.43
C_4HN_2	5.10	0.11
C_5HO	5.45	0.32
C_5H_3N	5.83	0.14
C_6H_5	6.56	0.18
78		
CH_2O_4	1.27	0.80
CH_4NO_3	1 64	0.61
$CH_6N_2O_2$	2.02	0.41
$C_2H_6O_3$	2.38	0.62
$C_4H_2N_2$	5.12	0.11
C_5H_2O	5.47	0.32
C_5H_4N	5.49	0.14
C_6H_6	6.58	0.18
79		
CH_3O_4	1.29	0.80
CH_5NO_3	1.66	0.61
C_3HN_3	4.40	0.08
C_4HNO	4.76	0.29
$C_4H_3N_2$	5.13	0.11
C_5H_3O	5.49	0.32
C_5H_5N	5.87	0.14
C_6H_7	6.60	0.18
80		
CH_4O_4	1.30	0.80
$C_3H_2N_3$	4.42	0.08
C_4H_2NO	4.78	0.29
$C_4H_4N_2$	5.15	0.11
C_5H_4O	5.51	0.32
C_5H_6N	5.88	0.14
C_6H_8	6.61	0 18
81		
C_2HN_4	3.70	0.05
C_3HN_2O	4.06	0 26

	M + 1	*M* + 2
$C_3H_3N_3$	4.43	0.08
C_4HO_2	4.42	0.48
C_4H_3NO	4.79	0.29
$C_4H_5N_2$	5.17	0.11
C_5H_5O	5.52	0.32
C_5H_7N	5.90	0.14
C_6H_9	6.63	0.18
82		
$C_2H_2N_4$	3.72	0.05
$C_3H_2N_2O$	4.08	0.36
$C_3H_4N_3$	4.45	0.08
$C_4H_2O_2$	4.43	0.48
C_4H_4NO	4.81	0.29
$C_4H_6N_2$	4.18	0.11
C_5H_6O	5.54	0.32
C_5H_8N	5.91	0.14
C_6H_{10}	6.64	0.19
83		
C_2HN_3O	3,36	0.24
$C_2H_3N_4$	3.74	0.06
C_3HNO_2	3.72	0.45
$C_3H_3N_2O$	4.09	0.27
$C_3H_5N_3$	4.47	0.08
$C_4H_3O_2$	4.45	0.48
C_4H_5NO	4.82	0.29
$C_4H_7N_2$	5.20	0.11
C_5H_7O	5.55	0.33
C_5H_9N	5.93	0.15
C_6H_{11}	6.66	0.19
84		
$C_2H_2N_3O$	3.38	0.24
$C_2H_4N_4$	3.75	0.06
$C_3H_2NO_2$	3.73	0.45
$C_3H_4N_2O$	4.11	0.27
$C_3H_6N_3$	4.48	0.81
$C_4H_4O_2$	4.47	0.48
C_4H_6NO	4.84	0.29
$C_4H_8N_2$	5.21	0.11
C_5H_8O	5.57	0.33
$C_5H_{10}N$	5.95	0.15
C_6H_{12}	6.68	0.19
85		
CHN_4O	2.66	0.23
$C_2HN_2O_2$	3.02	0.43
$C_2H_3N_3O$	3.39	0.24
$C_2H_5N_4$	3.77	0.06
C_3HO_3	3.38	0.64
$C_3H_3NO_2$	3.75	0.45
$C_3H_5N_2O$	4.12	0.27
$C_3H_7N_3$	4.50	0.08
$C_4H_5O_2$	4.48	0.48
C_4H_7NO	4.86	0.29
$C_4H_9N_2$	5.23	0.11
C_5H_9O	5.59	0.33
$C_5H_{11}N$	5.96	0.15
C_6H_{13}	6.69	0.19
C_7H	7.58	0.25
86		
CH_2N_4O	2.68	0.23
$C_2H_2N_2O_2$	3.03	0.43
$C_2H_4N_3O$	3.41	0.24
$C_2H_6N_4$	3.78	0.06
$C_3H_2O_3$	3.39	0.64
$C_3H_4NO_2$	3.77	0.45
$C_3H_6N_2O$	4.14	0.27
$C_3H_8N_3$	4.51	0.08
$C_4H_6O_2$	4.50	0.48
C_4H_8NO	4.87	0.30
$C_4H_{10}N_2$	5.25	0.11
$C_5H_{10}O$	5.60	0.33
$C_5H_{12}N$	5.98	0.15
C_6H_{14}	6.71	0.19
C_7H_2	7.60	0.25
87		
CHN_3O_2	2.32	0.42
CH_3N_4O	2.69	0.23
C_2HNO_3	2.68	0.62
$C_2H_3N_2O_2$	3.05	0.43
$C_2H_5N_3O$	3.43	0.25
$C_2H_7N_4$	3.80	0.06
$C_3H_3O_3$	3.41	0.64
$C_3H_5NO_2$	3.78	0.45
$C_3H_7N_2O$	4.16	0.27
$C_3H_9N_3$	4.53	0.08
$C_4H_7O_2$	4.51	0.48
C_4H_9NO	4.89	0.30
$C_4H_{11}N_2$	5.26	0.11
$C_5H_{11}O$	5.62	0.33
$C_5H_{13}N$	5.99	0.15
C_6HN	6.88	0.20
C_7H_3	7.61	0.25
88		
$CH_2N_3O_2$	2.34	0.42
CH_4N_4O	2.71	0.23
$C_2H_2NO_3$	2.69	0.63
$C_2H_4N_2O_2$	3.07	0.43
$C_2H_6N_3O$	3.44	0.25
$C_2H_8N_4$	3.82	0.06
$C_3H_4O_3$	3.42	0.64
$C_3H_6NO_2$	3.80	0.45
$C_3H_8N_2O$	4.17	0.27
$C_3H_{10}N_3$	4.55	0.08
$C_4H_8O_2$	4.53	0.48
$C_4H_{10}NO$	4.90	0.30
$C_4H_{12}N_2$	5.28	0.11
$C_5H_{12}O$	5.63	0.33
C_6H_2N	6.90	0.20
C_7H_4	7.63	0.25
89		
CHN_2O_3	1.98	0.61
$CH_3N_3O_2$	2.35	0.42
CH_5N_4O	2.73	0.23
C_2HO_4	2.33	0.82
$C_2H_3NO_3$	2.71	0.63
$C_2H_5N_2O_2$	3.08	0.44
$C_2H_7N_3O$	3.46	0.25
$C_2H_9N_4$	3.83	0.06
$C_3H_5O_3$	3.44	0.64
$C_3H_7NO_2$	3.81	0.46
$C_3H_9N_2O$	4.19	0.27
$C_3H_{11}N_3$	4.56	0.84
$C_4H_9O_2$	4.55	0.48
$C_4H_{11}NO$	4.92	0.30
C_5HN_2	6.18	0.16
C_6HO	6.54	0.38
C_6H_3N	6.91	0.20
C_7H_5	7.64	0.25
90		
$CH_2N_2O_3$	1.99	0.61
$CH_4N_3O_2$	2.37	0.42
CH_6N_4O	2.74	0.23
$C_2H_2O_4$	2.35	0.82
$C_2H_4NO_3$	2.72	0.63
$C_2H_6N_2O_2$	3.10	0.44
$C_2H_8N_3O$	3.47	0.25
$C_2H_{10}N_4$	3.85	0.06
$C_3H_6O_3$	3.46	0.64
$C_3H_8NO_2$	3.83	0.46
$C_3H_{10}N_2O$	4.20	0.27
$C_4H_{10}O_2$	4.56	0.48
$C_5H_2N_2$	6.20	0.16
C_6H_2O	6.56	0.38
C_6H_4N	6.93	0.20
C_7H_6	7.66	0.25
91		
$CHNO_4$	1.63	0.81
$CH_3N_2O_3$	2.01	0.61
$CH_5N_3O_2$	2.38	0.42
CH_7N_4O	2.76	0.23
$C_2H_3O_4$	2.37	0.82
$C_2H_5NO_3$	2.74	0.63
$C_2H_7N_2O_2$	3.11	0.44
$C_2H_9N_3O$	3.49	0.25
$C_3H_7O_3$	3.47	0.64
$C_3H_9NO_2$	3.85	0.46
C_4HN_3	5.48	0.12
C_5HNO	5.84	0.34
$C_5H_3N_2$	6.21	0.16
C_6H_3O	6.57	0.38
C_6H_5N	6.95	0.21
C_7H_7	7.68	0.25
92		
CH_2NO_4	1.65	0.81
$CH_4N_2O_3$	2.03	0.61
$CH_6N_3O_2$	2.40	0.42
CH_8N_4O	2.77	0.23
$C_2H_4O_4$	2.38	0.82
$C_2H_6NO_3$	2.76	0.63
$C_2H_8N_2O_2$	3.13	0.44
$C_3H_8O_3$	3.49	0.64
$C_4H_2N_3$	5.50	0.13
C_5H_2NO	5.86	0.34
$C_5H_4N_2$	6.23	0.16
C_6H_4O	6.59	0.38
C_6H_6N	6.96	0.21
C_7H_8	7.69	0.26
N_2O_4	9.19	0.80
93		
CH_3NO_4	1.67	0.81
$CH_5N_2O_3$	2.04	0.61
$CH_7N_3O_2$	2.42	0.42
$C_2H_5O_4$	2.40	0.82
$C_2H_7NO_3$	2.77	0.63
C_3HN_4	4.78	0.09
C_4HN_2O	5.14	0.31
$C_4H_3N_3$	5.52	0.13
C_5HO_2	5.50	0.52
C_5H_3NO	5.87	0.34
$C_5H_5N_2$	6.25	0.16
C_6H_5O	6.60	0.38
C_6H_7N	6.98	0.21
C_7H_9	7.71	0.26
94		
CH_4NO_4	1.68	0.81
$CH_6N_2O_3$	2.06	0.62
$C_2H_6O_4$	2.41	0.82
$C_3H_2N_4$	4.80	0.09
$C_4H_2N_2O$	5.16	0.31
$C_4H_4N_3$	5.53	0.13
$C_5H_2O_2$	5.51	0.52
C_5H_4NO	5.89	0.34
$C_5H_6N_2$	6.26	0.17
C_6H_6O	6.62	0.38
C_6H_8N	6.99	0.21
C_7H_{10}	7.72	0.26
95		
CH_5NO_4	1.70	0.81
C_3HN_3O	4.44	0.28
$C_3H_3N_4$	4.82	0.10
C_4HNO_2	4.80	0.49
$C_4H_3N_2O$	5.17	0.31
$C_4H_5N_3$	5.55	0.13
$C_5H_3O_2$	5.53	0.52
C_5H_5NO	5.90	0.34
$C_5H_7N_2$	6.28	0.17
C_6H_7O	6.64	0.39
C_6H_9N	7.01	0.21
C_7H_{11}	7.74	0.26
96		
$C_3H_2N_3O$	4.46	0.28
$C_3H_4N_4$	4.83	0.10
$C_4H_2NO_2$	4.81	0.49
$C_4H_4N_2O$	5.19	0.31
$C_4H_6N_3$	5.56	0.13
$C_5H_4O_2$	5.55	0.53
C_5H_6NO	5.92	0.35
$C_5H_8N_2$	6.29	0.17
C_6H_8O	6.65	0.39
$C_6H_{10}N$	7.03	0.21
C_7H_{12}	7.76	0.26
97		
C_2HN_4O	3.74	0.26
$C_3HN_2O_2$	4.10	0.47
$C_3H_3N_3O$	4.47	0.28
$C_3H_5N_4$	4.85	0.10
C_4HO_3	4.46	0.68
$C_4H_3NO_2$	4.83	0.49
$C_4H_5N_2O$	5.20	0.31
$C_4H_7N_3$	5.58	0.13
$C_5H_5O_2$	5.56	0.53

	M+1	M+2
C_5H_7NO	5.94	0.35
$C_5H_9N_2$	6.31	0.17
C_6H_9O	6.67	0.39
$C_6H_{11}N$	7.04	0.21
C_7H_{13}	7.77	0.26
C_8H	8.66	0.33
98		
$C_2H_2N_4O$	3.76	0.26
$C_3H_2N_2O_2$	4.12	0.47
$C_3H_4N_3O$	4.49	0.28
$C_3H_6N_4$	4.86	0.10
$C_4H_2O_3$	4.47	0.68
$C_4H_4NO_2$	4.85	0.49
$C_4H_6N_2O$	5.22	0.31
$C_4H_8N_3$	5.60	0.13
$C_5H_6O_2$	5.58	0.53
C_5H_8NO	5.95	0.35
$C_5H_{10}N_2$	6.33	0.17
$C_6H_{10}O$	6.68	0.39
$C_6H_{12}N$	7.06	0.21
C_7H_{14}	7.79	0.26
C_8H_2	8.68	0.33
99		
$C_2HN_3O_2$	3.40	0.44
$C_2H_3N_4O$	3.77	0.26
C_3HNO_3	3.76	0.65
$C_3H_3N_2O_2$	4.13	0.47
$C_3H_5N_3O$	4.51	0.28
$C_3H_7N_4$	4.88	0.10
$C_4H_3O_3$	4.49	0.68
$C_4H_5NO_2$	4.86	0.50
$C_4H_7N_2O$	5.24	0.31
$C_4H_9N_3$	5.61	0.13
$C_5H_7O_2$	5.59	0.53
C_5H_9NO	5.97	0.35
$C_5H_{11}N_2$	6.34	0.17
$C_6H_{11}O$	6.70	0.39
$C_6H_{13}N$	7.07	0.21
C_7HN	7.96	0.28
C_7H_{15}	7.80	0.26
C_8H_3	8.69	0.33
100		
$C_2H_2N_3O_2$	3.42	0.45
$C_2H_4N_4O$	3.79	0.26
$C_3H_2NO_3$	3.77	0.65
$C_3H_4N_2O_2$	4.15	0.47
$C_3H_6N_3O$	4.52	0.28
$C_3H_8N_4$	4.90	0.10
$C_4H_4O_3$	4.50	0.68
$C_4H_6NO_2$	4.88	0.50
$C_4H_8N_2O$	5.25	0.31
$C_4H_{10}N_3$	5.63	0.13
$C_5H_8O_2$	5.61	0.53
$C_5H_{10}NO$	5.98	0.35
$C_5H_{12}N_2$	6.36	0.17
$C_6H_{12}O$	6.72	0.39
$C_6H_{14}N$	7.09	0.22
C_7H_2N	7.98	0.28
C_7H_{16}	7.82	0.26
C_8H_4	8.71	0.33

	M+1	+2
101		
CHN_4O_2	2.70	0.43
$C_2HN_2O_3$	3.06	0.64
$C_2H_3N_3O_2$	3.43	0.45
$C_2H_5N_4O$	3.81	0.26
C_3HO_4	3.41	0.84
$C_3H_3NO_3$	3.79	0.65
$C_3H_5N_2O_2$	4.16	0.47
$C_3H_7N_3O$	4.54	0.28
$C_3H_9N_4$	4.91	0.10
$C_4H_5O_3$	4.52	0.68
$C_4H_7NO_2$	4.89	0.50
$C_4H_9N_2O$	5.27	0.31
$C_4H_{11}N_3$	5.64	0.13
$C_5H_9O_2$	5.63	0.53
$C_5H_{11}NO$	6.00	0.35
$C_5H_{13}N_2$	6.37	0.17
C_6HN_2	7.26	0.23
$C_6H_{13}O$	6.73	0.39
$C_6H_{15}N$	7.11	0.22
C_7HO	7.62	0.45
C_7H_3N	7.99	0.28
C_8H_5	8.73	0.33
102		
$CH_2N_4O_2$	2.72	0.43
$C_2H_2N_2O_3$	3.07	0.64
$C_2H_4N_3O_2$	3.45	0.45
$C_2H_6N_4O$	3.82	0.26
$C_3H_2O_4$	3.43	0.84
$C_3H_4NO_3$	3.81	0.66
$C_3H_6N_2O_2$	4.18	0.47
$C_3H_8N_3O$	4.55	0.28
$C_3H_{10}N_4$	4.93	0.10
$C_4H_6O_3$	4.54	0.68
$C_4H_8NO_2$	4.91	0.50
$C_4H_{10}N_2O$	5.28	0.32
$C_4H_{12}N_3$	5.66	0.13
$C_5H_{10}O_2$	5.64	0.53
$C_5H_{12}NO$	6.02	0.35
$C_5H_{14}N_2$	6.39	0.17
$C_6H_2N_2$	7.28	0.23
$C_6H_{14}O$	6.75	0.39
C_7H_2O	7.64	0.45
C_7H_4N	8.01	0.28
C_8H_6	8.74	0.34
103		
CHN_3O_3	2.36	0.62
$CH_3N_4O_2$	2.73	0.43
C_2HNO_4	2.72	0.83
$C_2H_3N_2O_3$	3.09	0.64
$C_2H_5N_3O_2$	3.46	0.45
$C_2H_7N_4O$	3.84	0.26
$C_3H_3O_4$	3.45	0.84
$C_3H_5NO_3$	3.82	0.66
$C_3H_7N_2O_2$	4.20	0.47
$C_3H_9N_3O$	4.57	0.29
$C_3H_{11}N_4$	4.94	0.10
$C_4H_7O_3$	4.55	0.68
$C_4H_9NO_2$	4.93	0.50
$C_4H_{11}N_2O$	5.30	0.32
$C_4H_{13}N_3$	5.68	0.14

	M+1	M+2
C_5HN_3	6.56	0.18
$C_5H_{11}O_2$	5.66	0.53
$C_5H_{13}NO$	6.03	0.35
C_6HNO	6.92	0.40
$C_6H_3N_2$	7.30	0.23
C_7H_3O	7.65	0.45
C_7H_5N	8.03	0.28
C_8H_7	8.76	0.34
104		
$CH_2N_3O_3$	2.37	0.62
$CH_4N_4O_2$	2.75	0.43
$C_2H_2NO_4$	2.73	0.83
$C_2H_4N_2O_3$	3.11	0.64
$C_2H_6N_3O_2$	3.48	0.45
$C_2H_8N_4O$	3.85	0.26
$C_3H_4O_4$	3.46	0.84
$C_3H_6NO_3$	3.84	0.66
$C_3H_8N_2O_2$	4.21	0.47
$C_3H_{10}N_3O$	4.59	0.29
$C_3H_{12}N_4$	4.96	0.10
$C_4H_6O_3$	4.57	0.68
$C_4H_{10}NO_2$	4.94	0.50
$C_4H_{12}N_2O$	5.32	0.32
$C_5H_2N_3$	6.58	0.19
$C_5H_{12}O_2$	5.67	0.53
C_6H_2NO	6.94	0.41
$C_6H_4N_2$	7.31	0.23
C_7H_4O	7.67	0.45
C_7H_6N	8.04	0.28
C_8H_8	8.77	0.34
105		
CHN_2O_4	2.02	0.81
$CH_3N_3O_3$	2.39	0.62
$CH_5N_4O_2$	2.77	0.43
$C_2H_3NO_4$	2.75	0.83
$C_2H_5N_2O_3$	3.12	0.64
$C_2H_7N_3O_2$	3.50	0.45
$C_2H_9N_4O$	3.87	0.26
$C_3H_5O_4$	3.48	0.84
$C_3H_7NO_3$	3.85	0.66
$C_3H_9N_2O_2$	4.23	0.47
$C_3H_{11}N_3O$	4.60	0.29
C_4HN_4	5.86	0.15
$C_4H_9O_3$	4.58	0.68
$C_4H_{11}NO_2$	4.96	0.50
C_5HN_2O	6.22	0.36
$C_5H_3N_3$	6.60	0.19
C_6HO_2	6.58	0.58
C_6H_3NO	6.95	0.41
$C_6H_5N_2$	7.33	0.23
C_7H_5O	7.68	0.45
C_7H_7N	8.06	0.28
C_8H_9	8.79	0.34
106		
$CH_2N_2O_4$	2.03	0.82
$CH_4N_3O_3$	2.41	0.62
$CH_6N_4O_2$	2.78	0.43
$C_2H_4NO_4$	2.76	0.83
$C_2H_6N_2O_3$	3.14	0.64
$C_2H_8N_3O_2$	3.51	0.45
$C_2H_{10}N_4O$	3.89	0.26

	M+1	M+2
$C_3H_6O_4$	3.49	0.85
$C_3H_8NO_3$	3.87	0.66
$C_3H_{10}N_2O_2$	4.24	0.47
$C_4H_2N_4$	5.88	0.15
$C_4H_{10}O_3$	4.60	0.68
$C_5H_2N_2O$	6.24	0.36
$C_5H_4N_3$	6.61	0.19
$C_6H_2O_2$	6.59	0.58
C_6H_4NO	6.97	0.41
$C_6H_6N_2$	7.34	0.23
C_7H_6O	7.70	0.46
C_7H_8N	8.07	0.28
C_8H_{10}	8.81	0.34
107		
$CH_3N_2O_4$	2.05	0.82
$CH_5N_3O_3$	2.42	0.62
$CH_7N_4O_2$	2.80	0.43
$C_2H_5NO_4$	2.78	0.83
$C_2H_7N_2O_3$	3.15	0.64
$C_2H_9N_3O_2$	3.53	0.45
$C_3H_7O_4$	3.51	0.85
$C_3H_9NO_3$	3.89	0.66
C_4HN_3O	5.52	0.33
$C_4H_3N_4$	5.90	0.15
C_5HNO_2	5.88	0.54
$C_5H_3N_2O$	6.25	0.37
$C_5H_5N_3$	6.63	0.19
$C_6H_3O_2$	6.61	0.58
C_6H_5NO	6.98	0.41
$C_6H_7N_2$	7.36	0.23
C_7H_7O	7.72	0.46
C_7H_9N	8.09	0.29
C_8H_{11}	8.82	0.34
108		
$CH_4N_2O_4$	2.06	0.82
$CH_6N_3O_3$	2.44	0.62
$CH_8N_4O_2$	2.81	0.43
$C_2H_6NO_4$	2.80	0.83
$C_2H_8N_2O_3$	3.17	0.64
$C_3H_8O_4$	3.53	0.85
$C_4H_2N_3O$	5.54	0.33
$C_4H_4N_4$	5.91	0.15
$C_5H_2NO_2$	5.90	0.54
$C_5H_4N_2O$	6.27	0.37
$C_5H_6N_3$	6.64	0.19
$C_6H_4O_2$	6.63	0.59
C_6H_6NO	7.00	0.41
$C_6H_8N_2$	7.38	0.24
C_7H_8O	7.73	0.46
$C_7H_{10}N$	8.11	0.29
C_8H_{12}	8.84	0.34
109		
$CH_5N_2O_4$	2.08	0.82
$CH_7N_3O_3$	2.45	0.62
$C_2H_7NO_4$	2.81	0.83
C_3HN_4O	4.82	0.30
$C_4HN_2O_2$	5.18	0.51
$C_4H_3N_3O$	5.55	0.33
$C_4H_5N_4$	5.93	0.15
C_5HO_3	5.54	0.73

	M + 1	M + 2
C5H3NO2	5.91	0.55
C5H5N2O	6.29	0.37
C5H7N3	6.66	0.19
C6H5O2	6.64	0.59
C6H7NO	7.02	0.41
C6H9N2	7.39	0.24
C7H9O	7.75	0.46
C7H11N	8.12	0.29
C8H13	8.85	0.35
C9H	9.74	0.42

110

	M + 1	M + 2
CH6N2O4	2.10	0.82
C3H2N4O	4.84	0.30
C4H2N2O2	5.20	0.51
C4H4N3O	5.57	0.33
C4H6N4	5.94	0.15
C5H2O3	5.55	0.73
C5H4NO2	5.93	0.55
C5H6N2O	6.30	0.37
C5H8N3	6.68	0.19
C6H6O2	6.66	0.59
C6H8NO	7.03	0.41
C6H10N2	7.41	0.24
C7H10O	7.76	0.46
C7H12N	8.14	0.29
C8H14	8.87	0.35
C9H2	9.76	0.42

111

	M + 1	M + 2
C3HN3O2	4.48	0.48
C3H3N4O	4.86	0.30
C4HNO3	4.84	0.69
C4H3N2O2	5.21	0.51
C4H5N3O	5.59	0.33
C4H7N4	5.96	0.15
C5H3O3	5.57	0.73
C5H5NO2	5.94	0.55
C5H7N2O	6.32	0.37
C5H9N3	6.69	0.19
C6H7O2	6.67	0.59
C6H9NO	7.05	0.41
C6H11N2	7.42	0.24
C7H11O	7.78	0.46
C7H13N	8.15	0.29
C8HN	9.04	0.36
C8H15	8.89	0.35
C9H3	9.77	0.43

112

	M + 1	M + 2
C3H2N3O2	4.50	0.48
C3H4N4O	4.87	0.30
C4H2NO3	4.85	0.70
C4H4N2O2	5.23	0.51
C4H6N3O	5.60	0.33
C4H8N4	5.98	0.15
C5H4O3	5.58	0.73
C5H6NO2	5.96	0.55
C5H8N2O	6.33	0.37
C5H10N3	6.71	0.19
C6H8O2	6.69	0.59
C6H10NO	7.06	0.41
C6H12N2	7.44	0.24
C7H12O	7.80	0.46

	M + 1	M + 2
C7H14N	8.17	0.29
C8H2N	9.06	0.36
C8H16	8.90	0.35
C9H4	9.79	0.43

113

	M + 1	M + 2
C2HN4O2	3.78	0.46
C3HN2O3	4.14	0.67
C3H3N3O2	4.51	0.48
C3H5N4O	4.89	0.30
C4HO4	4.50	0.88
C4H3NO3	4.87	0.70
C4H5N2O2	5.24	0.51
C4H7N3O	5.62	0.33
C4H9N4	5.99	0.15
C5H5O3	5.60	0.73
C5H7NO2	5.98	0.55
C5H9N2O	6.35	0.37
C5H11N3	6.72	0.19
C6H9O2	6.71	0.59
C6H11NO	7.08	0.42
C6H13N2	7.46	0.24
C7HN2	8.34	0.31
C7H13O	7.81	0.46
C7H15N	8.19	0.29
C8HO	8.70	0.53
C8H3N	9.07	0.36
C8H17	8.92	0.35
C9H5	9.81	0.43

114

	M + 1	M + 2
C2H2N4O2	3.80	0.46
C3H2N2O3	4.15	0.67
C3H4N3O2	4.53	0.48
C3H6N4O	4.90	0.30
C4H2O4	4.51	0.88
C4H4NO3	4.89	0.70
C4H6N2O2	5.26	0.51
C4H8N3O	5.63	0.33
C4H10N4	6.01	0.15
C5H6O3	5.62	0.73
C5H8NO2	5.99	0.55
C5H10N2O	6.37	0.37
C5H12N3	6.74	0.20
C6H10O2	6.72	0.59
C6H12NO	7.10	0.42
C6H14N2	7.47	0.24
C7H2N2	8.36	0.31
C7H14O	7.83	0.47
C7H16N	8.20	0.29
C8H2O	8.72	0.53
C8H4N	9.09	0.37
C8H18	8.93	0.35
C9H6	9.82	0.43

115

	M + 1	M + 2
C2HN3O3	3.44	0.65
C2H3N4O2	3.81	0.46
C3HNO4	3.80	0.86
C3H3N2O3	4.17	0.67
C3H5N3O2	4.54	0.48
C3H7N4O	4.92	0.30
C4H3O4	4.53	0.88
C4H5NO3	4.90	0.70

	M + 1	M + 2
C4H7N2O2	5.28	0.52
C4H9N3O	5.65	0.33
C4H11N4	6.02	0.16
C5H7O3	5.63	0.73
C5H9NO2	6.01	0.55
C5H11N2O	6.38	0.37
C5H13N3	6.76	0.20
C6HN3	7.64	0.25
C6H11O2	6.74	0.59
C6H13NO	7.11	0.42
C6H15N2	7.49	0.24
C7HNO	8.00	0.48
C7H3N2	8.38	0.31
C7H15O	7.84	0.47
C7H17N	8.22	0.30
C8H3O	8.73	0.54
C8H5N	9.11	0.37
C9H7	9.84	0.43

116

	M + 1	M + 2
C2H2N3O3	3.46	0.65
C2H4N4O2	3.83	0.46
C3H2NO4	3.81	0.86
C3H4N2O3	4.19	0.67
C3H6N3O2	4.56	0.49
C3H8N4O	4.94	0.30
C4H4O4	4.54	0.88
C4H6NO3	4.92	0.70
C4H8N2O2	5.29	0.52
C4H10N3O	5.67	0.34
C4H12N4	6.04	0.16
C5H8O3	5.65	0.73
C5H10NO2	6.02	0.55
C5H12N2O	6.40	0.37
C5H14N3	6.77	0.20
C6H2N3	7.66	0.26
C6H12O2	6.75	0.59
C6H14NO	7.13	0.42
C6H16N2	7.50	0.24
C7H2NO	8.02	0.48
C7H4N2	8.39	0.31
C7H16O	7.86	0.47
C8H4O	8.75	0.54
C8H6N	9.12	0.37
C9H8	9.85	0.43

117

	M + 1	M + 2
C2HN2O4	3.10	0.84
C2H3N3O3	3.47	0.65
C2H5N4O2	3.85	0.46
C3H3NO4	3.83	0.86
C3H5N2O3	4.20	0.67
C3H7N3O2	4.58	0.49
C3H9N4O	4.95	0.30
C4H5O4	4.56	0.88
C4H7NO3	4.93	0.70
C4H9N2O2	5.31	0.52
C4H11N3O	5.68	0.34
C4H13N4	6.06	0.16
C5HN4	6.95	0.21
C5H9O3	5.66	0.73
C5H11NO2	6.04	0.55
C5H13N2O	6.41	0.38
C5H15N3	6.79	0.20

	M + 1	M + 2
C6HN2O	7.30	0.43
C6H3N3	7.68	0.26
C6H13O2	6.77	0.60
C6H15NO	7.14	0.42
C7HO2	7.66	0.65
C7H3NO	8.03	0.48
C7H5N2	8.41	0.31
C8H5O	8.76	0.54
C8H7N	9.14	0.37
C9H9	9.87	0.43

118

	M + 1	M + 2
C2H2N2O4	3.11	0.84
C2H4N3O3	3.49	0.65
C2H6N4O2	3.86	0.46
C3H4NO4	3.84	0.86
C3H6N2O3	4.22	0.67
C3H8N3O2	4.59	0.49
C3H10N4O	4.97	0.30
C4H6O4	4.58	0.88
C4H8NO3	4.95	0.70
C4H10N2O2	5.32	0 52
C4H12N3O	5.70	0.34
C4H14N4	6.07	0.16
C5H2N4	6.96	0.21
C5H10O3	5.68	0.73
C5H12NO2	6.06	0.55
C5H14N2O	6.43	0.38
C6H2N2O	7.32	0.43
C6H4N3	7.69	0.26
C6H14O2	6.79	0.60
C7H2O2	7.67	0.65
C7H4NO	8.05	0.48
C7H6N2	8.42	0.31
C8H6O	8.78	0.54
C8H8N	9.15	0.37
C9H10	9.89	0.44

119

	M + 1	M + 2
C2H3N2O4	3.13	0.84
C2H5N3O3	3.50	0.65
C2H7N4O2	3.88	0.46
C3H5NO4	3.86	0.86
C3H7N2O3	4.23	0.67
C3H9N3O2	4.61	0.49
C3H11N4O	4.98	0.30
C4H7O4	4.59	0.88
C4H9NO3	4.97	0.70
C4H11N2O2	5.34	0.52
C4H13N3O	5.71	0.34
C5HN3O	6.60	0.39
C5H3N4	6.98	0.21
C5H11O3	5.70	0.73
C5H13NO2	6.07	0.56
C6HNO2	6.96	0.61
C6H3N2O	7.33	0.43
C6H5N3	7.71	0.26
C7H3O2	7.69	0.66
C7H5NO	8.07	0.48
C7H7N2	8.44	0.31
C8H7O	8.80	0.54
C8H9N	9.17	0.37
C9H11	9.90	0.44

	M + 1	M + 2
120		
$C_2H_4N_2O_4$	3.15	0.84
$C_2H_6N_3O_3$	3.52	0.65
$C_2H_8N_4O_2$	3.89	0.46
$C_3H_6NO_4$	3.88	0.86
$C_3H_8N_2O_3$	4.25	0.67
$C_3H_{10}N_3O_2$	4.62	0.49
$C_3H_{12}N_4O$	5.00	0.31
$C_4H_8O_4$	4.61	0.88
$C_4H_{10}NO_3$	4.98	0.70
$C_4H_{12}N_2O_2$	5.36	0.52
$C_5H_2N_3O$	6.62	0.39
$C_5H_4N_4$	6.99	0.21
$C_5H_{12}O_3$	5.71	0.74
$C_6H_2NO_2$	6.98	0.61
$C_6H_4N_2O$	7.35	0.43
$C_6H_6N_3$	7.72	0.26
$C_7H_4O_2$	7.71	0.66
C_7H_6NO	8.08	0.49
$C_7H_8N_2$	8.46	0.32
C_8H_8O	8.81	0.54
$C_8H_{10}N$	9.19	0.37
C_9H_{12}	9.92	0.44
121		
$C_2H_5N_2O_4$	3.16	0.84
$C_2H_7N_3O_3$	3.54	0.65
$C_2H_9N_4O_2$	3.91	0.46
$C_3H_7NO_4$	3.89	0.86
$C_3H_9N_2O_3$	4.27	0.67
$C_3H_{11}N_3O_2$	4.64	0.49
C_4HN_4O	5.90	0.35
$C_4H_9O_4$	4.62	0.89
$C_4H_{11}NO_3$	5.00	0.70
$C_5HN_2O_2$	6.26	0.57
$C_5H_3N_3O$	6.64	0.39
$C_5H_5N_4$	7.01	0.21
C_6HO_3	6.62	0.79
$C_6H_3NO_2$	6.99	0.61
$C_6H_5N_2O$	7.37	0.44
$C_6H_7N_3$	7.74	0.26
$C_7H_5O_2$	7.72	0.66
C_7H_7NO	8.10	0.49
$C_7H_9N_2$	8.47	0.32
C_8H_9O	8.83	0.54
$C_8H_{11}N$	9.20	0.38
C_9H_{13}	9.93	0.44
$C_{10}H$	10.82	0.53
122		
$C_2H_6N_2O_4$	3.18	0.84
$C_2H_8N_3O_3$	3.55	0.65
$C_2H_{10}N_4O_2$	3.93	0.46
$C_3H_8NO_4$	3.91	0.86
$C_3H_{10}N_2O_3$	4.28	0.67
$C_4H_2N_4O$	5.92	0.35
$C_4H_{10}O_4$	4.64	0.89
$C_5H_2N_2O_2$	6.28	0.57
$C_5H_4N_3O$	6.65	0.39
$C_5H_6N_4$	7.03	0.21
$C_6H_2O_3$	6.63	0.79
$C_6H_4NO_2$	7.01	0.61
$C_6H_6N_2O$	7.38	0.44
$C_6H_8N_3$	7.76	0.26
$C_7H_6O_2$	7.74	0.66
C_7H_8NO	8.11	0.49
$C_7H_{10}N_2$	8.49	0.32
$C_8H_{10}O$	8.84	0.54
$C_8H_{12}N$	9.22	0.38
C_9H_{14}	9.95	0.44
$C_{10}H_2$	10.84	0.53
123		
$C_2H_7N_2O_4$	3.19	0.84
$C_2H_9N_3O_3$	3.57	0.65
$C_3H_9NO_4$	3.92	0.86
$C_4HN_3O_2$	5.56	0.53
$C_4H_3N_4O$	5.94	0.35
C_5HNO_3	5.92	0.75
$C_5H_3N_2O_2$	6.29	0.57
$C_5H_5N_3O$	6.67	0.39
$C_5H_7N_4$	7.04	0.22
$C_6H_3O_3$	6.65	0.79
$C_6H_5NO_2$	7.02	0.61
$C_6H_7N_2O$	7.40	0.44
$C_6H_9N_3$	7.77	0.26
$C_7H_7O_2$	7.75	0.66
C_7H_9NO	8.13	0.49
$C_7H_{11}N_2$	8.50	0.32
$C_8H_{11}O$	8.86	0.55
$C_8H_{13}N$	9.23	0.38
C_9HN	10.12	0.46
C_9H_{15}	9.97	0.44
$C_{10}H_3$	10.85	0.53
124		
$C_2H_8N_2O_4$	3.21	0.84
$C_4H_2N_3O_2$	5.58	0.53
$C_4H_4N_4O$	5.95	0.35
$C_5H_2NO_3$	5.93	0.75
$C_5H_4N_2O_2$	6.31	0.57
$C_5H_6N_3O$	6.68	0.39
$C_5H_8N_4$	7.06	0.22
$C_6H_4O_3$	6.67	0.79
$C_6H_6NO_2$	7.04	0.61
$C_6H_8N_2O$	7.41	0.44
$C_6H_{10}N_3$	7.79	0.27
$C_7H_8O_2$	7.77	0.66
$C_7H_{10}NO$	8.15	0.49
$C_7H_{12}N_2$	8.52	0.32
$C_8H_{12}O$	8.88	0.55
$C_8H_{14}N$	9.25	0.38
C_9H_2N	10.14	0.46
C_9H_{16}	9.98	0.45
$C_{10}H_4$	10.87	0.53
125		
$C_3HN_4O_2$	4.86	0.50
$C_4HN_2O_3$	5.22	0.71
$C_4H_3N_3O_2$	5.59	0.53
$C_4H_5N_4O$	5.97	0.35
C_5HO_4	5.58	0.93
$C_5H_3NO_3$	5.95	0.75
$C_5H_5N_2O_2$	6.32	0.57
$C_5H_7N_3O$	6.70	0.39
$C_5H_9N_4$	7.07	0.22
$C_6H_5O_3$	6.68	0.79
$C_6H_7NO_2$	7.06	0.61
$C_6H_9N_2O$	7.43	0.44
$C_6H_{11}N_3$	7.80	0.27
$C_7H_9O_2$	7.79	0.66
$C_7H_{11}NO$	8.16	0.49
$C_7H_{13}N_2$	8.54	0.32
C_8HN_2	9.42	0.40
$C_8H_{13}O$	8.89	0.55
$C_8H_{15}N$	9.27	0.38
C_9HO	9.78	0.63
C_9H_3N	10.16	0.46
C_9H_{17}	10.00	0.45
$C_{10}H_5$	10.89	0.53
126		
$C_3H_2N_4O_2$	4.88	0.50
$C_4H_2N_2O_3$	5.24	0.71
$C_4H_4N_3O_2$	5.61	0.53
$C_4H_6N_4O$	5.98	0.35
$C_5H_2O_4$	5.59	0.93
$C_5H_4NO_3$	5.97	0.75
$C_5H_6N_2O_2$	6.34	0.57
$C_5H_8N_3O$	6.72	0.35
$C_5H_{10}N_4$	7.09	0.22
$C_6H_6O_3$	6.70	0.79
$C_6H_8NO_2$	7.07	0.62
$C_6H_{10}N_2O$	7.45	0.44
$C_6H_{12}N_3$	7.82	0.27
$C_7H_{10}O_2$	7.80	0.66
$C_7H_{12}NO$	8.18	0.49
$C_7H_{14}N_2$	8.55	0.32
$C_8H_2N_2$	9.44	0.40
$C_8H_{14}O$	8.91	0.55
$C_8H_{16}N$	9.28	0.38
C_9H_2O	9.80	0.63
C_9H_4N	10.17	0.46
C_9H_{18}	10.01	0.45
$C_{10}H_6$	10.90	0.54
127		
$C_3HN_3O_3$	4.52	0.68
$C_3H_3N_4O_2$	4.89	0.50
C_4HNO_4	4.88	0.90
$C_4H_3N_2O_3$	5.25	0.71
$C_4H_5N_3O_2$	5.63	0.53
$C_4H_7N_4O$	6.00	0.35
$C_5H_3O_4$	5.61	0.93
$C_5H_5NO_3$	5.98	0.75
$C_5H_7N_2O_2$	6.36	0.57
$C_5H_9N_3O$	6.73	0.40
$C_5H_{11}N_4$	7.11	0.22
$C_6H_7O_3$	6.71	0.79
$C_6H_9NO_2$	7.09	0.62
$C_6H_{11}N_2O$	7.46	0.44
$C_6H_{13}N_3$	7.84	0.27
C_7HN_3	8.73	0.34
$C_7H_{11}O_2$	7.82	0.67
$C_7H_{13}NO$	8.19	0.49
$C_7H_{15}N_2$	8.57	0.32
C_8HNO	9.08	0.57
$C_8H_3N_2$	9.46	0.40
$C_8H_{15}O$	8.92	0.55
$C_8H_{17}N$	9.30	0.38
C_9H_3O	9.81	0.63
C_9H_5N	10.19	0.47
C_9H_{19}	10.03	0.45
$C_{10}H_7$	10.92	0.54
128		
$C_3H_2N_3O_3$	4.54	0.68
$C_3H_4N_4O_2$	4.91	0.50
$C_4H_2NO_4$	4.89	0.90
$C_4H_4N_2O_3$	5.27	0.72
$C_4H_6N_3O_2$	5.64	0.53
$C_4H_8N_4O$	6.02	0.36
$C_5H_4O_4$	5.62	0.93
$C_5H_6NO_3$	6.00	0.75
$C_5H_8N_2O_2$	6.37	0.57
$C_5H_{10}N_3O$	6.75	0.40
$C_5H_{12}N_4$	7.12	0.22
$C_6H_8O_3$	6.73	0.79
$C_6H_{10}NO_2$	7.10	0.62
$C_6H_{12}N_2O$	7.48	0.44
$C_6H_{14}N_3$	7.85	0.27
$C_7H_2N_3$	8.74	0.34
$C_7H_{12}O_2$	7.83	0.67
$C_7H_{14}NO$	8.21	0.50
$C_7H_{16}N_2$	8.58	0.33
C_8H_2NO	9.10	0.57
$C_8H_4N_2$	9.47	0.40
$C_8H_{16}O$	8.94	0.55
$C_8H_{18}N$	9.31	0.39
C_9H_4O	9.83	0.63
C_9H_6N	10.20	0.47
C_9H_{20}	10.05	0.45
$C_{10}H_8$	10.94	0.54
129		
$C_3HN_2O_4$	4.18	0.87
$C_3H_3N_3O_3$	4.55	0.69
$C_3H_5N_4O_2$	4.93	0.50
$C_4H_3NO_4$	4.91	0.90
$C_4H_5N_2O_3$	5.28	0.72
$C_4H_7N_3O_2$	5.66	0.54
$C_4H_9N_4O$	6.03	0.36
$C_5H_5O_4$	5.64	0.93
$C_5H_7NO_3$	6.01	0.75
$C_5H_9N_2O_2$	6.39	0.57
$C_5H_{11}N_3O$	6.76	0.40
$C_5H_{13}N_4$	7.14	0.22
C_6HN_4	8.03	0.28
$C_6H_9O_3$	6.75	0.79
$C_6H_{11}NO_2$	7.12	0.62
$C_6H_{13}N_2O$	7.49	0.44
$C_6H_{15}N_3$	7.87	0.27
C_7HN_2O	8.38	0.51
$C_7H_3N_3$	8.76	0.34
$C_7H_{13}O_2$	7.85	0.67
$C_7H_{15}NO$	8.23	0.50
$C_7H_{17}N_2$	8.60	0.33
C_8HO_2	8.74	0.74
C_8H_3NO	9.11	0.57
$C_8H_5N_2$	9.49	0.40
$C_8H_{17}O$	8.96	0.55
$C_8H_{19}N$	9.33	0.39
C_9H_5O	9.85	0.63
C_9H_7N	10.22	0.47
$C_{10}H_9$	10.95	0.54

	M + 1	M + 2
130		
$C_3H_2N_2O_4$	4.19	0.87
$C_3H_4N_3O_3$	4.57	0.69
$C_3H_6N_4O_2$	4.94	0.50
$C_4H_4NO_4$	4.92	0.90
$C_4H_6N_2O_3$	5.30	0.72
$C_4H_8N_3O_2$	5.67	0.54
$C_4H_{10}N_4O$	6.05	0.36
$C_5H_6O_4$	5.66	0.93
$C_5H_8NO_3$	6.03	0.75
$C_5H_{10}N_2O_2$	6.40	0.58
$C_5H_{12}N_3O$	6.78	0.40
$C_5H_{14}N_4$	7.15	0.22
$C_6H_2N_4$	8.04	0.29
$C_6H_{10}O_3$	6.76	0.79
$C_6H_{12}NO_2$	7.14	0.62
$C_6H_{14}N_2O$	7.51	0.45
$C_6H_{16}N_3$	7.88	0.27
$C_7H_2N_2O$	8.40	0.51
$C_7H_4N_3$	8.77	0.34
$C_7H_{14}O_2$	7.87	0.67
$C_7H_{16}NO$	8.24	0.50
$C_7H_{18}N_2$	8.62	0.33
$C_8H_2O_2$	8.76	0.74
C_8H_4NO	9.13	0.57
$C_8H_6N_2$	9.50	0.40
$C_8H_{18}O$	8.97	0.56
C_9H_6O	9.86	0.63
C_9H_8N	10.24	0.47
$C_{10}H_{10}$	10.97	0.54
131		
$C_3H_3N_2O_4$	4.21	0.87
$C_3H_5N_3O_3$	4.58	0.69
$C_3H_7N_4O_2$	4.96	0.50
$C_4H_5NO_4$	4.94	0.90
$C_4H_7N_2O_3$	5.32	0.72
$C_4H_9N_3O_2$	5.69	0.54
$C_4H_{11}N_4O$	6.06	0.36
$C_5H_7O_4$	5.67	0.93
$C_5H_9NO_3$	6.05	0.75
$C_5H_{11}N_2O_2$	6.42	0.58
$C_5H_{13}N_3O$	6.80	0.40
$C_5H_{15}N_4$	7.17	0.22
C_6HN_3O	7.68	0.46
$C_6H_3N_4$	8.06	0.29
$C_6H_{11}O_3$	6.78	0.80
$C_6H_{13}NO_2$	7.15	0.62
$C_6H_{15}N_2O$	7.53	0.45
$C_6H_{17}N_3$	7.90	0.27
C_7HNO_2	8.04	0.68
$C_7H_3N_2O$	8.41	0.51
$C_7H_5N_3$	8.79	0.34
$C_7H_{15}O_2$	7.88	0.67
$C_7H_{17}NO$	8.26	0.50
$C_8H_3O_2$	8.77	0.74
C_8H_5NO	9.15	0.57
$C_8H_7N_2$	9.52	0.41
C_9H_7O	9.88	0.64
C_9H_9N	10.25	0.47
$C_{10}H_{11}$	10.98	0.54
132		
$C_3H_4N_2O_4$	4.23	0.87
$C_3H_6N_3O_3$	4.60	0.69
$C_3H_8N_4O_2$	4.97	0.50
$C_4H_6NO_4$	4.96	0.90
$C_4H_8N_2O_3$	5.33	0.72
$C_4H_{10}N_3O_2$	5.71	0.54
$C_4H_{12}N_4O$	6.08	0.36
$C_5H_8O_4$	5.69	0.93
$C_5H_{10}NO_3$	6.06	0.76
$C_5H_{12}N_2O_2$	6.44	0.58
$C_5H_{14}N_3O$	6.81	0.40
$C_5H_{16}N_4$	7.19	0.23
$C_6H_2N_3O$	7.70	0.46
$C_6H_4N_4$	8.07	0.29
$C_6H_{12}O_3$	6.97	0.80
$C_6H_{14}NO_2$	7.17	0.62
$C_6H_{16}N_2O$	7.54	0.45
$C_7H_2NO_2$	8.06	0.68
$C_7H_4N_2O$	8.43	0.51
$C_7H_6N_3$	8.81	0.34
$C_7H_{16}O_2$	7.90	0.67
$C_8H_4O_2$	8.79	0.74
C_8H_6NO	9.16	0.57
$C_8H_8N_2$	9.54	0.41
C_9H_8O	9.89	0.64
$C_9H_{10}N$	10.27	0.47
$C_{10}H_{12}$	11.00	0.55
133		
$C_3H_5N_2O_4$	4.24	0.87
$C_3H_7N_3O_3$	4.62	0.69
$C_3H_9N_4O_2$	4.99	0.51
$C_4H_7NO_4$	4.97	0.90
$C_4H_9N_2O_3$	5.35	0.72
$C_4H_{11}N_3O_2$	5.72	0.54
$C_4H_{13}N_4O$	6.10	0.36
C_5HN_4O	6.98	0.41
$C_5H_9O_4$	5.70	0.94
$C_5H_{11}NO_3$	6.08	0.76
$C_5H_{13}N_2O_2$	6.45	0.58
$C_5H_{15}N_3O$	6.83	0.40
$C_6HN_2O_2$	7.34	0.63
$C_6H_3N_3O$	7.72	0.46
$C_6H_5N_4$	8.09	0.29
$C_6H_{13}O_3$	6.81	0.80
$C_6H_{15}NO_2$	7.18	0.62
C_7HO_3	7.70	0.86
$C_7H_3NO_2$	8.07	0.69
$C_7H_5N_2O$	8.45	0.51
$C_7H_7N_3$	8.82	0.35
$C_8H_5O_2$	8.80	0.74
C_8H_7NO	9.18	0.57
$C_8H_9N_2$	9.55	0.41
C_9H_9O	9.91	0.64
$C_9H_{11}N$	10.28	0.48
$C_{10}H_{13}$	11.01	0.55
$C_{11}H$	11.90	0.64
134		
$C_3H_6N_2O_4$	4.26	0.87
$C_3H_8N_3O_3$	4.63	0.69
$C_3H_{10}N_4O_2$	5.01	0.51
$C_4H_8NO_4$	4.99	0.90
$C_4H_{10}N_2O_3$	5.36	0.72
$C_4H_{12}N_3O_2$	5.74	0.54
$C_4H_{14}N_4O$	6.11	0.36
$C_5H_2N_4O$	7.00	0.41
$C_5H_{10}O_4$	5.72	0.94
$C_5H_{12}NO_3$	6.09	0.76
$C_5H_{14}N_2O_2$	6.47	0.58
$C_6H_2N_2O_2$	7.36	0.64
$C_6H_4N_3O$	7.73	0.46
$C_6H_6N_4$	8.11	0.29
$C_6H_{14}O_3$	6.83	0.80
$C_7H_2O_3$	7.71	0.86
$C_7H_4NO_2$	8.09	0.69
$C_7H_6N_2O$	8.46	0.52
$C_7H_8N_3$	8.84	0.35
$C_8H_6O_2$	8.82	0.74
C_8H_8NO	9.19	0.58
$C_8H_{10}N_2$	9.57	0.41
$C_9H_{10}O$	9.93	0.64
$C_9H_{12}N$	10.30	0.48
$C_{10}H_{14}$	11.03	0.55
$C_{11}H_2$	11.92	0.65
135		
$C_3H_7N_2O_4$	4.27	0.87
$C_3H_9N_3O_3$	4.65	0.69
$C_3H_{11}N_4O_2$	5.02	0.51
$C_4H_9NO_4$	5.00	0.90
$C_4H_{11}N_2O_3$	5.38	0.72
$C_4H_{13}N_3O_2$	5.75	0.54
$C_5HN_3O_2$	6.64	0.59
$C_5H_3N_4O$	7.02	0.41
$C_5H_{11}O_4$	5.74	0.94
$C_5H_{13}NO_3$	6.11	0.76
C_6HNO_3	7.00	0.81
$C_6H_3N_2O_2$	7.37	0.64
$C_6H_5N_3O$	7.75	0.46
$C_6H_7N_4$	8.12	0.29
$C_7H_3O_3$	7.73	0.86
$C_7H_5NO_2$	8.10	0.69
$C_7H_7N_2O$	8.48	0.52
$C_7H_9N_3$	8.85	0.35
$C_8H_7O_2$	8.84	0.74
C_8H_9NO	9.21	0.58
$C_8H_{11}N_2$	9.58	0.41
$C_9H_{11}O$	9.94	0.64
$C_9H_{13}N$	10.32	0.48
$C_{10}HN$	11.20	0.57
$C_{10}H_{15}$	11.05	0.55
$C_{11}H_3$	11.94	0.65
136		
$C_3H_8N_2O_4$	4.29	0.87
$C_3H_{10}N_3O_3$	4.66	0.69
$C_3H_{12}N_4O_2$	5.04	0.51
$C_4H_{10}NO_4$	5.02	0.90
$C_4H_{12}N_2O_3$	5.40	0.72
$C_5H_2N_3O_2$	6.66	0.59
$C_5H_4N_4O$	7.03	0.42
$C_5H_{12}O_4$	5.75	0.94
$C_6H_2NO_3$	7.01	0.81
$C_6H_4N_2O_2$	7.39	0.64
$C_6H_6N_3O$	7.76	0.46
$C_6H_8N_4$	8.14	0.29
$C_7H_4O_3$	7.75	0.86
$C_7H_6NO_2$	8.12	0.69
$C_7H_8N_2O$	8.49	0.52
$C_7H_{10}N_3$	8.87	0.35
$C_8H_8O_2$	8.85	0.75
$C_8H_{10}NO$	9.23	0.58
$C_8H_{12}N_2$	9.60	0.41
$C_8H_{12}O$	9.96	0.64
$C_9H_{14}N$	10.33	0.48
$C_{10}H_2N$	11.22	0.57
$C_{10}H_{16}$	11.06	0.55
$C_{11}H_4$	11.95	0.65
137		
$C_3H_9N_2O_4$	4.31	0.88
$C_3H_{11}N_3O_3$	4.68	0.69
$C_4HN_4O_2$	5.94	0.55
$C_4H_{11}NO_4$	5.04	0.90
$C_5HN_2O_3$	6.30	0.77
$C_5H_3N_3O_2$	6.67	0.59
$C_5H_5N_4O$	7.05	0.42
C_6HO_4	6.66	0.99
$C_6H_3NO_3$	7.03	0.81
$C_6H_5N_2O_2$	7.41	0.64
$C_6H_7N_3O$	7.78	0.47
$C_6H_9N_4$	8.15	0.29
$C_7H_5O_3$	7.76	0.86
$C_7H_7NO_2$	8.14	0.69
$C_7H_9N_2O$	8.51	0.52
$C_7H_{11}N_3$	8.89	0.35
$C_8H_9O_2$	8.87	0.75
$C_8H_{11}NO$	9.24	0.58
$C_8H_{13}N_2$	9.62	0.41
C_9HN_2	10.50	0.50
$C_9H_{13}O$	9.97	0.65
$C_9H_{15}N$	10.35	0.48
$C_{10}HO$	10.86	0.73
$C_{10}H_3N$	11.24	0.57
$C_{10}H_{17}$	11.08	0.56
$C_{11}H_5$	11.97	0.65
138		
$C_3H_{10}N_2O_4$	4.32	0.88
$C_4H_2N_4O_2$	5.96	0.55
$C_5H_2N_2O_3$	6.32	0.77
$C_5H_4N_3O_2$	6.69	0.59
$C_5H_6N_4O$	7.06	0.42
$C_6H_2O_4$	6.67	0.99
$C_6H_4NO_3$	7.05	0.81
$C_6H_6N_2O_2$	7.42	0.64
$C_6H_8N_3O$	7.80	0.47
$C_6H_{10}N_4$	8.17	0.30
$C_7H_6O_3$	7.78	0.86
$C_7H_8NO_2$	8.15	0.69
$C_7H_{10}N_2O$	8.53	0.52
$C_7H_{12}N_3$	8.90	0.35
$C_8H_{10}O_2$	8.88	0.75
$C_8H_{12}NO$	9.26	0.58
$C_8H_{14}N_2$	9.63	0.42
$C_9H_2N_2$	10.52	0.50
$C_9H_{14}O$	9.99	0.65
$C_9H_{16}N$	10.36	0.48
$C_{10}H_2O$	10.88	0.73

	$M+1$	$M+2$
$C_{10}H_4N$	11.25	0.57
$C_{10}H_{18}$	11.09	0.56
$C_{11}H_6$	11.98	0.65

139

	$M+1$	$M+2$
$C_4H_3N_3O_3$	5.60	0.73
$C_4H_3N_4O_2$	5.97	0.55
C_5HNO_4	5.96	0.95
$C_5H_3N_2O_3$	6.33	0.77
$C_5H_5N_3O_2$	6.71	0.59
$C_5H_7N_4O$	7.03	0.42
$C_6H_3O_4$	6.69	0.99
$C_6H_5NO_3$	7.06	0.82
$C_6H_7N_2O_2$	7.44	0.64
$C_6H_9N_3O$	7.81	0.47
$C_6H_{11}N_4$	8.19	0.30
$C_7H_7O_3$	7.79	0.86
$C_7H_9NO_2$	8.17	0.69
$C_7H_{11}N_2O$	8.54	0.52
$C_7H_{13}N_3$	8.92	0.35
C_8HN_3	9.81	0.43
$C_8H_{11}O_2$	8.90	0.75
$C_8H_{13}NO$	9.27	0.58
$C_8H_{15}N_2$	9.65	0.42
C_9HNO	10.16	0.66
$C_9H_3N_2$	10.54	0.50
$C_9H_{15}O$	10.01	0.65
$C_9H_{17}N$	10.38	0.49
$C_{10}H_3O$	10.89	0.74
$C_{10}H_5N$	11.27	0.58
$C_{10}H_{19}$	11.11	0.56
$C_{11}H_7$	12.00	0.66

140

	$M+1$	$M+2$
$C_4H_2N_3O_3$	5.62	0.73
$C_4H_4N_4O_2$	5.99	0.55
$C_5H_2NO_4$	5.97	0.95
$C_5H_4N_2O_3$	6.35	0.77
$C_5H_6N_3O_2$	6.72	0.60
$C_5H_8N_4O$	7.10	0.42
$C_6H_4O_4$	6.70	0.99
$C_6H_6NO_3$	7.08	0.82
$C_6H_8N_2O_2$	7.45	0.64
$C_6H_{10}N_3O$	7.83	0.47
$C_6H_{12}N_4$	8.20	0.30
$C_7H_8O_3$	7.81	0.87
$C_7H_{10}NO_2$	8.18	0.69
$C_7H_{12}N_2O$	8.56	0.52
$C_7H_{14}N_3$	8.93	0.36
$C_8H_2N_3$	9.82	0.43
$C_8H_{12}O_2$	8.92	0.75
$C_8H_{14}NO$	9.29	0.58
$C_8H_{16}N_2$	9.66	0.42
C_9H_2NO	10.18	0.67
$C_9H_4N_2$	10.55	0.50
$C_9H_{16}O$	10.02	0.65
$C_9H_{18}N$	10.40	0.49
$C_{10}H_4O$	10.91	0.74
$C_{10}H_6N$	11.28	0.58
$C_{10}H_{20}$	11.13	0.56
$C_{11}H_8$	12.02	0.66

141

	$M+1$	$M+2$
$C_4HN_2O_4$	5.26	0.92
$C_4H_3N_3O_3$	5.63	0.73
$C_4H_5N_4O_2$	6.01	0.56
$C_5H_3NO_4$	5.99	0.95
$C_5H_5N_2O_3$	6.36	0.77
$C_5H_7N_3O_2$	6.74	0.60
$C_5H_9N_4O$	7.11	0.42
$C_6H_5O_4$	6.72	0.99
$C_6H_7NO_3$	7.09	0.82
$C_6H_9N_2O_2$	7.47	0.64
$C_6H_{11}N_3O$	7.84	0.47
$C_6H_{13}N_4$	8.22	0.30
C_7HN_4	9.11	0.37
$C_7H_9O_3$	7.83	0.87
$C_7H_{11}NO_2$	8.20	0.70
$C_7H_{13}N_2O$	8.57	0.53
$C_7H_{15}N_3$	8.95	0.36
C_8HN_2O	9.46	0.60
$C_8H_3N_3$	9.84	0.44
$C_8H_{13}O_2$	8.93	0.75
$C_8H_{15}NO$	9.31	0.59
$C_8H_{17}N_2$	9.68	0.42
C_9HO_2	9.82	0.83
C_9H_3NO	10.19	0.67
$C_9H_5N_2$	10.57	0.50
$C_9H_{17}O$	10.04	0.65
$C_9H_{19}N$	10.41	0.49
$C_{10}H_5O$	10.93	0.74
$C_{10}H_7N$	11.30	0.58
$C_{10}H_{21}$	11.14	0.56
$C_{11}H_9$	12.03	0.66

142

	$M+1$	$M+2$
$C_4H_2N_2O_4$	5.27	0.92
$C_4H_4N_3O_3$	5.65	0.74
$C_4H_6N_4O_2$	6.02	0.56
$C_5H_4NO_4$	6.01	0.95
$C_5H_6N_2O_3$	6.38	0.77
$C_5H_8N_3O_2$	6.75	0.60
$C_5H_{10}N_4O$	7.13	0.42
$C_6H_6O_4$	6.74	0.99
$C_6H_8NO_3$	7.11	0.82
$C_6H_{10}N_2O_2$	7.49	0.64
$C_6H_{12}N_3O$	7.86	0.47
$C_6H_{14}N_4$	8.23	0.30
$C_7H_2N_4$	9.12	0.37
$C_7H_{10}O_3$	7.84	0.87
$C_7H_{12}NO_2$	8.22	0.70
$C_7H_{14}N_2O$	8.59	0.53
$C_7H_{16}N_3$	8.97	0.36
$C_8H_2N_2O$	9.48	0.60
$C_8H_4N_3$	9.85	0.44
$C_8H_{14}O_2$	8.95	0.75
$C_8H_{16}NO$	9.32	0.59
$C_8H_{18}N_2$	9.70	0.42
$C_9H_2O_2$	9.84	0.83
C_9H_4NO	10.21	0.67
$C_9H_6N_2$	10.58	0.51
$C_9H_{18}O$	10.05	0.65
$C_9H_{20}N$	10.43	0.49
$C_{10}H_6O$	10.94	0.74
$C_{10}H_8N$	11.32	0.58
$C_{10}H_{22}$	11.16	0.56
$C_{11}H_{10}$	12.05	0.66

143

	$M+1$	$M+2$
$C_4H_3N_2O_4$	5.29	0.92
$C_4H_5N_3O_3$	5.66	0.74
$C_4H_7N_4O_2$	6.04	0.56
$C_5H_5NO_4$	6.02	0.95
$C_5H_7N_2O_3$	6.40	0.78
$C_5H_9N_3O_2$	6.77	0.60
$C_5H_{11}N_4O$	7.14	0.42
$C_6H_7O_4$	6.75	0.99
$C_6H_9NO_3$	7.13	0.82
$C_6H_{11}N_2O_2$	7.50	0.65
$C_6H_{13}N_3O$	7.88	0.47
$C_6H_{15}N_4$	8.25	0.30
C_7HN_3O	8.76	0.54
$C_7H_3N_4$	9.14	0.37
$C_7H_{11}O_3$	7.86	0.87
$C_7H_{13}NO_2$	8.23	0.70
$C_7H_{15}N_2O$	8.61	0.53
$C_7H_{17}N_3$	8.98	0.36
C_8HNO_2	9.12	0.77
$C_8H_3N_2O$	9.50	0.60
$C_8H_5N_3$	9.87	0.44
$C_8H_{15}O_2$	8.96	0.76
$C_8H_{17}NO$	9.34	0.59
$C_8H_{19}N_2$	9.71	0.42
$C_9H_3O_2$	9.85	0.83
C_9H_5NO	10.23	0.67
$C_9H_7N_2$	10.60	0.51
$C_9H_{19}O$	10.07	0.65
$C_9H_{21}N$	10.44	0.49
$C_{10}H_7O$	10.96	0.74
$C_{10}H_9N$	11.33	0.58
$C_{11}H_{11}$	12.06	0.66

144

	$M+1$	$M+2$
$C_4H_4N_2O_4$	5.31	0.92
$C_4H_6N_3O_3$	5.68	0.74
$C_4H_8N_4O_2$	6.05	0.56
$C_5H_6NO_4$	6.04	0.95
$C_5H_8N_2O_3$	6.41	0.78
$C_5H_{10}N_3O_2$	6.79	0.60
$C_5H_{12}N_4O$	7.16	0.42
$C_6H_8O_4$	6.77	1.00
$C_6H_{10}NO_3$	7.14	0.82
$C_6H_{12}N_2O_2$	7.52	0.65
$C_6H_{14}N_3O$	7.89	0.47
$C_6H_{16}N_4$	8.27	0.30
$C_7H_2N_3O$	8.78	0.54
$C_7H_4N_4$	9.15	0.38
$C_7H_{12}O_3$	7.87	0.87
$C_7H_{14}NO_2$	8.25	0.70
$C_7H_{16}N_2O$	8.62	0.53
$C_7H_{18}N_3$	9.00	0.36
$C_8H_2NO_2$	9.14	0.77
$C_8H_4N_2O$	9.51	0.60
$C_8H_6N_3$	9.89	0.44
$C_8H_{16}O_2$	8.98	0.76
$C_8H_{18}NO$	9.35	0.59
$C_8H_{20}N_2$	9.73	0.43
$C_9H_4O_2$	9.87	0.84
C_9H_6NO	10.24	0.67
$C_9H_8N_2$	10.62	0.51
$C_9H_{20}O$	10.09	0.66
$C_{10}H_8O$	10.97	0.74
$C_{10}H_{10}N$	11.35	0.58
$C_{11}H_{12}$	12.08	0.67

145

	$M+1$	$M+2$
$C_4H_5N_2O_4$	5.32	0.92
$C_4H_7N_3O_3$	5.70	0.74
$C_4H_9N_4O_2$	6.07	0.56
$C_5H_7NO_4$	6.05	0.96
$C_5H_9N_2O_3$	6.43	0.78
$C_5H_{11}N_3O_2$	6.80	0.60
$C_5H_{13}N_4O$	7.18	0.43
C_6HN_4O	8.07	0.49
$C_6H_9O_4$	6.78	1.00
$C_6H_{11}NO_3$	7.16	0.82
$C_6H_{13}N_2O_2$	7.53	0.65
$C_6H_{15}N_3O$	7.91	0.48
$C_6H_{17}N_4$	8.28	0.31
$C_7HN_2O_2$	8.42	0.71
$C_7H_3N_3O$	8.80	0.54
$C_7H_5N_4$	9.17	0.38
$C_7H_{13}O_3$	7.89	0.87
$C_7H_{15}NO_2$	8.26	0.70
$C_7H_{17}N_2O$	8.64	0.53
$C_7H_{19}N_3$	9.01	0.36
C_8HO_3	8.78	0.94
$C_8H_3NO_2$	9.15	0.77
$C_8H_5N_2O$	9.53	0.61
$C_8H_7N_3$	9.90	0.44
$C_8H_{17}O_2$	9.00	0.76
$C_8H_{19}NO$	9.37	0.59
$C_9H_5O_2$	9.88	0.84
C_9H_7NO	10.26	0.67
$C_9H_9N_2$	10.63	0.51
$C_{10}H_9O$	10.99	0.75
$C_{10}H_{11}N$	11.36	0.59
$C_{11}H_{13}$	12.10	0.67
$C_{12}H$	12.98	0.77

146

	$M+1$	$M+2$
$C_4H_6N_2O_4$	5.34	0.92
$C_4H_8N_3O_3$	5.71	0.74
$C_4H_{10}N_4O_2$	6.09	0.56
$C_5H_8NO_4$	6.07	0.96
$C_5H_{10}N_2O_3$	6.44	0.78
$C_5H_{12}N_3O_2$	6.82	0.60
$C_5H_{14}N_4O$	7.19	0.43
$C_6H_2N_4O$	8.08	0.49
$C_6H_{10}O_4$	6.80	1.00
$C_6H_{12}NO_3$	7.17	0.82
$C_6H_{14}N_2O_2$	7.55	0.65
$C_6H_{16}N_3O$	7.92	0.48
$C_6H_{18}N_4$	8.30	0.31
$C_7H_2N_2O_2$	8.44	0.71
$C_7H_4N_3O$	8.81	0.55
$C_7H_6N_4$	9.19	0.38
$C_7H_{14}O_3$	7.91	0.87
$C_7H_{16}NO_2$	8.28	0.70
$C_7H_{18}N_2O$	8.65	0.53
$C_9H_2O_3$	8.79	0.94
$C_8H_4NO_2$	9.17	0.77
$C_8H_6N_2O$	9.54	0.61
$C_8H_8N_3$	9.92	0.44
$C_8H_{18}O_2$	9.01	0.76
$C_9H_6O_2$	9.90	0.84

	M+1	M+2		M+1	M+2		M+1	M+2		M+1	M+2
C9H8NO	10.27	0.68	C9H12N2	10.68	0.52	C10H2N2	11.60	0.61	C10H18N	11.48	0.60
C9H10N2	10.65	0.51	C10H12O	11.04	0.75	C10H14O	11.07	0.75	C11H4O	11.99	0.86
C10H10O	11.01	0.75	C10H14N	11.41	0.59	C10H16N	11.44	0.60	C11H6N	12.36	0.70
C10H12N	11.38	0.59	C11H2N	12.30	0.69	C11H2O	11.96	0.85	C11H20	12.21	0.68
C11H14	12.11	0.67	C11H16	12.14	0.67	C11H4N	12.33	0.70	C12H8	13.10	0.79
C12H2	13.00	0.77	C12H4	13.03	0.78	C11H18	12.18	0.68			
						C12H6	13.06	0.78	**153**		
147			**149**						C5HN2O4	6.34	0.97
C4H7N2O4	5.35	0.92	C4H9N2O4	5.39	0.92	**151**			C5H3N3O3	6.71	0.80
C4H9N3O3	5.73	0.74	C4H11N3O3	5.76	0.74	C4H11N2O4	5.42	0.92	C5H5N4O2	7.09	0.62
C4H11N4O2	6.10	0.56	C4H13N4O2	6.13	0.56	C4H13N3O3	5.79	0.74	C6H3NO4	7.07	1.02
C5H9NO4	6.09	0.96	C5HN4O2	7.02	0.62	C5HN3O3	6.68	0.79	C6H5N2O3	7.44	0.84
C5H11N2O3	6.46	0.78	C5H11NO4	6.12	0.96	C5H3N4O2	7.06	0.62	C6H7N3O2	7.82	0.67
C5H13N3O2	6.83	0.60	C5H13N2O3	6.49	0.78	C5H13NO4	6.15	0.96	C6H9N4O	8.19	0.50
C5H15N4O	7.21	0.43	C5H15N3O2	6.87	0.61	C6HNO4	7.04	1.01	C7H5O4	7.80	1.07
C6HN3O2	7.72	0.66	C6HN2O3	7.38	0.84	C6H3N2O3	7.41	0.84	C7H7NO3	8.18	0.89
C6H3N4O	8.10	0.49	C6H3N3O2	7.75	0.66	C6H5N3O2	7.79	0.67	C7H9N2O2	8.55	0.72
C6H11O4	6.82	1.00	C6H5N4O	8.13	0.49	C6H7N4O	8.16	0.50	C7H11N3O	8.92	0.56
C6H13NO3	7.19	0.82	C6H13O4	6.85	1.00	C7H3O4	7.77	1.06	C7H13N4	9.30	0.39
C6H15N2O2	7.57	0.65	C6H15NO3	7.22	0.83	C7H5NO3	8.14	0.89	C8HN4	10.19	0.47
C6H17N3O	7.94	0.48	C7HO4	7.74	1.06	C7H7N2O2	8.52	0.72	C8H9O3	8.91	0.95
C7HNO3	8.08	0.89	C7H3NO3	8.11	0.89	C7H9N3O	8.89	0.55	C8H11NO2	9.28	0.78
C7H3N2O2	8.45	0.72	C7H5N2O2	8.49	0.72	C7H11N4	9.27	0.39	C8H13N2O	9.66	0.62
C7H5N3O	8.83	0.55	C7H7N3O	8.86	0.55	C8H7O3	8.87	0.95	C8H15N3	10.03	0.45
C7H7N4	9.20	0.38	C7H9N4	9.23	0.38	C8H9NO2	9.25	0.78	C9HN2O	10.54	0.70
C7H15O3	7.92	0.87	C8H5O3	8.84	0.95	C8H11N2O	9.62	0.62	C9H3N3	10.92	0.54
C7H17NO2	8.30	0.70	C8H7NO2	9.22	0.78	C8H13N3	10.00	0.45	C9H13O2	10.01	0.85
C8H3O3	8.81	0.94	C8H9N2O	9.59	0.61	C9HN3	10.89	0.54	C9H15NO	10.39	0.69
C8H5NO2	9.19	0.78	C8H11N3	9.97	0.45	C9H11O2	9.98	0.85	C9H17N2	10.76	0.52
C8H7N2O	9.56	0.61	C9H9O2	9.95	0.84	C9H13NO	10.36	0.68	C10HO2	10.90	0.94
C8H9N3	9.93	0.44	C9H11NO	10.32	0.68	C9H15N2	10.73	0.52	C10H3NO	11.28	0.78
C9H7O2	9.92	0.84	C9H13N2	10.70	0.52	C10HNO	11.24	0.77	C10H5N2	11.65	0.62
C9H9NO	10.29	0.68	C10HN2	11.59	0.61	C10H3N2	11.62	0.61	C10H17O	11.12	0.76
C9H11N2	10.66	0.51	C10H13O	11.05	0.75	C10H15O	11.09	0.76	C10H19N	11.49	0.60
C10H11O	11.02	0.75	C10H15N	11.43	0.59	C10H17N	11.46	0.60	C11H5O	12.01	0.86
C10H13N	11.40	0.59	C11HO	11.94	0.85	C11H3O	11.97	0.85	C11H7N	12.38	0.70
C11HN	12.28	0.69	C11H3N	12.32	0.69	C11H5N	12.35	0.70	C11H21	12.22	0.68
C11H15	12.13	0.67	C11H17	12.16	0.67	C11H19	12.19	0.68	C12H9	13.11	0.79
C12H3	13.02	0.78	C12H5	13.05	0.78	C12H7	13.08	0.79			
									154		
148						**152**			C5H2N2O4	6.35	0.97
C4H8N2O4	5.37	0.92	**150**			C4H12N2O4	5.43	0.92	C5H4N3O3	6.73	0.80
C4H10N3O3	5.74	0.74	C4H10N2O4	5.40	0.92	C5H2N3O3	6.70	0.79	C5H6N4O2	7.10	0.62
C4H12N4O2	6.12	0.56	C4H12N3O3	5.78	0.74	C5H4N4O2	7.07	0.62	C6H4NO4	7.09	1.02
C5H10NO4	6.10	0.96	C4H14N4O2	6.15	0.56	C6H2NO4	7.05	1.01	C6H6N2O3	7.46	0.84
C5H12N2O3	6.48	0.78	C5H2N4O2	7.04	0.62	C6H4N2O3	7.43	0.84	C6H8N3O2	7.83	0.67
C5H14N3O2	6.85	0.60	C5H12NO4	6.13	0.96	C6H6N3O2	7.80	0.67	C6H10N4O	8.21	0.50
C5H16N4O	7.22	0.43	C5H14N2O3	6.51	0.78	C6H8N4O	8.18	0.50	C7H6O4	7.82	1.07
C6H2N3O2	7.74	0.66	C6H2N2O3	7.40	0.84	C7H4O4	7.79	1.06	C7H8NO3	8.19	0.90
C6H4N4O	8.11	0.49	C6H4N3O2	7.77	0.67	C7H6NO3	8.16	0.89	C7H10N2O2	8.57	0.73
C6H12O4	6.83	1.00	C6H6N4O	8.15	0.49	C7H8N2O2	8.53	0.72	C7H12N3O	8.94	0.56
C6H14NO3	7.21	0.83	C6H14O4	6.86	1.00	C7H10N3O	8.91	0.55	C7H14N4	9.31	0.39
C6H16N2O2	7.58	0.65	C7H2O4	7.75	1.06	C7H12N4	9.28	0.39	C8H2N4	10.20	0.47
C7H2NO3	8.10	0.89	C7H4NO3	8.13	0.89	C8H8O3	8.89	0.95	C8H10O3	8.92	0.95
C7H4N2O2	8.47	0.72	C7H6N2O2	8.50	0.72	C8H10NO2	9.27	0.78	C8H12NO2	9.30	0.79
C7H6N3O	8.84	0.55	C7H8N3O	8.88	0.55	C8H12N2O	9.64	0.62	C8H14N2O	9.67	0.62
C7H8N4	9.22	0.38	C7H10N4	9.25	0.38	C8H14N3	10.01	0.45	C8H16N3	10.05	0.46
C7H16O3	7.94	0.88	C8H6O3	8.86	0.95	C9H2N3	10.90	0.54	C9H2N2O	10.56	0.70
C8H4O3	8.83	0.94	C8H8NO2	9.23	0.78	C9H12O2	10.00	0.85	C9H4N3	10.93	0.54
C8H6NO2	9.20	0.78	C8H10N2O	9.61	0.61	C9H14NO	10.37	0.68	C9H14O2	10.03	0.85
C8H8N2O	9.58	0.61	C8H12N3	9.98	0.45	C9H16N2	10.74	0.52	C9H16NO	10.40	0.69
C8H10N3	9.95	0.45	C9H10O2	9.96	0.84	C10H2NO	11.26	0.78	C9H18N2	10.78	0.53
C9H8O2	9.93	0.84	C9H12NO	10.34	0.68	C10H4N2	11.63	0.62	C10H2O2	10.92	0.94
C9H10NO	10.31	0.68	C9H14N2	10.71	0.52	C10H16O	11.10	0.76	C10H4NO	11.29	0.78

	M + 1	M + 2		M + 1	M + 2		M + 1	M + 2		M + 1	M + 2
$C_{10}H_8N_2$	11.67	0.62	$C_9H_6N_3$	10.97	0.55	$C_7H_{14}N_2O_2$	8.63	0.73	**160**		
$C_{10}H_{18}O$	11.13	0.76	$C_9H_{16}O_2$	10.06	0.85	$C_7H_{16}N_3O$	9.00	0.56	$C_5H_8N_2O_4$	6.45	0.98
$C_{10}H_{20}N$	11.51	0.60	$C_9H_{18}NO$	10.43	0.69	$C_7H_{18}N_4$	9.38	0.40	$C_5H_{10}N_3O_3$	6.83	0.80
$C_{11}H_6O$	12.02	0.86	$C_9H_{20}N_2$	10.81	0.53	$C_8H_2N_2O_2$	9.52	0.81	$C_5H_{12}N_4O_2$	7.20	0.63
$C_{11}H_8N$	12.40	0.70	$C_{10}H_4O_2$	10.95	0.94	$C_8H_4N_3O$	9.89	0.64	$C_6H_{10}NO_4$	7.18	1.02
$C_{11}H_{22}$	12.24	0.68	$C_{10}H_6NO$	11.32	0.78	$C_8H_6N_4$	10.27	0.48	$C_6H_{12}N_2O_3$	7.56	0.85
$C_{12}H_{10}$	13.13	0.79	$C_{10}H_8N_2$	11.70	0.62	$C_8H_{14}O_3$	8.99	0.96	$C_6H_{14}N_3O_2$	7.93	0.68
			$C_{10}H_{20}O$	11.17	0.77	$C_8H_{16}NO_2$	9.36	0.79	$C_6H_{16}N_4O$	8.31	0.51
155			$C_{10}H_{22}N$	11.54	0.61	$C_8H_{18}N_2O$	9.74	0.63	$C_7H_2N_3O_2$	8.82	0.75
$C_5H_3N_2O_4$	6.37	0.97	$C_{11}H_8O$	12.05	0.86	$C_8H_{20}N_3$	10.11	0.46	$C_7H_4N_4O$	9.19	0.58
$C_5H_5N_3O_3$	6.75	0.80	$C_{11}H_{10}N$	12.43	0.71	$C_9H_2O_3$	9.88	1.04	$C_7H_{12}O_4$	7.91	1.07
$C_5H_7N_4O_2$	7.12	0.62	$C_{11}H_{24}$	12.27	0.69	$C_9H_4NO_2$	10.25	0.87	$C_7H_{14}NO_3$	8.29	0.90
$C_6H_5NO_4$	7.10	1.02	$C_{12}H_{12}$	13.16	0.80	$C_9H_6N_2O$	10.62	0.71	$C_7H_{16}N_2O_2$	8.66	0.73
$C_6H_7N_2O_3$	7.48	0.84				$C_9H_8N_3$	11.00	0.55	$C_7H_{18}N_3O$	9.04	0.57
$C_6H_9N_3O_2$	7.85	0.67	**157**			$C_9H_{18}O_2$	10.09	0.86	$C_7H_{20}N_4$	9.41	0.40
$C_6H_{11}N_4O$	8.23	0.50	$C_5H_5N_2O_4$	6.40	0.98	$C_9H_{20}NO$	10.47	0.69	$C_8H_2NO_3$	9.18	0.97
$C_7H_7O_4$	7.83	1.07	$C_5H_7N_3O_3$	6.78	0.80	$C_9H_{22}N_2$	10.84	0.53	$C_8H_4N_2O_2$	9.55	0.81
$C_7H_9NO_3$	8.21	0.90	$C_5H_9N_4O_2$	7.15	0.62	$C_{10}H_6O_2$	10.98	0.95	$C_8H_6N_3O$	9.92	0.64
$C_7H_{11}N_2O_2$	8.58	0.73	$C_6H_7NO_4$	7.13	1.02	$C_{10}H_8NO$	11.36	0.79	$C_8H_8N_4$	10.30	0.48
$C_7H_{13}N_3O$	8.96	0.56	$C_6H_9N_2O_3$	7.51	0.85	$C_{10}H_{10}N_2$	11.73	0.63	$C_8H_{16}O_3$	9.02	0.96
$C_7H_{15}N_4$	9.33	0.39	$C_6H_{11}N_3O_2$	7.88	0.67	$C_{10}H_{22}O$	11.20	0.77	$C_8H_{18}NO_2$	9.39	0.79
C_8HN_3O	9.84	0.64	$C_6H_{13}N_4O$	8.26	0.50	$C_{11}H_{10}O$	12.09	0.87	$C_8H_{20}N_2O$	9.77	0.63
$C_8H_3N_4$	10.22	0.47	C_7HN_4O	9.15	0.57	$C_{11}H_{12}N$	12.46	0.71	$C_9H_4O_3$	9.91	1.04
$C_8H_{11}O_3$	8.94	0.95	$C_7H_9O_4$	7.87	1.07	$C_{12}H_{14}$	13.19	0.80	$C_9H_6NO_2$	10.28	0.88
$C_8H_{13}NO_2$	9.31	0.79	$C_7H_{11}NO_3$	8.24	0.90	$C_{13}H_2$	14.08	0.92	$C_9H_8N_2O$	10.66	0.71
$C_8H_{15}N_2O$	9.69	0.62	$C_7H_{13}N_2O_2$	8.61	0.73				$C_9H_{10}N_3$	11.03	0.55
$C_8H_{17}N_3$	10.06	0.46	$C_7H_{15}N_3O$	8.99	0.56	**159**			$C_9H_{20}O_2$	10.12	0.86
C_9HNO_2	10.20	0.87	$C_7H_{17}N_4$	9.36	0.39	$C_5H_7N_2O_4$	6.43	0.98	$C_{10}H_8O_2$	11.01	0.95
$C_9H_3N_2O$	10.58	0.71	$C_8HN_2O_2$	9.50	0.80	$C_5H_9N_3O_3$	6.81	0.80	$C_{10}H_{10}NO$	11.39	0.79
$C_9H_5N_3$	10.95	0.54	$C_8H_3N_3O$	9.88	0.64	$C_5H_{11}N_4O_2$	7.18	0.63	$C_{10}H_{12}N_2$	11.76	0.63
$C_9H_{15}O_2$	10.04	0.85	$C_8H_5N_4$	10.25	0.48	$C_6H_9NO_4$	7.17	1.02	$C_{11}H_{12}O$	12.12	0.87
$C_9H_{17}NO$	10.42	0.69	$C_8H_{13}O_3$	8.97	0.96	$C_6H_{11}N_2O_3$	7.54	0.85	$C_{11}H_{14}N$	12.49	0.72
$C_9H_{19}N_2$	10.79	0.53	$C_8H_{15}NO_2$	9.35	0.79	$C_6H_{13}N_3O_2$	7.91	0.68	$C_{12}H_2N$	13.38	0.82
$C_{10}H_3O_2$	10.93	0.94	$C_8H_{17}N_2O$	9.72	0.62	$C_6H_{15}N_4O$	8.29	0.51	$C_{12}H_{16}$	13.22	0.80
$C_{10}H_5NO$	11.31	0.78	$C_8H_{19}N_3$	10.09	0.46	$C_7HN_3O_2$	8.80	0.75	$C_{13}H_4$	14.11	0.92
$C_{10}H_7N_2$	11.68	0.62	C_9HO_3	9.86	1.03	$C_7H_3N_4O$	9.18	0.58			
$C_{10}H_{19}O$	11.15	0.76	$C_9H_3NO_2$	10.23	0.87	$C_7H_{11}O_4$	7.90	1.07	**161**		
$C_{10}H_{21}N$	11.52	0.60	$C_9H_5N_2O$	10.61	0.71	$C_7H_{13}NO_3$	8.27	0.90	$C_5H_9N_2O_4$	6.47	0.98
$C_{11}H_7O$	12.04	0.86	$C_9H_7N_3$	10.98	0.55	$C_7H_{15}N_2O_2$	8.65	0.73	$C_5H_{11}N_3O_3$	6.84	0.80
$C_{11}H_9N$	12.41	0.71	$C_9H_{17}O_2$	10.08	0.86	$C_7H_{17}N_3O$	9.02	0.56	$C_5H_{13}N_4O_2$	7.22	0.63
$C_{11}H_{23}$	12.26	0.69	$C_9H_{19}NO$	10.45	0.69	$C_7H_{19}N_4$	9.39	0.40	$C_6HN_4O_2$	8.10	0.69
$C_{12}H_{11}$	13.14	0.79	$C_9H_{21}N_2$	10.82	0.53	C_8HNO_3	9.16	0.97	$C_6H_{11}NO_4$	7.20	1.03
			$C_{10}H_5O_2$	10.96	0.94	$C_8H_3N_2O_2$	9.53	0.81	$C_6H_{13}N_2O_3$	7.57	0.85
156			$C_{10}H_7NO$	11.34	0.78	$C_8H_5N_3O$	9.91	0.64	$C_6H_{15}N_3O_2$	7.95	0.68
$C_5H_4N_2O_4$	6.39	0.98	$C_{10}H_9N_2$	11.71	0.63	$C_8H_7N_4$	10.28	0.48	$C_6H_{17}N_4O$	8.32	0.51
$C_5H_6N_3O_3$	6.76	0.80	$C_{10}H_{21}O$	11.18	0.77	$C_8H_{15}O_3$	9.00	0.96	$C_7HN_2O_3$	8.46	0.92
$C_5H_8N_4O_2$	7.14	0.62	$C_{10}H_{23}N$	11.56	0.61	$C_8H_{17}NO_2$	9.38	0.79	$C_7H_3N_3O_2$	8.84	0.75
$C_6H_6NO_4$	7.12	1.02	$C_{11}H_9O$	12.07	0.86	$C_8H_{19}N_2O$	9.75	0.63	$C_7H_5N_4O$	9.21	0.58
$C_6H_8N_2O_3$	7.49	0.85	$C_{11}H_{11}N$	12.44	0.71	$C_8H_{21}N_3$	10.13	0.46	$C_7H_{13}O_4$	7.93	1.08
$C_6H_{10}N_3O_2$	7.87	0.67	$C_{12}H_{13}$	13.18	0.80	$C_9H_3O_3$	9.89	1.04	$C_7H_{15}NO_3$	8.30	0.90
$C_6H_{12}N_4O$	8.24	0.50	$C_{13}H$	14.06	0.91	$C_9H_5NO_2$	10.27	0.87	$C_7H_{17}N_2O_2$	8.68	0.74
$C_7H_8O_4$	7.85	1.07				$C_9H_7N_2O$	10.64	0.71	$C_7H_{19}N_3O$	9.05	0.57
$C_7H_{10}NO_3$	8.22	0.90	**158**			$C_9H_9N_3$	11.01	0.55	C_8HO_4	8.82	1.14
$C_7H_{12}N_2O_2$	8.60	0.73	$C_5H_6N_2O_4$	6.42	0.98	$C_9H_{19}O_2$	10.11	0.86	$C_8H_3NO_3$	9.19	0.98
$C_7H_{14}N_3O$	8.97	0.56	$C_5H_8N_3O_3$	6.79	0.80	$C_9H_{21}NO$	10.48	0.70	$C_8H_5N_2O_2$	9.57	0.81
$C_7H_{16}N_4$	9.35	0.39	$C_5H_{10}N_4O_2$	7.17	0.63	$C_{10}H_7O_2$	11.00	0.95	$C_8H_7N_3O$	9.94	0.65
$C_8H_2N_3O$	9.86	0.64	$C_6H_8NO_4$	7.15	1.02	$C_{10}H_9NO$	11.37	0.79	$C_8H_9N_4$	10.32	0.48
$C_8H_4N_4$	10.24	0.47	$C_6H_{10}N_2O_3$	7.52	0.85	$C_{10}H_{11}N_2$	11.75	0.63	$C_8H_{17}O_3$	9.03	0.96
$C_8H_{12}O_3$	8.95	0.96	$C_6H_{12}N_3O_2$	7.90	0.68	$C_{11}H_{11}O$	12.10	0.87	$C_8H_{19}NO_2$	9.41	0.80
$C_8H_{14}NO_2$	9.33	0.79	$C_6H_{14}N_4O$	8.27	0.50	$C_{11}H_{13}N$	12.48	0.71	$C_9H_5O_3$	9.92	1.04
$C_8H_{16}N_2O$	9.70	0.62	$C_7H_2N_4O$	9.16	0.58	$C_{12}HN$	13.37	0.82	$C_9H_7NO_2$	10.30	0.88
$C_8H_{18}N_3$	10.08	0.46	$C_7H_{10}O_4$	7.88	1.07	$C_{12}H_{15}$	13.21	0.80	$C_9H_9N_2O$	10.67	0.72
$C_9H_2NO_2$	10.22	0.87	$C_7H_{12}NO_3$	8.26	0.90	$C_{13}H_3$	14.10	0.92	$C_9H_{11}N_3$	11.05	0.56
$C_9H_4N_2O$	10.59	0.71									

	M + 1	M + 2
$C_{10}H_9O_2$	11.03	0.95
$C_{10}H_{11}NO$	11.40	0.79
$C_{10}H_{13}N_2$	11.78	0.63
$C_{11}HN_2$	12.67	0.74
$C_{11}H_{13}O$	12.13	0.87
$C_{11}H_{15}N$	12.51	0.72
$C_{12}HO$	13.02	0.98
$C_{12}H_3N$	13.40	0.83
$C_{12}H_{17}$	13.24	0.81
$C_{13}H_5$	14.13	0.92
162		
$C_5H_{10}N_2O_4$	6.48	0.98
$C_5H_{12}N_3O_3$	6.86	0.81
$C_5H_{14}N_4O_2$	7.23	0.63
$C_6H_2N_4O_2$	8.12	0.69
$C_6H_{12}NO_4$	7.21	1.03
$C_6H_{14}N_2O_3$	7.59	0.85
$C_6H_{16}N_3O_2$	7.96	0.68
$C_6H_{18}N_4O$	8.34	0.51
$C_7H_2N_2O_3$	8.48	0.92
$C_7H_4N_3O_2$	8.85	0.75
$C_7H_6N_4O$	9.23	0.58
$C_7H_{14}O_4$	7.95	1.08
$C_7H_{16}NO_3$	8.32	0.91
$C_7H_{18}N_2O_2$	8.69	0.74
$C_8H_2O_4$	8.83	1.15
$C_8H_4NO_3$	9.21	0.98
$C_8H_6N_2O_2$	9.58	0.81
$C_8H_8N_3O$	9.96	0.65
$C_8H_{10}N_4$	10.33	0.48
$C_8H_{18}O_3$	9.05	0.96
$C_9H_6O_3$	9.94	1.04
$C_9H_8NO_2$	10.31	0.88
$C_9H_{10}N_2O$	10.69	0.72
$C_9H_{12}N_3$	11.06	0.56
$C_{10}H_{10}O_2$	11.04	0.95
$C_{10}H_{12}NO$	11.42	0.79
$C_{10}H_{14}N_2$	11.79	0.64
$C_{11}H_2N_2$	12.68	0.74
$C_{11}H_{14}O$	12.15	0.87
$C_{11}H_{16}N$	12.52	0.72
$C_{12}H_2O$	13.04	0.98
$C_{12}H_4N$	13.41	0.83
$C_{12}H_{18}$	13.26	0.81
$C_{13}H_6$	14.14	0.92
163		
$C_5H_{11}N_2O_4$	6.50	0.98
$C_5H_{13}N_3O_3$	6.87	0.81
$C_5H_{15}N_4O_2$	7.25	0.63
$C_6HN_3O_3$	7.76	0.87
$C_6H_3N_4O_2$	8.14	0.69
$C_6H_{13}NO_4$	7.23	1.03
$C_6H_{15}N_2O_3$	7.60	0.85
$C_6H_{17}N_3O_2$	7.98	0.68
C_7HNO_4	8.12	1.09
$C_7H_3N_2O_3$	8.49	0.92
$C_7H_5N_3O_2$	8.87	0.75
$C_7H_7N_4O$	9.24	0.58
$C_7H_{15}O_4$	7.96	1.08
$C_7H_{17}NO_3$	8.34	0.91
$C_8H_3O_4$	8.85	1.15
$C_8H_5NO_3$	9.22	0.98

	M + 1	M + 2
$C_8H_7N_2O_2$	9.60	0.81
$C_8H_9N_3O$	9.97	0.65
$C_8H_{11}N_4$	10.35	0.49
$C_9H_7O_3$	9.96	1.04
$C_9H_9NO_2$	10.33	0.88
$C_9H_{11}N_2O$	10.70	0.72
$C_9H_{13}N_3$	11.08	0.56
$C_{10}HN_3$	11.97	0.66
$C_{10}H_{11}O_2$	11.06	0.95
$C_{10}H_{13}NO$	11.44	0.80
$C_{10}H_{15}N_2$	11.81	0.64
$C_{11}HNO$	12.32	0.89
$C_{11}H_3N_2$	12.70	0.74
$C_{11}H_{15}O$	12.17	0.88
$C_{11}H_{17}N$	12.54	0.72
$C_{12}H_3O$	13.05	0.98
$C_{12}H_5N$	13.43	0.83
$C_{12}H_{19}$	13.27	0.81
$C_{13}H_7$	14.16	0.93
164		
$C_5H_{12}N_2O_4$	6.51	0.98
$C_5H_{14}N_3O_3$	6.89	0.81
$C_5H_{16}N_4O_2$	7.26	0.63
$C_6H_2N_3O_3$	7.78	0.87
$C_6H_4N_4O_2$	8.15	0.70
$C_6H_{14}NO_4$	7.25	1.03
$C_6H_{16}N_2O_3$	7.62	0.86
$C_7H_2NO_4$	8.13	1.09
$C_7H_4N_2O_3$	8.51	0.92
$C_7H_6N_3O_2$	8.88	0.75
$C_7H_8N_4O$	9.26	0.59
$C_7H_{16}O_4$	7.98	1.08
$C_8H_4O_4$	8.87	1.15
$C_8H_6NO_3$	9.24	0.98
$C_8H_8N_2O_2$	9.61	0.81
$C_8H_{10}N_3O$	9.99	0.65
$C_8H_{12}N_4$	10.36	0.49
$C_9H_8O_3$	9.97	1.05
$C_9H_{10}NO_2$	10.35	0.88
$C_9H_{12}N_2O$	10.72	0.72
$C_9H_{14}N_3$	11.09	0.56
$C_{10}H_2N_3$	11.98	0.66
$C_{10}H_{12}O_2$	11.08	0.96
$C_{10}H_{14}NO$	11.45	0.80
$C_{10}H_{16}N_2$	11.83	0.64
$C_{11}H_2NO$	12.34	0.90
$C_{11}H_4N_2$	12.71	0.74
$C_{11}H_{16}O$	12.18	0.88
$C_{11}H_{18}N$	12.56	0.72
$C_{12}H_4O$	13.07	0.98
$C_{12}H_6N$	13.45	0.83
$C_{12}H_{20}$	13.29	0.81
$C_{13}H_8$	14.18	0.93
165		
$C_5H_{13}N_2O_4$	6.53	0.98
$C_5H_{15}N_3O_3$	6.91	0.81
$C_6HN_2O_4$	7.42	1.04
$C_6H_3N_3O_3$	7.79	0.87
$C_6H_5N_4O_2$	8.17	0.70
$C_6H_{15}NO_4$	7.26	1.03
$C_7H_3NO_4$	8.15	1.09

	M + 1	M + 2
$C_7H_5N_2O_3$	8.52	0.92
$C_7H_7N_3O_2$	8.90	0.75
$C_7H_9N_4O$	9.27	0.59
$C_8H_5O_4$	8.88	1.15
$C_8H_7NO_3$	9.26	0.98
$C_8H_9N_2O_2$	9.63	0.82
$C_8H_{11}N_3O$	10.00	0.65
$C_8H_{13}N_4$	10.38	0.49
C_9HN_4	11.27	0.58
$C_9H_9O_3$	9.99	1.05
$C_9H_{11}NO_2$	10.36	0.88
$C_9H_{13}N_2O$	10.74	0.72
$C_9H_{15}N_3$	11.11	0.56
$C_{10}HN_2O$	11.62	0.82
$C_{10}H_3N_3$	12.00	0.66
$C_{10}H_{13}O_2$	11.09	0.96
$C_{10}H_{15}NO$	11.47	0.80
$C_{10}H_{17}N_2$	11.84	0.64
$C_{11}HO_2$	11.98	1.05
$C_{11}H_3NO$	12.36	0.90
$C_{11}H_5N_2$	12.73	0.74
$C_{11}H_{17}O$	12.20	0.88
$C_{11}H_{19}N$	12.57	0.73
$C_{12}H_5O$	13.09	0.99
$C_{12}H_7N$	13.46	0.84
$C_{12}H_{21}$	13.30	0.81
$C_{13}H_9$	14.19	0.93
166		
$C_5H_{14}N_2O_4$	6.55	0.99
$C_6H_2N_2O_4$	7.44	1.04
$C_6H_4N_3O_3$	7.81	0.87
$C_6H_6N_4O_2$	8.18	0.70
$C_7H_4NO_4$	8.17	1.09
$C_7H_6N_2O_3$	8.54	0.92
$C_7H_8N_3O_2$	8.92	0.76
$C_7H_{10}N_4O$	9.29	0.59
$C_8H_6O_4$	8.90	1.15
$C_8H_8NO_3$	9.27	0.98
$C_8H_{10}N_2O_2$	9.65	0.82
$C_8H_{12}N_3O$	10.02	0.65
$C_8H_{14}N_4$	10.40	0.49
$C_9H_2N_4$	11.28	0.58
$C_9H_{10}O_3$	10.00	1.05
$C_9H_{12}NO_2$	10.38	0.89
$C_9H_{14}N_2O$	10.75	0.72
$C_9H_{16}N_3$	11.13	0.56
$C_{10}H_2N_2O$	11.64	0.82
$C_{10}H_4N_3$	12.01	0.66
$C_{10}H_{14}O_2$	11.11	0.96
$C_{10}H_{16}NO$	11.48	0.80
$C_{10}H_{18}N_2$	11.86	0.64
$C_{11}H_2O_2$	12.00	1.06
$C_{11}H_4NO$	12.37	0.90
$C_{11}H_6N_2$	12.75	0.75
$C_{11}H_{18}O$	12.21	0.88
$C_{11}H_{20}N$	12.59	0.73
$C_{12}H_6O$	13.10	0.99
$C_{12}H_8N$	13.48	0.84
$C_{12}H_{22}$	13.32	0.82
$C_{13}H_{10}$	14.21	0.93
167		
$C_6H_3N_2O_4$	7.45	1.04

	M + 1	M + 2
$C_6H_5N_3O_3$	7.83	0.87
$C_6H_7N_4O_2$	8.20	0.70
$C_7H_5NO_4$	8.18	1.10
$C_7H_7N_2O_3$	8.56	0.93
$C_7H_9N_3O_2$	8.93	0.76
$C_7H_{11}N_4O$	9.31	0.59
$C_8H_7O_4$	8.91	1.15
$C_8H_9NO_3$	9.29	0.99
$C_8H_{11}N_2O_2$	9.66	0.82
$C_8H_{13}N_3O$	10.04	0.66
$C_8H_{15}N_4$	10.41	0.49
C_9HN_3O	10.93	0.74
$C_9H_3N_4$	11.30	0.58
$C_9H_{11}O_3$	10.02	1.05
$C_9H_{13}NO_2$	10.39	0.89
$C_9H_{15}N_2O$	10.77	0.73
$C_9H_{17}N_3$	11.14	0.57
$C_{10}HNO_2$	11.28	0.98
$C_{10}H_3N_2O$	11.66	0.82
$C_{10}H_5N_3$	12.03	0.66
$C_{10}H_{15}O_2$	11.12	0.96
$C_{10}H_{17}NO$	11.50	0.80
$C_{10}H_{19}N_2$	11.87	0.65
$C_{11}H_3O_2$	12.01	1.06
$C_{11}H_5NO$	12.39	0.90
$C_{11}H_7N_2$	12.76	0.75
$C_{11}H_{19}O$	12.23	0.88
$C_{11}H_{21}N$	12.60	0.73
$C_{12}H_7O$	13.12	0.99
$C_{12}H_9N$	13.49	0.84
$C_{12}H_{23}$	13.34	0.82
$C_{13}H_{11}$	14.22	0.94
168		
$C_6H_4N_2O_4$	7.47	1.04
$C_6H_6N_3O_3$	7.84	0.87
$C_6H_8N_4O_2$	8.22	0.70
$C_7H_6NO_4$	8.20	1.10
$C_7H_8N_2O_3$	8.57	0.93
$C_7H_{10}N_3O_2$	8.95	0.76
$C_7H_{12}N_4O$	9.32	0.59
$C_8H_8O_4$	8.93	1.15
$C_8H_{10}NO_3$	9.30	0.99
$C_8H_{12}N_2O_2$	9.68	0.82
$C_8H_{14}N_3O$	10.05	0.66
$C_8H_{16}N_4$	10.43	0.49
$C_9H_2N_3O$	10.94	0.74
$C_9H_4N_4$	11.32	0.58
$C_9H_{12}O_3$	10.04	1.05
$C_9H_{14}NO_2$	10.41	0.89
$C_9H_{16}N_2O$	10.78	0.73
$C_9H_{18}N_3$	11.16	0.57
$C_{10}H_2NO_2$	11.30	0.98
$C_{10}H_4N_2O$	11.67	0.82
$C_{10}H_6N_3$	12.05	0.67
$C_{10}H_{16}O_2$	11.14	0.96
$C_{10}H_{18}NO$	11.52	0.80
$C_{10}H_{20}N_2$	11.89	0.65
$C_{11}H_4O_2$	12.03	1.06
$C_{11}H_6NO$	12.40	0.90
$C_{11}H_8N_2$	12.78	0.75
$C_{11}H_{20}O$	12.25	0.89
$C_{11}H_{22}N$	12.62	0.73

	M+1	M+2
$C_{12}H_9O$	13.13	0.99
$C_{12}H_{10}N$	13.51	0.84
$C_{12}H_{24}$	13.35	0.82
$C_{13}H_{12}$	14.24	0.94

169

	M+1	M+2
$C_6H_5N_2O_4$	7.48	1.05
$C_6H_7N_3O_3$	7.86	0.87
$C_6H_9N_4O_2$	8.23	0.70
$C_7H_7NO_4$	8.21	1.10
$C_7H_9N_2O_3$	8.59	0.93
$C_7H_{11}N_3O_2$	8.96	0.76
$C_7H_{13}N_4O$	9.34	0.59
C_8HN_4O	10.23	0.67
$C_8H_9O_4$	8.95	1.16
$C_8H_{11}NO_3$	9.32	0.99
$C_8H_{13}N_2O_2$	9.69	0.82
$C_8H_{15}N_3O$	10.07	0.66
$C_8H_{17}N_4$	10.44	0.50
$C_9HN_2O_2$	10.58	0.91
$C_9H_3N_3O$	10.96	0.75
$C_9H_5N_4$	11.33	0.59
$C_9H_{13}O_3$	10.05	1.05
$C_9H_{15}NO_2$	10.43	0.89
$C_9H_{17}N_2O$	10.80	0.73
$C_9H_{19}N_3$	11.17	0.57
$C_{10}HO_3$	10.94	1.14
$C_{10}H_3NO_2$	11.31	0.98
$C_{10}H_5N_2O$	11.69	0.82
$C_{10}H_7N_3$	12.06	0.67
$C_{10}H_{17}O_2$	11.16	0.96
$C_{10}H_{19}NO$	11.53	0.81
$C_{10}H_{21}N_2$	11.91	0.65
$C_{11}H_5O_2$	12.05	1.06
$C_{11}H_7NO$	12.42	0.91
$C_{11}H_9N_2$	12.79	0.75
$C_{11}H_{21}O$	12.26	0.89
$C_{11}H_{23}N$	12.64	0.73
$C_{12}H_9O$	13.15	1.00
$C_{12}H_{11}N$	13.53	0.84
$C_{12}H_{25}$	13.37	0.82
$C_{13}H_{13}$	14.26	0.94
$C_{14}H$	15.14	1.07

170

	M+1	M+2
$C_6H_6N_2O_4$	7.50	1.05
$C_6H_8N_3O_3$	7.87	0.87
$C_6H_{10}N_4O_2$	8.25	0.70
$C_7H_8NO_4$	8.23	1.10
$C_7H_{10}N_2O_3$	8.60	0.93
$C_7H_{12}N_3O_2$	8.98	0.76
$C_7H_{14}N_4O$	9.35	0.59
$C_8H_2N_4O$	10.24	0.68
$C_8H_{10}O_4$	8.96	1.16
$C_8H_{12}NO_3$	9.34	0.99
$C_8H_{14}N_2O_2$	9.71	0.82
$C_8H_{16}N_3O$	10.08	0.66
$C_8H_{18}N_4$	10.46	0.50
$C_9H_2N_2O_2$	10.60	0.91
$C_9H_4N_3O$	10.97	0.75
$C_9H_6N_4$	11.35	0.59
$C_9H_{14}O_3$	10.07	1.06
$C_9H_{16}NO_2$	10.44	0.89
$C_9H_{18}N_2O$	10.82	0.73

	M+1	M+2
$C_9H_{20}N_3$	11.19	0.57
$C_{10}H_2O_3$	10.96	1.14
$C_{10}H_4NO_2$	11.33	0.98
$C_{10}H_6N_2O$	11.70	0.83
$C_{10}H_8N_3$	12.08	0.67
$C_{10}H_{18}O_2$	11.17	0.97
$C_{10}H_{20}NO$	11.55	0.81
$C_{10}H_{22}N_2$	11.92	0.65
$C_{11}H_6O_2$	12.06	1.06
$C_{11}H_8NO$	12.44	0.91
$C_{11}H_{10}N_2$	12.81	0.75
$C_{11}H_{22}O$	12.28	0.89
$C_{11}H_{24}N$	12.65	0.74
$C_{12}H_{10}O$	13.17	1.00
$C_{12}H_{12}N$	13.54	0.85
$C_{12}H_{26}$	13.38	0.83
$C_{13}H_{14}$	14.27	0.94
$C_{14}H_2$	15.16	1.07

171

	M+1	M+2
$C_6H_7N_2O_4$	7.52	1.05
$C_6H_9N_3O_3$	7.89	0.88
$C_6H_{11}N_4O_2$	8.26	0.70
$C_7H_9NO_4$	8.25	1.10
$C_7H_{11}N_2O_3$	8.62	0.93
$C_7H_{13}N_3O_2$	9.00	0.76
$C_7H_{15}N_4O$	9.37	0.60
$C_8HN_3O_2$	9.88	0.84
$C_8H_3N_4O$	10.26	0.68
$C_8H_{11}O_4$	8.98	1.16
$C_8H_{13}NO_3$	9.35	0.99
$C_8H_{15}N_2O_2$	9.73	0.83
$C_8H_{17}N_3O$	10.10	0.66
$C_8H_{19}N_4$	10.48	0.50
C_8HNO_3	10.24	1.07
$C_9H_3N_2O_2$	10.61	0.91
$C_9H_5N_3O$	10.99	0.75
$C_9H_7N_4$	11.36	0.59
$C_9H_{15}O_3$	10.08	1.06
$C_9H_{17}NO_2$	10.46	0.89
$C_9H_{19}N_2O$	10.83	0.73
$C_9H_{21}N_3$	11.21	0.57
$C_{10}H_3O_3$	10.97	1.14
$C_{10}H_5NO_2$	11.35	0.99
$C_{10}H_7N_2O$	11.72	0.83
$C_{10}H_9N_3$	12.09	0.67
$C_{10}H_{19}O_2$	11.19	0.97
$C_{10}H_{21}NO$	11.56	0.81
$C_{10}H_{23}N_2$	11.94	0.65
$C_{11}H_7O_2$	12.08	1.07
$C_{11}H_9NO$	12.45	0.91
$C_{11}H_{11}N_2$	12.83	0.76
$C_{11}H_{23}O$	12.29	0.89
$C_{11}H_{25}N$	12.67	0.74
$C_{12}H_{11}O$	13.18	1.00
$C_{12}H_{13}N$	13.56	0.85
$C_{13}HN$	14.45	0.97
$C_{13}H_{15}$	14.29	0.94
$C_{14}H_3$	15.18	1.07

172

	M+1	M+2
$C_6H_8N_2O_4$	7.53	1.05
$C_6H_{10}N_3O_3$	7.91	0.88

	M+1	M+2
$C_6H_{12}N_4O_2$	8.28	0.71
$C_7H_{10}NO_4$	8.26	1.10
$C_7H_{12}N_2O_3$	8.64	0.93
$C_7H_{14}N_3O_2$	9.01	0.76
$C_7H_{16}N_4O$	9.39	0.60
$C_8H_2N_3O_2$	9.90	0.84
$C_8H_4N_4O$	10.27	0.68
$C_8H_{12}O_4$	8.99	1.16
$C_8H_{14}NO_3$	9.37	0.99
$C_8H_{16}N_2O_2$	9.74	0.83
$C_8H_{18}N_3O$	10.12	0.66
$C_8H_{20}N_4$	10.49	0.50
C_9NO_3	10.26	1.07
$C_9H_4N_2O_2$	10.63	0.91
$C_9H_6N_3O$	11.01	0.75
$C_9H_8N_4$	11.38	0.59
$C_9H_{16}O_3$	10.10	1.06
$C_9H_{18}NO_2$	10.47	0.90
$C_9H_{20}N_2O$	10.85	0.73
$C_9H_{22}N_3$	11.22	0.57
$C_{10}H_4O_3$	10.99	1.15
$C_{10}H_6NO_2$	11.36	0.99
$C_{10}H_8N_2O$	11.74	0.83
$C_{10}H_{10}N_3$	12.11	0.67
$C_{10}H_{20}O_2$	11.20	0.97
$C_{10}H_{22}NO$	11.58	0.81
$C_{10}H_{24}N_2$	11.95	0.65
$C_{11}H_8O_2$	12.09	1.07
$C_{11}H_{10}NO$	12.47	0.91
$C_{11}H_{12}N_2$	12.84	0.76
$C_{11}H_{24}O$	12.31	0.89
$C_{12}H_{12}O$	13.20	1.00
$C_{12}H_{14}N$	13.57	0.85
$C_{13}H_2N$	14.46	0.97
$C_{13}H_{16}$	14.30	0.95
$C_{14}H_4$	15.19	1.07

173

	M+1	M+2
$C_6H_9N_2O_4$	7.55	1.05
$C_6H_{11}N_3O_3$	7.92	0.88
$C_6H_{13}N_4O_2$	8.30	0.71
$C_7HN_4O_2$	9.18	0.78
$C_7H_{11}NO_4$	8.28	1.10
$C_7H_{13}N_2O_3$	8.65	0.93
$C_7H_{15}N_3O_2$	9.03	0.77
$C_7H_{17}N_4O$	9.40	0.60
$C_8HN_2O_3$	9.54	1.01
$C_8H_3N_3O_2$	9.92	0.84
$C_8H_5N_4O$	10.29	0.68
$C_8H_{13}O_4$	9.01	1.16
$C_8H_{15}NO_3$	9.38	0.99
$C_8H_{17}N_2O_2$	9.76	0.83
$C_8H_{19}N_3O$	10.13	0.66
$C_9H_{21}N_4$	10.51	0.50
C_9HO_4	9.90	1.24
$C_9H_3NO_3$	10.27	1.08
$C_9H_5N_2O_2$	10.65	0.91
$C_9H_7N_3O$	11.02	0.75
$C_9H_9N_4$	11.40	0.59
$C_9H_{17}O_3$	10.12	1.06
$C_9H_{19}NO_2$	10.49	0.90
$C_9H_{21}N_2O$	10.86	0.74
$C_9H_{23}N_3$	11.24	0.58

	M+1	M+2
$C_{10}H_5O_3$	11.00	1.15
$C_{10}H_7NO_2$	11.38	0.99
$C_{10}H_9N_2O$	11.75	0.83
$C_{10}H_{11}N_3$	12.13	0.67
$C_{10}H_{21}O_2$	11.22	0.97
$C_{10}H_{23}NO$	11.60	0.81
$C_{11}H_9O_2$	12.11	1.07
$C_{11}H_{11}NO$	12.48	0.91
$C_{11}H_{13}N_2$	12.86	0.76
$C_{12}HN_2$	13.75	0.87
$C_{12}H_{13}O$	13.21	1.00
$C_{12}H_{15}N$	13.59	0.85
$C_{13}HO$	14.10	1.12
$C_{13}H_3N$	14.48	0.97
$C_{13}H_{17}$	14.32	0.95
$C_{14}H_5$	15.21	1.07

174

	M+1	M+2
$C_6H_{10}N_2O_4$	7.56	1.05
$C_6H_{12}N_3O_3$	7.94	0.88
$C_6H_{14}N_4O_2$	8.31	0.71
$C_7H_2N_4O_2$	9.20	0.78
$C_7H_{12}NO_4$	8.29	1.10
$C_7H_{14}N_2O_3$	8.67	0.93
$C_7H_{16}N_3O_2$	9.04	0.77
$C_7H_{18}N_4O$	9.42	0.60
$C_8H_2N_2O_3$	9.56	1.01
$C_8H_4N_3O_2$	9.93	0.85
$C_8H_6N_4O$	10.31	0.68
$C_8H_{14}O_4$	9.03	1.16
$C_8H_{16}NO_3$	9.40	1.00
$C_8H_{18}N_2O_2$	9.77	0.83
$C_8H_{20}N_3O$	10.15	0.67
$C_8H_{22}N_4$	10.52	0.50
$C_9H_2O_4$	9.91	1.24
$C_9H_4NO_3$	10.29	1.08
$C_9H_6N_2O_2$	10.66	0.92
$C_9H_8N_3O$	11.04	0.75
$C_9H_{10}N_4$	11.41	0.60
$C_9H_{18}O_3$	10.13	1.06
$C_9H_{20}NO_2$	10.51	0.90
$C_9H_{22}N_2O$	10.88	0.74
$C_{10}H_6O_3$	11.02	1.15
$C_{10}H_8NO_2$	11.39	0.99
$C_{10}H_{10}N_2O$	11.77	0.83
$C_{10}H_{12}N_3$	12.14	0.68
$C_{10}H_{22}O_2$	11.24	0.97
$C_{11}H_{10}O_2$	12.13	1.07
$C_{11}H_{12}NO$	12.50	0.92
$C_{11}H_{14}N_2$	12.87	0.76
$C_{12}H_2N_2$	13.76	0.88
$C_{12}H_{14}O$	13.23	1.01
$C_{12}H_{16}N$	13.61	0.85
$C_{13}H_2O$	14.12	1.12
$C_{13}H_4N$	14.49	0.97
$C_{13}H_{18}$	14.34	0.95
$C_{14}H_6$	15.22	1.08

175

	M+1	M+2
$C_6H_{11}N_2O_4$	7.58	1.05
$C_6H_{13}N_3O_3$	7.95	0.88
$C_6H_{15}N_4O_2$	8.33	0.71
$C_7HN_3O_3$	8.84	0.95

	M+1	M+2		M+1	M+2		M+1	M+2		M+1	M+2
$C_7H_3N_4O_2$	9.22	0.78	$C_{10}H_{14}N_3$	12.17	0.68	$C_7H_{18}N_2O_3$	8.73	0.94	$C_{12}H_{21}N$	13.69	0.87
$C_7H_{13}NO_4$	8.31	1.11	$C_{11}H_2N_3$	13.06	0.79	$C_8H_4NO_4$	9.25	1.18	$C_{13}H_7O$	14.20	1.13
$C_7H_{15}N_2O_3$	8.68	0.94	$C_{11}H_{12}O_2$	12.16	1.08	$C_8H_6N_2O_3$	9.62	1.02	$C_{13}H_9N$	14.57	0.99
$C_7H_{17}N_3O_2$	9.06	0.77	$C_{11}H_{14}NO$	12.53	0.92	$C_8H_8N_3O_2$	10.00	0.85	$C_{13}H_{23}$	14.42	0.96
$C_7H_{19}N_4O$	9.43	0.60	$C_{11}H_{16}N_2$	12.91	0.77	$C_8H_{10}N_4O$	10.37	0.69	$C_{14}H_{11}$	15.30	1.09
C_8HNO_4	9.20	1.18	$C_{12}H_2NO$	13.42	1.03	$C_8H_{18}O_4$	9.09	1.17			
$C_8H_3N_2O_3$	9.57	1.01	$C_{12}H_4N_2$	13.79	0.88	$C_9H_6O_4$	9.98	1.25	**180**		
$C_8H_5N_3O_2$	9.95	0.85	$C_{12}H_{16}O$	13.26	1.01	$C_9H_8NO_3$	10.35	1.08	$C_6H_{16}N_2O_4$	7.66	1.06
$C_8H_7N_4O$	10.32	0.68	$C_{12}H_{18}N$	13.64	0.86	$C_9H_{10}N_2O_2$	10.73	0.92	$C_7H_4N_2O_4$	8.55	1.12
$C_8H_{17}NO_3$	9.42	1.00	$C_{13}H_4O$	14.15	1.13	$C_9H_{12}N_3O$	11.10	0.76	$C_7H_6N_3O_3$	8.92	0.96
$C_8H_{19}N_2O_2$	9.79	0.83	$C_{13}H_6N$	14.53	0.98	$C_9H_{14}N_4$	11.48	0.60	$C_7H_8N_4O_2$	9.30	0.79
$C_8H_{21}N_3O$	10.16	0.67	$C_{13}H_{20}$	14.37	0.96	$C_{10}H_2N_4$	12.36	0.70	$C_8H_6NO_4$	9.28	1.18
$C_8H_{15}O_4$	9.04	1.16	$C_{14}H_8$	15.26	1.08	$C_{10}H_{10}O_3$	11.08	1.16	$C_8H_8N_2O_3$	9.65	1.02
$C_9H_3O_4$	9.93	1.24				$C_{10}H_{12}NO_2$	11.46	1.00	$C_8H_{10}N_3O_2$	10.03	0.85
$C_9H_5NO_3$	10.30	1.08	**177**			$C_{10}H_{14}N_2O$	11.83	0.84	$C_8H_{12}N_4O$	10.40	0.69
$C_9H_7N_2O_2$	10.68	0.92	$C_6H_{13}N_2O_4$	7.61	1.06	$C_{10}H_{16}N_3$	12.21	0.68	$C_9H_8O_4$	10.01	1.25
$C_9H_9N_3O$	11.05	0.76	$C_6H_{15}N_3O_3$	7.99	0.88	$C_{11}H_2N_2O$	12.72	0.94	$C_9H_{10}NO_3$	10.38	1.09
$C_9H_{11}N_4$	11.43	0.60	$C_6H_{17}N_4O_2$	8.36	0.71	$C_{11}H_4N_3$	13.10	0.79	$C_9H_{12}N_2O_2$	10.76	0.93
$C_9H_{19}O_3$	10.15	1.06	$C_7HN_2O_4$	8.50	1.12	$C_{11}H_{14}O_2$	12.19	1.08	$C_9H_{14}N_3O$	11.13	0.77
$C_9H_{21}NO_2$	10.52	0.90	$C_7H_3N_3O_3$	8.87	0.95	$C_{11}H_{16}NO$	12.56	0.92	$C_9H_{16}N_4$	11.51	0.61
$C_{10}H_7O_3$	11.04	1.15	$C_7H_5N_4O_2$	9.25	0.78	$C_{11}H_{18}N_2$	12.94	0.77	$C_{10}H_2N_3O$	12.02	0.86
$C_{10}H_9NO_2$	11.41	0.99	$C_7H_{15}NO_4$	8.34	1.11	$C_{12}H_2O_2$	13.08	1.19	$C_{10}H_4N_4$	12.40	0.71
$C_{10}H_{11}N_2O$	11.78	0.83	$C_7H_{17}N_2O_3$	8.72	0.94	$C_{12}H_4NO$	13.45	1.03	$C_{10}H_{12}O_3$	11.12	1.16
$C_{10}H_{13}N_3$	12.16	0.68	$C_7H_{19}N_3O_2$	9.09	0.77	$C_{12}H_6N_2$	13.83	0.88	$C_{10}H_{14}NO_2$	11.49	1.00
$C_{11}HN_3$	13.05	0.78	$C_8H_3NO_4$	9.23	1.18	$C_{12}H_{18}O$	13.29	1.01	$C_{10}H_{16}N_2O$	11.86	0.84
$C_{11}H_{11}O_2$	12.14	1.07	$C_8H_5N_2O_3$	9.61	1.01	$C_{12}H_{20}N$	13.67	0.86	$C_{10}H_{18}N_3$	12.24	0.69
$C_{11}H_{13}NO$	12.52	0.92	$C_8H_7N_3O_2$	9.98	0.85	$C_{13}H_6O$	14.18	1.13	$C_{11}H_2NO_2$	12.38	1.10
$C_{11}H_{15}N_2$	12.89	0.77	$C_8H_9N_4O$	10.35	0.69	$C_{13}H_8N$	14.56	0.98	$C_{11}H_4N_2O$	12.75	0.95
$C_{12}HNO$	13.40	1.03	$C_8H_{17}O_4$	9.07	1.17	$C_{13}H_{22}$	14.40	0.96	$C_{11}H_6N_3$	13.13	0.80
$C_{12}H_3N_2$	13.78	0.88	$C_8H_{19}NO_3$	9.45	1.00	$C_{14}H_{10}$	15.29	1.09	$C_{11}H_{16}O_2$	12.22	1.08
$C_{12}H_{15}O$	13.25	1.01	$C_9H_5O_4$	9.96	1.25				$C_{11}H_{18}NO$	12.60	0.93
$C_{12}H_{17}N$	13.62	0.86	$C_9H_7NO_3$	10.34	1.08	**179**			$C_{11}H_{20}N_2$	12.97	0.78
$C_{13}H_3O$	14.14	1.12	$C_9H_9N_2O_2$	10.71	0.92	$C_6H_{15}N_2O_4$	7.64	1.06	$C_{12}H_4O_2$	13.11	1.19
$C_{13}H_5N$	14.51	0.98	$C_9H_{11}N_3O$	11.09	0.76	$C_6H_{17}N_3O_3$	8.02	0.89	$C_{12}H_6NO$	13.48	1.04
$C_{13}H_{19}$	14.35	0.95	$C_9H_{13}N_4$	11.46	0.60	$C_7H_3N_2O_4$	8.53	1.12	$C_{12}H_8N_2$	13.86	0.89
$C_{14}H_7$	15.24	1.08	$C_{10}HN_4$	12.35	0.70	$C_7H_5N_3O_3$	8.91	0.95	$C_{12}H_{20}O$	13.33	1.02
			$C_{10}H_9O_3$	11.07	1.16	$C_7H_7N_4O_2$	9.28	0.79	$C_{12}H_{22}N$	13.70	0.87
176			$C_{10}H_{11}NO_2$	11.44	1.00	$C_7H_{17}NO_4$	8.37	1.11	$C_{13}H_8O$	14.22	1.13
$C_6H_{12}N_2O_4$	7.60	1.05	$C_{10}H_{13}N_2O$	11.82	0.84	$C_8H_5NO_4$	9.26	1.18	$C_{13}H_{10}N$	14.59	0.99
$C_6H_{14}N_3O_3$	7.97	0.88	$C_{10}H_{15}N_3$	12.19	0.68	$C_8H_7N_2O_3$	9.64	1.02	$C_{13}H_{24}$	14.43	0.97
$C_6H_{16}N_4O_2$	8.34	0.71	$C_{11}HN_2O$	12.71	0.94	$C_8H_9N_3O_2$	10.01	0.85	$C_{14}H_{12}$	15.32	1.09
$C_7H_2N_3O_3$	8.86	0.95	$C_{11}H_3N_3$	13.08	0.79	$C_8H_{11}N_4O$	10.39	0.69			
$C_7H_4N_4O_2$	9.23	0.78	$C_{11}H_{13}O_2$	12.17	1.08	$C_9H_7O_4$	9.99	1.25	**181**		
$C_7H_{14}NO_4$	8.33	1.11	$C_{11}H_{15}NO$	12.55	0.92	$C_9H_9NO_3$	10.37	1.09	$C_7H_5N_2O_4$	8.56	1.13
$C_7H_{16}N_2O_3$	8.70	0.94	$C_{11}H_{17}N_2$	12.92	0.77	$C_9H_{11}N_2O_2$	10.74	0.92	$C_7H_7N_3O_3$	8.94	0.96
$C_7H_{18}N_3O_2$	9.08	0.77	$C_{12}HO_2$	13.06	1.18	$C_9H_{13}N_3O$	11.12	0.76	$C_7H_9N_4O_2$	9.31	0.79
$C_7H_{20}N_4O$	9.45	0.60	$C_{12}H_3NO$	13.44	1.03	$C_9H_{15}N_4$	11.49	0.60	$C_8H_7NO_4$	9.30	1.19
$C_8H_2NO_4$	9.22	1.18	$C_{12}H_5N_2$	13.81	0.88	$C_{10}HN_3O$	12.01	0.86	$C_8H_9N_2O_3$	9.67	1.02
$C_8H_4N_2O_3$	9.59	1.01	$C_{12}H_{17}O$	13.28	1.01	$C_{10}H_3N_4$	12.38	0.71	$C_8H_{11}N_3O_2$	10.04	0.86
$C_8H_6N_3O_2$	9.96	0.85	$C_{12}H_{19}N$	13.65	0.86	$C_{10}H_{11}O_3$	11.10	1.16	$C_8H_{13}N_4O$	10.42	0.69
$C_8H_8N_4O$	10.34	0.69	$C_{13}H_5O$	14.17	1.13	$C_{10}H_{13}NO_2$	11.47	1.00	C_9HN_4O	11.31	0.78
$C_8H_{16}O_4$	9.06	1.17	$C_{13}H_7N$	14.54	0.98	$C_{10}H_{15}N_2O$	11.85	0.84	$C_9H_9O_4$	10.03	1.25
$C_8H_{18}NO_3$	9.43	1.00	$C_{13}H_{21}$	14.38	0.96	$C_{10}H_{17}N_3$	12.22	0.69	$C_9H_{11}NO_3$	10.40	1.09
$C_8H_{20}N_2O_2$	9.81	0.83	$C_{14}H_9$	15.27	1.08	$C_{11}HNO_2$	12.36	1.10	$C_9H_{13}N_2O_2$	10.78	0.93
$C_9H_4O_4$	9.95	1.24				$C_{11}H_3N_2O$	12.74	0.95	$C_9H_{15}N_3O$	11.15	0.77
$C_9H_6NO_3$	10.32	1.08	**178**			$C_{11}H_5N_3$	13.11	0.79	$C_9H_{17}N_4$	11.52	0.61
$C_9H_8N_2O_2$	10.70	0.92	$C_6H_{14}N_2O_4$	7.63	1.06	$C_{11}H_{15}O_2$	12.21	1.08	$C_{10}HN_2O_2$	11.66	1.02
$C_9H_{10}N_3O$	11.07	0.76	$C_6H_{16}N_3O_3$	8.00	0.88	$C_{11}H_{17}NO$	12.58	0.93	$C_{10}H_3N_3O$	12.04	0.86
$C_9H_{12}N_4$	11.44	0.60	$C_6H_{18}N_4O_2$	8.38	0.71	$C_{11}H_{19}N_2$	12.95	0.77	$C_{10}H_5N_4$	12.41	0.71
$C_9H_{20}O_3$	10.16	1.07	$C_7H_2N_2O_4$	8.52	1.12	$C_{12}H_3O_2$	13.09	1.19	$C_{10}H_{13}O_3$	11.13	1.16
$C_{10}H_6O_3$	11.05	1.15	$C_7H_4N_3O_3$	8.89	0.95	$C_{12}H_5NO$	13.47	1.04	$C_{10}H_{15}NO_2$	11.51	1.00
$C_{10}H_{10}NO_2$	11.43	0.99	$C_7H_6N_4O_2$	9.26	0.79	$C_{12}H_7N_2$	13.84	0.89	$C_{10}H_{17}N_2O$	11.88	0.85
$C_{10}H_{12}N_2O$	11.80	0.84	$C_7H_{16}NO_4$	8.36	1.11	$C_{12}H_{19}O$	13.31	1.02	$C_{10}H_{19}N_3$	12.25	0.69

Formula	M+1	M+2
$C_{11}HO_3$	12.02	1.26
$C_{11}H_3NO_2$	12.39	1.10
$C_{11}H_5N_2O$	12.77	0.95
$C_{11}H_7N_3$	13.14	0.80
$C_{11}H_{17}O_2$	12.24	1.09
$C_{11}H_{19}NO$	12.61	0.93
$C_{11}H_{21}N_2$	12.99	0.78
$C_{12}H_5O_2$	13.13	1.19
$C_{12}H_7NO$	13.50	1.04
$C_{12}H_9N_2$	13.87	0.89
$C_{12}H_{21}O$	13.34	1.02
$C_{12}H_{23}N$	13.72	0.87
$C_{13}H_9O$	14.23	1.14
$C_{13}H_{11}N$	14.61	0.99
$C_{13}H_{25}$	14.45	0.97
$C_{14}H_{13}$	15.34	1.09
$C_{15}H$	16.23	1.23

182

Formula	M+1	M+2
$C_7H_6N_2O_4$	8.58	1.13
$C_7H_8N_3O_3$	8.95	0.96
$C_7H_{10}N_4O_2$	9.33	0.79
$C_8H_8NO_4$	9.31	1.19
$C_8H_{10}N_2O_3$	9.69	1.02
$C_8H_{12}N_3O_2$	10.06	0.86
$C_8H_{14}N_4O$	10.43	0.70
$C_9H_2N_4O$	11.32	0.79
$C_9H_{10}O_4$	10.04	1.25
$C_9H_{12}NO_3$	10.42	1.09
$C_9H_{14}N_2O_2$	10.79	0.93
$C_9H_{16}N_3O$	11.17	0.77
$C_9H_{18}N_4$	11.54	0.61
$C_{10}H_2N_2O_2$	11.68	1.02
$C_{10}H_4N_3O$	12.05	0.87
$C_{10}H_6N_4$	12.43	0.71
$C_{10}H_{14}O_3$	11.15	1.16
$C_{10}H_{16}NO_2$	11.52	1.01
$C_{10}H_{18}N_2O$	11.90	0.85
$C_{10}H_{20}N_3$	12.27	0.69
$C_{11}H_2O_3$	12.04	1.26
$C_{11}H_4NO_2$	12.41	1.11
$C_{11}H_6N_2O$	12.79	0.95
$C_{11}H_8N_3$	13.16	0.80
$C_{11}H_{18}O_2$	12.25	1.09
$C_{11}H_{20}NO$	12.63	0.93
$C_{11}H_{22}N_2$	13.00	0.78
$C_{12}H_6O_2$	13.14	1.19
$C_{12}H_8NO$	13.52	1.04
$C_{12}H_{10}N_2$	13.89	0.89
$C_{12}H_{22}O$	13.36	1.02
$C_{12}H_{24}N$	13.73	0.87
$C_{13}H_{10}O$	14.25	1.14
$C_{13}H_{12}N$	14.62	0.99
$C_{13}H_{26}$	14.46	0.97
$C_{14}H_{14}$	15.35	1.10
$C_{15}H_2$	16.24	1.21

183

Formula	M+1	M+2
$C_7H_7N_2O_4$	8.60	1.13
$C_7H_9N_3O_3$	8.97	0.96
$C_7H_{11}N_4O_2$	9.34	0.79
$C_8H_9NO_4$	9.33	1.19
$C_8H_{11}N_2O_3$	9.70	1.02
$C_8H_{13}N_3O_2$	10.08	0.86
$C_8H_{15}N_4O$	10.45	0.70
$C_9HN_3O_2$	10.96	0.95
$C_9H_3N_4O$	11.34	0.79
$C_9H_{11}O_4$	10.06	1.26
$C_9H_{13}NO_3$	10.43	1.09
$C_9H_{15}N_2O_2$	10.81	0.93
$C_9H_{17}N_3O$	11.18	0.77
$C_9H_{19}N_4$	11.56	0.61
$C_{10}HNO_3$	11.32	1.18
$C_{10}H_3N_2O_2$	11.70	1.03
$C_{10}H_5N_3O$	12.07	0.87
$C_{10}H_7N_4$	12.44	0.71
$C_{10}H_{15}O_3$	11.16	1.17
$C_{10}H_{17}NO_2$	11.54	1.01
$C_{10}H_{19}N_2O$	11.91	0.85
$C_{10}H_{21}N_3$	12.29	0.69
$C_{11}H_3O_3$	12.05	1.26
$C_{11}H_5NO_2$	12.43	1.11
$C_{11}H_7N_2O$	12.80	0.95
$C_{11}H_9N_3$	13.18	0.80
$C_{11}H_{19}O_2$	12.27	1.09
$C_{11}H_{21}NO$	12.64	0.93
$C_{11}H_{23}N_2$	13.02	0.78
$C_{12}H_7O_2$	13.16	1.20
$C_{12}H_9NO$	13.53	1.05
$C_{12}H_{11}N_2$	13.91	0.90
$C_{12}H_{23}O$	13.37	1.02
$C_{12}H_{25}N$	13.75	0.87
$C_{13}H_{11}O$	14.26	1.14
$C_{13}H_{13}N$	14.64	0.99
$C_{13}H_{27}$	14.48	0.97
$C_{14}HN$	15.53	1.12
$C_{14}H_{15}$	15.37	1.10
$C_{15}H_3$	16.26	1.23

184

Formula	M+1	M+2
$C_7H_8N_2O_4$	8.61	1.13
$C_7H_{10}N_3O_3$	8.99	0.96
$C_7H_{12}N_4O_2$	9.36	0.80
$C_8H_{10}NO_4$	9.34	1.19
$C_8H_{12}N_2O_3$	9.72	1.03
$C_8H_{14}N_3O_2$	10.09	0.86
$C_8H_{16}N_4O$	10.47	0.70
$C_9H_2N_3O_2$	10.98	0.95
$C_9H_4N_4O$	11.35	0.79
$C_9H_{12}O_4$	10.07	1.26
$C_9H_{14}NO_3$	10.45	1.09
$C_9H_{16}N_2O_2$	10.82	0.93
$C_9H_{18}N_3O$	11.20	0.77
$C_9H_{20}N_4$	11.57	0.61
$C_{10}H_2NO_3$	11.34	1.18
$C_{10}H_4N_2O_2$	11.71	1.03
$C_{10}H_6N_3O$	12.09	0.87
$C_{10}H_8N_4$	12.46	0.71
$C_{10}H_{16}O_3$	11.18	1.17
$C_{10}H_{18}NO_2$	11.55	1.01
$C_{10}H_{20}N_2O$	11.93	0.85
$C_{10}H_{22}N_3$	12.30	0.70
$C_{11}H_4O_3$	12.07	1.27
$C_{11}H_6NO_2$	12.44	1.11
$C_{11}H_8N_2O$	12.82	0.96
$C_{11}H_{10}N_3$	13.19	0.80
$C_{11}H_{20}O_2$	12.29	1.09
$C_{11}H_{22}NO$	12.66	0.94
$C_{11}H_{24}N_2$	13.03	0.78
$C_{12}H_8O_2$	13.17	1.20
$C_{12}H_{10}NO$	13.55	1.05
$C_{12}H_{12}N_2$	13.92	0.90
$C_{12}H_{24}O$	13.39	1.03
$C_{12}H_{26}N$	13.77	0.88
$C_{13}H_{12}O$	14.28	1.14
$C_{13}H_{14}N$	14.65	1.00
$C_{13}H_{28}$	14.50	0.97
$C_{14}H_2N$	15.54	1.13
$C_{14}H_{16}$	15.38	1.10
$C_{15}H_4$	16.27	1.24

185

Formula	M+1	M+2
$C_7H_9N_2O_4$	8.63	1.13
$C_7H_{11}N_3O_3$	9.00	0.96
$C_7H_{13}N_4O_2$	9.38	0.80
$C_8HN_4O_2$	10.27	0.88
$C_8H_{11}NO_4$	9.36	1.19
$C_8H_{13}N_2O_3$	9.73	1.03
$C_8H_{15}N_3O_2$	10.11	0.86
$C_8H_{17}N_4O$	10.48	0.70
$C_9HN_2O_3$	10.62	1.11
$C_9H_3N_3O_2$	11.00	0.95
$C_9H_5N_4O$	11.37	0.79
$C_9H_{13}O_4$	10.09	1.26
$C_9H_{15}NO_3$	10.46	1.10
$C_9H_{17}N_2O_2$	10.84	0.93
$C_9H_{19}N_3O$	11.21	0.77
$C_9H_{21}N_4$	11.59	0.62
$C_{10}HO_4$	10.98	1.35
$C_{10}H_3NO_3$	11.35	1.19
$C_{10}H_5N_2O_2$	11.73	1.03
$C_{10}H_7N_3O$	12.10	0.87
$C_{10}H_9N_4$	12.48	0.72
$C_{10}H_{17}O_3$	11.20	1.17
$C_{10}H_{19}NO_2$	11.57	1.01
$C_{10}H_{21}N_2O$	11.94	0.85
$C_{10}H_{23}N_3$	12.32	0.70
$C_{11}H_5O_3$	12.08	1.27
$C_{11}H_7NO_2$	12.46	1.11
$C_{11}H_9N_2O$	12.83	0.96
$C_{11}H_{11}N_3$	13.21	0.81
$C_{11}H_{21}O_2$	12.30	1.09
$C_{11}H_{23}NO$	12.68	0.94
$C_{11}H_{25}N_2$	13.05	0.79
$C_{12}H_9O_2$	13.19	1.20
$C_{12}H_{11}NO$	13.56	1.05
$C_{12}H_{13}N_2$	13.94	0.90
$C_{12}H_{25}O$	13.41	1.03
$C_{12}H_{27}N$	13.78	0.88
$C_{13}HN_2$	14.83	1.02
$C_{13}H_{13}O$	14.30	1.15
$C_{13}H_{15}N$	14.67	1.00
$C_{14}HO$	15.18	1.27
$C_{14}H_3N$	15.56	1.13
$C_{14}H_{17}$	15.40	1.10
$C_{15}H_5$	16.29	1.24

186

Formula	M+1	M+2
$C_7H_{10}N_2O_4$	8.64	1.13
$C_7H_{12}N_3O_3$	9.02	0.97
$C_7H_{14}N_4O_2$	9.39	0.80
$C_8H_2N_4O_2$	10.28	0.88
$C_8H_{12}NO_4$	9.38	1.19
$C_8H_{14}N_2O_3$	9.75	1.03
$C_8H_{16}N_3O_2$	10.12	0.86
$C_8H_{18}N_4O$	10.50	0.70
$C_9H_2N_2O_3$	10.64	1.11
$C_9H_4N_3O_2$	11.01	0.95
$C_9H_6N_4O$	11.39	0.79
$C_9H_{14}O_4$	10.11	1.26
$C_9H_{16}NO_3$	10.48	1.10
$C_9H_{18}N_2O_2$	10.86	0.94
$C_9H_{20}N_3O$	11.23	0.78
$C_9H_{22}N_4$	11.60	0.62
$C_{10}H_2O_4$	10.99	1.35
$C_{10}H_4NO_3$	11.37	1.19
$C_{10}H_6N_2O_2$	11.74	1.03
$C_{10}H_8N_3O$	12.12	0.87
$C_{10}H_{10}N_4$	12.49	0.72
$C_{10}H_{18}O_3$	11.21	1.17
$C_{10}H_{20}NO_2$	11.59	1.01
$C_{10}H_{22}N_2O$	11.96	0.86
$C_{10}H_{24}N_3$	12.33	0.70
$C_{11}H_6O_3$	12.10	1.27
$C_{11}H_8NO_2$	12.47	1.11
$C_{11}H_{10}N_2O$	12.85	0.96
$C_{11}H_{12}N_3$	13.22	0.81
$C_{11}H_{22}O_2$	12.32	1.10
$C_{11}H_{24}NO$	12.69	0.94
$C_{11}H_{26}N_2$	13.07	0.79
$C_{12}H_{10}O_2$	13.21	1.20
$C_{12}H_{12}NO$	13.58	1.05
$C_{12}H_{14}N_2$	13.95	0.90
$C_{12}H_{26}O$	13.42	1.03
$C_{13}H_2N_2$	14.84	1.02
$C_{13}H_{14}O$	14.31	1.15
$C_{13}H_{16}N$	14.69	1.00
$C_{14}H_2O$	15.20	1.27
$C_{14}H_4N$	15.57	1.13
$C_{14}H_{18}$	15.42	1.11
$C_{15}H_6$	16.31	1.24

187

Formula	M+1	M+2
$C_7H_{11}N_2O_4$	8.66	1.13
$C_7H_{13}N_3O_3$	9.03	0.97
$C_7H_{15}N_4O_2$	9.41	0.80
$C_8HN_3O_3$	9.92	1.04
$C_8H_3N_4O_2$	10.30	0.88
$C_8H_{13}NO_4$	9.39	1.20
$C_8H_{15}N_2O_3$	9.77	1.03
$C_8H_{17}N_3O_2$	10.14	0.87
$C_8H_{19}N_4O$	10.51	0.70
C_9HNO_4	10.28	1.28
$C_9H_3N_2O_3$	10.65	1.11
$C_9H_5N_3O_2$	11.03	0.95
$C_9H_7N_4O$	11.40	0.80
$C_9H_{15}O_4$	10.12	1.26
$C_9H_{17}NO_3$	10.50	1.10
$C_9H_{19}N_2O_2$	10.87	0.94
$C_9H_{21}N_3O$	11.25	0.78
$C_9H_{23}N_4$	11.62	0.62
$C_{10}H_3O_4$	11.01	1.35
$C_{10}H_5NO_3$	11.39	1.19

	M + 1	M + 2
$C_{10}H_7N_2O_2$	11.76	1.03
$C_{10}H_9N_3O$	12.13	0.88
$C_{10}H_{11}N_4$	12.51	0.72
$C_{10}H_{19}O_3$	11.23	1.17
$C_{10}H_{21}NO_2$	11.60	1.01
$C_{10}H_{23}N_2O$	11.98	0.86
$C_{10}H_{25}N_3$	12.35	0.70
$C_{11}H_7O_3$	12.12	1.27
$C_{11}H_9NO_2$	12.49	1.12
$C_{11}H_{11}N_2O$	12.87	0.96
$C_{11}H_{13}N_3$	13.24	0.81
$C_{11}H_{23}O_2$	12.33	1.10
$C_{11}H_{25}NO$	12.71	0.94
$C_{12}HN_3$	14.13	0.93
$C_{12}H_{11}O_2$	13.22	1.20
$C_{12}H_{13}NO$	13.60	1.05
$C_{12}H_{15}N_2$	13.97	0.90
$C_{13}HNO$	14.48	1.17
$C_{13}H_3N_2$	14.86	1.03
$C_{13}H_{15}O$	14.33	1.15
$C_{13}H_{17}N$	14.70	1.00
$C_{14}H_3O$	15.22	1.28
$C_{14}H_5N$	15.59	1.13
$C_{14}H_{19}$	15.43	1.11
$C_{15}H_7$	16.32	1.24

188

	M + 1	M + 2
$C_7H_{12}N_2O_4$	8.68	1.14
$C_7H_{14}N_3O_3$	9.05	0.97
$C_7H_{16}N_4O_2$	9.42	0.80
$C_8H_2N_3O_3$	9.94	1.05
$C_8H_4N_4O_2$	10.31	0.88
$C_8H_{14}NO_4$	9.41	1.20
$C_8H_{16}N_2O_3$	9.78	1.03
$C_8H_{18}N_3O_2$	10.16	0.87
$C_8H_{20}N_4O$	10.53	0.71
$C_9H_2NO_4$	10.30	1.28
$C_9H_4N_2O_3$	10.67	1.12
$C_9H_6N_3O_2$	11.04	0.96
$C_9H_8N_4O$	11.42	0.80
$C_9H_{16}O_4$	10.14	1.26
$C_9H_{18}NO_3$	10.51	1.10
$C_9H_{20}N_2O_2$	10.89	0.94
$C_9H_{22}N_3O$	11.26	0.78
$C_9H_{24}N_4$	11.64	0.62
$C_{10}H_4O_4$	11.03	1.35
$C_{10}H_6NO_3$	11.40	1.19
$C_{10}H_8N_2O_2$	11.78	1.03
$C_{10}H_{10}N_3O$	12.15	0.88
$C_{10}H_{12}N_4$	12.52	0.72
$C_{10}H_{20}O_3$	11.24	1.18
$C_{10}H_{22}NO_2$	11.62	1.02
$C_{10}H_{24}N_2O$	11.99	0.86
$C_{11}H_8O_3$	12.13	1.27
$C_{11}H_{10}NO_2$	12.51	1.12
$C_{11}H_{12}N_2O$	12.88	0.96
$C_{11}H_{14}N_3$	13.26	0.81
$C_{11}H_{24}O_2$	12.35	1.10
$C_{12}H_2N_3$	14.14	0.93
$C_{12}H_{12}O_2$	13.24	1.21
$C_{12}H_{14}NO$	13.61	1.06
$C_{12}H_{16}N_2$	13.99	0.91
$C_{13}H_2NO$	14.50	1.18

	M + 1	M + 2
$C_{13}H_4N_2$	14.88	1.03
$C_{13}H_{16}O$	14.34	1.15
$C_{13}H_{18}N$	14.72	1.01
$C_{14}H_4O$	15.23	1.28
$C_{14}H_6N$	15.61	1.14
$C_{14}H_{20}$	15.45	1.11
$C_{15}H_8$	16.34	1.25

189

	M + 1	M + 2
$C_7H_{13}N_2O_4$	8.69	1.14
$C_7H_{15}N_3O_3$	9.07	0.97
$C_7H_{17}N_4O_2$	9.44	0.80
$C_8HN_2O_4$	9.58	1.21
$C_8H_3N_3O_3$	9.95	1.05
$C_8H_5N_4O_2$	10.33	0.88
$C_8H_{15}NO_4$	9.42	1.20
$C_8H_{17}N_2O_3$	9.80	1.03
$C_8H_{19}N_3O_2$	10.17	0.87
$C_8H_{21}N_4O$	10.55	0.71
$C_9H_3NO_4$	10.31	1.28
$C_9H_5N_2O_3$	10.69	1.12
$C_9H_7N_3O_2$	11.06	0.96
$C_9H_9N_4O$	11.43	0.80
$C_9H_{17}O_4$	10.15	1.26
$C_9H_{19}NO_3$	10.53	1.10
$C_9H_{21}N_2O_2$	10.90	0.94
$C_9H_{23}N_3O$	11.28	0.78
$C_{10}H_5O_4$	11.04	1.35
$C_{10}H_7NO_3$	11.42	1.19
$C_{10}H_9N_2O_2$	11.79	1.04
$C_{10}H_{11}N_3O$	12.17	0.88
$C_{10}H_{13}N_4$	12.54	0.72
$C_{10}H_{21}O_3$	11.26	1.18
$C_{10}H_{23}NO_2$	11.63	1.02
$C_{11}HN_4$	13.43	0.83
$C_{11}H_9O_3$	12.15	1.28
$C_{11}H_{11}NO_2$	12.52	1.12
$C_{11}H_{13}N_2O$	12.90	0.97
$C_{11}H_{15}N_3$	13.27	0.81
$C_{12}HN_2O$	13.79	1.08
$C_{12}H_3N_3$	14.16	0.93
$C_{12}H_{13}O_2$	13.25	1.21
$C_{12}H_{15}NO$	13.63	1.06
$C_{12}H_{17}N_2$	14.00	0.91
$C_{13}HO_2$	14.14	1.33
$C_{13}H_3NO$	14.52	1.18
$C_{13}H_5N_2$	14.89	1.03
$C_{13}H_{17}O$	14.36	1.16
$C_{13}H_{19}N$	14.73	1.01
$C_{14}H_5O$	15.25	1.28
$C_{14}H_7N$	15.62	1.14
$C_{14}H_{21}$	15.46	1.11
$C_{15}H_9$	16.35	1.25

190

	M + 1	M + 2
$C_7H_{14}N_2O_4$	8.71	1.14
$C_7H_{16}N_3O_3$	9.08	0.97
$C_7H_{18}N_4O_2$	9.46	0.80
$C_8H_2N_2O_4$	9.60	1.21
$C_8H_4N_3O_3$	9.97	1.05
$C_8H_6N_4O_2$	10.35	0.89
$C_8H_{16}NO_4$	9.44	1.20
$C_8H_{18}N_2O_3$	9.81	1.03

	M + 1	M + 2
$C_8H_{20}N_3O_2$	10.19	0.87
$C_8H_{22}N_4O$	10.56	0.71
$C_9H_4NO_4$	10.33	1.28
$C_9H_6N_2O_3$	10.70	1.12
$C_9H_8N_3O_2$	11.08	0.96
$C_9H_{10}N_4O$	11.45	0.80
$C_9H_{18}O_4$	10.17	1.27
$C_9H_{20}NO_3$	10.54	1.10
$C_9H_{22}N_2O_2$	10.92	0.94
$C_{10}H_6O_4$	11.06	1.35
$C_{10}H_8NO_3$	11.43	1.20
$C_{10}H_{10}N_2O_2$	11.81	1.03
$C_{10}H_{12}N_3O$	12.18	0.88
$C_{10}H_{14}N_4$	12.56	0.73
$C_{10}H_{22}O_3$	11.28	1.18
$C_{11}H_2N_4$	13.44	0.84
$C_{11}H_{10}O_3$	12.16	1.28
$C_{11}H_{12}NO_2$	12.54	1.12
$C_{11}H_{14}N_2O$	12.91	0.97
$C_{11}H_{16}N_3$	13.29	0.82
$C_{12}H_2N_2O$	13.80	1.08
$C_{12}H_4N_3$	14.18	0.93
$C_{12}H_{14}O_2$	13.27	1.21
$C_{12}H_{16}NO$	13.64	1.06
$C_{12}H_{18}N_2$	14.02	0.91
$C_{13}H_2O_2$	14.16	1.33
$C_{13}H_4NO$	14.53	1.18
$C_{13}H_6N_2$	14.91	1.03
$C_{13}H_{18}O$	14.38	1.16
$C_{13}H_{20}N$	14.75	1.01
$C_{14}H_6O$	15.26	1.28
$C_{14}H_8N$	15.64	1.14
$C_{14}H_{22}$	15.48	1.12
$C_{15}H_{10}$	16.37	1.25

191

	M + 1	M + 2
$C_7H_{15}N_2O_4$	8.72	1.14
$C_7H_{17}N_3O_3$	9.10	0.97
$C_7H_{19}N_4O_2$	9.47	0.81
$C_8H_3N_2O_4$	9.61	1.22
$C_8H_5N_3O_3$	9.99	1.05
$C_8H_7N_4O_2$	10.36	0.89
$C_8H_{17}NO_4$	9.46	1.20
$C_8H_{19}N_2O_3$	9.83	1.04
$C_8H_{21}N_3O_2$	10.20	0.87
$C_9H_5NO_4$	10.34	1.28
$C_9H_7N_2O_3$	10.72	1.12
$C_9H_9N_3O_2$	11.09	0.96
$C_9H_{11}N_4O$	11.47	0.80
$C_9H_{19}O_4$	10.19	1.27
$C_9H_{21}NO_3$	10.56	1.11
$C_{10}H_7O_4$	11.07	1.36
$C_{10}H_9NO_3$	11.45	1.20
$C_{10}H_{11}N_2O_2$	11.82	1.04
$C_{10}H_{13}N_3O$	12.20	0.88
$C_{10}H_{15}N_4$	12.57	0.73
$C_{11}HN_3O$	13.09	0.99
$C_{11}H_3N_4$	13.46	0.84
$C_{11}H_{11}O_3$	12.18	1.28
$C_{11}H_{13}NO_2$	12.55	1.12
$C_{11}H_{15}N_2O$	12.93	0.97
$C_{11}H_{17}N_3$	13.30	0.82
$C_{12}HNO_2$	13.44	1.23

	M + 1	M + 2
$C_{12}H_3N_2O$	13.82	1.08
$C_{12}H_5N_3$	14.19	0.93
$C_{12}H_{15}O_2$	13.29	1.21
$C_{12}H_{17}NO$	13.66	1.06
$C_{12}H_{19}N_2$	14.03	0.91
$C_{13}H_3O_2$	14.17	1.33
$C_{13}H_5NO$	14.55	1.18
$C_{13}H_7N_2$	14.92	1.04
$C_{13}H_{19}O$	14.39	1.16
$C_{13}H_{21}N$	14.77	1.01
$C_{14}H_7O$	15.28	1.29
$C_{14}H_9N$	15.65	1.14
$C_{14}H_{23}$	15.50	1.12
$C_{15}H_{11}$	16.39	1.25

192

	M + 1	M + 2
$C_7H_{16}N_2O_4$	8.74	1.14
$C_7H_{18}N_3O_3$	9.11	0.97
$C_7H_{20}N_4O_2$	9.49	0.81
$C_8H_4N_2O_4$	9.63	1.22
$C_8H_6N_3O_3$	10.00	1.05
$C_8H_8N_4O_2$	10.38	0.89
$C_8H_{18}NO_4$	9.47	1.20
$C_8H_{20}N_2O_3$	9.85	1.04
$C_9H_6NO_4$	10.36	1.29
$C_9H_8N_2O_3$	10.73	1.12
$C_9H_{10}N_3O_2$	11.11	0.96
$C_9H_{12}N_4O$	11.48	0.80
$C_9H_{20}O_4$	10.20	1.27
$C_{10}H_8O_4$	11.09	1.36
$C_{10}H_{10}NO_3$	11.47	1.20
$C_{10}H_{12}N_2O_2$	11.84	1.04
$C_{10}H_{14}N_3O$	12.21	0.89
$C_{10}H_{16}N_4$	12.59	0.73
$C_{11}H_2N_3O$	13.10	0.99
$C_{11}H_4N_4$	13.48	0.84
$C_{11}H_{12}O_3$	12.20	1.28
$C_{11}H_{14}NO_2$	12.57	1.13
$C_{11}H_{16}N_2O$	12.95	0.97
$C_{11}H_{18}N_3$	13.32	0.82
$C_{12}H_2NO_2$	13.46	1.24
$C_{12}H_4N_2O$	13.83	1.09
$C_{12}H_6N_3$	14.21	0.94
$C_{12}H_{16}O_2$	13.30	1.22
$C_{12}H_{18}NO$	13.68	1.06
$C_{12}H_{20}N_2$	14.05	0.92
$C_{13}H_4O_2$	14.19	1.33
$C_{13}H_6NO$	14.56	1.18
$C_{13}H_8N_2$	14.94	1.04
$C_{13}H_{20}O$	14.41	1.16
$C_{13}H_{22}N$	14.78	1.02
$C_{14}H_8O$	15.30	1.29
$C_{14}H_{10}N$	15.67	1.15
$C_{14}H_{24}$	15.51	1.12
$C_{15}H_{12}$	16.40	1.26

193

	M + 1	M + 2
$C_7H_{17}N_2O_4$	8.76	1.14
$C_7H_{19}N_3O_3$	9.13	0.98
$C_8H_5N_2O_4$	9.64	1.22
$C_8H_7N_3O_3$	10.02	1.05
$C_8H_9N_4O_2$	10.39	0.89
$C_8H_{19}NO_4$	9.49	1.20

	M + 1	M + 2
$C_9H_7NO_4$	10.38	1.29
$C_9H_9N_2O_3$	10.75	1.13
$C_9H_{11}N_3O_2$	11.12	0.96
$C_9H_{13}N_4O$	11.50	0.81
$C_{10}HN_4O$	12.39	0.91
$C_{10}H_9O_4$	11.11	1.36
$C_{10}H_{11}NO_3$	11.48	1.20
$C_{10}H_{13}N_2O_2$	11.86	1.04
$C_{10}H_{15}N_3O$	12.23	0.89
$C_{10}H_{17}N_4$	12.60	0.73
$C_{11}HN_2O_2$	12.74	1.15
$C_{11}H_3N_3O$	13.12	0.99
$C_{11}H_5N_4$	13.49	0.84
$C_{11}H_{13}O_3$	12.21	1.28
$C_{11}H_{15}NO_2$	12.59	1.13
$C_{11}H_{17}N_2O$	12.96	0.97
$C_{11}H_{19}N_3$	13.34	0.82
$C_{12}HO_3$	13.10	1.39
$C_{12}H_3NO_2$	13.48	1.24
$C_{12}H_5N_2O$	13.85	1.09
$C_{12}H_7N_3$	14.22	0.94
$C_{12}H_{17}O_2$	13.32	1.22
$C_{12}H_{19}NO$	13.69	1.07
$C_{12}H_{21}N_2$	14.07	0.92
$C_{13}H_5O_2$	14.21	1.33
$C_{13}H_7NO$	14.58	1.19
$C_{13}H_9N_2$	14.96	1.04
$C_{13}H_{21}O$	14.42	1.16
$C_{13}H_{23}N$	14.80	1.02
$C_{14}H_9O$	15.31	1.29
$C_{14}H_{11}N$	15.69	1.15
$C_{14}H_{25}$	15.53	1.12
$C_{15}H_{13}$	16.42	1.26
$C_{16}H$	17.31	1.40

194

	M + 1	M + 2
$C_7H_{18}N_2O_4$	8.77	1.14
$C_8H_6N_2O_4$	9.66	1.22
$C_8H_8N_3O_3$	10.03	1.06
$C_8H_{10}N_4O_2$	10.41	0.89
$C_9H_8NO_4$	10.39	1.29
$C_9H_{10}N_2O_3$	10.77	1.13
$C_9H_{12}N_3O_2$	11.14	0.97
$C_9H_{14}N_4O$	11.51	0.81
$C_{10}H_6N_4O$	12.40	0.91
$C_{10}H_{10}O_4$	11.12	1.36
$C_{10}H_{12}NO_3$	11.50	1.20
$C_{10}H_{14}N_2O_2$	11.87	1.05
$C_{10}H_{16}N_3O$	12.25	0.89
$C_{10}H_{18}N_4$	12.62	0.74
$C_{11}H_2N_2O_2$	12.76	1.15
$C_{11}H_4N_3O$	13.13	1.00
$C_{11}H_6N_4$	13.51	0.85
$C_{11}H_{14}O_3$	12.23	1.28
$C_{11}H_{16}NO_2$	12.60	1.13
$C_{11}H_{18}N_2O$	12.98	0.98
$C_{11}H_{20}N_3$	13.35	0.82
$C_{12}H_2O_3$	13.12	1.39
$C_{12}H_4NO_2$	13.49	1.24
$C_{12}H_6N_2O$	13.87	1.09
$C_{12}H_8N_3$	14.24	0.94
$C_{12}H_{18}O_2$	13.33	1.22
$C_{12}H_{20}NO$	13.71	1.07
$C_{12}H_{22}N_2$	14.08	0.92

	M + 1	M + 2
$C_{13}H_6O_2$	14.22	1.34
$C_{13}H_8NO$	14.60	1.19
$C_{13}H_{10}N_2$	14.97	1.04
$C_{13}H_{22}O$	14.44	1.17
$C_{13}H_{24}N$	14.81	1.02
$C_{14}H_{10}O$	15.33	1.29
$C_{14}H_{12}N$	15.70	1.15
$C_{14}H_{26}$	15.54	1.13
$C_{15}H_{14}$	16.43	1.26
$C_{16}H_2$	17.32	1.41

195

	M + 1	M + 2
$C_8H_7N_2O_4$	9.68	1.22
$C_8H_9N_3O_3$	10.05	1.06
$C_8H_{11}N_4O_2$	10.43	0.89
$C_9H_9NO_4$	10.41	1.29
$C_9H_{11}N_2O_3$	10.78	1.13
$C_9H_{13}N_3O_2$	11.16	0.97
$C_9H_{15}N_4O$	11.53	0.81
$C_{10}HN_3O_2$	12.05	1.07
$C_{10}H_3N_4O$	12.42	0.91
$C_{10}H_{11}O_4$	11.14	1.36
$C_{10}H_{13}NO_3$	11.51	1.21
$C_{10}H_{15}N_2O_2$	11.89	1.05
$C_{10}H_{17}N_3O$	12.26	0.89
$C_{10}H_{19}N_4$	12.64	0.74
$C_{11}HNO_3$	12.40	1.31
$C_{11}H_3N_2O_2$	12.78	1.15
$C_{11}H_5N_3O$	13.15	1.00
$C_{11}H_7N_4$	13.52	0.85
$C_{11}H_{15}O_3$	12.24	1.29
$C_{11}H_{17}NO_2$	12.62	1.13
$C_{11}H_{19}N_2O$	12.99	0.98
$C_{11}H_{21}N_3$	13.37	0.83
$C_{12}H_3O_3$	13.13	1.39
$C_{12}H_5NO_2$	13.51	1.24
$C_{12}H_7N_2O$	13.88	1.09
$C_{12}H_9N_3$	14.26	0.94
$C_{12}H_{19}O_2$	13.35	1.22
$C_{12}H_{21}NO$	13.72	1.07
$C_{12}H_{23}N_2$	14.10	0.92
$C_{13}H_7O_2$	14.24	1.34
$C_{13}H_9NO$	14.61	1.19
$C_{13}H_{11}N_2$	14.99	1.05
$C_{13}H_{23}O$	14.46	1.17
$C_{13}H_{25}N$	14.83	1.02
$C_{14}H_{11}O$	15.34	1.30
$C_{14}H_{13}N$	15.72	1.15
$C_{14}H_{27}$	15.56	1.13
$C_{15}HN$	16.61	1.29
$C_{15}H_{15}$	16.45	1.27
$C_{16}H_3$	17.34	1.41

196

	M + 1	M + 2
$C_8H_8N_2O_4$	9.69	1.22
$C_8H_{10}N_3O_3$	10.07	1.06
$C_8H_{12}N_4O_2$	10.44	0.90
$C_9H_{10}NO_4$	10.42	1.29
$C_9H_{12}N_2O_3$	10.80	1.13
$C_9H_{14}N_3O_2$	11.17	0.97
$C_9H_{16}N_4O$	11.55	0.81
$C_{10}H_2N_3O_2$	12.06	1.07
$C_{10}H_4N_4O$	12.44	0.91

	M + 1	M + 2
$C_{10}H_{12}O_4$	11.15	1.37
$C_{10}H_{14}NO_3$	11.53	1.21
$C_{10}H_{16}N_2O_2$	11.90	1.05
$C_{10}H_{18}N_3O$	12.28	0.89
$C_{10}H_{20}N_4$	12.65	0.74
$C_{11}H_2NO_3$	12.42	1.31
$C_{11}H_4N_2O_2$	12.79	1.15
$C_{11}H_6N_3O$	13.17	1.00
$C_{11}H_8N_4$	13.54	0.85
$C_{11}H_{16}O_3$	12.26	1.29
$C_{11}H_{18}NO_2$	12.63	1.13
$C_{11}H_{20}N_2O$	13.01	0.98
$C_{11}H_{22}N_3$	13.38	0.83
$C_{12}H_4O_3$	13.15	1.40
$C_{12}H_6NO_2$	13.52	1.24
$C_{12}H_8N_2O$	13.90	1.09
$C_{12}H_{10}N_3$	14.27	0.95
$C_{12}H_{20}O_2$	13.37	1.22
$C_{12}H_{22}NO$	13.74	1.07
$C_{12}H_{24}N_2$	14.11	0.92
$C_{13}H_8O_2$	14.25	1.34
$C_{13}H_{10}NO$	14.63	1.19
$C_{13}H_{12}N_2$	15.00	1.05
$C_{13}H_{24}O$	14.47	1.17
$C_{13}H_{26}N$	14.85	1.03
$C_{14}H_{12}O$	15.36	1.30
$C_{14}H_{14}N$	15.73	1.16
$C_{14}H_{28}$	15.58	1.13
$C_{15}H_2N$	16.62	1.29
$C_{15}H_{16}$	16.47	1.27
$C_{16}H_4$	17.35	1.41

197

	M + 1	M + 2
$C_8H_9N_2O_4$	9.71	1.23
$C_8H_{11}N_3O_3$	10.08	1.06
$C_8H_{13}N_4O_2$	10.46	0.90
$C_9HN_4O_2$	11.35	0.99
$C_9H_{11}NO_4$	10.44	1.29
$C_9H_{13}N_2O_3$	10.81	1.13
$C_9H_{15}N_3O_2$	11.19	0.97
$C_9H_{17}N_4O$	11.56	0.81
$C_{10}HN_2O_3$	11.70	1.23
$C_{10}H_3N_3O_2$	12.08	1.07
$C_{10}H_5N_4O$	12.45	0.91
$C_{10}H_{13}O_4$	11.17	1.37
$C_{10}H_{15}NO_3$	11.55	1.21
$C_{10}H_{17}N_2O_2$	11.92	1.05
$C_{10}H_{19}N_3O$	12.29	0.90
$C_{10}H_{21}N_4$	12.67	0.74
$C_{11}HO_4$	12.06	1.46
$C_{11}H_3NO_3$	12.43	1.31
$C_{11}H_5N_2O_2$	12.81	1.16
$C_{11}H_7N_3O$	13.18	1.00
$C_{11}H_9N_4$	13.56	0.85
$C_{11}H_{17}O_3$	12.28	1.29
$C_{11}H_{19}NO_2$	12.65	1.14
$C_{11}H_{21}N_2O$	13.03	0.98
$C_{11}H_{23}N_3$	13.40	0.83
$C_{12}H_5O_3$	13.16	1.40
$C_{12}H_7NO_2$	13.54	1.25
$C_{12}H_9N_2O$	13.91	1.10
$C_{12}H_{11}N_3$	14.29	0.95
$C_{12}H_{21}O_2$	13.38	1.23

	M + 1	M + 2
$C_{12}H_{23}NO$	13.76	1.08
$C_{12}H_{25}N_2$	14.13	0.93
$C_{13}H_9O_2$	14.27	1.34
$C_{13}H_{11}NO$	14.64	1.20
$C_{13}H_{13}N_2$	15.02	1.05
$C_{13}H_{25}O$	14.49	1.17
$C_{13}H_{27}N$	14.86	1.03
$C_{14}HN_2$	15.91	1.18
$C_{14}H_{13}O$	15.38	1.30
$C_{14}H_{15}N$	15.75	1.16
$C_{14}H_{29}$	15.59	1.13
$C_{15}HO$	16.26	1.44
$C_{15}H_3N$	16.64	1.30
$C_{15}H_{17}$	16.48	1.27
$C_{16}H_5$	17.37	1.42

198

	M + 1	M + 2
$C_8H_{10}N_2O_4$	9.72	1.23
$C_8H_{12}N_3O_3$	10.10	1.06
$C_8H_{14}N_4O$	10.47	0.90
$C_9H_2N_4O_2$	11.36	0.99
$C_9H_{12}NO_4$	10.46	1.30
$C_9H_{14}N_2O_3$	10.83	1.13
$C_9H_{16}N_3O_2$	11.20	0.97
$C_9H_{18}N_4O$	11.58	0.82
$C_{10}H_2N_2O_3$	11.72	1.23
$C_{10}H_4N_3O_2$	12.09	1.07
$C_{10}H_6N_4O$	12.47	0.92
$C_{10}H_{14}O_4$	11.19	1.37
$C_{10}H_{16}NO_3$	11.56	1.21
$C_{10}H_{18}N_2O_2$	11.94	1.05
$C_{10}H_{20}N_3O$	12.31	0.90
$C_{10}H_{22}N_4$	12.68	0.74
$C_{11}H_2O_4$	12.08	1.47
$C_{11}H_4NO_3$	12.45	1.31
$C_{11}H_6N_2O_2$	12.82	1.16
$C_{11}H_8N_3O$	13.20	1.01
$C_{11}H_{10}N_4$	13.57	0.85
$C_{11}H_{18}O_3$	12.29	1.29
$C_{11}H_{20}NO_2$	12.67	1.14
$C_{11}H_{22}N_2O$	13.04	0.99
$C_{11}H_{24}N_3$	13.42	0.83
$C_{12}H_6O_3$	13.18	1.40
$C_{12}H_8NO_2$	13.56	1.25
$C_{12}H_{10}N_2O$	13.93	1.10
$C_{12}H_{12}N_3$	14.30	0.95
$C_{12}H_{22}O_2$	13.40	1.23
$C_{12}H_{24}NO$	13.77	1.08
$C_{12}H_{26}N_2$	14.15	0.93
$C_{13}H_{10}O_2$	14.29	1.35
$C_{13}H_{12}NO$	14.66	1.20
$C_{13}H_{14}N_2$	15.04	1.05
$C_{13}H_{26}O$	14.50	1.18
$C_{13}H_{28}N$	14.88	1.03
$C_{14}H_2N_2$	15.92	1.18
$C_{14}H_{14}O$	15.39	1.30
$C_{14}H_{16}N$	15.77	1.16
$C_{14}H_{30}$	15.61	1.14
$C_{15}H_2O$	16.28	1.44
$C_{15}H_4N$	16.65	1.30
$C_{15}H_{18}$	16.50	1.27
$C_{16}H_6$	17.39	1.42

199

	M + 1	M + 2
$C_8H_{11}N_2O_4$	9.74	1.23
$C_8H_{13}N_3O_3$	10.11	1.06
$C_8H_{15}N_4O_2$	10.49	0.90
$C_9HN_3O_3$	11.00	1.15
$C_9H_3N_4O_2$	11.38	0.99
$C_9H_{13}NO_4$	10.47	1.30
$C_9H_{15}N_2O_3$	10.85	1.14
$C_9H_{17}N_3O_2$	11.22	0.98
$C_9H_{19}N_4O$	11.59	0.82
$C_{10}HNO_4$	11.36	1.39
$C_{10}H_3N_2O_3$	11.73	1.23
$C_{10}H_5N_3O_2$	12.11	1.07
$C_{10}H_7N_4O$	12.48	0.92
$C_{10}H_{15}O_4$	11.20	1.37
$C_{10}H_{17}NO_3$	11.58	1.21
$C_{10}H_{19}N_2O_2$	11.95	1.01
$C_{10}H_{21}N_3O$	12.33	0.90
$C_{10}H_{23}N_4$	12.70	0.75
$C_{11}H_3O_4$	12.09	1.47
$C_{11}H_5NO_3$	12.47	1.31
$C_{11}H_7N_2O_2$	12.84	1.16
$C_{11}H_9N_3O$	13.21	1.01
$C_{11}H_{11}N_4$	13.59	0.86
$C_{11}H_{19}O_3$	12.31	1.29
$C_{11}H_{21}NO_2$	12.68	1.14
$C_{11}H_{23}N_2O$	13.06	0.99
$C_{11}H_{25}N_3$	13.43	0.84
$C_{12}H_7O_3$	13.20	1.40
$C_{12}H_9NO_2$	13.57	1.25
$C_{12}H_{11}N_2O$	13.95	1.10
$C_{12}H_{13}N_3$	14.32	0.95
$C_{12}H_{23}O_2$	13.41	1.23
$C_{12}H_{25}NO$	13.79	1.08
$C_{12}H_{27}N_2$	14.16	0.93
$C_{13}HN_3$	15.21	1.08
$C_{13}H_{11}O_2$	14.30	1.35
$C_{13}H_{13}NO$	14.68	1.20
$C_{13}H_{15}N_2$	15.05	1.06
$C_{13}H_{27}O$	14.52	1.18
$C_{13}H_{29}N$	14.89	1.03
$C_{14}HNO$	15.57	1.33
$C_{14}H_3N_2$	15.94	1.19
$C_{14}H_{15}O$	15.41	1.31
$C_{14}H_{17}N$	15.78	1.16
$C_{15}H_3O$	16.30	1.44
$C_{15}H_5N$	16.67	1.30
$C_{15}H_{19}$	16.51	1.28
$C_{16}H_7$	17.40	1.42

200

	M + 1	M + 2
$C_8H_{12}N_2O_4$	9.76	1.23
$C_8H_{14}N_3O_3$	10.13	1.07
$C_8H_{16}N_4O_2$	10.51	0.90
$C_9H_2N_3O_3$	11.02	1.15
$C_9H_4N_4O_2$	11.39	0.99
$C_9H_{14}NO_4$	10.49	1.30
$C_9H_{16}N_2O_3$	10.86	1.14
$C_9H_{18}N_3O_2$	11.24	0.98
$C_9H_{20}N_4O$	11.61	0.82
$C_{10}H_2NO_4$	11.38	1.39
$C_{10}H_4N_2O_3$	11.75	1.23
$C_{10}H_6N_3O_2$	12.13	1.08
$C_{10}H_8N_4O$	12.50	0.92
$C_{10}H_{16}O_4$	11.22	1.37
$C_{10}H_{18}NO_3$	11.59	1.21
$C_{10}H_{20}N_2O_2$	11.97	1.06
$C_{10}H_{22}N_3O$	12.34	0.90
$C_{10}H_{24}N_4$	12.72	0.75
$C_{11}H_4O_4$	12.11	1.47
$C_{11}H_6NO_3$	12.48	1.32
$C_{11}H_8N_2O_2$	12.86	1.16
$C_{11}H_{10}N_3O$	13.23	1.01
$C_{11}H_{12}N_4$	13.60	0.86
$C_{11}H_{20}O_3$	12.32	1.30
$C_{11}H_{22}NO_2$	12.70	1.14
$C_{11}H_{24}N_2O$	13.07	0.99
$C_{11}H_{26}N_3$	13.45	0.84
$C_{12}H_8O_3$	13.21	1.40
$C_{12}H_{10}NO_2$	13.59	1.25
$C_{12}H_{12}N_2O$	13.96	1.10
$C_{12}H_{14}N_3$	13.34	0.96
$C_{12}H_{24}O_2$	13.43	1.23
$C_{12}H_{26}NO$	13.80	1.08
$C_{12}H_{28}N_2$	14.18	0.93
$C_{13}H_2N_3$	15.22	1.08
$C_{13}H_{12}O_2$	14.32	1.35
$C_{13}H_{14}NO$	14.69	1.20
$C_{13}H_{16}N_2$	15.07	1.06
$C_{13}H_{28}O$	14.54	1.18
$C_{14}H_2NO$	15.58	1.33
$C_{14}H_4N_2$	15.96	1.19
$C_{14}H_{16}O$	15.42	1.31
$C_{14}H_{18}N$	15.80	1.17
$C_{15}H_4O$	16.31	1.44
$C_{15}H_6N$	16.69	1.30
$C_{15}H_{20}$	16.53	1.28
$C_{16}H_8$	17.42	1.42

201

	M + 1	M + 2
$C_8H_{13}N_2O_4$	9.77	1.23
$C_8H_{15}N_3O_3$	10.15	1.07
$C_8H_{17}N_4O_2$	10.52	0.90
$C_9HN_2O_4$	10.66	1.32
$C_9H_3N_3O_3$	11.04	1.16
$C_9H_5N_4O_2$	11.41	1.00
$C_9H_{15}NO_4$	10.50	1.30
$C_9H_{17}N_2O_3$	10.88	1.14
$C_9H_{19}N_3O_2$	11.25	0.98
$C_9H_{21}N_4O$	11.63	0.82
$C_{10}H_3NO_4$	11.39	1.39
$C_{10}H_5N_2O_3$	11.77	1.23
$C_{10}H_7N_3O_2$	12.14	1.08
$C_{10}H_9N_4O$	12.52	0.92
$C_{10}H_{17}O_4$	11.23	1.37
$C_{10}H_{19}NO_3$	11.61	1.22
$C_{10}H_{21}N_2O_2$	11.98	1.06
$C_{10}H_{23}N_3O$	12.36	0.90
$C_{10}H_{25}N_4$	12.73	0.75
$C_{11}H_5O_4$	12.12	1.47
$C_{11}H_7NO_3$	12.50	1.32
$C_{11}H_9N_2O_2$	12.87	1.16
$C_{11}H_{11}N_3O$	13.25	1.01
$C_{11}H_{13}N_4$	13.62	0.86
$C_{11}H_{21}O_3$	12.34	1.30
$C_{11}H_{23}NO_2$	12.71	1.14
$C_{11}H_{25}N_2O$	13.09	0.99
$C_{11}H_{27}N_3$	13.46	0.84
$C_{12}HN_4$	14.51	0.98
$C_{12}H_9O_3$	13.23	1.41
$C_{12}H_{11}NO_2$	13.60	1.26
$C_{12}H_{13}N_2O$	13.98	1.11
$C_{12}H_{15}N_3$	14.35	0.96
$C_{12}H_{25}O_2$	13.45	1.23
$C_{12}H_{27}NO$	13.82	1.08
$C_{13}HN_2O$	14.87	1.23
$C_{13}H_3N_3$	15.24	1.08
$C_{13}H_{13}O_2$	14.33	1.35
$C_{13}H_{15}NO$	14.71	1.21
$C_{13}H_{17}N_2$	15.08	1.06
$C_{14}HO_2$	15.22	1.48
$C_{14}H_3NO$	15.60	1.33
$C_{14}H_5N_2$	15.97	1.19
$C_{14}H_{17}O$	15.44	1.31
$C_{14}H_{19}N$	15.81	1.17
$C_{15}H_5O$	16.33	1.45
$C_{15}H_7N$	16.70	1.31
$C_{15}H_{21}$	16.55	1.28
$C_{16}H_9$	17.43	1.43

202

	M + 1	M + 2
$C_8H_{14}N_2O_4$	9.79	1.23
$C_8H_{16}N_3O_3$	10.16	1.07
$C_8H_{18}N_4O_2$	10.54	0.91
$C_9H_2N_2O_4$	10.68	1.32
$C_9H_4N_3O_3$	11.05	1.16
$C_9H_6N_4O_2$	11.43	1.00
$C_9H_{16}NO_4$	10.52	1.30
$C_9H_{18}N_2O_3$	10.89	1.14
$C_9H_{20}N_3O_2$	11.27	0.98
$C_9H_{22}N_4O$	11.64	0.82
$C_{10}H_4NO_4$	11.41	1.39
$C_{10}H_6N_2O_3$	11.78	1.24
$C_{10}H_8N_3O_2$	12.16	1.08
$C_{10}H_{10}N_4O$	12.53	0.92
$C_{10}H_{18}O_4$	11.25	1.38
$C_{10}H_{20}NO_3$	11.63	1.22
$C_{10}H_{22}N_2O_2$	12.00	1.06
$C_{10}H_{24}N_3O$	12.37	0.91
$C_{10}H_{26}N_4$	12.75	0.75
$C_{11}H_6O_4$	12.14	1.47
$C_{11}H_8NO_3$	12.51	1.32
$C_{11}H_{10}N_2O_2$	12.89	1.17
$C_{11}H_{12}N_3O$	13.26	1.01
$C_{11}H_{14}N_4$	13.64	0.86
$C_{11}H_{22}O_3$	12.36	1.30
$C_{11}H_{24}NO_2$	12.73	1.15
$C_{11}H_{26}N_2O$	13.11	0.99
$C_{12}H_2N_4$	14.53	0.98
$C_{12}H_{10}O_3$	13.25	1.41
$C_{12}H_{12}NO_2$	13.62	1.26
$C_{12}H_{14}N_2O$	13.99	1.11
$C_{12}H_{16}N_3$	14.37	0.96
$C_{12}H_{26}O_2$	13.46	1.24
$C_{13}H_2N_2O$	14.88	1.23
$C_{13}H_4N_3$	15.26	1.09
$C_{13}H_{14}O_2$	14.35	1.35
$C_{13}H_{16}NO$	14.72	1.21
$C_{13}H_{18}N_2$	15.10	1.06
$C_{14}H_2O_2$	15.24	1.48
$C_{14}H_4NO$	15.61	1.34
$C_{14}H_6N_2$	15.99	1.19
$C_{14}H_{18}O$	15.46	1.31
$C_{14}H_{20}N$	15.83	1.17
$C_{15}H_6O$	16.34	1.45
$C_{15}H_8N$	16.72	1.31
$C_{15}H_{22}$	16.56	1.28
$C_{16}H_{10}$	17.45	1.43

203

	M + 1	M + 2
$C_8H_{15}N_2O_4$	9.80	1.23
$C_8H_{17}N_3O_3$	10.18	1.07
$C_8H_{19}N_4O_2$	10.55	0.91
$C_9H_3N_2O_4$	10.69	1.32
$C_9H_5N_3O_3$	11.07	1.16
$C_5H_7N_4O_2$	11.44	1.00
$C_9H_{17}NO_4$	10.54	1.30
$C_9H_{19}N_2O_3$	10.91	1.14
$C_9H_{21}N_3O_2$	11.28	0.98
$C_9H_{23}N_4O$	11.66	0.82
$C_{10}H_5NO_4$	11.42	1.40
$C_{10}H_7N_2O_3$	11.80	1.24
$C_{10}H_9N_3O_2$	12.17	1.08
$C_{10}H_{11}N_4O$	12.55	0.93
$C_{10}H_{19}O_4$	11.27	1.38
$C_{10}H_{21}NO_3$	11.64	1.22
$C_{10}H_{23}N_2O_2$	12.02	1.06
$C_{10}H_{25}N_3O$	12.39	0.91
$C_{11}H_7O_4$	12.16	1.48
$C_{11}H_9NO_3$	12.53	1.32
$C_{11}H_{11}N_2O_2$	12.90	1.17
$C_{11}H_{13}N_3O$	13.28	1.02
$C_{11}H_{15}N_4$	13.65	0.86
$C_{11}H_{23}O_3$	12.37	1.30
$C_{11}H_{25}NO_2$	12.75	1.15
$C_{12}HN_3O$	14.17	1.13
$C_{12}H_3N_4$	14.54	0.98
$C_{12}H_{11}O_3$	13.26	1.41
$C_{12}H_{13}NO_2$	13.64	1.26
$C_{12}H_{15}N_2O$	14.01	1.11
$C_{12}H_{17}N_3$	14.38	0.96
$C_{13}HNO_2$	14.52	1.38
$C_{13}H_3N_2O$	14.90	1.23
$C_{13}H_5N_3$	15.27	1.09
$C_{13}H_{15}O_2$	14.37	1.36
$C_{13}H_{17}NO$	14.74	1.21
$C_{13}H_{19}N_2$	15.12	1.06
$C_{14}H_3O_2$	15.26	1.48
$C_{14}H_5NO$	15.63	1.34
$C_{14}H_7N_2$	16.00	1.20
$C_{14}H_{19}O$	15.47	1.32
$C_{14}H_{21}N$	15.85	1.17
$C_{15}H_7O$	16.36	1.45
$C_{15}H_9N$	16.73	1.31
$C_{15}H_{23}$	16.58	1.29
$C_{16}H_{11}$	17.47	1.43

204

	M + 1	M + 2
$C_8H_{16}N_2O_4$	9.82	1.24
$C_8H_{18}N_3O_3$	10.19	1.07
$C_8H_{20}N_4O_2$	10.57	0.91
$C_9H_4N_2O_4$	10.71	1.32

	M+1	M+2		M+1	M+2		M+1	M+2		M+1	M+2
C$_9$H$_6$N$_3$O$_3$	11.08	1.16	C$_{11}$H$_{17}$N$_4$	13.68	0.87	C$_{14}$H$_{22}$O	15.52	1.32	C$_{11}$H$_4$N$_4$O	13.52	1.05
C$_9$H$_8$N$_4$O$_2$	11.46	1.00	C$_{12}$HN$_2$O$_2$	13.82	1.29	C$_{14}$H$_{24}$N	15.89	1.18	C$_{11}$H$_{12}$O$_4$	12.24	1.49
C$_9$H$_{18}$NO$_4$	10.55	1.31	C$_{12}$H$_3$N$_3$O	14.20	1.14	C$_{15}$H$_{10}$O	16.41	1.46	C$_{11}$H$_{14}$NO$_3$	12.61	1.33
C$_9$H$_{20}$N$_2$O$_3$	10.93	1.14	C$_{12}$H$_5$N$_4$	14.57	0.99	C$_{15}$H$_{12}$N	16.78	1.32	C$_{11}$H$_{16}$N$_2$O$_2$	12.98	1.18
C$_9$H$_{22}$N$_3$O$_2$	11.30	0.98	C$_{12}$H$_{13}$O$_3$	13.29	1.41	C$_{15}$H$_{26}$	16.63	1.29	C$_{11}$H$_{18}$N$_3$O	13.36	1.03
C$_9$H$_{24}$N$_4$O	11.67	0.83	C$_{12}$H$_{15}$NO$_2$	13.67	1.26	C$_{16}$H$_{14}$	17.51	1.44	C$_{11}$H$_{20}$N$_4$	13.73	0.88
C$_{10}$H$_6$NO$_4$	11.44	1.40	C$_{12}$H$_{17}$N$_2$O	14.04	1.11	C$_{17}$H$_2$	18.40	1.59	C$_{12}$H$_2$NO$_3$	13.50	1.44
C$_{10}$H$_8$N$_2$O$_3$	11.81	1.24	C$_{12}$H$_{19}$N$_3$	14.42	0.97				C$_{12}$H$_4$N$_2$O$_2$	13.87	1.29
C$_{10}$H$_{10}$N$_3$O$_2$	12.19	1.08	C$_{13}$HO$_3$	14.18	1.53	**207**			C$_{12}$H$_6$N$_3$O	14.25	1.14
C$_{10}$H$_{12}$N$_4$O	12.56	0.93	C$_{13}$H$_3$NO$_2$	14.56	1.38	C$_8$H$_{19}$N$_2$O$_4$	9.87	1.24	C$_{12}$H$_8$N$_4$	14.62	1.00
C$_{10}$H$_{20}$O$_4$	11.28	1.38	C$_{13}$H$_5$N$_2$O	14.93	1.24	C$_8$H$_{21}$N$_3$O$_3$	10.24	1.08	C$_{12}$H$_{16}$O$_3$	13.34	1.42
C$_{10}$H$_{22}$NO$_3$	11.66	1.22	C$_{13}$H$_7$N$_3$	15.30	1.09	C$_9$H$_7$N$_2$O$_4$	10.76	1.33	C$_{12}$H$_{18}$NO$_2$	13.72	1.27
C$_{10}$H$_{24}$N$_2$O$_2$	12.03	1.06	C$_{13}$H$_{17}$O$_2$	14.40	1.36	C$_9$H$_9$N$_3$O$_3$	11.13	1.17	C$_{12}$H$_{20}$N$_2$O	14.09	1.12
C$_{11}$H$_8$O$_4$	12.17	1.48	C$_{13}$H$_{19}$NO	14.77	1.21	C$_9$H$_{11}$N$_4$O$_2$	11.51	1.01	C$_{12}$H$_{22}$N$_3$	14.46	0.97
C$_{11}$H$_{10}$NO$_3$	12.55	1.32	C$_{13}$H$_{21}$N$_2$	15.15	1.07	C$_9$H$_{21}$NO$_4$	10.60	1.31	C$_{13}$H$_4$O$_3$	14.23	1.54
C$_{11}$H$_{12}$N$_2$O$_2$	12.92	1.17	C$_{14}$H$_5$O$_2$	15.29	1.49	C$_{10}$H$_9$NO$_4$	11.49	1.40	C$_{13}$H$_6$NO$_2$	14.60	1.39
C$_{11}$H$_{14}$N$_3$O	13.29	1.02	C$_{14}$H$_7$NO	15.66	1.34	C$_{10}$H$_{11}$N$_2$O$_3$	11.86	1.25	C$_{13}$H$_8$N$_2$O	14.98	1.24
C$_{11}$H$_{16}$N$_4$	13.67	0.87	C$_{14}$H$_9$N$_2$	16.04	1.20	C$_{10}$H$_{13}$N$_3$O$_2$	12.24	1.09	C$_{13}$H$_{10}$N$_3$	15.35	1.10
C$_{11}$H$_{24}$O$_3$	12.39	1.30	C$_{14}$H$_{21}$O	15.50	1.32	C$_{11}$H$_{15}$N$_4$O	12.61	0.93	C$_{13}$H$_{20}$O$_2$	14.45	1.37
C$_{12}$H$_2$N$_3$O	14.18	1.13	C$_{14}$H$_{23}$N	15.88	1.18	C$_{11}$HN$_3$O$_2$	13.13	1.20	C$_{13}$H$_{22}$NO	14.82	1.22
C$_{12}$H$_4$N$_4$	14.56	0.99	C$_{15}$H$_9$O	16.39	1.46	C$_{11}$H$_3$N$_4$O	13.50	1.04	C$_{13}$H$_{24}$N$_2$	15.20	1.08
C$_{12}$H$_{12}$O$_3$	13.28	1.41	C$_{15}$H$_{11}$N	16.77	1.32	C$_{11}$H$_{11}$O$_4$	12.22	1.48	C$_{14}$H$_8$O$_2$	15.34	1.50
C$_{12}$H$_{14}$NO$_2$	13.65	1.26	C$_{15}$H$_{25}$	16.61	1.29	C$_{11}$H$_{13}$NO$_3$	12.59	1.33	C$_{14}$H$_{10}$NO	15.71	1.35
C$_{12}$H$_{16}$N$_2$O	14.03	1.11	C$_{16}$H$_{13}$	17.50	1.44	C$_{11}$H$_{15}$N$_2$O$_2$	12.97	1.18	C$_{14}$H$_{12}$N$_2$	16.08	1.21
C$_{12}$H$_{18}$N$_3$	14.40	0.96	C$_{17}$H	18.39	1.59	C$_{11}$H$_{17}$N$_3$O	13.34	1.02	C$_{14}$H$_{24}$O	15.55	1.33
C$_{13}$H$_2$NO$_2$	14.54	1.38				C$_{11}$H$_{19}$N$_4$	13.72	0.87	C$_{14}$H$_{26}$N	15.93	1.19
C$_{13}$H$_4$N$_2$O	14.91	1.24	**206**			C$_{12}$HNO$_3$	13.48	1.44	C$_{15}$H$_{12}$O	16.44	1.46
C$_{13}$H$_6$N$_3$	15.29	1.09	C$_8$H$_{18}$N$_2$O$_4$	9.85	1.24	C$_{12}$H$_3$N$_2$O$_2$	13.86	1.29	C$_{15}$H$_{14}$N	16.81	1.33
C$_{13}$H$_{16}$O$_2$	14.38	1.36	C$_8$H$_{20}$N$_3$O$_3$	10.23	1.08	C$_{12}$H$_5$N$_3$O	14.23	1.14	C$_{15}$H$_{28}$	16.66	1.30
C$_{13}$H$_{18}$NO	14.76	1.21	C$_8$H$_{22}$N$_4$O$_2$	10.60	0.91	C$_{12}$H$_7$N$_4$	14.61	0.99	C$_{16}$H$_2$N	17.70	1.47
C$_{13}$H$_{20}$N$_2$	15.13	1.07	C$_9$H$_6$N$_2$O$_4$	10.74	1.32	C$_{12}$H$_{15}$O$_3$	13.33	1.42	C$_{16}$H$_{16}$	17.55	1.45
C$_{14}$H$_4$O$_2$	15.27	1.49	C$_9$H$_8$N$_3$O$_3$	11.12	1.16	C$_{12}$H$_{17}$NO$_2$	13.70	1.27	C$_{17}$H$_4$	18.43	1.60
C$_{14}$H$_6$NO	15.65	1.34	C$_9$H$_{10}$N$_4$O$_2$	11.49	1.01	C$_{12}$H$_{19}$N$_2$O	14.07	1.12			
C$_{14}$H$_8$N$_2$	16.02	1.20	C$_9$H$_{20}$NO$_4$	10.58	1.31	C$_{12}$H$_{21}$N$_3$	14.45	0.97	**209**		
C$_{14}$H$_{20}$O	15.49	1.32	C$_9$H$_{22}$N$_2$O$_3$	10.96	1.15	C$_{13}$H$_3$O$_3$	14.21	1.54	C$_9$H$_9$N$_2$O$_4$	10.79	1.33
C$_{14}$H$_{22}$N	15.86	1.18	C$_{10}$H$_8$NO$_4$	11.47	1.40	C$_{13}$H$_5$NO$_2$	14.59	1.39	C$_9$H$_{11}$N$_3$O$_3$	11.16	1.17
C$_{15}$H$_8$O	16.38	1.45	C$_{10}$H$_{10}$N$_2$O$_3$	11.85	1.24	C$_{13}$H$_7$N$_2$O	14.96	1.24	C$_9$H$_{13}$N$_4$O$_2$	11.54	1.01
C$_{15}$H$_{10}$N	16.75	1.31	C$_{10}$H$_{12}$N$_3$O$_2$	12.22	1.09	C$_{13}$H$_9$N$_3$	15.34	1.10	C$_{10}$HN$_4$O$_2$	12.43	1.11
C$_{15}$H$_{24}$	16.59	1.29	C$_{10}$H$_{14}$N$_4$O	12.60	0.93	C$_{13}$H$_{19}$O$_2$	14.43	1.37	C$_{10}$H$_{11}$NO$_4$	11.52	1.41
C$_{16}$H$_{12}$	17.48	1.43	C$_{10}$H$_{22}$O$_4$	11.31	1.38	C$_{13}$H$_{21}$NO	14.80	1.22	C$_{10}$H$_{13}$N$_2$O$_3$	11.89	1.25
			C$_{11}$H$_2$N$_4$O	13.48	1.04	C$_{13}$H$_{23}$N$_2$	15.18	1.07	C$_{10}$H$_{15}$N$_3$O$_2$	12.27	1.09
			C$_{11}$H$_{10}$O$_4$	12.20	1.48	C$_{14}$H$_7$O$_2$	15.32	1.49	C$_{10}$H$_{17}$N$_4$O	12.64	0.94
205			C$_{11}$H$_{12}$NO$_3$	12.58	1.33	C$_{14}$H$_9$NO	15.69	1.35	C$_{11}$HN$_2$O$_3$	12.78	1.35
C$_8$H$_{17}$N$_2$O$_4$	9.84	1.24	C$_{11}$H$_{14}$N$_2$O$_2$	12.95	1.17	C$_{14}$H$_{11}$N$_2$	16.07	1.21	C$_{11}$H$_3$N$_3$O$_2$	13.16	1.20
C$_8$H$_{19}$N$_3$O$_3$	10.21	1.07	C$_{11}$H$_{16}$N$_3$O	13.33	1.02	C$_{14}$H$_{23}$O	15.54	1.33	C$_{11}$H$_5$N$_4$O	13.53	1.05
C$_8$H$_{21}$N$_4$O$_2$	10.59	0.91	C$_{11}$H$_{18}$N$_4$	13.70	0.87	C$_{14}$H$_{25}$N	15.91	1.18	C$_{11}$H$_{13}$O$_4$	12.25	1.49
C$_9$H$_5$N$_2$O$_4$	10.73	1.32	C$_{12}$H$_2$N$_2$O$_2$	13.84	1.29	C$_{15}$H$_{11}$O	16.42	1.46	C$_{11}$H$_{15}$NO$_3$	12.63	1.33
C$_9$H$_7$N$_3$O$_3$	11.10	1.16	C$_{12}$H$_4$N$_3$O	14.22	1.14	C$_{15}$H$_{13}$N	16.80	1.32	C$_{11}$H$_{17}$N$_2$O$_2$	13.00	1.18
C$_9$H$_9$N$_4$O$_2$	11.47	1.00	C$_{12}$H$_6$N$_4$	14.59	0.99	C$_{15}$H$_{27}$	16.64	1.30	C$_{11}$H$_{19}$N$_3$O	13.37	1.03
C$_9$H$_{19}$NO$_4$	10.57	1.31	C$_{12}$H$_{14}$O$_3$	13.31	1.42	C$_{16}$HN	17.69	1.47	C$_{11}$H$_{21}$N$_4$	13.75	0.88
C$_9$H$_{21}$N$_2$O$_3$	10.94	1.15	C$_{12}$H$_{16}$NO$_2$	13.68	1.27	C$_{16}$H$_{15}$	17.53	1.44	C$_{12}$HO$_4$	13.14	1.60
C$_9$H$_{23}$N$_3$O$_2$	11.32	0.99	C$_{12}$H$_{18}$N$_2$O	14.06	1.12	C$_{17}$H$_3$	18.42	1.60	C$_{12}$H$_3$NO$_3$	13.51	1.44
C$_{10}$H$_7$NO$_4$	11.46	1.40	C$_{12}$H$_{20}$N$_3$	14.43	0.97				C$_{12}$H$_5$N$_2$O$_2$	13.89	1.29
C$_{10}$H$_9$N$_2$O$_3$	11.83	1.24	C$_{13}$H$_2$O$_3$	14.20	1.53	**208**			C$_{12}$H$_7$N$_3$O	14.26	1.15
C$_{10}$H$_{11}$N$_3$O$_2$	12.21	1.09	C$_{13}$H$_4$NO$_2$	14.57	1.39	C$_8$H$_{20}$N$_2$O$_4$	9.88	1.24	C$_{12}$H$_9$N$_4$	14.64	1.00
C$_{10}$H$_{13}$N$_4$O	12.58	0.93	C$_{13}$H$_6$N$_2$O	14.95	1.24	C$_9$H$_8$N$_2$O$_4$	10.77	1.33	C$_{12}$H$_{17}$O$_3$	13.36	1.42
C$_{10}$H$_{21}$O$_4$	11.30	1.38	C$_{13}$H$_8$N$_3$	15.32	1.10	C$_9$H$_{10}$N$_3$O$_3$	11.15	1.17	C$_{12}$H$_{19}$NO$_2$	13.73	1.27
C$_{10}$H$_{23}$NO$_3$	11.67	1.22	C$_{13}$H$_{18}$O$_2$	14.41	1.36	C$_9$H$_{12}$N$_4$O$_2$	11.52	1.01	C$_{12}$H$_{21}$N$_2$O	14.11	1.12
C$_{11}$HN$_4$O	13.47	1.04	C$_{13}$H$_{20}$NO	14.79	1.22	C$_{10}$H$_{10}$NO$_4$	11.50	1.40	C$_{12}$H$_{23}$N$_3$	14.48	0.98
C$_{11}$H$_9$O$_4$	12.19	1.48	C$_{13}$H$_{22}$N$_2$	15.16	1.07	C$_{10}$H$_{12}$N$_2$O$_3$	11.88	1.25	C$_{13}$H$_5$O$_3$	14.25	1.54
C$_{11}$H$_{11}$NO$_3$	12.56	1.33	C$_{14}$H$_6$O$_2$	15.30	1.49	C$_{10}$H$_{14}$N$_3$O$_2$	12.25	1.09	C$_{13}$H$_7$NO$_2$	14.62	1.39
C$_{11}$H$_{13}$N$_2$O$_2$	12.94	1.17	C$_{14}$H$_8$NO	15.68	1.35	C$_{10}$H$_{16}$N$_4$O	12.63	0.94	C$_{13}$H$_9$N$_2$O	14.99	1.25
C$_{11}$H$_{15}$N$_3$O	13.31	1.02	C$_{14}$H$_{10}$N$_2$	16.05	1.21	C$_{11}$H$_2$N$_3$O$_2$	13.14	1.20	C$_{13}$H$_{11}$N$_3$	15.37	1.10

	M + 1	M + 2
$C_{13}H_{21}O_2$	14.46	1.37
$C_{13}H_{23}NO$	14.84	1.22
$C_{13}H_{25}N_2$	15.21	1.08
$C_{14}H_9O_2$	15.35	1.50
$C_{14}H_{11}NO$	15.73	1.35
$C_{14}H_{13}N_2$	16.10	1.21
$C_{14}H_{25}O$	15.57	1.33
$C_{14}H_{27}N$	15.94	1.19
$C_{15}HN_2$	16.99	1.35
$C_{15}H_{13}O$	16.46	1.47
$C_{15}H_{15}N$	16.83	1.33
$C_{15}H_{29}$	16.67	1.30
$C_{16}HO$	17.35	1.61
$C_{16}H_3N$	17.72	1.48
$C_{16}H_{17}$	17.56	1.45
$C_{17}H_5$	18.45	1.60

210

	M + 1	M + 2
$C_9H_{10}N_2O_4$	10.81	1.33
$C_9H_{12}N_3O_3$	11.18	1.17
$C_9H_{14}N_4O_2$	11.55	1.01
$C_{10}H_2N_4O_2$	12.44	1.11
$C_{10}H_{12}NO_4$	11.54	1.41
$C_{10}H_{14}N_2O_3$	11.91	1.25
$C_{10}H_{16}N_3O_2$	12.29	1.09
$C_{10}H_{18}N_4O$	12.66	0.94
$C_{11}H_2N_2O_3$	12.80	1.35
$C_{11}H_4N_3O_2$	13.17	1.20
$C_{11}H_6N_4O$	13.55	1.05
$C_{11}H_{14}O_4$	12.27	1.49
$C_{11}H_{16}NO_3$	12.64	1.34
$C_{11}H_{18}N_2O_2$	13.02	1.18
$C_{11}H_{20}N_3O$	13.39	1.03
$C_{11}H_{22}N_4$	13.76	0.88
$C_{12}H_2O_4$	13.16	1.60
$C_{12}H_4NO_3$	13.53	1.45
$C_{12}H_6N_2O_2$	13.90	1.30
$C_{12}H_8N_3O$	14.28	1.15
$C_{12}H_{10}N_4$	14.65	1.00
$C_{12}H_{18}O_3$	13.37	1.43
$C_{12}H_{20}NO_2$	13.75	1.28
$C_{12}H_{22}N_2O$	14.12	1.13
$C_{12}H_{24}N_3$	14.50	0.98
$C_{13}H_6O_3$	14.26	1.54
$C_{13}H_8NO_2$	14.64	1.40
$C_{13}H_{10}N_2O$	15.01	1.25
$C_{13}H_{12}N_3$	15.38	1.11
$C_{13}H_{22}O_2$	14.48	1.37
$C_{13}H_{24}NO$	14.85	1.23
$C_{13}H_{26}N_2$	15.23	1.08
$C_{14}H_{10}O_2$	15.37	1.50
$C_{14}H_{12}NO$	15.74	1.36
$C_{14}H_{14}N_2$	16.12	1.22
$C_{14}H_{26}O$	15.58	1.33
$C_{14}H_{28}N$	15.96	1.19
$C_{15}H_2N_2$	17.00	1.36
$C_{15}H_{14}O$	16.47	1.47
$C_{15}H_{16}N$	16.85	1.33
$C_{15}H_{30}$	16.69	1.31
$C_{16}H_2O$	17.36	1.61
$C_{16}H_4N$	17.74	1.48
$C_{16}H_{18}$	17.58	1.45
$C_{17}H_6$	18.47	1.61

211

	M + 1	M + 2
$C_9H_{11}N_2O_4$	10.82	1.33
$C_9H_{13}N_3O_3$	11.20	1.17
$C_9H_{15}N_4O_2$	11.57	1.01
$C_{10}HN_3O_3$	12.08	1.27
$C_{10}H_3N_4O_2$	12.46	1.12
$C_{10}H_{13}NO_4$	11.55	1.41
$C_{10}H_{15}N_2O_3$	11.93	1.25
$C_{10}H_{17}N_3O_2$	12.30	1.10
$C_{10}H_{19}N_4O$	12.68	0.94
$C_{11}HNO_4$	12.44	1.51
$C_{11}H_3N_2O_3$	12.82	1.36
$C_{11}H_5N_3O_2$	13.19	1.20
$C_{11}H_7N_4O$	13.56	1.05
$C_{11}H_{15}O_4$	12.28	1.49
$C_{11}H_{17}NO_3$	12.66	1.34
$C_{11}H_{19}N_2O_2$	13.03	1.18
$C_{11}H_{21}N_3O$	13.41	1.03
$C_{11}H_{23}N_4$	13.78	0.88
$C_{12}H_3O_4$	13.17	1.60
$C_{12}H_5NO_3$	13.55	1.45
$C_{12}H_7N_2O_2$	13.92	1.30
$C_{12}H_9N_3O$	14.30	1.15
$C_{12}H_{11}N_4$	14.67	1.60
$C_{12}H_{19}O_3$	13.39	1.43
$C_{12}H_{21}NO_2$	13.76	1.28
$C_{12}H_{23}N_2O$	14.14	1.13
$C_{12}H_{25}N_3$	14.51	0.98
$C_{13}H_7O_3$	14.28	1.54
$C_{13}H_9NO_2$	14.65	1.40
$C_{13}H_{11}N_2O$	15.03	1.25
$C_{13}H_{13}N_3$	15.40	1.11
$C_{13}H_{23}O_2$	14.49	1.38
$C_{13}H_{25}NO$	14.87	1.23
$C_{13}H_{27}N_2$	15.24	1.08
$C_{14}HN_3$	16.29	1.24
$C_{14}H_{11}O_2$	15.38	1.50
$C_{14}H_{13}NO$	15.76	1.36
$C_{14}H_{15}N_2$	16.13	1.22
$C_{14}H_{27}O$	15.60	1.34
$C_{14}H_{29}N$	15.97	1.19
$C_{15}HNO$	16.65	1.50
$C_{15}H_3N_2$	17.02	1.36
$C_{15}H_{15}O$	16.49	1.47
$C_{15}H_{17}N$	16.86	1.33
$C_{15}H_{31}$	16.71	1.31
$C_{16}H_3O$	17.38	1.62
$C_{16}H_5N$	17.75	1.48
$C_{16}H_{19}$	17.59	1.45
$C_{17}H_7$	18.48	1.61

212

	M + 1	M + 2
$C_9H_{12}N_2O_4$	10.84	1.34
$C_9H_{14}N_3O_3$	11.21	1.18
$C_9H_{16}N_4O_2$	11.59	1.02
$C_{10}H_2N_3O_3$	12.10	1.27
$C_{10}H_4N_4O_2$	12.47	1.12
$C_{10}H_{14}NO_4$	11.57	1.41
$C_{10}H_{16}N_2O_3$	11.94	1.25
$C_{10}H_{18}N_3O_2$	12.32	1.10
$C_{10}H_{20}N_4O$	12.69	0.94
$C_{11}H_2NO_4$	12.46	1.51
$C_{11}H_4N_2O_3$	12.83	1.36
$C_{11}H_6N_3O_2$	13.21	1.21
$C_{11}H_8N_4O$	13.58	1.06
$C_{11}H_{16}O_4$	12.30	1.49
$C_{11}H_{18}NO_3$	12.67	1.34
$C_{11}H_{20}N_2O_2$	13.05	1.19
$C_{11}H_{22}N_3O$	13.42	1.03
$C_{11}H_{24}N_4$	13.80	0.88
$C_{12}H_4O_4$	13.19	1.60
$C_{12}H_6NO_3$	13.56	1.45
$C_{12}H_8N_2O_2$	13.94	1.30
$C_{12}H_{10}N_3O$	14.31	1.15
$C_{12}H_{12}N_4$	14.69	1.01
$C_{12}H_{20}O_3$	13.41	1.43
$C_{12}H_{22}NO_2$	13.78	1.28
$C_{12}H_{24}N_2O$	14.15	1.13
$C_{12}H_{26}N_3$	14.53	0.98
$C_{13}H_8O_3$	14.29	1.55
$C_{13}H_{10}NO_2$	14.67	1.40
$C_{13}H_{12}N_2O$	15.04	1.25
$C_{13}H_{14}N_3$	15.42	1.11
$C_{13}H_{24}O_2$	14.51	1.38
$C_{13}H_{26}NO$	14.88	1.23
$C_{13}H_{28}N_2$	15.26	1.09
$C_{14}H_2N_3$	16.31	1.25
$C_{14}H_{12}O_2$	15.40	1.50
$C_{14}H_{14}NO$	15.77	1.36
$C_{14}H_{16}N_2$	16.15	1.22
$C_{14}H_{28}O$	15.62	1.34
$C_{14}H_{30}N$	15.99	1.20
$C_{15}H_2NO$	16.66	1.50
$C_{15}H_4N_2$	17.04	1.36
$C_{15}H_{16}O$	16.50	1.47
$C_{15}H_{18}N$	16.88	1.34
$C_{15}H_{32}$	16.72	1.31
$C_{16}H_4O$	17.39	1.62
$C_{16}H_6N$	17.77	1.48
$C_{16}H_{20}$	17.61	1.46
$C_{17}H_8$	18.50	1.61

213

	M + 1	M + 2
$C_9H_{13}N_2O_4$	10.86	1.34
$C_9H_{15}N_3O_3$	11.23	1.18
$C_9H_{17}N_4O_2$	11.60	1.02
$C_{10}HN_2O_4$	11.74	1.43
$C_{10}H_3N_3O_3$	12.12	1.27
$C_{10}H_5N_4O_2$	12.49	1.12
$C_{10}H_{15}NO_4$	11.58	1.41
$C_{10}H_{17}N_2O_3$	11.96	1.26
$C_{10}H_{19}N_3O_2$	12.33	1.10
$C_{10}H_{21}N_4O$	12.71	0.95
$C_{11}H_3NO_4$	12.47	1.51
$C_{11}H_5N_2O_3$	12.85	1.36
$C_{11}H_7N_3O_2$	13.22	1.21
$C_{11}H_9N_4O$	13.60	1.06
$C_{11}H_{17}O_4$	12.32	1.50
$C_{11}H_{19}NO_3$	12.69	1.34
$C_{11}H_{21}N_2O_2$	13.06	1.19
$C_{11}H_{23}N_3O$	13.44	1.04
$C_{11}H_{25}N_4$	13.81	0.89
$C_{12}H_5O_4$	13.20	1.60
$C_{12}H_7NO_3$	13.58	1.45
$C_{12}H_9N_2O_2$	13.95	1.30
$C_{12}H_{11}N_3O$	14.33	1.15

214

	M + 1	M + 2
$C_{12}H_{13}N_4$	14.70	1.01
$C_{12}H_{21}O_3$	13.42	1.43
$C_{12}H_{23}NO_2$	13.80	1.28
$C_{12}H_{25}N_2O$	14.17	1.13
$C_{12}H_{27}N_3$	14.54	0.99
$C_{13}HN_4$	15.59	1.14
$C_{13}H_9O_3$	14.31	1.55
$C_{13}H_{11}NO_2$	14.68	1.40
$C_{13}H_{13}N_2O$	15.06	1.26
$C_{13}H_{15}N_3$	15.43	1.11
$C_{13}H_{25}O_2$	14.53	1.38
$C_{13}H_{27}NO$	14.90	1.23
$C_{13}H_{29}N_2$	15.28	1.09
$C_{14}HN_2O$	15.95	1.39
$C_{14}H_3N_3$	16.32	1.25
$C_{14}H_{13}O_2$	15.42	1.51
$C_{14}H_{15}NO$	15.79	1.36
$C_{14}H_{17}N_2$	16.16	1.22
$C_{14}H_{29}O$	15.63	1.34
$C_{14}H_{31}N$	16.01	1.20
$C_{15}HO_2$	16.30	1.64
$C_{15}H_3NO$	16.68	1.50
$C_{15}H_5N_2$	17.05	1.36
$C_{15}H_{17}O$	16.52	1.48
$C_{15}H_{19}N$	16.90	1.34
$C_{16}H_5O$	17.41	1.62
$C_{16}H_7N$	17.78	1.49
$C_{16}H_{21}$	17.63	1.46
$C_{17}H_9$	18.51	1.61

	M + 1	M + 2
$C_9H_{14}N_2O_4$	10.87	1.34
$C_9H_{16}N_3O_3$	11.24	1.18
$C_9H_{18}N_4O_2$	11.62	1.02
$C_{10}H_2N_2O_4$	11.76	1.43
$C_{10}H_4N_3O_3$	12.13	1.28
$C_{10}H_6N_4O_2$	12.51	1.12
$C_{10}H_{16}NO_4$	11.60	1.42
$C_{10}H_{18}N_2O_3$	11.97	1.26
$C_{10}H_{20}N_3O_2$	12.35	1.10
$C_{10}H_{22}N_4O$	12.72	0.95
$C_{11}H_4NO_4$	12.49	1.52
$C_{11}H_6N_2O_3$	12.86	1.36
$C_{11}H_8N_3O_2$	13.24	1.21
$C_{11}H_{10}N_4O$	13.61	1.06
$C_{11}H_{18}O_4$	12.33	1.50
$C_{11}H_{20}NO_3$	12.71	1.34
$C_{11}H_{22}N_2O_2$	13.08	1.19
$C_{11}H_{24}N_3O$	13.45	1.04
$C_{11}H_{26}N_4$	13.83	0.89
$C_{12}H_6O_4$	13.22	1.61
$C_{12}H_8NO_3$	13.59	1.45
$C_{12}H_{10}N_2O_2$	13.97	1.31
$C_{12}H_{12}N_3O$	14.34	1.16
$C_{12}H_{14}N_4$	14.72	1.01
$C_{12}H_{22}O_3$	13.44	1.43
$C_{12}H_{24}NO_2$	13.81	1.28
$C_{12}H_{26}N_2O$	14.19	1.14
$C_{12}H_{28}N_3$	14.56	0.99
$C_{13}H_2N_4$	15.61	1.14
$C_{13}H_{10}O_3$	14.33	1.55
$C_{13}H_{12}NO_2$	14.70	1.40
$C_{13}H_{14}N_2O$	15.07	1.26

	M + 1	M + 2
C$_{13}$H$_{16}$N$_3$	15.45	1.12
C$_{13}$H$_{26}$O$_2$	14.54	1.38
C$_{13}$H$_{28}$NO	14.92	1.24
C$_{13}$H$_{30}$N$_2$	15.29	1.09
C$_{14}$H$_2$N$_2$O	15.96	1.39
C$_{14}$H$_4$N$_3$	16.34	1.25
C$_{14}$H$_{14}$O$_2$	15.43	1.51
C$_{14}$H$_{16}$NO	15.81	1.37
C$_{14}$H$_{18}$N$_2$	16.18	1.23
C$_{14}$H$_{30}$O	15.65	1.34
C$_{15}$H$_2$O$_2$	16.32	1.64
C$_{15}$H$_4$NO	16.69	1.51
C$_{15}$H$_6$N$_2$	17.07	1.37
C$_{15}$H$_{18}$O	16.54	1.48
C$_{15}$H$_{20}$N	16.91	1.34
C$_{16}$H$_6$O	17.43	1.63
C$_{16}$H$_8$N	17.80	1.49
C$_{16}$H$_{22}$	17.64	1.46
C$_{17}$H$_{10}$	18.53	1.62

215

	M + 1	M + 2
C$_9$H$_{15}$N$_2$O$_4$	10.89	1.34
C$_9$H$_{17}$N$_3$O$_3$	11.26	1.18
C$_9$H$_{19}$N$_4$O$_2$	11.63	1.02
C$_{10}$H$_3$N$_2$O$_4$	11.77	1.44
C$_{10}$H$_5$N$_3$O$_3$	12.15	1.28
C$_{10}$H$_7$N$_4$O$_2$	12.52	1.12
C$_{10}$H$_{17}$NO$_4$	11.62	1.42
C$_{10}$H$_{19}$N$_2$O$_3$	11.99	1.26
C$_{10}$H$_{21}$N$_3$O$_2$	12.37	1.10
C$_{10}$H$_{23}$N$_4$O	12.74	0.95
C$_{11}$H$_5$NO$_4$	12.50	1.52
C$_{11}$H$_7$N$_2$O$_3$	12.88	1.37
C$_{11}$H$_9$N$_3$O$_2$	13.25	1.21
C$_{11}$H$_{11}$N$_4$O	13.63	1.06
C$_{11}$H$_{19}$O$_4$	12.35	1.50
C$_{11}$H$_{21}$NO$_3$	12.72	1.35
C$_{11}$H$_{23}$N$_2$O$_2$	13.10	1.19
C$_{11}$H$_{25}$N$_3$O	13.47	1.04
C$_{11}$H$_{27}$N$_4$	13.84	0.89
C$_{12}$H$_7$O$_4$	13.24	1.61
C$_{12}$H$_9$NO$_3$	13.61	1.46
C$_{12}$H$_{11}$N$_2$O$_2$	13.98	1.31
C$_{12}$H$_{13}$N$_3$O	14.36	1.16
C$_{12}$H$_{15}$N$_4$	14.73	1.01
C$_{12}$H$_{23}$O$_3$	13.45	1.44
C$_{12}$H$_{25}$NO$_2$	13.83	1.29
C$_{12}$H$_{27}$N$_2$O	14.20	1.14
C$_{12}$H$_{29}$N$_3$	14.58	0.99
C$_{13}$HN$_3$O	15.25	1.28
C$_{13}$H$_3$N$_4$	15.62	1.14
C$_{13}$H$_{11}$O$_3$	14.34	1.55
C$_{13}$H$_{13}$NO$_2$	14.72	1.41
C$_{13}$H$_{15}$N$_2$O	15.09	1.26
C$_{13}$H$_{17}$N$_3$	15.46	1.12
C$_{13}$H$_{27}$O$_2$	14.56	1.38
C$_{13}$H$_{29}$NO	14.93	1.24
C$_{14}$HNO$_2$	15.60	1.54
C$_{14}$H$_3$N$_2$O	15.98	1.39
C$_{14}$H$_5$N$_3$	16.35	1.25
C$_{14}$H$_{15}$O$_2$	15.45	1.51
C$_{14}$H$_{17}$NO	15.82	1.37
C$_{14}$H$_{19}$N$_2$	16.20	1.23

	M + 1	M + 2
C$_{15}$H$_3$O$_2$	16.34	1.65
C$_{15}$H$_5$NO	16.71	1.51
C$_{15}$H$_7$N$_2$	17.08	1.37
C$_{15}$H$_{19}$O	16.55	1.48
C$_{15}$H$_{21}$N	16.93	1.34
C$_{16}$H$_7$O	17.44	1.63
C$_{16}$H$_9$N	17.82	1.49
C$_{16}$H$_{23}$	17.66	1.47
C$_{17}$H$_{11}$	18.55	1.62

216

	M + 1	M + 2
C$_9$H$_{16}$N$_2$O$_4$	10.90	1.34
C$_9$H$_{18}$N$_3$O$_3$	11.28	1.18
C$_9$H$_{20}$N$_4$O$_2$	11.65	1.02
C$_{10}$H$_4$N$_2$O$_4$	11.79	1.44
C$_{10}$H$_6$N$_3$O$_3$	12.16	1.28
C$_{10}$H$_8$N$_4$O$_2$	12.54	1.13
C$_{10}$H$_{18}$NO$_4$	11.63	1.42
C$_{10}$H$_{20}$N$_2$O$_3$	12.01	1.26
C$_{10}$H$_{22}$N$_3$O$_2$	12.38	1.11
C$_{10}$H$_{24}$N$_4$O	12.76	0.95
C$_{11}$H$_6$NO$_4$	12.52	1.52
C$_{11}$H$_8$N$_2$O$_3$	12.90	1.37
C$_{11}$H$_{10}$N$_3$O$_2$	13.27	1.21
C$_{11}$H$_{12}$N$_4$O	13.64	1.06
C$_{11}$H$_{20}$O$_4$	12.36	1.50
C$_{11}$H$_{22}$NO$_3$	12.74	1.35
C$_{11}$H$_{24}$N$_2$O$_2$	13.11	1.19
C$_{11}$H$_{26}$N$_3$O	13.49	1.04
C$_{11}$H$_{28}$N$_4$	13.86	0.89
C$_{12}$H$_8$O$_4$	13.25	1.61
C$_{12}$H$_{10}$NO$_3$	13.63	1.46
C$_{12}$H$_{12}$N$_2$O$_2$	14.00	1.31
C$_{12}$H$_{14}$N$_3$O	14.38	1.16
C$_{12}$H$_{16}$N$_4$	14.75	1.01
C$_{12}$H$_{24}$O$_3$	13.47	1.44
C$_{12}$H$_{26}$NO$_2$	13.84	1.29
C$_{12}$H$_{28}$N$_2$O	14.22	1.14
C$_{13}$H$_2$N$_3$O	15.26	1.29
C$_{13}$H$_4$N$_4$	15.64	1.14
C$_{13}$H$_{12}$O$_3$	14.36	1.56
C$_{13}$H$_{14}$NO$_2$	14.73	1.41
C$_{13}$H$_{16}$N$_2$O	15.11	1.26
C$_{13}$H$_{18}$N$_3$	15.48	1.12
C$_{13}$H$_{28}$O$_2$	14.57	1.39
C$_{14}$H$_2$NO$_2$	15.62	1.54
C$_{14}$H$_4$N$_2$O	15.99	1.40
C$_{14}$H$_6$N$_3$	16.37	1.26
C$_{14}$H$_{16}$O$_2$	15.46	1.51
C$_{14}$H$_{18}$NO	15.84	1.37
C$_{14}$H$_{20}$N$_2$	16.21	1.23
C$_{15}$H$_4$O$_2$	16.35	1.65
C$_{15}$H$_6$NO	16.73	1.51
C$_{15}$H$_8$N$_2$	17.10	1.37
C$_{15}$H$_{20}$O	16.57	1.49
C$_{15}$H$_{22}$N	16.94	1.35
C$_{16}$H$_8$O	17.46	1.63
C$_{16}$H$_{10}$N	17.83	1.50
C$_{16}$H$_{24}$	17.67	1.47
C$_{17}$H$_{12}$	18.56	1.62

217

	M + 1	M + 2
C$_9$H$_{17}$N$_2$O$_4$	10.92	1.34
C$_9$H$_{19}$N$_3$O$_3$	11.29	1.18
C$_9$H$_{21}$N$_4$O$_2$	11.67	1.03
C$_{10}$H$_5$N$_2$O$_4$	11.81	1.44
C$_{10}$H$_7$N$_3$O$_3$	12.18	1.28
C$_{10}$H$_9$N$_4$O$_2$	12.55	1.13
C$_{10}$H$_{19}$NO$_4$	11.65	1.42
C$_{10}$H$_{21}$N$_2$O$_3$	12.02	1.26
C$_{10}$H$_{23}$N$_3$O$_2$	12.40	1.11
C$_{10}$H$_{25}$N$_4$O	12.77	0.95
C$_{11}$H$_7$NO$_4$	12.54	1.52
C$_{11}$H$_9$N$_2$O$_3$	12.91	1.37
C$_{11}$H$_{11}$N$_3$O$_2$	13.29	1.22
C$_{11}$H$_{13}$N$_4$O	13.66	1.07
C$_{11}$H$_{21}$O$_4$	12.38	1.50
C$_{11}$H$_{23}$NO$_3$	12.75	1.35
C$_{11}$H$_{25}$N$_2$O$_2$	13.13	1.20
C$_{11}$H$_{27}$N$_3$O	13.50	1.05
C$_{12}$HN$_4$O	14.55	1.19
C$_{12}$H$_9$O$_4$	13.27	1.61
C$_{12}$H$_{11}$NO$_3$	13.64	1.46
C$_{12}$H$_{13}$N$_2$O$_2$	14.02	1.31
C$_{12}$H$_{15}$N$_3$O	14.39	1.16
C$_{12}$H$_{17}$N$_4$	14.77	1.02
C$_{12}$H$_{25}$O$_3$	13.49	1.44
C$_{12}$H$_{27}$NO$_2$	13.86	1.29
C$_{13}$HN$_2$O$_2$	14.91	1.43
C$_{13}$H$_3$N$_3$O	15.28	1.29
C$_{13}$H$_5$N$_4$	15.65	1.15
C$_{13}$H$_{13}$O$_3$	14.37	1.56
C$_{13}$H$_{15}$NO$_2$	14.75	1.41
C$_{13}$H$_{17}$N$_2$O	15.12	1.27
C$_{13}$H$_{19}$N$_3$	15.50	1.12
C$_{14}$HO$_3$	15.26	1.68
C$_{14}$H$_3$NO$_2$	15.64	1.54
C$_{14}$H$_5$N$_2$O	16.01	1.40
C$_{14}$H$_7$N$_3$	16.39	1.26
C$_{14}$H$_{17}$O$_2$	15.48	1.52
C$_{14}$H$_{19}$NO	15.58	1.37
C$_{14}$H$_{21}$N$_2$	16.23	1.23
C$_{15}$H$_5$O$_2$	16.37	1.65
C$_{15}$H$_7$NO	16.74	1.51
C$_{15}$H$_9$N$_2$	17.12	1.38
C$_{15}$H$_{21}$O	16.58	1.49
C$_{15}$H$_{23}$N	16.96	1.35
C$_{16}$H$_9$O	17.47	1.63
C$_{16}$H$_{11}$N	17.85	1.50
C$_{16}$H$_{25}$	17.69	1.47
C$_{17}$H$_{13}$	18.58	1.63
C$_{18}$H	19.47	1.79

218

	M + 1	M + 2
C$_9$H$_{18}$N$_2$O$_4$	10.93	1.35
C$_9$H$_{20}$N$_3$O$_3$	11.31	1.19
C$_9$H$_{22}$N$_4$O$_2$	11.68	1.03
C$_{10}$H$_6$N$_2$O$_4$	11.82	1.44
C$_{10}$H$_8$N$_3$O$_3$	12.20	1.28
C$_{10}$H$_{10}$N$_4$O$_2$	12.57	1.13
C$_{10}$H$_{20}$NO$_4$	11.66	1.42
C$_{10}$H$_{22}$N$_2$O$_3$	12.04	1.27
C$_{10}$H$_{24}$N$_3$O$_2$	12.41	1.11
C$_{10}$H$_{26}$N$_4$O	12.79	0.96
C$_{11}$H$_8$NO$_4$	12.55	1.52
C$_{11}$H$_{10}$N$_2$O$_3$	12.93	1.37

	M + 1	M + 2
C$_{11}$H$_{12}$N$_3$O$_2$	13.30	1.22
C$_{11}$H$_{14}$N$_4$O	13.68	1.07
C$_{11}$H$_{22}$O$_4$	12.40	1.51
C$_{11}$H$_{24}$NO$_3$	12.77	1.35
C$_{11}$H$_{26}$N$_2$O$_2$	13.14	1.20
C$_{12}$H$_2$N$_4$O	14.56	1.19
C$_{12}$H$_{10}$O$_4$	13.28	1.61
C$_{12}$H$_{12}$NO$_3$	13.66	1.46
C$_{12}$H$_{14}$N$_2$O$_2$	14.03	1.31
C$_{12}$H$_{16}$N$_3$O	14.41	1.17
C$_{12}$H$_{18}$N$_4$	14.78	1.02
C$_{12}$H$_{26}$O$_3$	13.50	1.44
C$_{13}$H$_2$N$_2$O$_2$	14.92	1.44
C$_{13}$H$_4$N$_3$O	15.30	1.29
C$_{13}$H$_6$N$_4$	15.67	1.15
C$_{13}$H$_{14}$O$_3$	14.39	1.56
C$_{13}$H$_{16}$NO$_2$	14.76	1.41
C$_{13}$H$_{18}$N$_2$O	15.14	1.27
C$_{13}$H$_{20}$N$_3$	15.51	1.13
C$_{14}$H$_2$O$_3$	15.28	1.69
C$_{14}$H$_4$NO$_2$	15.65	1.54
C$_{14}$H$_6$N$_2$O	16.03	1.40
C$_{14}$H$_8$N$_3$	16.40	1.26
C$_{14}$H$_{18}$O$_2$	15.50	1.52
C$_{14}$H$_{20}$NO	15.87	1.38
C$_{14}$H$_{22}$N$_2$	16.24	1.24
C$_{15}$H$_6$O$_2$	16.38	1.66
C$_{15}$H$_8$NO	16.76	1.52
C$_{15}$H$_{10}$N$_2$	17.13	1.38
C$_{15}$H$_{22}$O	16.60	1.49
C$_{15}$H$_{24}$N	16.98	1.35
C$_{16}$H$_{10}$O	17.49	1.64
C$_{16}$H$_{12}$N	17.86	1.50
C$_{16}$H$_{26}$	17.71	1.47
C$_{17}$H$_{14}$	18.59	1.63
C$_{18}$H$_2$	19.48	1.79

219

	M + 1	M + 2
C$_9$H$_{19}$N$_2$O$_4$	10.95	1.35
C$_9$H$_{21}$N$_3$O$_3$	11.32	1.19
C$_9$H$_{23}$N$_4$O$_2$	11.70	1.03
C$_{10}$H$_7$N$_2$O$_4$	11.84	1.44
C$_{10}$H$_9$N$_3$O$_3$	12.21	1.29
C$_{10}$H$_{11}$N$_4$O$_2$	12.59	1.13
C$_{10}$H$_{21}$NO$_4$	11.68	1.42
C$_{10}$H$_{23}$N$_2$O$_3$	12.05	1.27
C$_{10}$H$_{25}$N$_3$O$_2$	12.43	1.11
C$_{11}$H$_9$NO$_4$	12.57	1.53
C$_{11}$H$_{11}$N$_2$O$_3$	12.94	1.37
C$_{11}$H$_{13}$N$_3$O$_2$	13.32	1.22
C$_{11}$H$_{15}$N$_4$O	13.69	1.07
C$_{11}$H$_{23}$O$_4$	12.41	1.51
C$_{11}$H$_{25}$NO$_3$	12.79	1.35
C$_{12}$HN$_3$O$_2$	14.21	1.34
C$_{12}$H$_3$N$_4$O	14.58	1.19
C$_{12}$H$_{11}$O$_4$	13.30	1.62
C$_{12}$H$_{13}$NO$_3$	13.67	1.47
C$_{12}$H$_{15}$N$_2$O$_2$	14.05	1.32
C$_{12}$H$_{17}$N$_3$O	14.42	1.17
C$_{12}$H$_{19}$N$_4$	14.80	1.02
C$_{13}$HNO$_3$	14.56	1.59
C$_{13}$H$_3$N$_2$O$_2$	14.94	1.44
C$_{13}$H$_5$N$_3$O	15.31	1.29

	M + 1	M + 2
$C_{13}H_7N_4$	15.69	1.15
$C_{13}H_{15}O_3$	14.41	1.56
$C_{13}H_{17}NO_2$	14.78	1.42
$C_{13}H_{19}N_2O$	15.15	1.27
$C_{13}H_{21}N_3$	15.53	1.13
$C_{14}H_3O_3$	15.29	1.69
$C_{14}H_5NO_2$	15.67	1.55
$C_{14}H_7N_2O$	16.04	1.40
$C_{14}H_9N_3$	16.42	1.26
$C_{14}H_{19}O_2$	15.51	1.52
$C_{14}H_{21}NO$	15.89	1.38
$C_{14}H_{23}N_2$	16.26	1.24
$C_{15}H_7O_2$	16.40	1.66
$C_{15}H_9NO$	16.77	1.52
$C_{15}H_{11}N_2$	17.15	1.38
$C_{15}H_{23}O$	16.62	1.49
$C_{15}H_{25}N$	16.99	1.36
$C_{16}H_{11}O$	17.51	1.64
$C_{16}H_{13}N$	17.88	1.50
$C_{16}H_{27}$	17.72	1.48
$C_{17}HN$	18.77	1.66
$C_{17}H_{15}$	18.61	1.63
$C_{18}H_3$	19.50	1.80

220

	M + 1	M + 2
$C_9H_{20}N_2O_4$	10.97	1.35
$C_9H_{22}N_3O_3$	11.34	1.19
$C_9H_{24}N_4O_2$	11.71	1.03
$C_{10}H_8N_2O_4$	11.85	1.44
$C_{10}H_{10}N_3O_3$	12.23	1.29
$C_{10}H_{12}N_4O_2$	12.60	1.13
$C_{10}H_{22}NO_4$	11.70	1.43
$C_{10}H_{24}N_2O_3$	12.07	1.27
$C_{11}H_{10}NO_4$	12.58	1.53
$C_{11}H_{12}N_2O_3$	12.96	1.38
$C_{11}H_{14}N_3O_2$	13.33	1.22
$C_{11}H_{16}N_4O$	13.71	1.07
$C_{11}H_{24}O_4$	12.43	1.51
$C_{12}H_2N_3O_2$	14.22	1.34
$C_{12}H_4N_4O$	14.60	1.19
$C_{12}H_{12}O_4$	13.32	1.62
$C_{12}H_{14}NO_3$	13.69	1.47
$C_{12}H_{16}N_2O_2$	14.06	1.32
$C_{12}H_{18}N_3O$	14.44	1.17
$C_{12}H_{20}N_4$	14.81	1.02
$C_{13}H_2NO_3$	14.58	1.59
$C_{13}H_4N_2O_2$	14.95	1.44
$C_{13}H_6N_3O$	15.33	1.30
$C_{13}H_8N_4$	15.70	1.15
$C_{13}H_{16}O_3$	14.42	1.57
$C_{13}H_{18}NO_2$	14.80	1.42
$C_{13}H_{20}N_2O$	15.17	1.27
$C_{13}H_{22}N_3$	15.54	1.13
$C_{14}H_4O_3$	15.31	1.69
$C_{14}H_6NO_2$	15.68	1.55
$C_{14}H_8N_2O$	16.06	1.41
$C_{14}H_{10}N_3$	16.43	1.27
$C_{14}H_{20}O_2$	15.53	1.52
$C_{14}H_{22}NO$	15.90	1.38
$C_{14}H_{24}N_2$	16.28	1.24
$C_{15}H_8O_2$	16.42	1.66
$C_{15}H_{10}NO$	16.79	1.52
$C_{15}H_{12}N_2$	17.16	1.38
$C_{15}H_{24}O$	16.63	1.50
$C_{15}H_{26}N$	17.01	1.36
$C_{16}H_{12}O$	17.52	1.64
$C_{16}H_{14}N$	17.90	1.51
$C_{16}H_{28}$	17.74	1.48
$C_{17}H_2N$	18.78	1.66
$C_{17}H_{16}$	18.63	1.64
$C_{18}H_4$	19.52	1.80

221

	M + 1	M + 2
$C_9H_{21}N_2O_4$	10.98	1.35
$C_9H_{23}N_3O_3$	11.36	1.19
$C_{10}H_9N_2O_4$	11.87	1.45
$C_{10}H_{11}N_3O_3$	12.24	1.29
$C_{10}H_{13}N_4O_2$	12.62	1.14
$C_{10}H_{23}NO_4$	11.71	1.43
$C_{11}HN_4O_2$	13.51	1.25
$C_{11}H_{11}NO_4$	12.60	1.53
$C_{11}H_{13}N_2O_3$	12.98	1.38
$C_{11}H_{15}N_3O_2$	13.35	1.23
$C_{11}H_{17}N_4O$	13.72	1.08
$C_{12}HN_2O_3$	13.86	1.49
$C_{12}H_3N_3O_2$	14.24	1.34
$C_{12}H_5N_4O$	14.61	1.20
$C_{12}H_{13}O_4$	13.33	1.62
$C_{12}H_{15}NO_3$	13.71	1.47
$C_{12}H_{17}N_2O_2$	14.08	1.32
$C_{12}H_{19}N_3O$	14.46	1.17
$C_{12}H_{21}N_4$	14.83	1.03
$C_{13}HO_4$	14.22	1.74
$C_{13}H_3NO_3$	14.60	1.59
$C_{13}H_5N_2O_2$	14.97	1.44
$C_{13}H_7N_3O$	15.34	1.30
$C_{13}H_9N_4$	15.72	1.16
$C_{13}H_{17}O_3$	14.44	1.57
$C_{13}H_{19}NO_2$	14.81	1.42
$C_{13}H_{21}N_2O$	15.19	1.28
$C_{13}H_{23}N_3$	15.56	1.13
$C_{14}H_5O_3$	15.33	1.69
$C_{14}H_7NO_2$	15.70	1.55
$C_{14}H_9N_2O$	16.07	1.41
$C_{14}H_{11}N_3$	16.45	1.27
$C_{14}H_{21}O_2$	15.54	1.53
$C_{14}H_{23}NO$	15.92	1.38
$C_{14}H_{25}N_2$	16.29	1.24
$C_{15}H_9O_2$	16.43	1.66
$C_{15}H_{11}NO$	16.81	1.52
$C_{15}H_{13}N_2$	17.18	1.39
$C_{15}H_{25}O$	16.65	1.50
$C_{15}H_{27}N$	17.02	1.36
$C_{16}HN_2$	18.07	1.54
$C_{16}H_{13}O$	17.54	1.64
$C_{16}H_{15}N$	17.91	1.51
$C_{16}H_{29}$	17.75	1.48
$C_{17}HO$	18.43	1.80
$C_{17}H_3N$	18.80	1.67
$C_{17}H_{17}$	18.64	1.64
$C_{18}H_5$	19.53	1.80

222

	M + 1	M + 2
$C_9H_{22}N_2O_4$	11.00	1.35
$C_{10}H_{10}N_2O_4$	11.89	1.45
$C_{10}H_{12}N_3O_3$	12.26	1.29
$C_{10}H_{14}N_4O_2$	12.63	1.14
$C_{11}H_2N_4O_2$	13.52	1.25
$C_{11}H_{12}NO_4$	12.62	1.53
$C_{11}H_{14}N_2O_3$	12.99	1.38
$C_{11}H_{16}N_3O_2$	13.37	1.23
$C_{11}H_{18}N_4O$	13.74	1.08
$C_{12}H_2N_2O_3$	13.88	1.49
$C_{12}H_4N_3O_2$	14.25	1.34
$C_{12}H_6N_4O$	14.63	1.20
$C_{12}H_{14}O_4$	13.35	1.62
$C_{12}H_{16}NO_3$	13.72	1.47
$C_{12}H_{18}N_2O_2$	14.10	1.32
$C_{12}H_{20}N_3O$	14.47	1.18
$C_{12}H_{22}N_4$	14.85	1.03
$C_{13}H_2O_4$	14.24	1.74
$C_{13}H_4NO_3$	14.61	1.59
$C_{13}H_6N_2O_2$	14.99	1.45
$C_{13}H_8N_3O$	15.36	1.30
$C_{13}H_{10}N_4$	15.73	1.16
$C_{13}H_{18}O_3$	14.45	1.57
$C_{13}H_{20}NO_2$	14.83	1.42
$C_{13}H_{22}N_2O$	15.20	1.28
$C_{13}H_{24}N_3$	15.58	1.14
$C_{14}H_6O_3$	15.34	1.70
$C_{14}H_8NO_2$	15.72	1.55
$C_{14}H_{10}N_2O$	16.09	1.41
$C_{14}H_{12}N_3$	16.47	1.27
$C_{14}H_{22}O_2$	15.56	1.53
$C_{14}H_{24}NO$	15.93	1.39
$C_{14}H_{26}N_2$	16.31	1.25
$C_{15}H_{10}O_2$	16.45	1.67
$C_{15}H_{12}NO$	16.82	1.53
$C_{15}H_{14}N_2$	17.20	1.39
$C_{15}H_{26}O$	16.66	1.50
$C_{15}H_{28}N$	17.04	1.36
$C_{16}H_2N_2$	18.09	1.54
$C_{16}H_{14}O$	17.55	1.65
$C_{16}H_{16}N$	17.93	1.51
$C_{16}H_{30}$	17.77	1.49
$C_{17}H_2O$	18.44	1.80
$C_{17}H_4N$	18.82	1.67
$C_{17}H_{18}$	18.66	1.64
$C_{18}H_6$	19.55	1.81

223

	M + 1	M + 2
$C_{10}H_{11}N_2O_4$	11.90	1.45
$C_{10}H_{13}N_3O_3$	12.28	1.29
$C_{10}H_{15}N_4O_2$	12.65	1.14
$C_{11}HN_3O_3$	13.16	1.40
$C_{11}H_3N_4O_2$	13.54	1.25
$C_{11}H_{13}NO_4$	12.63	1.53
$C_{11}H_{15}N_2O_3$	13.01	1.38
$C_{11}H_{17}N_3O_2$	13.38	1.23
$C_{11}H_{19}N_4O$	13.76	1.08
$C_{12}HNO_4$	13.52	1.65
$C_{12}H_3N_2O_3$	13.90	1.50
$C_{12}H_5N_3O_2$	14.27	1.35
$C_{12}H_7N_4O$	14.64	1.20
$C_{12}H_{15}O_4$	13.36	1.62
$C_{12}H_{17}NO_3$	13.74	1.47
$C_{12}H_{19}N_2O_2$	14.11	1.33
$C_{12}H_{21}N_3O$	14.49	1.18
$C_{12}H_{23}N_4$	14.86	1.03
$C_{13}H_3O_4$	14.25	1.74
$C_{13}H_5NO_3$	14.63	1.59
$C_{13}H_7N_2O_2$	15.00	1.45
$C_{13}H_9N_3O$	15.38	1.30
$C_{13}H_{11}N_4$	15.75	1.16
$C_{13}H_{19}O_3$	14.47	1.57
$C_{13}H_{21}NO_2$	14.84	1.43
$C_{13}H_{23}N_2O$	15.22	1.28
$C_{13}H_{25}N_3$	15.59	1.14
$C_{14}H_7O_3$	15.36	1.70
$C_{14}H_9NO_2$	15.73	1.56
$C_{14}H_{11}N_2O$	16.11	1.41
$C_{14}H_{13}N_3$	16.48	1.27
$C_{14}H_{23}O_2$	15.58	1.53
$C_{14}H_{25}NO$	15.95	1.39
$C_{14}H_{27}N_2$	16.32	1.25
$C_{15}HN_3$	17.37	1.42
$C_{15}H_{11}O_2$	16.46	1.67
$C_{15}H_{13}NO$	16.84	1.53
$C_{15}H_{15}N_2$	17.21	1.39
$C_{15}H_{27}O$	16.68	1.50
$C_{15}H_{29}N$	17.06	1.37
$C_{16}HNO$	17.73	1.68
$C_{16}H_3N_2$	18.10	1.54
$C_{16}H_{15}O$	17.57	1.65
$C_{16}H_{17}N$	17.94	1.52
$C_{16}H_{31}$	17.79	1.49
$C_{17}H_3O$	18.46	1.80
$C_{17}H_5N$	18.83	1.67
$C_{17}H_{19}$	18.67	1.64
$C_{18}H_7$	19.56	1.81

224

	M + 1	M + 2
$C_{10}H_{12}N_2O_4$	11.92	1.45
$C_{10}H_{14}N_3O_3$	12.29	1.30
$C_{10}H_{16}N_4O_2$	12.67	1.14
$C_{11}H_2N_3O_3$	13.18	1.40
$C_{11}H_4N_4O_2$	13.56	1.25
$C_{11}H_{14}NO_4$	12.65	1.54
$C_{11}H_{16}N_2O_3$	13.02	1.38
$C_{11}H_{18}N_3O_2$	13.40	1.23
$C_{11}H_{20}N_4O$	13.77	1.08
$C_{12}H_2NO_4$	13.54	1.65
$C_{12}H_4N_2O_3$	13.91	1.50
$C_{12}H_6N_3O_2$	14.29	1.35
$C_{12}H_8N_4O$	14.66	1.20
$C_{12}H_{16}O_4$	13.38	1.63
$C_{12}H_{18}NO_3$	13.75	1.48
$C_{12}H_{20}N_2O_2$	14.13	1.33
$C_{12}H_{22}N_3O$	14.50	1.18
$C_{12}H_{24}N_4$	14.88	1.03
$C_{13}H_4O_4$	14.27	1.74
$C_{13}H_6NO_3$	14.64	1.60
$C_{13}H_8N_2O_2$	15.02	1.45
$C_{13}H_{10}N_3O$	15.39	1.31
$C_{13}H_{12}N_4$	15.77	1.16
$C_{13}H_{20}O_3$	14.49	1.57
$C_{13}H_{22}NO_2$	14.86	1.43
$C_{13}H_{24}N_2O$	15.23	1.28
$C_{13}H_{26}N_3$	15.61	1.14
$C_{14}H_8O_3$	15.37	1.70
$C_{14}H_{10}NO_2$	15.75	1.56
$C_{14}H_{12}N_2O$	16.12	1.42
$C_{14}H_{14}N_3$	16.50	1.28
$C_{14}H_{24}O_2$	15.59	1.53

	M + 1	M + 2
C$_{14}$H$_{26}$NO	15.97	1.39
C$_{14}$H$_{28}$N$_2$	16.34	1.25
C$_{15}$H$_2$N$_3$	17.39	1.42
C$_{15}$H$_{12}$O$_2$	16.48	1.67
C$_{15}$H$_{14}$NO	16.85	1.53
C$_{15}$H$_{16}$N$_2$	17.23	1.40
C$_{15}$H$_{28}$O	16.70	1.51
C$_{15}$H$_{30}$N	17.07	1.37
C$_{16}$H$_2$NO	17.74	1.68
C$_{16}$H$_4$N$_2$	18.12	1.55
C$_{16}$H$_{16}$O	17.59	1.65
C$_{16}$H$_{18}$N	17.96	1.52
C$_{16}$H$_{32}$	17.80	1.49
C$_{17}$H$_4$O	18.47	1.81
C$_{17}$H$_6$N	18.85	1.68
C$_{17}$H$_{20}$	18.69	1.65
C$_{18}$H$_8$	19.58	1.81
225		
C$_{10}$H$_{13}$N$_2$O$_4$	11.93	1.45
C$_{10}$H$_{15}$N$_3$O$_3$	12.31	1.30
C$_{10}$H$_{17}$N$_4$O$_2$	12.68	1.14
C$_{11}$HN$_2$O$_4$	12.82	1.56
C$_{11}$H$_3$N$_3$O$_3$	13.20	1.41
C$_{11}$H$_5$N$_4$O$_2$	13.57	1.25
C$_{11}$H$_{15}$NO$_4$	12.66	1.54
C$_{11}$H$_{17}$N$_2$O$_3$	13.04	1.39
C$_{11}$H$_{19}$N$_3$O$_2$	13.41	1.23
C$_{11}$H$_{21}$N$_4$O	13.79	1.08
C$_{12}$H$_3$NO$_4$	13.55	1.65
C$_{12}$H$_5$N$_2$O$_3$	13.93	1.50
C$_{12}$H$_7$N$_3$O$_2$	14.30	1.35
C$_{12}$H$_9$N$_4$O	14.68	1.20
C$_{12}$H$_{17}$O$_4$	13.40	1.63
C$_{12}$H$_{19}$NO$_3$	13.77	1.48
C$_{12}$H$_{21}$N$_2$O$_2$	14.14	1.33
C$_{12}$H$_{23}$N$_3$O	14.52	1.18
C$_{12}$H$_{25}$N$_4$	14.89	1.04
C$_{13}$H$_5$O$_4$	14.28	1.75
C$_{13}$H$_7$NO$_3$	14.66	1.60
C$_{13}$H$_9$N$_2$O$_2$	15.03	1.45
C$_{13}$H$_{11}$N$_3$O	15.41	1.31
C$_{13}$H$_{13}$N$_4$	15.78	1.17
C$_{13}$H$_{21}$O$_3$	14.50	1.58
C$_{13}$H$_{23}$NO$_2$	14.88	1.43
C$_{13}$H$_{25}$N$_2$O	15.25	1.29
C$_{13}$H$_{27}$N$_3$	15.62	1.14
C$_{14}$HN$_4$	16.67	1.30
C$_{14}$H$_9$O$_3$	15.39	1.70
C$_{14}$H$_{11}$NO$_2$	15.76	1.56
C$_{14}$H$_{13}$N$_2$O	16.14	1.42
C$_{14}$H$_{15}$N$_3$	16.51	1.28
C$_{14}$H$_{25}$O$_2$	15.61	1.54
C$_{14}$H$_{27}$NO	15.98	1.39
C$_{14}$H$_{29}$N$_2$	16.36	1.25
C$_{15}$HN$_2$O	17.03	1.56
C$_{15}$H$_3$N$_3$	17.40	1.42
C$_{15}$H$_{13}$O$_2$	16.50	1.67
C$_{15}$H$_{15}$NO	16.87	1.54
C$_{15}$H$_{17}$N$_2$	17.24	1.40
C$_{15}$H$_{29}$O	16.71	1.51
C$_{15}$H$_{31}$N	17.09	1.37
C$_{16}$HO$_2$	17.38	1.82
C$_{16}$H$_3$NO	17.76	1.68
C$_{16}$H$_5$N$_2$	18.13	1.55
C$_{16}$H$_{17}$O	17.60	1.66
C$_{16}$H$_{19}$N	17.98	1.52
C$_{16}$H$_{33}$	17.82	1.49
C$_{17}$H$_5$O	18.49	1.81
C$_{17}$H$_7$N	18.86	1.68
C$_{17}$H$_{21}$	18.71	1.65
C$_{18}$H$_9$	19.60	1.81
226		
C$_{10}$H$_{14}$N$_2$O$_4$	11.95	1.46
C$_{10}$H$_{16}$N$_3$O$_3$	12.32	1.30
C$_{10}$H$_{18}$N$_4$O$_2$	12.70	1.15
C$_{11}$H$_2$N$_2$O$_4$	12.84	1.56
C$_{11}$H$_4$N$_3$O$_3$	13.21	1.41
C$_{11}$H$_6$N$_4$O$_2$	13.59	1.26
C$_{11}$H$_{16}$NO$_4$	12.68	1.54
C$_{11}$H$_{18}$N$_2$O$_3$	13.06	1.39
C$_{11}$H$_{20}$N$_3$O$_2$	13.43	1.24
C$_{11}$H$_{22}$N$_4$O	13.80	1.09
C$_{12}$H$_4$NO$_4$	13.57	1.65
C$_{12}$H$_6$N$_2$O$_3$	13.94	1.50
C$_{12}$H$_8$N$_3$O$_2$	14.32	1.35
C$_{12}$H$_{10}$N$_4$O	14.69	1.21
C$_{12}$H$_{18}$O$_4$	13.41	1.63
C$_{12}$H$_{20}$NO$_3$	13.79	1.48
C$_{12}$H$_{22}$N$_2$O$_2$	14.16	1.33
C$_{12}$H$_{24}$N$_3$O	14.54	1.18
C$_{12}$H$_{26}$N$_4$	14.91	1.04
C$_{13}$H$_6$O$_4$	14.30	1.75
C$_{13}$H$_8$NO$_3$	14.68	1.60
C$_{13}$H$_{10}$N$_2$O$_2$	15.05	1.46
C$_{13}$H$_{12}$N$_3$O	15.42	1.31
C$_{13}$H$_{14}$N$_4$	15.80	1.17
C$_{13}$H$_{22}$O$_3$	14.52	1.58
C$_{13}$H$_{24}$NO$_2$	14.89	1.43
C$_{13}$H$_{26}$N$_2$O	15.27	1.29
C$_{13}$H$_{28}$N$_3$	15.64	1.15
C$_{14}$H$_2$N$_4$	16.69	1.31
C$_{14}$H$_{10}$O$_3$	15.41	1.71
C$_{14}$H$_{12}$NO$_2$	15.78	1.56
C$_{14}$H$_{14}$N$_2$O	16.15	1.42
C$_{14}$H$_{16}$N$_3$	16.53	1.28
C$_{14}$H$_{26}$O$_2$	15.62	1.54
C$_{14}$H$_{28}$O$_2$	16.00	1.40
C$_{14}$H$_{30}$N$_2$	16.37	1.26
C$_{15}$H$_2$N$_2$O	17.04	1.56
C$_{15}$H$_4$N$_3$	17.42	1.43
C$_{15}$H$_{14}$O$_2$	16.51	1.68
C$_{15}$H$_{16}$NO	16.89	1.54
C$_{15}$H$_{18}$N$_2$	17.26	1.40
C$_{15}$H$_{30}$O	16.73	1.51
C$_{15}$H$_{32}$N	17.10	1.37
C$_{16}$H$_2$O$_2$	17.40	1.82
C$_{16}$H$_4$NO	17.77	1.69
C$_{16}$H$_6$N$_2$	18.15	1.55
C$_{16}$H$_{18}$O	17.62	1.66
C$_{16}$H$_{20}$N	17.99	1.52
C$_{16}$H$_{34}$	17.83	1.50
C$_{17}$H$_6$O	18.51	1.81
C$_{17}$H$_8$N	18.88	1.68
C$_{17}$H$_{22}$	18.72	1.65
C$_{18}$H$_{10}$	19.61	1.82
227		
C$_{10}$H$_{15}$N$_2$O$_4$	11.97	1.46
C$_{10}$H$_{17}$N$_3$O$_3$	12.34	1.30
C$_{10}$H$_{19}$N$_4$O$_2$	12.71	1.15
C$_{11}$H$_3$N$_2$O$_4$	12.85	1.56
C$_{11}$H$_5$N$_3$O$_3$	13.23	1.41
C$_{11}$H$_7$N$_4$O$_2$	13.60	1.26
C$_{11}$H$_{17}$NO$_4$	12.70	1.54
C$_{11}$H$_{19}$N$_2$O$_3$	13.07	1.39
C$_{11}$H$_{21}$N$_3$O$_2$	13.45	1.24
C$_{11}$H$_{23}$N$_4$O	13.82	1.09
C$_{12}$H$_5$NO$_4$	13.59	1.65
C$_{12}$H$_7$N$_2$O$_3$	13.96	1.50
C$_{12}$H$_9$N$_3$O$_2$	14.33	1.36
C$_{12}$H$_{11}$N$_4$O	14.71	1.21
C$_{12}$H$_{19}$O$_4$	13.43	1.63
C$_{12}$H$_{21}$NO$_3$	13.80	1.48
C$_{12}$H$_{23}$N$_2$O$_2$	14.18	1.33
C$_{12}$H$_{25}$N$_3$O	14.55	1.19
C$_{12}$H$_{27}$N$_4$	14.93	1.04
C$_{13}$H$_7$O$_4$	14.32	1.75
C$_{13}$H$_9$NO$_3$	14.69	1.60
C$_{13}$H$_{11}$N$_2$O$_2$	15.07	1.46
C$_{13}$H$_{13}$N$_3$O	15.44	1.31
C$_{13}$H$_{15}$N$_4$	15.81	1.17
C$_{13}$H$_{23}$O$_3$	14.53	1.58
C$_{13}$H$_{25}$NO$_2$	14.91	1.44
C$_{13}$H$_{27}$N$_2$O	15.28	1.29
C$_{13}$H$_{29}$N$_3$	15.66	1.15
C$_{14}$HN$_3$O	16.33	1.45
C$_{14}$H$_3$N$_4$	16.70	1.31
C$_{14}$H$_{11}$O$_3$	15.42	1.71
C$_{14}$H$_{13}$NO$_2$	15.80	1.57
C$_{14}$H$_{15}$N$_2$O	16.17	1.42
C$_{14}$H$_{17}$N$_3$	16.55	1.28
C$_{14}$H$_{27}$O$_2$	15.64	1.54
C$_{14}$H$_{29}$NO	16.01	1.40
C$_{14}$H$_{31}$N$_2$	16.39	1.26
C$_{15}$HNO$_2$	16.69	1.70
C$_{15}$H$_3$N$_2$O	17.06	1.57
C$_{15}$H$_5$N$_3$	17.43	1.43
C$_{15}$H$_{15}$O$_2$	16.53	1.68
C$_{15}$H$_{17}$NO	16.90	1.54
C$_{15}$H$_{19}$N$_2$	17.28	1.40
C$_{15}$H$_{31}$O	16.74	1.51
C$_{15}$H$_{33}$N	17.12	1.38
C$_{16}$H$_3$O$_2$	17.42	1.82
C$_{16}$H$_5$NO	17.79	1.69
C$_{16}$H$_7$N$_2$	18.17	1.55
C$_{16}$H$_{19}$O	17.63	1.66
C$_{16}$H$_{21}$N	18.01	1.53
C$_{17}$H$_7$O	18.52	1.82
C$_{17}$H$_9$N	18.90	1.69
C$_{17}$H$_{23}$	18.74	1.66
C$_{18}$H$_{11}$	19.63	1.82
228		
C$_{10}$H$_{16}$N$_2$O$_4$	11.98	1.46
C$_{10}$H$_{18}$N$_3$O$_3$	12.36	1.30
C$_{10}$H$_{20}$N$_4$O$_2$	12.73	1.15
C$_{11}$H$_4$N$_2$O$_4$	12.87	1.56
C$_{11}$H$_6$N$_3$O$_3$	13.24	1.41
C$_{11}$H$_8$N$_4$O$_2$	13.62	1.26
C$_{11}$H$_{18}$NO$_4$	12.71	1.55
C$_{11}$H$_{20}$N$_2$O$_3$	13.09	1.39
C$_{11}$H$_{22}$N$_3$O$_2$	13.46	1.24
C$_{11}$H$_{24}$N$_4$O	13.84	1.09
C$_{12}$H$_6$NO$_4$	13.60	1.66
C$_{12}$H$_8$N$_2$O$_3$	13.98	1.51
C$_{12}$H$_{10}$N$_3$O$_2$	14.35	1.36
C$_{12}$H$_{12}$N$_4$O	14.72	1.21
C$_{12}$H$_{20}$O$_4$	13.44	1.64
C$_{12}$H$_{22}$NO$_3$	13.82	1.49
C$_{12}$H$_{24}$N$_2$O$_2$	14.19	1.34
C$_{12}$H$_{26}$N$_3$O	14.57	1.19
C$_{12}$H$_{28}$N$_4$	14.94	1.04
C$_{13}$H$_8$O$_4$	14.33	1.75
C$_{13}$H$_{10}$NO$_3$	14.71	1.61
C$_{13}$H$_{12}$N$_2$O$_2$	15.08	1.46
C$_{13}$H$_{14}$N$_3$O	15.46	1.32
C$_{13}$H$_{16}$N$_4$	15.83	1.17
C$_{13}$H$_{24}$O$_3$	14.55	1.58
C$_{13}$H$_{26}$NO$_2$	14.92	1.44
C$_{13}$H$_{28}$N$_2$O	15.30	1.29
C$_{13}$H$_{30}$N$_3$	15.67	1.15
C$_{14}$H$_2$N$_3$O	16.34	1.45
C$_{14}$H$_4$N$_4$	16.72	1.31
C$_{14}$H$_{12}$O$_3$	15.44	1.71
C$_{14}$H$_{14}$NO$_2$	15.81	1.57
C$_{14}$H$_{16}$N$_2$O	16.19	1.43
C$_{14}$H$_{18}$N$_3$	16.56	1.29
C$_{14}$H$_{28}$O$_2$	15.66	1.54
C$_{14}$H$_{30}$NO	16.03	1.40
C$_{14}$H$_{32}$N$_2$	16.40	1.26
C$_{15}$H$_2$NO$_2$	16.70	1.71
C$_{15}$H$_4$N$_2$O	17.08	1.57
C$_{15}$H$_6$N$_3$	17.45	1.43
C$_{15}$H$_{16}$O$_2$	16.54	1.68
C$_{15}$H$_{18}$NO	16.92	1.54
C$_{15}$H$_{20}$N$_2$	17.29	1.41
C$_{15}$H$_{32}$O	16.76	1.52
C$_{16}$H$_4$O$_2$	17.43	1.83
C$_{16}$H$_6$NO	17.81	1.69
C$_{16}$H$_8$N$_2$	18.18	1.56
C$_{16}$H$_{20}$O	17.65	1.66
C$_{16}$H$_{22}$N	18.02	1.53
C$_{17}$H$_8$O	18.54	1.82
C$_{17}$H$_{10}$N	18.91	1.69
C$_{17}$H$_{24}$	18.75	1.66
C$_{18}$H$_{12}$	19.64	1.82
229		
C$_{10}$H$_{17}$N$_2$O$_4$	12.00	1.46
C$_{10}$H$_{19}$N$_3$O$_3$	12.37	1.31
C$_{10}$H$_{21}$N$_4$O$_2$	12.75	1.15
C$_{11}$H$_5$N$_2$O$_4$	12.89	1.57
C$_{11}$H$_7$N$_3$O$_3$	13.26	1.41
C$_{11}$H$_9$N$_4$O$_2$	13.64	1.26
C$_{11}$H$_{19}$NO$_4$	12.73	1.55
C$_{11}$H$_{21}$N$_2$O$_3$	13.10	1.39
C$_{11}$H$_{23}$N$_3$O$_2$	13.48	1.24
C$_{11}$H$_{25}$N$_4$O	13.85	1.09
C$_{12}$H$_7$NO$_4$	13.62	1.66
C$_{12}$H$_9$N$_2$O$_3$	13.99	1.51

	M+1	M+2		M+1	M+2		M+1	M+2		M+1	M+2
$C_{12}H_{11}N_3O_2$	14.37	1.36	$C_{12}H_{30}N_4$	14.97	1.05	$C_{14}HNO_3$	15.64	1.74	$C_{15}H_8N_2O$	17.14	1.58
$C_{12}H_{13}N_4O$	14.74	1.21	$C_{13}H_2N_4O$	15.65	1.35	$C_{14}H_3N_2O_2$	16.02	1.60	$C_{15}H_{10}N_3$	17.51	1.44
$C_{12}H_{21}O_4$	13.46	1.64	$C_{13}H_{10}O_4$	14.36	1.76	$C_{14}H_5N_3O$	16.39	1.46	$C_{15}H_{20}O_2$	16.61	1.69
$C_{12}H_{23}NO_3$	13.83	1.49	$C_{13}H_{12}NO_3$	14.74	1.61	$C_{14}H_7N_4$	16.77	1.32	$C_{15}H_{22}NO$	16.98	1.55
$C_{12}H_{25}N_2O_2$	14.21	1.34	$C_{13}H_{14}N_2O_2$	15.11	1.47	$C_{14}H_{15}O_3$	15.49	1.72	$C_{15}H_{24}N_2$	17.36	1.42
$C_{12}H_{27}N_3O$	14.58	1.19	$C_{13}H_{16}N_3O$	15.49	1.32	$C_{14}H_{17}NO_2$	15.86	1.58	$C_{16}H_8O_2$	17.50	1.84
$C_{12}H_{29}N_4$	14.96	1.05	$C_{13}H_{18}N_4$	15.86	1.18	$C_{14}H_{19}N_2O$	16.23	1.44	$C_{16}H_{10}NO$	17.87	1.70
$C_{13}HN_4O$	15.63	1.34	$C_{13}H_{26}O_3$	14.58	1.59	$C_{14}H_{21}N_3$	16.61	1.30	$C_{16}H_{12}N_2$	18.25	1.57
$C_{13}H_9O_4$	14.35	1.76	$C_{13}H_{28}NO_2$	14.96	1.44	$C_{15}H_3O_3$	16.37	1.85	$C_{16}H_{24}O$	17.71	1.68
$C_{13}H_{11}NO_3$	14.72	1.61	$C_{13}H_{30}N_2O$	15.33	1.30	$C_{15}H_5NO_2$	16.75	1.72	$C_{16}H_{26}N$	18.09	1.54
$C_{13}H_{13}N_2O_2$	15.10	1.46	$C_{14}H_2N_2O_2$	16.00	1.60	$C_{15}H_7N_2O$	17.12	1.58	$C_{17}H_{12}O$	18.60	1.83
$C_{13}H_{15}N_3O$	15.47	1.32	$C_{14}H_4N_3O$	16.38	1.46	$C_{15}H_9N_3$	17.50	1.44	$C_{17}H_{14}N$	18.98	1.70
$C_{13}H_{17}N_4$	15.85	1.18	$C_{14}H_6N_4$	16.75	1.32	$C_{15}H_{19}O_2$	16.59	1.69	$C_{17}H_{28}$	18.82	1.67
$C_{13}H_{25}O_3$	14.57	1.59	$C_{14}H_{14}O_3$	15.47	1.72	$C_{15}H_{21}NO$	16.97	1.55	$C_{18}H_2N$	19.86	1.87
$C_{13}H_{27}NO_2$	14.94	1.44	$C_{14}H_{16}NO_2$	15.84	1.57	$C_{15}H_{23}N_2$	17.34	1.41	$C_{18}H_{16}$	19.71	1.84
$C_{13}H_{29}N_2O$	15.31	1.30	$C_{14}H_{18}N_2O$	16.22	1.43	$C_{16}H_7O_2$	17.48	1.84	$C_{19}H_4$	20.60	2.01
$C_{13}H_{31}N_3$	15.69	1.15	$C_{14}H_{20}N_3$	16.59	1.29	$C_{16}H_9NO$	17.85	1.70			
$C_{14}HN_2O_2$	15.99	1.60	$C_{14}H_{30}O_2$	15.69	1.55	$C_{16}H_{11}N_2$	18.23	1.57	**233**		
$C_{14}H_3N_3O$	16.36	1.45	$C_{15}H_2O_3$	16.36	1.85	$C_{16}H_{23}O$	17.70	1.67	$C_{10}H_{21}N_2O_4$	12.06	1.47
$C_{14}H_5N_4$	16.73	1.32	$C_{15}H_4NO_2$	16.73	1.71	$C_{16}H_{25}N$	18.07	1.54	$C_{10}H_{23}N_3O_3$	12.44	1.31
$C_{14}H_{13}O_3$	15.45	1.71	$C_{15}H_6N_2O$	17.11	1.57	$C_{17}H_{11}O$	18.59	1.83	$C_{10}H_{25}N_4O_2$	12.81	1.16
$C_{14}H_{15}NO_2$	15.83	1.57	$C_{15}H_8N_3$	17.48	1.44	$C_{17}H_{13}N$	18.96	1.70	$C_{11}H_9N_2O_4$	12.95	1.57
$C_{14}H_{17}N_2O$	16.20	1.43	$C_{15}H_{18}O_2$	16.58	1.69	$C_{17}H_{27}$	18.80	1.67	$C_{11}H_{11}N_3O_3$	13.32	1.42
$C_{14}H_{19}N_3$	16.58	1.29	$C_{15}H_{20}NO$	16.95	1.55	$C_{18}HN$	19.85	1.86	$C_{11}H_{13}N_4O_2$	13.70	1.27
$C_{14}H_{29}O_2$	15.67	1.55	$C_{15}H_{22}N_2$	17.32	1.41	$C_{18}H_{15}$	19.69	1.83	$C_{11}H_{23}NO_4$	12.79	1.56
$C_{14}H_{31}NO$	16.05	1.41	$C_{16}H_6O_2$	17.46	1.83	$C_{19}H_3$	20.58	2.01	$C_{11}H_{25}N_2O_3$	13.17	1.40
$C_{15}HO_3$	16.34	1.85	$C_{16}H_8NO$	17.84	1.70				$C_{11}H_{27}N_3O_2$	13.54	1.25
$C_{15}H_3NO_2$	16.72	1.71	$C_{16}H_{10}N_2$	18.21	1.56	**232**			$C_{12}HN_4O_2$	14.59	1.39
$C_{15}H_5N_2O$	17.09	1.57	$C_{16}H_{22}O$	17.68	1.67	$C_{10}H_{20}N_2O_4$	12.05	1.47	$C_{12}H_{11}NO_4$	13.68	1.67
$C_{15}H_7N_3$	17.47	1.44	$C_{16}H_{24}N$	18.06	1.54	$C_{10}H_{22}N_3O_3$	12.42	1.31	$C_{12}H_{13}N_2O_3$	14.06	1.52
$C_{15}H_{17}O_2$	16.56	1.68	$C_{17}H_{10}O$	18.57	1.83	$C_{10}H_{24}N_4O_2$	12.79	1.16	$C_{12}H_{15}N_3O_2$	14.43	1.37
$C_{15}H_{19}NO$	16.93	1.55	$C_{17}H_{12}N$	18.94	1.69	$C_{11}H_8N_2O_4$	12.93	1.57	$C_{12}H_{17}N_4O$	14.80	1.22
$C_{15}H_{21}N_2$	17.31	1.41	$C_{17}H_{26}$	18.79	1.67	$C_{11}H_{10}N_3O_3$	13.31	1.42	$C_{12}H_{25}O_4$	13.52	1.65
$C_{16}H_5O_2$	17.45	1.83	$C_{18}H_{14}$	19.68	1.83	$C_{11}H_{12}N_4O_2$	13.68	1.27	$C_{12}H_{27}NO_3$	13.90	1.50
$C_{16}H_7NO$	17.82	1.69	$C_{19}H_2$	20.56	2.00	$C_{11}H_{22}NO_4$	12.78	1.55	$C_{13}HN_2O_3$	14.94	1.64
$C_{16}H_9N_2$	18.20	1.56				$C_{11}H_{24}N_2O_3$	13.15	1.40	$C_{13}H_3N_3O_2$	15.32	1.50
$C_{16}H_{21}O$	17.67	1.67	**231**			$C_{11}H_{26}N_3O_2$	13.53	1.25	$C_{13}H_5N_4O$	15.69	1.35
$C_{16}H_{23}N$	18.04	1.53	$C_{10}H_{19}N_2O_4$	12.03	1.47	$C_{11}H_{28}N_4O$	13.90	1.10	$C_{13}H_{13}O_4$	14.41	1.76
$C_{17}H_9O$	18.55	1.82	$C_{10}H_{21}N_3O_3$	12.40	1.31	$C_{12}H_{10}NO_4$	13.67	1.66	$C_{13}H_{15}NO_3$	14.79	1.62
$C_{17}H_{11}N$	18.93	1.69	$C_{10}H_{23}N_4O_2$	12.78	1.16	$C_{12}H_{12}N_2O_3$	14.04	1.52	$C_{13}H_{17}N_2O_2$	15.16	1.47
$C_{17}H_{25}$	18.77	1.66	$C_{11}H_7N_2O_4$	12.92	1.57	$C_{12}H_{14}N_3O_2$	14.41	1.37	$C_{13}H_{19}N_3O$	15.54	1.33
$C_{18}H_{13}$	19.66	1.83	$C_{11}H_9N_3O_3$	13.29	1.42	$C_{12}H_{16}N_4O$	14.79	1.22	$C_{13}H_{21}N_4$	15.91	1.19
$C_{19}H$	20.55	2.00	$C_{11}H_{11}N_4O_2$	13.67	1.27	$C_{12}H_{24}O_4$	13.51	1.64	$C_{14}HO_4$	15.30	1.89
			$C_{11}H_{21}NO_4$	12.76	1.55	$C_{12}H_{26}NO_3$	13.88	1.49	$C_{14}H_3NO_3$	15.68	1.75
230			$C_{11}H_{23}N_2O_3$	13.14	1.40	$C_{12}H_{28}N_2O_2$	14.26	1.35	$C_{14}H_5N_2O_2$	16.05	1.61
$C_{10}H_{18}N_2O_4$	12.01	1.46	$C_{11}H_{25}N_3O_2$	13.51	1.25	$C_{13}H_2N_3O_2$	15.30	1.49	$C_{14}H_7N_3O$	16.42	1.47
$C_{10}H_{20}N_3O_3$	12.39	1.31	$C_{11}H_{27}N_4O$	13.88	1.10	$C_{13}H_4N_4O$	15.68	1.35	$C_{14}H_9N_4$	16.80	1.33
$C_{10}H_{22}N_4O_2$	12.76	1.15	$C_{12}H_9NO_4$	13.65	1.66	$C_{13}H_{12}O_4$	14.40	1.76	$C_{14}H_{17}O_3$	15.52	1.72
$C_{11}H_6N_2O_4$	12.90	1.57	$C_{12}H_{11}N_2O_3$	14.02	1.51	$C_{13}H_{14}NO_3$	14.77	1.62	$C_{14}H_{19}NO_2$	15.89	1.58
$C_{11}H_8N_3O_3$	13.28	1.42	$C_{12}H_{13}N_3O_2$	14.40	1.37	$C_{13}H_{16}N_2O_2$	15.15	1.47	$C_{14}H_{21}N_2O$	16.27	1.44
$C_{11}H_{10}N_4O_2$	13.65	1.27	$C_{12}H_{15}N_4O$	14.77	1.22	$C_{13}H_{18}N_3O$	15.52	1.33	$C_{14}H_{23}N_3$	16.64	1.30
$C_{11}H_{20}NO_4$	12.74	1.55	$C_{12}H_{23}O_4$	13.49	1.64	$C_{13}H_{20}N_4$	15.89	1.18	$C_{15}H_5O_3$	16.41	1.86
$C_{11}H_{22}N_2O_3$	13.12	1.40	$C_{12}H_{25}NO_3$	13.87	1.49	$C_{13}H_{28}O_3$	14.61	1.59	$C_{15}H_7NO_2$	16.78	1.72
$C_{11}H_{24}N_3O_2$	13.49	1.24	$C_{12}H_{27}N_2O_2$	14.24	1.34	$C_{14}H_2NO_3$	15.66	1.75	$C_{15}H_9N_2O$	17.16	1.58
$C_{11}H_{26}N_4O$	13.87	1.09	$C_{12}H_{29}N_3O$	14.62	1.20	$C_{14}H_4N_2O_2$	16.03	1.60	$C_{15}H_{11}N_3$	17.53	1.45
$C_{12}H_8NO_4$	13.63	1.66	$C_{13}HN_3O_2$	15.29	1.49	$C_{14}H_6N_3O$	16.41	1.46	$C_{15}H_{21}O_2$	16.62	1.70
$C_{12}H_{10}N_2O_3$	14.01	1.51	$C_{13}H_3N_4O$	15.66	1.35	$C_{14}H_8N_4$	16.78	1.32	$C_{15}H_{23}NO$	17.00	1.56
$C_{12}H_{12}N_3O_2$	14.38	1.36	$C_{13}H_{11}O_4$	14.38	1.76	$C_{14}H_{16}O_3$	15.50	1.72	$C_{15}H_{25}N_2$	17.37	1.42
$C_{12}H_{14}N_4O$	14.76	1.22	$C_{13}H_{13}NO_3$	14.76	1.61	$C_{14}H_{18}NO_2$	15.88	1.58	$C_{16}H_9O_2$	17.51	1.84
$C_{12}H_{22}O_4$	13.48	1.64	$C_{13}H_{15}N_2O_2$	15.13	1.47	$C_{14}H_{20}N_2O$	16.25	1.44	$C_{16}H_{11}NO$	17.89	1.71
$C_{12}H_{24}NO_3$	13.85	1.49	$C_{13}H_{17}N_3O$	15.50	1.32	$C_{14}H_{22}N_3$	16.63	1.30	$C_{16}H_{13}N_2$	18.26	1.57
$C_{12}H_{26}N_2O_2$	14.22	1.34	$C_{13}H_{19}N_4$	15.88	1.18	$C_{15}H_4O_3$	16.39	1.86	$C_{16}H_{25}O$	17.73	1.68
$C_{12}H_{28}N_3O$	14.60	1.19	$C_{13}H_{27}O_3$	14.60	1.59	$C_{15}H_6NO_2$	16.77	1.72	$C_{16}H_{27}N$	18.10	1.54
			$C_{13}H_{29}NO_2$	14.97	1.45						

Formula	M + 1	M + 2
C$_{17}$HN$_2$	19.15	1.73
C$_{17}$H$_{13}$O	18.62	1.83
C$_{17}$H$_{15}$N	18.99	1.70
C$_{17}$H$_{29}$	18.83	1.67
C$_{18}$HO	19.51	2.00
C$_{18}$H$_3$N	19.88	1.87
C$_{18}$H$_{17}$	19.72	1.84
C$_{19}$H$_5$	20.61	2.01

234

Formula	M + 1	M + 2
C$_{10}$H$_{22}$N$_2$O$_4$	12.08	1.47
C$_{10}$H$_{24}$N$_3$O$_3$	12.45	1.32
C$_{10}$H$_{26}$N$_4$O$_2$	12.83	1.16
C$_{11}$H$_{10}$N$_2$O$_4$	12.97	1.58
C$_{11}$H$_{12}$N$_3$O$_3$	13.34	1.42
C$_{11}$H$_{14}$N$_4$O$_2$	13.72	1.27
C$_{11}$H$_{24}$NO$_4$	12.81	1.56
C$_{11}$H$_{26}$N$_2$O$_3$	13.18	1.40
C$_{12}$H$_2$N$_4$O$_2$	14.60	1.39
C$_{12}$H$_{12}$NO$_4$	13.70	1.67
C$_{12}$H$_{14}$N$_2$O$_3$	14.07	1.52
C$_{12}$H$_{16}$N$_3$O$_2$	14.45	1.37
C$_{12}$H$_{18}$N$_4$O	14.82	1.23
C$_{12}$H$_{26}$O$_4$	13.54	1.65
C$_{13}$H$_2$N$_2$O$_3$	14.96	1.64
C$_{13}$H$_4$N$_3$O$_2$	15.33	1.50
C$_{13}$H$_6$N$_4$O	15.71	1.36
C$_{13}$H$_{14}$O$_4$	14.43	1.77
C$_{13}$H$_{16}$NO$_3$	14.80	1.62
C$_{13}$H$_{18}$N$_2$O$_2$	15.18	1.48
C$_{13}$H$_{20}$N$_3$O	15.55	1.33
C$_{13}$H$_{22}$N$_4$	15.93	1.19
C$_{14}$H$_2$O$_4$	15.32	1.89
C$_{14}$H$_4$NO$_3$	15.69	1.75
C$_{14}$H$_6$N$_2$O$_2$	16.07	1.61
C$_{14}$H$_8$N$_3$O	16.44	1.47
C$_{14}$H$_{10}$N$_4$	16.81	1.33
C$_{14}$H$_{18}$O$_3$	15.53	1.73
C$_{14}$H$_{20}$NO$_2$	15.91	1.58
C$_{14}$H$_{22}$N$_2$O	16.28	1.44
C$_{14}$H$_{24}$N$_3$	16.66	1.30
C$_{15}$H$_6$O$_3$	16.42	1.86
C$_{15}$H$_8$NO$_2$	16.80	1.72
C$_{15}$H$_{10}$N$_2$O	17.17	1.59
C$_{15}$H$_{12}$N$_3$	17.55	1.45
C$_{15}$H$_{22}$O$_2$	16.64	1.70
C$_{15}$H$_{24}$NO	17.01	1.56
C$_{15}$H$_{26}$N$_2$	17.39	1.42
C$_{16}$H$_{10}$O$_2$	17.53	1.84
C$_{16}$H$_{12}$NO	17.90	1.71
C$_{16}$H$_{14}$N$_2$	18.28	1.58
C$_{16}$H$_{26}$O	17.75	1.68
C$_{16}$H$_{28}$N	18.12	1.55
C$_{17}$H$_2$N$_2$	19.17	1.74
C$_{17}$H$_{14}$O	18.63	1.84
C$_{17}$H$_{16}$N	19.01	1.71
C$_{17}$H$_{30}$	18.85	1.68
C$_{18}$H$_2$O	19.52	2.00
C$_{18}$H$_4$N	19.90	1.87
C$_{18}$H$_{18}$	19.74	1.84
C$_{19}$H$_6$	20.63	2.02

235

Formula	M + 1	M + 2
C$_{10}$H$_{23}$N$_2$O$_4$	12.09	1.47
C$_{10}$H$_{25}$N$_3$O$_3$	12.47	1.32
C$_{11}$H$_{11}$N$_2$O$_4$	12.98	1.58
C$_{11}$H$_{13}$N$_3$O$_3$	13.36	1.43
C$_{11}$H$_{15}$N$_4$O$_2$	13.73	1.28
C$_{11}$H$_{25}$NO$_4$	12.82	1.56
C$_{12}$HN$_3$O$_3$	14.25	1.54
C$_{12}$H$_3$N$_4$O$_2$	14.62	1.40
C$_{12}$H$_{13}$NO$_4$	13.71	1.67
C$_{12}$H$_{15}$N$_2$O$_3$	14.09	1.52
C$_{12}$H$_{17}$N$_3$O$_2$	14.46	1.37
C$_{12}$H$_{19}$N$_4$O	14.84	1.23
C$_{13}$HNO$_4$	14.60	1.79
C$_{13}$H$_3$N$_2$O$_3$	14.98	1.65
C$_{13}$H$_5$N$_3$O$_2$	15.35	1.50
C$_{13}$H$_7$N$_4$O	15.73	1.36
C$_{13}$H$_{15}$O$_4$	14.44	1.77
C$_{13}$H$_{17}$NO$_3$	14.82	1.62
C$_{13}$H$_{19}$N$_2$O$_2$	15.19	1.48
C$_{13}$H$_{21}$N$_3$O	15.57	1.33
C$_{13}$H$_{23}$N$_4$	15.94	1.19
C$_{14}$H$_3$O$_4$	15.33	1.90
C$_{14}$H$_5$NO$_3$	15.71	1.75
C$_{14}$H$_7$N$_2$O$_2$	16.08	1.61
C$_{14}$H$_9$N$_3$O	16.46	1.47
C$_{14}$H$_{11}$N$_4$	16.83	1.33
C$_{14}$H$_{19}$O$_3$	15.55	1.73
C$_{14}$H$_{21}$NO$_2$	15.92	1.59
C$_{14}$H$_{23}$N$_2$O	16.30	1.45
C$_{14}$H$_{25}$N$_3$	16.67	1.31
C$_{15}$H$_7$O$_3$	16.44	1.86
C$_{15}$H$_9$NO$_2$	16.81	1.73
C$_{15}$H$_{11}$N$_2$O	17.19	1.59
C$_{15}$H$_{13}$N$_3$	17.56	1.44
C$_{15}$H$_{23}$O$_2$	16.66	1.70
C$_{15}$H$_{25}$NO	17.03	1.56
C$_{15}$H$_{27}$N$_2$	17.40	1.43
C$_{16}$HN$_3$	18.45	1.61
C$_{16}$H$_{11}$O$_2$	17.54	1.85
C$_{16}$H$_{13}$NO	17.92	1.71
C$_{16}$H$_{15}$N$_2$	18.29	1.58
C$_{16}$H$_{27}$O	17.76	1.68
C$_{16}$H$_{29}$N	18.14	1.55
C$_{17}$HNO	18.81	1.87
C$_{17}$H$_3$N$_2$	19.18	1.74
C$_{17}$H$_5$O	18.65	1.84
C$_{17}$H$_{17}$N	19.02	1.71
C$_{17}$H$_{31}$	18.87	1.68
C$_{18}$H$_3$O	19.54	2.00
C$_{18}$H$_5$N	19.91	1.88
C$_{18}$H$_{19}$	19.76	1.85
C$_{19}$H$_7$	20.64	2.02

236

Formula	M + 1	M + 2
C$_{10}$H$_{24}$N$_2$O$_4$	12.11	1.48
C$_{11}$H$_{12}$N$_2$O$_4$	13.00	1.58
C$_{11}$H$_{14}$N$_3$O$_3$	13.37	1.43
C$_{11}$H$_{16}$N$_4$O$_2$	13.75	1.28
C$_{12}$H$_2$N$_3$O$_3$	14.26	1.55
C$_{12}$H$_4$N$_4$O$_2$	14.64	1.40
C$_{12}$H$_{14}$NO$_4$	13.73	1.67
C$_{12}$H$_{16}$N$_2$O$_3$	14.10	1.52
C$_{12}$H$_{18}$N$_3$O$_2$	14.48	1.38
C$_{12}$H$_{20}$N$_4$O	14.85	1.23
C$_{13}$H$_2$NO$_4$	14.62	1.79
C$_{13}$H$_4$N$_2$O$_3$	14.99	1.65
C$_{13}$H$_6$N$_3$O$_2$	15.37	1.50
C$_{13}$H$_8$N$_4$O	15.74	1.36
C$_{13}$H$_{16}$O$_4$	14.46	1.77
C$_{13}$H$_{18}$NO$_3$	14.84	1.63
C$_{13}$H$_{20}$N$_2$O$_2$	15.21	1.48
C$_{13}$H$_{22}$N$_3$O	15.58	1.34
C$_{13}$H$_{24}$N$_4$	15.96	1.19
C$_{14}$H$_4$O$_4$	15.35	1.90
C$_{14}$H$_6$NO$_3$	15.72	1.76
C$_{14}$H$_8$N$_2$O$_2$	16.10	1.61
C$_{14}$H$_{10}$N$_3$O	16.47	1.47
C$_{14}$H$_{12}$N$_4$	16.85	1.33
C$_{14}$H$_{20}$O$_3$	15.57	1.73
C$_{14}$H$_{22}$NO$_2$	15.94	1.59
C$_{14}$H$_{24}$N$_2$O	16.31	1.45
C$_{14}$H$_{26}$N$_3$	16.69	1.31
C$_{15}$H$_8$O$_3$	16.45	1.87
C$_{15}$H$_{10}$NO$_2$	16.83	1.73
C$_{15}$H$_{12}$N$_2$O	17.20	1.59
C$_{15}$H$_{14}$N$_3$	17.58	1.46
C$_{15}$H$_{24}$O$_2$	16.67	1.70
C$_{15}$H$_{26}$NO	17.05	1.56
C$_{15}$H$_{28}$N$_2$	17.42	1.43
C$_{16}$H$_2$N$_3$	18.47	1.61
C$_{16}$H$_{12}$O$_2$	17.56	1.85
C$_{16}$H$_{14}$NO	17.93	1.71
C$_{16}$H$_{16}$N$_2$	18.31	1.58
C$_{16}$H$_{28}$O	17.78	1.69
C$_{16}$H$_{30}$N	18.15	1.55
C$_{17}$H$_2$NO	18.82	1.87
C$_{17}$H$_4$N$_2$	19.20	1.74
C$_{17}$H$_{16}$O	18.67	1.84
C$_{17}$H$_{18}$N	19.04	1.71
C$_{17}$H$_{32}$	18.88	1.68
C$_{18}$H$_4$O	19.55	2.01
C$_{18}$H$_6$N	19.93	1.88
C$_{18}$H$_{20}$	19.77	1.85
C$_{19}$H$_8$	20.66	2.02

237

Formula	M + 1	M + 2
C$_{11}$H$_{13}$N$_2$O$_4$	13.01	1.58
C$_{11}$H$_{15}$N$_3$O$_3$	13.39	1.43
C$_{11}$H$_{17}$N$_4$O$_2$	13.76	1.28
C$_{12}$HN$_2$O$_4$	13.90	1.70
C$_{12}$H$_3$N$_3$O$_3$	14.28	1.55
C$_{12}$H$_5$N$_4$O$_2$	14.65	1.40
C$_{12}$H$_{15}$NO$_4$	13.75	1.68
C$_{12}$H$_{17}$N$_2$O$_3$	14.12	1.53
C$_{12}$H$_{19}$N$_3$O$_2$	14.49	1.38
C$_{12}$H$_{21}$N$_4$O	14.87	1.23
C$_{13}$H$_3$NO$_4$	14.63	1.80
C$_{13}$H$_5$N$_2$O$_3$	15.01	1.65
C$_{13}$H$_7$N$_3$O$_2$	15.38	1.51
C$_{13}$H$_9$N$_4$O	15.76	1.36
C$_{13}$H$_{17}$O$_4$	14.48	1.77
C$_{13}$H$_{19}$NO$_3$	14.85	1.63
C$_{13}$H$_{21}$N$_2$O$_2$	15.23	1.48
C$_{13}$H$_{23}$N$_3$O	15.60	1.34
C$_{13}$H$_{25}$N$_4$	15.97	1.20
C$_{14}$H$_5$O$_4$	15.37	1.90
C$_{14}$H$_7$NO$_3$	15.74	1.76
C$_{14}$H$_9$N$_2$O$_2$	16.11	1.62
C$_{14}$H$_{11}$N$_3$O	16.49	1.48
C$_{14}$H$_{13}$N$_4$	16.86	1.34
C$_{14}$H$_{21}$O$_3$	15.58	1.73
C$_{14}$H$_{23}$NO$_2$	15.96	1.59
C$_{14}$H$_{25}$N$_2$O	16.33	1.45
C$_{14}$H$_{27}$N$_3$	16.71	1.31
C$_{15}$HN$_4$	17.75	1.49
C$_{15}$H$_9$O$_3$	16.47	1.87
C$_{15}$H$_{11}$NO$_2$	16.85	1.73
C$_{15}$H$_{13}$N$_2$O	17.22	1.59
C$_{15}$H$_{15}$N$_3$	17.59	1.46
C$_{15}$H$_{25}$O$_2$	16.69	1.71
C$_{15}$H$_{27}$NO	17.06	1.57
C$_{15}$H$_{29}$N$_2$	17.44	1.43
C$_{16}$HN$_2$O	18.11	1.75
C$_{16}$H$_3$N$_3$	18.48	1.61
C$_{16}$H$_{13}$O$_2$	17.58	1.85
C$_{16}$H$_{15}$NO	17.95	1.72
C$_{16}$H$_{17}$N$_2$	18.33	1.58
C$_{16}$H$_{29}$O	17.79	1.69
C$_{16}$H$_{31}$N	18.17	1.56
C$_{17}$HO$_2$	18.46	2.01
C$_{17}$H$_3$NO	18.84	1.88
C$_{17}$H$_5$N$_2$	19.21	1.75
C$_{17}$H$_{17}$O	18.68	1.85
C$_{17}$H$_{19}$N	19.06	1.72
C$_{17}$H$_{33}$	18.90	1.69
C$_{18}$H$_5$O	19.57	2.01
C$_{18}$H$_7$N	19.94	1.88
C$_{18}$H$_{21}$	19.79	1.85
C$_{19}$H$_9$	20.68	2.03

238

Formula	M + 1	M + 2
C$_{11}$H$_{14}$N$_2$O$_4$	13.03	1.59
C$_{11}$H$_{16}$N$_3$O$_3$	13.40	1.43
C$_{11}$H$_{18}$N$_4$O$_2$	13.78	1.28
C$_{12}$H$_2$N$_2$O$_4$	13.92	1.70
C$_{12}$H$_4$N$_3$O$_3$	14.29	1.55
C$_{12}$H$_6$N$_4$O$_2$	14.67	1.40
C$_{12}$H$_{16}$NO$_4$	13.76	1.68
C$_{12}$H$_{18}$N$_2$O$_3$	14.14	1.53
C$_{12}$H$_{20}$N$_3$O$_2$	14.51	1.38
C$_{12}$H$_{22}$N$_4$O	14.88	1.24
C$_{13}$H$_4$NO$_4$	14.65	1.80
C$_{13}$H$_6$N$_2$O$_3$	15.02	1.65
C$_{13}$H$_8$N$_3$O$_2$	15.40	1.51
C$_{13}$H$_{10}$N$_4$O	15.77	1.37
C$_{13}$H$_{18}$O$_4$	14.49	1.78
C$_{13}$H$_{20}$NO$_3$	14.87	1.63
C$_{13}$H$_{22}$N$_2$O$_2$	15.24	1.49
C$_{13}$H$_{24}$N$_3$O	15.62	1.34
C$_{13}$H$_{26}$N$_4$	15.99	1.20
C$_{14}$H$_6$O$_4$	15.38	1.90
C$_{14}$H$_8$NO$_3$	15.76	1.76
C$_{14}$H$_{10}$N$_2$O$_2$	16.13	1.62
C$_{14}$H$_{12}$N$_3$O	16.50	1.48
C$_{14}$H$_{14}$N$_4$	16.88	1.34
C$_{14}$H$_{22}$O$_3$	15.60	1.74
C$_{14}$H$_{24}$NO$_2$	15.97	1.59
C$_{14}$H$_{26}$N$_2$O	16.35	1.45
C$_{14}$H$_{28}$N$_3$	16.72	1.31
C$_{15}$H$_2$N$_4$	17.77	1.49
C$_{15}$H$_{10}$O$_3$	16.49	1.87

	M + 1	M + 2
$C_{15}H_{12}NO_2$	16.86	1.73
$C_{15}H_{14}N_2O$	17.24	1.60
$C_{15}H_{16}N_3$	17.61	1.46
$C_{15}H_{28}O_2$	16.70	1.71
$C_{15}H_{28}NO$	17.08	1.57
$C_{15}H_{30}N_2$	17.45	1.43
$C_{16}H_2N_2O$	18.12	1.75
$C_{16}H_4N_3$	18.50	1.62
$C_{16}H_{14}O_2$	17.59	1.85
$C_{16}H_{16}NO$	17.97	1.72
$C_{16}H_{18}N_2$	18.34	1.59
$C_{16}H_{30}O$	17.81	1.69
$C_{16}H_{32}N$	18.18	1.56
$C_{17}H_2O_2$	18.48	2.01
$C_{17}H_4NO$	18.86	1.88
$C_{17}H_6N_2$	19.23	1.75
$C_{17}H_{18}O$	18.70	1.85
$C_{17}H_{20}N$	19.07	1.72
$C_{17}H_{34}$	18.91	1.69
$C_{18}H_6O$	19.59	2.01
$C_{18}H_8N$	19.96	1.89
$C_{18}H_{22}$	19.80	1.86
$C_{19}H_{10}$	20.69	2.03

239

	M + 1	M + 2
$C_{11}H_{15}N_2O_4$	13.05	1.59
$C_{11}H_{17}N_3O_3$	13.42	1.44
$C_{11}H_{19}N_4O_2$	13.80	1.29
$C_{12}H_3N_2O_4$	13.93	1.70
$C_{12}H_5N_3O_3$	14.31	1.55
$C_{12}H_7N_4O_2$	14.68	1.41
$C_{12}H_{17}NO_4$	13.78	1.68
$C_{12}H_{19}N_2O_3$	14.15	1.53
$C_{12}H_{21}N_3O_2$	14.53	1.38
$C_{12}H_{23}N_4O$	14.90	1.24
$C_{13}H_5NO_4$	14.67	1.80
$C_{13}H_7N_2O_3$	15.04	1.66
$C_{13}H_9N_3O_2$	15.41	1.51
$C_{13}H_{11}N_4O$	15.79	1.37
$C_{13}H_{19}O_4$	14.51	1.78
$C_{13}H_{21}NO_3$	14.88	1.63
$C_{13}H_{23}N_2O_2$	15.26	1.49
$C_{13}H_{25}N_3O$	15.63	1.34
$C_{13}H_{27}N_4$	16.01	1.20
$C_{14}H_7O_4$	15.40	1.91
$C_{14}H_9NO_3$	15.77	1.76
$C_{14}H_{11}N_2O_2$	16.15	1.62
$C_{14}H_{13}N_3O$	16.52	1.48
$C_{14}H_{15}N_4$	16.89	1.34
$C_{14}H_{23}O_3$	15.61	1.74
$C_{14}H_{25}NO_2$	15.99	1.60
$C_{14}H_{27}N_2O$	16.36	1.46
$C_{14}H_{29}N_3$	16.74	1.32
$C_{15}HN_3O$	17.41	1.63
$C_{15}H_3N_4$	17.78	1.49
$C_{15}H_{11}O_3$	16.50	1.88
$C_{15}H_{13}NO_2$	16.88	1.74
$C_{15}H_{15}N_2O$	17.25	1.60
$C_{15}H_{17}N_3$	17.63	1.46
$C_{15}H_{27}O_2$	16.72	1.71
$C_{15}H_{29}NO$	17.09	1.57
$C_{15}H_{31}N_2$	17.47	1.44
$C_{16}HNO_2$	17.77	1.88
$C_{16}H_3N_2O$	18.14	1.75
$C_{16}H_5N_3$	18.51	1.62
$C_{16}H_{15}O_2$	17.61	1.86
$C_{16}H_{17}NO$	17.98	1.72
$C_{16}H_{19}N_2$	18.36	1.59
$C_{16}H_{31}O$	17.83	1.70
$C_{16}H_{33}N$	18.20	1.56
$C_{17}H_3O_2$	18.50	2.01
$C_{17}H_5NO$	18.87	1.88
$C_{17}H_7N_2$	19.25	1.75
$C_{17}H_{19}O$	18.71	1.85
$C_{17}H_{21}N$	19.09	1.72
$C_{17}H_{35}$	18.93	1.69
$C_{18}H_7O$	19.60	2.02
$C_{18}H_9N$	19.98	1.89
$C_{18}H_{23}$	19.82	1.86
$C_{19}H_{11}$	20.71	2.03

240

	M + 1	M + 2
$C_{11}H_{16}N_2O_4$	13.06	1.59
$C_{11}H_{18}N_3O_3$	13.44	1.44
$C_{11}H_{20}N_4O_2$	13.81	1.29
$C_{12}H_4N_2O_4$	13.95	1.70
$C_{12}H_6N_3O_3$	14.33	1.56
$C_{12}H_8N_4O_2$	14.70	1.41
$C_{12}H_{18}NO_4$	13.79	1.68
$C_{12}H_{20}N_2O_3$	14.17	1.53
$C_{12}H_{22}N_3O_2$	14.54	1.39
$C_{12}H_{24}N_4O$	14.92	1.24
$C_{13}H_6NO_4$	14.68	1.80
$C_{13}H_8N_2O_3$	15.06	1.66
$C_{13}H_{10}N_3O_2$	15.43	1.51
$C_{13}H_{12}N_4O$	15.81	1.37
$C_{13}H_{20}O_4$	14.52	1.78
$C_{13}H_{22}NO_3$	14.90	1.63
$C_{13}H_{24}N_2O_2$	15.27	1.49
$C_{13}H_{26}N_3O$	15.65	1.35
$C_{13}H_{28}N_4$	16.02	1.20
$C_{14}H_8O_4$	15.41	1.91
$C_{14}H_{10}NO_3$	15.79	1.77
$C_{14}H_{12}N_2O_2$	16.16	1.62
$C_{14}H_{14}N_3O$	16.54	1.48
$C_{14}H_{16}N_4$	16.91	1.35
$C_{14}H_{24}O_3$	15.63	1.74
$C_{14}H_{26}NO_2$	16.00	1.60
$C_{14}H_{28}N_2O$	16.38	1.46
$C_{14}H_{30}N_3$	16.75	1.32
$C_{15}H_2N_3O$	17.42	1.63
$C_{15}H_4N_4$	17.80	1.49
$C_{15}H_{12}O_3$	16.52	1.88
$C_{15}H_{14}NO_2$	16.89	1.74
$C_{15}H_{16}N_2O$	17.27	1.60
$C_{15}H_{18}N_3$	17.64	1.47
$C_{15}H_{28}O_2$	16.74	1.71
$C_{15}H_{30}NO$	17.11	1.58
$C_{15}H_{32}N_2$	17.48	1.44
$C_{16}H_2NO_2$	17.78	1.89
$C_{16}H_4N_2O$	18.16	1.75
$C_{16}H_6N_3$	18.53	1.62
$C_{16}H_{16}O_2$	17.62	1.86
$C_{16}H_{18}NO$	18.00	1.73
$C_{16}H_{20}N_2$	18.37	1.59
$C_{16}H_{32}O$	17.84	1.70
$C_{16}H_{34}N$	18.22	1.56
$C_{17}H_4O_2$	18.51	2.02
$C_{17}H_6NO$	18.89	1.88
$C_{17}H_8N_2$	19.26	1.75
$C_{17}H_{20}O$	18.73	1.86
$C_{17}H_{22}N$	19.10	1.72
$C_{17}H_{36}$	18.95	1.70
$C_{18}H_8O$	19.62	2.02
$C_{18}H_{10}N$	19.99	1.89
$C_{18}H_{24}$	19.84	1.86
$C_{19}H_{12}$	20.72	2.04

241

	M + 1	M + 2
$C_{11}H_{17}N_2O_4$	13.08	1.59
$C_{11}H_{19}N_3O_3$	13.45	1.44
$C_{11}H_{21}N_4O_2$	13.83	1.29
$C_{12}H_5N_2O_4$	13.97	1.71
$C_{12}H_7N_3O_3$	14.34	1.56
$C_{12}H_9N_4O_2$	14.72	1.41
$C_{12}H_{19}NO_4$	13.81	1.68
$C_{12}H_{21}N_2O_3$	14.18	1.54
$C_{12}H_{23}N_3O_2$	14.56	1.39
$C_{12}H_{25}N_4O$	14.93	1.24
$C_{13}H_7NO_4$	14.70	1.81
$C_{13}H_9N_2O_3$	15.07	1.66
$C_{13}H_{11}N_3O_2$	15.45	1.52
$C_{13}H_{13}N_4O$	15.82	1.37
$C_{13}H_{21}O_4$	14.54	1.78
$C_{13}H_{23}NO_3$	14.92	1.64
$C_{13}H_{25}N_2O_2$	15.29	1.49
$C_{13}H_{27}N_3O$	15.66	1.35
$C_{13}H_{29}N_4$	16.04	1.21
$C_{14}HN_4O$	16.71	1.51
$C_{14}H_9O_4$	15.43	1.91
$C_{14}H_{11}NO_3$	15.80	1.77
$C_{14}H_{13}N_2O_2$	16.18	1.63
$C_{14}H_{15}N_3O$	16.55	1.49
$C_{14}H_{17}N_4$	16.93	1.35
$C_{14}H_{25}O_3$	15.65	1.74
$C_{14}H_{27}NO_2$	16.02	1.60
$C_{14}H_{29}N_2O$	16.39	1.46
$C_{14}H_{31}N_3$	16.77	1.32
$C_{15}HN_2O_2$	17.07	1.77
$C_{15}H_3N_3O$	17.44	1.63
$C_{15}H_5N_4$	17.82	1.50
$C_{15}H_{13}O_3$	16.53	1.88
$C_{15}H_{15}NO_2$	16.91	1.74
$C_{15}H_{17}N_2O$	17.28	1.60
$C_{15}H_{19}N_3$	17.66	1.47
$C_{15}H_{29}O_2$	16.75	1.72
$C_{15}H_{31}NO$	17.13	1.58
$C_{15}H_{33}N_2$	17.50	1.44
$C_{16}HO_3$	17.42	2.03
$C_{16}H_3NO_2$	17.80	1.89
$C_{16}H_5N_2O$	18.17	1.76
$C_{16}H_7N_3$	18.55	1.62
$C_{16}H_{17}O_2$	17.64	1.86
$C_{16}H_{19}NO$	18.01	1.73
$C_{16}H_{21}N_2$	18.39	1.60
$C_{16}H_{33}O$	17.86	1.70
$C_{16}H_{35}N$	18.23	1.57
$C_{17}H_5O_2$	18.53	2.02
$C_{17}H_7NO$	18.90	1.89
$C_{17}H_9N_2$	19.28	1.76
$C_{17}H_{21}O$	18.75	1.86
$C_{17}H_{23}N$	19.12	1.73
$C_{18}H_9O$	19.63	2.02
$C_{18}H_{11}N$	20.01	1.90
$C_{18}H_{25}$	19.85	1.87
$C_{19}H_{13}$	20.74	2.04
$C_{20}H$	21.63	2.22

242

	M + 1	M + 2
$C_{11}H_{18}N_2O_4$	13.09	1.59
$C_{11}H_{20}N_3O_3$	13.47	1.44
$C_{11}H_{22}N_4O_2$	13.84	1.29
$C_{12}H_6N_2O_4$	13.98	1.71
$C_{12}H_8N_3O_3$	14.36	1.56
$C_{12}H_{10}N_4O_2$	14.73	1.41
$C_{12}H_{20}NO_4$	13.83	1.69
$C_{12}H_{22}N_2O_3$	14.20	1.54
$C_{12}H_{24}N_3O_2$	14.57	1.39
$C_{12}H_{26}N_4O$	14.95	1.24
$C_{13}H_8NO_4$	14.71	1.81
$C_{13}H_{10}N_2O_3$	15.09	1.66
$C_{13}H_{12}N_3O_2$	15.46	1.52
$C_{13}H_{14}N_4O$	15.84	1.38
$C_{13}H_{22}O_4$	14.56	1.79
$C_{13}H_{24}NO_3$	14.93	1.64
$C_{13}H_{26}N_2O_2$	15.31	1.49
$C_{13}H_{28}N_3O$	15.68	1.35
$C_{13}H_{30}N_4$	16.05	1.21
$C_{14}H_2N_4O$	16.73	1.51
$C_{14}H_{10}O_4$	15.45	1.91
$C_{14}H_{12}NO_3$	15.82	1.77
$C_{14}H_{14}N_2O_2$	16.19	1.63
$C_{14}H_{16}N_3O$	16.57	1.49
$C_{14}H_{18}N_4$	16.94	1.35
$C_{14}H_{26}O_3$	15.66	1.75
$C_{14}H_{28}NO_2$	16.04	1.60
$C_{14}H_{30}N_2O$	16.41	1.46
$C_{14}H_{32}N_3$	16.79	1.32
$C_{15}H_2N_2O_2$	17.08	1.77
$C_{15}H_4N_3O$	17.46	1.63
$C_{15}H_6N_4$	17.83	1.50
$C_{15}H_{14}O_3$	16.55	1.88
$C_{15}H_{16}NO_2$	16.93	1.74
$C_{15}H_{18}N_2O$	17.30	1.61
$C_{15}H_{20}N_3$	17.67	1.47
$C_{15}H_{30}O_2$	16.77	1.72
$C_{15}H_{32}NO$	17.14	1.58
$C_{15}H_{34}N_2$	17.52	1.45
$C_{16}H_2O_3$	17.44	2.03
$C_{16}H_4NO_2$	17.81	1.89
$C_{16}H_6N_2O$	18.19	1.76
$C_{16}H_8N_3$	18.56	1.63
$C_{16}H_{18}O_2$	17.66	1.87
$C_{16}H_{20}NO$	18.03	1.73
$C_{16}H_{22}N_2$	18.41	1.60
$C_{16}H_{34}O$	17.87	1.70
$C_{17}H_6O_2$	18.54	2.02
$C_{17}H_8NO$	18.92	1.89
$C_{17}H_{10}N_2$	19.29	1.76
$C_{17}H_{22}O$	18.76	1.86
$C_{17}H_{24}N$	19.14	1.73
$C_{18}H_{10}O$	19.65	2.03
$C_{18}H_{12}N$	20.02	1.90

	M + 1	M + 2		M + 1	M + 2		M + 1	M + 2		M + 1	M + 2
$C_{18}H_{26}$	19.87	1.87	**244**			$C_{12}H_{11}N_3O_3$	14.41	1.57	$C_{12}H_{28}N_3O_2$	14.64	1.40
$C_{19}H_{14}$	20.76	2.04	$C_{11}H_{20}N_2O_4$	13.13	1.60	$C_{12}H_{13}N_4O_2$	14.78	1.42	$C_{12}H_{30}N_4O$	15.01	1.25
$C_{20}H_2$	21.64	2.23	$C_{11}H_{22}N_3O_3$	13.50	1.45	$C_{12}H_{23}NO_4$	13.87	1.69	$C_{13}H_2N_4O_2$	15.68	1.55
			$C_{11}H_{24}N_4O_2$	13.88	1.30	$C_{12}H_{25}N_2O_3$	14.25	1.54	$C_{13}H_{12}NO_4$	14.78	1.82
243			$C_{12}H_8N_2O_4$	14.01	1.71	$C_{12}H_{27}N_3O_2$	14.62	1.40	$C_{13}H_{14}N_2O_3$	15.15	1.67
$C_{11}H_{19}N_2O_4$	13.11	1.60	$C_{12}H_{10}N_3O_3$	14.39	1.56	$C_{12}H_{29}N_4O$	15.00	1.25	$C_{13}H_{16}N_3O_2$	15.53	1.53
$C_{11}H_{21}N_3O_3$	13.48	1.44	$C_{12}H_{12}N_4O_2$	14.76	1.42	$C_{13}HN_4O_2$	15.67	1.55	$C_{13}H_{18}N_4O$	15.90	1.39
$C_{11}H_{23}N_4O_2$	13.86	1.29	$C_{12}H_{22}NO_4$	13.86	1.69	$C_{13}H_{11}NO_4$	14.76	1.81	$C_{13}H_{26}O_4$	14.62	1.79
$C_{12}H_7N_2O_4$	14.00	1.71	$C_{12}H_{24}N_2O_3$	14.23	1.54	$C_{13}H_{13}N_2O_3$	15.14	1.67	$C_{13}H_{28}NO_3$	15.00	1.65
$C_{12}H_9N_3O_3$	14.37	1.56	$C_{12}H_{26}N_3O_2$	14.61	1.40	$C_{13}H_{15}N_3O_2$	15.51	1.53	$C_{13}H_{30}N_2O_2$	15.37	1.50
$C_{12}H_{11}N_4O_2$	14.75	1.42	$C_{12}H_{28}N_4O$	14.98	1.25	$C_{13}H_{17}N_4O$	15.89	1.38	$C_{14}H_2N_2O_3$	16.04	1.80
$C_{12}H_{21}NO_4$	13.84	1.69	$C_{13}H_{10}NO_4$	14.75	1.81	$C_{13}H_{25}O_4$	14.60	1.79	$C_{14}H_4N_3O_2$	16.42	1.66
$C_{12}H_{23}N_2O_3$	14.22	1.54	$C_{13}H_{12}N_2O_3$	15.12	1.67	$C_{13}H_{27}NO_3$	14.98	1.65	$C_{14}H_6N_4O$	16.79	1.53
$C_{12}H_{25}N_3O_2$	14.59	1.39	$C_{13}H_{14}N_3O_2$	15.49	1.52	$C_{13}H_{29}N_2O_2$	15.35	1.50	$C_{14}H_{14}O_4$	15.51	1.92
$C_{12}H_{27}N_4O$	14.96	1.25	$C_{13}H_{16}N_4O$	15.87	1.38	$C_{13}H_{31}N_3O$	15.73	1.36	$C_{14}H_{16}NO_3$	15.88	1.78
$C_{13}H_9NO_4$	14.73	1.81	$C_{13}H_{24}O_4$	14.59	1.79	$C_{14}HN_2O_3$	16.03	1.80	$C_{14}H_{18}N_2O_2$	16.26	1.64
$C_{13}H_{11}N_2O_3$	15.10	1.66	$C_{13}H_{26}NO_3$	14.96	1.64	$C_{14}H_3N_3O_2$	16.40	1.66	$C_{14}H_{20}N_3O$	16.63	1.50
$C_{13}H_{13}N_3O_2$	15.48	1.52	$C_{13}H_{28}N_2O_2$	15.34	1.50	$C_{14}H_5N_4O$	16.77	1.52	$C_{14}H_{22}N_4$	17.01	1.36
$C_{13}H_{15}N_4O$	15.85	1.38	$C_{13}H_{30}N_3O$	15.71	1.36	$C_{14}H_{13}O_4$	15.49	1.92	$C_{14}H_{30}O_3$	15.73	1.76
$C_{13}H_{23}O_4$	14.57	1.79	$C_{13}H_{32}N_4$	16.09	1.21	$C_{14}H_{15}NO_3$	15.87	1.78	$C_{15}H_2O_4$	16.40	2.06
$C_{13}H_{25}NO_3$	14.95	1.64	$C_{14}H_2N_3O_2$	16.38	1.66	$C_{14}H_{17}N_2O_2$	16.24	1.64	$C_{15}H_4NO_3$	16.77	1.92
$C_{13}H_{27}N_2O_2$	15.32	1.50	$C_{14}H_4N_4O$	16.76	1.52	$C_{14}H_{19}N_3O$	16.62	1.50	$C_{15}H_6N_2O_2$	17.15	1.78
$C_{13}H_{29}N_3O$	15.70	1.35	$C_{14}H_{12}O_4$	15.48	1.92	$C_{14}H_{21}N_4$	16.99	1.36	$C_{15}H_8N_3O$	17.52	1.65
$C_{13}H_{31}N_4$	16.07	1.21	$C_{14}H_{14}NO_3$	15.85	1.78	$C_{14}H_{29}O_3$	15.71	1.75	$C_{15}H_{10}N_4$	17.90	1.51
$C_{14}HN_3O_2$	16.37	1.66	$C_{14}H_{16}N_2O_2$	16.23	1.63	$C_{14}H_{31}NO_2$	16.08	1.61	$C_{15}H_{18}O_3$	16.61	1.89
$C_{14}H_3N_4O$	16.74	1.52	$C_{14}H_{18}N_3O$	16.60	1.49	$C_{15}HO_4$	16.38	2.06	$C_{15}H_{20}NO_2$	16.99	1.76
$C_{14}H_{11}O_4$	15.46	1.92	$C_{14}H_{20}N_4$	16.97	1.36	$C_{15}H_3NO_3$	16.76	1.92	$C_{15}H_{22}N_2O$	17.36	1.62
$C_{14}H_{13}NO_3$	15.84	1.77	$C_{14}H_{28}O_3$	15.69	1.75	$C_{15}H_5N_2O_2$	17.13	1.78	$C_{15}H_{24}N_3$	17.74	1.48
$C_{14}H_{15}N_2O_2$	16.21	1.63	$C_{14}H_{30}NO_2$	16.07	1.61	$C_{15}H_7N_3O$	17.50	1.64	$C_{16}H_6O_3$	17.50	2.04
$C_{14}H_{17}N_3O$	16.58	1.49	$C_{14}H_{32}N_2O$	16.44	1.47	$C_{15}H_9N_4$	17.88	1.51	$C_{16}H_8NO_2$	17.88	1.90
$C_{14}H_{19}N_4$	16.96	1.35	$C_{15}H_2NO_3$	16.74	1.91	$C_{15}H_{17}O_3$	16.60	1.89	$C_{16}H_{10}N_2O$	18.25	1.77
$C_{14}H_{27}O_3$	15.68	1.75	$C_{15}H_4N_2O_2$	17.11	1.78	$C_{15}H_{19}NO_2$	16.97	1.75	$C_{16}H_{12}N_3$	18.63	1.64
$C_{14}H_{29}NO_2$	16.05	1.61	$C_{15}H_6N_3O$	17.49	1.64	$C_{15}H_{21}N_2O$	17.35	1.62	$C_{16}H_{22}O_2$	17.72	1.88
$C_{14}H_{31}N_2O$	16.43	1.47	$C_{15}H_8N_4$	17.86	1.50	$C_{15}H_{23}N_3$	17.72	1.48	$C_{16}H_{24}NO$	18.09	1.74
$C_{14}H_{33}N_3$	16.80	1.33	$C_{15}H_{16}O_3$	16.58	1.89	$C_{16}H_5O_3$	17.49	2.04	$C_{16}H_{26}N_2$	18.47	1.61
$C_{15}HNO_3$	16.72	1.91	$C_{15}H_{18}NO_2$	16.96	1.75	$C_{16}H_7NO_2$	17.86	1.90	$C_{17}H_{10}O_2$	18.61	2.03
$C_{15}H_3N_2O_2$	17.10	1.77	$C_{15}H_{20}N_2O$	17.33	1.61	$C_{16}H_9N_2O$	18.24	1.77	$C_{17}H_{12}NO$	18.98	1.90
$C_{15}H_5N_3O$	17.47	1.64	$C_{15}H_{22}N_3$	17.71	1.48	$C_{16}H_{11}N_3$	18.61	1.64	$C_{17}H_{14}N_2$	19.36	1.77
$C_{15}H_7N_4$	17.85	1.50	$C_{15}H_{32}O_2$	16.80	1.72	$C_{16}H_{21}O_2$	17.70	1.87	$C_{17}H_{26}O$	18.83	1.87
$C_{15}H_{15}O_3$	16.57	1.89	$C_{16}H_4O_3$	17.47	2.03	$C_{16}H_{23}NO$	18.08	1.74	$C_{17}H_{28}N$	19.20	1.74
$C_{15}H_{17}NO_2$	16.94	1.75	$C_{16}H_6NO_2$	17.85	1.90	$C_{16}H_{25}N_2$	18.45	1.61	$C_{18}H_2N_2$	20.25	1.94
$C_{15}H_{19}N_2O$	17.32	1.61	$C_{16}H_8N_2O$	18.22	1.77	$C_{17}H_9O_2$	18.59	2.03	$C_{18}H_{14}O$	19.71	2.04
$C_{15}H_{21}N_3$	17.69	1.47	$C_{16}H_{10}N_3$	18.59	1.63	$C_{17}H_{11}NO$	18.97	1.90	$C_{18}H_{16}N$	20.09	1.91
$C_{15}H_{31}O_2$	16.78	1.72	$C_{16}H_{20}O_2$	17.69	1.87	$C_{17}H_{13}N_2$	19.34	1.77	$C_{18}H_{30}$	19.93	1.88
$C_{15}H_{33}NO$	17.16	1.58	$C_{16}H_{22}NO$	18.06	1.74	$C_{17}H_{25}O$	18.81	1.87	$C_{19}H_2O$	20.60	2.21
$C_{16}H_3O_3$	17.46	2.03	$C_{16}H_{24}N_2$	18.44	1.60	$C_{17}H_{27}N$	19.18	1.74	$C_{19}H_4N$	20.98	2.09
$C_{16}H_5NO_2$	17.83	1.90	$C_{17}H_8O_2$	18.58	2.03	$C_{18}HN_2$	20.23	1.94	$C_{19}H_{18}$	20.82	2.06
$C_{16}H_7N_2O$	18.20	1.76	$C_{17}H_{10}NO$	18.95	1.90	$C_{18}H_{13}O$	19.70	2.04	$C_{20}H_6$	21.71	2.24
$C_{16}H_9N_3$	18.58	1.63	$C_{17}H_{12}N_2$	19.33	1.77	$C_{18}H_{15}N$	20.07	1.91			
$C_{16}H_{19}O_2$	17.67	1.87	$C_{17}H_{24}O$	18.79	1.87	$C_{18}H_{29}$	19.92	1.88			
$C_{16}H_{21}NO$	18.05	1.73	$C_{17}H_{26}N$	19.17	1.74	$C_{19}H_{17}$	20.80	2.05	**247**		
$C_{16}H_{23}N_2$	18.42	1.60	$C_{18}H_{12}O$	19.68	2.03	$C_{19}HO$	20.59	2.21	$C_{11}H_{23}N_2O_4$	13.17	1.60
$C_{17}H_7O_2$	18.56	2.02	$C_{18}H_{14}N$	20.06	1.91	$C_{19}H_3N$	20.96	2.09	$C_{11}H_{25}N_3O_3$	13.55	1.45
$C_{17}H_9NO$	18.94	1.89	$C_{18}H_{28}$	19.90	1.87	$C_{20}H_5$	21.69	2.24	$C_{11}H_{27}N_4O_2$	13.92	1.30
$C_{17}H_{11}N_2$	19.31	1.76	$C_{19}H_2N$	20.95	2.08				$C_{12}H_{11}N_2O_4$	14.06	1.72
$C_{17}H_{23}O$	18.78	1.86	$C_{19}H_{16}$	20.79	2.05	**246**			$C_{12}H_{13}N_3O_3$	14.44	1.57
$C_{17}H_{25}N$	19.15	1.73	$C_{20}H_4$	21.68	2.23	$C_{11}H_{22}N_2O_4$	13.16	1.60	$C_{12}H_{15}N_4O_2$	14.81	1.42
$C_{18}H_{11}O$	19.67	2.03				$C_{11}H_{24}N_3O_3$	13.53	1.45	$C_{12}H_{25}NO_4$	13.91	1.70
$C_{18}H_{13}N$	20.04	1.90				$C_{11}H_{26}N_4O_2$	13.91	1.30	$C_{12}H_{27}N_2O_3$	14.28	1.55
$C_{18}H_{27}$	19.88	1.87	**245**			$C_{12}H_{10}N_2O_4$	14.05	1.72	$C_{12}H_{29}N_3O_2$	14.65	1.40
$C_{19}HN$	20.93	2.08	$C_{11}H_{21}N_2O_4$	13.14	1.60	$C_{12}H_{12}N_3O_3$	14.42	1.57	$C_{13}HN_3O_3$	15.33	1.70
$C_{19}H_{15}$	20.77	2.05	$C_{11}H_{23}N_3O_3$	13.52	1.45	$C_{12}H_{14}N_4O_2$	14.80	1.42	$C_{13}H_3N_4O_2$	15.70	1.55
$C_{20}H_3$	21.66	2.23	$C_{11}H_{25}N_4O_2$	13.89	1.30	$C_{12}H_{24}NO_4$	13.89	1.70	$C_{13}H_{13}NO_4$	14.79	1.82
			$C_{12}H_9N_2O_4$	14.03	1.71	$C_{12}H_{26}N_2O_3$	14.26	1.55	$C_{13}H_{15}N_2O_3$	15.17	1.67

	M+1	M+2
$C_{13}H_{17}N_3O_2$	15.54	1.53
$C_{13}H_{19}N_4O$	15.92	1.39
$C_{13}H_{27}O_4$	14.64	1.80
$C_{13}H_{29}NO_3$	15.01	1.65
$C_{14}HNO_4$	15.68	1.95
$C_{14}H_3N_2O_3$	16.06	1.81
$C_{14}H_5N_3O_2$	16.43	1.67
$C_{14}H_7N_4O$	16.81	1.53
$C_{14}H_{15}O_4$	15.53	1.93
$C_{14}H_{17}NO_3$	15.90	1.78
$C_{14}H_{19}N_2O_2$	16.27	1.64
$C_{14}H_{21}N_3O$	16.65	1.50
$C_{14}H_{23}N_4$	17.02	1.36
$C_{15}H_3O_4$	16.41	2.06
$C_{15}H_5NO_3$	16.79	1.92
$C_{15}H_7N_2O_2$	17.16	1.78
$C_{15}H_9N_3O$	17.54	1.65
$C_{15}H_{11}N_4$	17.91	1.51
$C_{15}H_{19}O_3$	16.63	1.90
$C_{15}H_{21}NO_2$	17.01	1.76
$C_{15}H_{23}N_2O$	17.38	1.62
$C_{15}H_{25}N_3$	17.75	1.49
$C_{16}H_7O_3$	17.52	2.04
$C_{16}H_9NO_2$	17.89	1.91
$C_{16}H_{11}N_2O$	18.27	1.77
$C_{16}H_{13}N_3$	18.64	1.64
$C_{16}H_{23}O_2$	17.74	1.88
$C_{16}H_{25}NO$	18.11	1.75
$C_{16}H_{27}N_2$	18.49	1.61
$C_{17}HN_3$	19.53	1.81
$C_{17}H_{11}O_2$	18.62	2.04
$C_{17}H_{13}NO$	19.00	1.91
$C_{17}H_{15}N_2$	19.37	1.78
$C_{17}H_{27}O$	18.84	1.88
$C_{17}H_{29}N$	19.22	1.75
$C_{18}HNO$	19.89	2.07
$C_{18}H_3N_2$	20.26	1.95
$C_{18}H_{15}O$	19.73	2.04
$C_{18}H_{17}N$	20.10	1.92
$C_{18}H_{31}$	19.95	1.88
$C_{19}H_3O$	20.62	2.22
$C_{19}H_5N$	20.99	2.09
$C_{19}H_{19}$	20.84	2.06
$C_{20}H_7$	21.72	2.24

248

	M+1	M+2
$C_{11}H_{24}N_2O_4$	13.19	1.61
$C_{11}H_{26}N_3O_3$	13.56	1.45
$C_{11}H_{28}N_4O_2$	13.94	1.31
$C_{12}H_{12}N_2O_4$	14.08	1.72
$C_{12}H_{14}N_3O_3$	14.45	1.57
$C_{12}H_{16}N_4O_2$	14.83	1.43
$C_{12}H_{26}NO_4$	13.92	1.70
$C_{12}H_{28}N_2O_3$	14.30	1.55

	M+1	M+2
$C_{13}H_2N_3O_3$	15.34	1.70
$C_{13}H_4N_4O_2$	15.72	1.56
$C_{13}H_{14}NO_4$	14.81	1.82
$C_{13}H_{16}N_2O_3$	15.18	1.68
$C_{13}H_{18}N_3O_2$	15.56	1.53
$C_{13}H_{20}N_4O$	15.93	1.39
$C_{13}H_{28}O_4$	14.65	1.80
$C_{14}H_2NO_4$	15.70	1.95
$C_{14}H_4N_2O_3$	16.07	1.81
$C_{14}H_6N_3O_2$	16.45	1.67
$C_{14}H_8N_4O$	16.82	1.53
$C_{14}H_{16}O_4$	15.54	1.93
$C_{14}H_{18}NO_3$	15.92	1.79
$C_{14}H_{20}N_2O_2$	16.29	1.64
$C_{14}H_{22}N_3O$	16.66	1.51
$C_{14}H_{24}N_4$	17.04	1.37
$C_{15}H_4O_4$	16.43	2.06
$C_{15}H_6NO_3$	16.80	1.92
$C_{15}H_8N_2O_2$	17.18	1.79
$C_{15}H_{10}N_3O$	17.55	1.65
$C_{15}H_{12}N_4$	17.93	1.52
$C_{15}H_{20}O_3$	16.65	1.90
$C_{15}H_{22}NO_2$	17.02	1.76
$C_{15}H_{24}N_2O$	17.40	1.62
$C_{15}H_{26}N_3$	17.77	1.49
$C_{16}H_8O_3$	17.54	2.05
$C_{16}H_{10}NO_2$	17.91	1.91
$C_{16}H_{12}N_2O$	18.28	1.78
$C_{16}H_{14}N_3$	18.66	1.65
$C_{16}H_{24}O_2$	17.75	1.88
$C_{16}H_{26}NO$	18.13	1.75
$C_{16}H_{28}N_2$	18.50	1.62
$C_{17}H_2N_3$	19.55	1.81
$C_{17}H_{12}O_2$	18.64	2.04
$C_{17}H_{14}NO$	19.02	1.91
$C_{17}H_{16}N_2$	19.39	1.78
$C_{17}H_{28}O$	18.86	1.88
$C_{17}H_{30}N$	19.23	1.75
$C_{18}H_2NO$	19.90	2.08
$C_{18}H_4N_2$	20.28	1.95
$C_{18}H_{16}O$	19.75	2.04
$C_{18}H_{18}N$	20.12	1.92
$C_{18}H_{32}$	19.96	1.89
$C_{19}H_4O$	20.64	2.22
$C_{19}H_6N$	21.01	2.10
$C_{19}H_{20}$	20.85	2.06
$C_{20}H_8$	21.74	2.25

249

	M+1	M+2
$C_{11}H_{25}N_2O_4$	13.21	1.61
$C_{11}H_{27}N_3O_3$	13.58	1.46
$C_{12}H_{13}N_2O_4$	14.10	1.72
$C_{12}H_{15}N_3O_3$	14.47	1.58
$C_{12}H_{17}N_4O_2$	14.84	1.43

	M+1	M+2
$C_{12}H_{27}NO_4$	13.94	1.70
$C_{13}HN_2O_4$	14.98	1.85
$C_{13}H_3N_3O_3$	15.36	1.70
$C_{13}H_5N_4O_2$	15.73	1.56
$C_{13}H_{15}NO_4$	14.83	1.82
$C_{13}H_{17}N_2O_3$	15.20	1.68
$C_{13}H_{19}N_3O_2$	15.57	1.54
$C_{13}H_{21}N_4O$	15.95	1.39
$C_{14}H_3NO_3$	15.71	1.95
$C_{14}H_5N_2O_3$	16.09	1.81
$C_{14}H_7N_3O_2$	16.46	1.67
$C_{14}H_9N_4O$	16.84	1.53
$C_{14}H_{17}O_4$	15.56	1.93
$C_{14}H_{19}NO_3$	15.93	1.79
$C_{14}H_{21}N_2O_2$	16.31	1.65
$C_{14}H_{23}N_3O$	16.68	1.51
$C_{14}H_{25}N_4$	17.05	1.37
$C_{15}H_5O_4$	16.45	2.07
$C_{15}H_7NO_3$	16.82	1.93
$C_{15}H_9N_2O_2$	17.19	1.79
$C_{15}H_{11}N_3O$	17.57	1.65
$C_{15}H_{13}N_4$	17.94	1.52
$C_{15}H_{21}O_3$	16.66	1.90
$C_{15}H_{23}NO_2$	17.04	1.76
$C_{15}H_{25}N_2O$	17.41	1.63
$C_{15}H_{27}N_3$	17.79	1.49
$C_{16}HN_4$	18.83	1.68
$C_{16}H_9O_3$	17.55	2.05
$C_{16}H_{11}NO_2$	17.93	1.91
$C_{16}H_{13}N_2O$	18.30	1.78
$C_{16}H_{15}N_3$	18.67	1.65
$C_{16}H_{25}O_2$	17.77	1.89
$C_{16}H_{27}NO$	18.14	1.75
$C_{16}H_{29}N_2$	18.52	1.62
$C_{17}HN_2O$	19.19	1.94
$C_{17}H_3N_3$	19.56	1.81
$C_{17}H_{13}O_2$	18.66	2.04
$C_{17}H_{15}NO$	19.03	1.91
$C_{17}H_{17}N_2$	19.41	1.78
$C_{17}H_{29}O$	18.87	1.88
$C_{17}H_{31}N$	19.25	1.75
$C_{18}HO_2$	19.55	2.21
$C_{18}H_3NO$	19.92	2.08
$C_{18}H_5N_2$	20.29	1.95
$C_{18}H_{17}O$	19.76	2.05
$C_{18}H_{19}N$	20.14	1.92
$C_{18}H_{33}$	19.98	1.89
$C_{19}H_5O$	20.65	2.22
$C_{19}H_7N$	21.03	2.10
$C_{19}H_{21}$	20.87	2.07
$C_{20}H_9$	21.76	2.25

250

	M+1	M+2
$C_{11}H_{26}N_2O_4$	13.22	1.61

	M+1	M+2
$C_{12}H_{14}N_2O_4$	14.11	1.73
$C_{12}H_{16}N_3O_3$	14.49	1.58
$C_{12}H_{18}N_4O_2$	14.86	1.43
$C_{13}H_2N_2O_4$	15.00	1.85
$C_{13}H_4N_3O_3$	15.37	1.71
$C_{13}H_6N_4O_2$	15.75	1.56
$C_{13}H_{16}NO_4$	14.84	1.83
$C_{13}H_{18}N_2O_3$	15.22	1.68
$C_{13}H_{20}N_3O_2$	15.59	1.54
$C_{13}H_{22}N_4O$	15.97	1.40
$C_{14}H_4NO_4$	15.73	1.96
$C_{14}H_6N_2O_3$	16.11	1.82
$C_{14}H_8N_3O_2$	16.48	1.67
$C_{14}H_{10}N_4O$	16.85	1.54
$C_{14}H_{18}O_4$	15.57	1.93
$C_{14}H_{20}NO_3$	15.95	1.79
$C_{14}H_{22}N_2O_2$	16.32	1.65
$C_{14}H_{24}N_3O$	16.70	1.51
$C_{14}H_{26}N_4$	17.07	1.37
$C_{15}H_6O_4$	16.46	2.07
$C_{15}H_8NO_3$	16.84	1.93
$C_{15}H_{10}N_2O_2$	17.21	1.79
$C_{15}H_{12}N_3O$	17.58	1.66
$C_{15}H_{14}N_4$	17.96	1.52
$C_{15}H_{22}O_3$	16.68	1.90
$C_{15}H_{24}NO_2$	17.05	1.77
$C_{15}H_{26}N_2O$	17.43	1.63
$C_{15}H_{28}N_3$	17.80	1.49
$C_{16}H_2N_4$	18.85	1.68
$C_{16}H_{10}O_3$	17.57	2.05
$C_{16}H_{12}NO_2$	17.94	1.92
$C_{16}H_{14}N_2O$	18.32	1.78
$C_{16}H_{16}N_3$	18.69	1.65
$C_{16}H_{26}O_2$	17.78	1.89
$C_{16}H_{28}NO$	18.16	1.75
$C_{16}H_{30}N_2$	18.53	1.62
$C_{17}H_2N_2O$	19.20	1.94
$C_{17}H_4N_3$	19.58	1.82
$C_{17}H_{14}O_2$	18.67	2.05
$C_{17}H_{16}NO$	19.05	1.91
$C_{17}H_{18}N_2$	19.42	1.79
$C_{17}H_{30}O$	18.89	1.89
$C_{17}H_{32}N$	19.26	1.76
$C_{18}H_2O_2$	19.56	2.21
$C_{18}H_4NO$	19.94	2.08
$C_{18}H_6N_2$	20.31	1.96
$C_{18}H_{18}O$	19.78	2.05
$C_{18}H_{20}N$	20.15	1.92
$C_{18}H_{34}$	20.00	1.89
$C_{19}H_6O$	20.67	2.23
$C_{19}H_8N$	21.04	2.10
$C_{19}H_{22}$	20.88	2.07
$C_{20}H_{10}$	21.77	2.25

appendix b

COMMON FRAGMENT IONS

Not all members of homologous and isomeric series are given. The list is meant to be suggestive rather than exhaustive. Appendix II of Hamming and Foster,[10f] Table A-7 of McLafferty's Interpretative book,[10c] and the high-resolution ion data of McLafferty are recommended as supplements.

m/e	Ions*
14	CH_2
15	CH_3
16	O
17	OH
18	H_2O, NH_4
19	F, H_3O
26	$C \equiv N$
27	C_2H_3
28	C_2H_4, CO, N_2 (air), $CH=NH$
29	C_2H_5, CHO
30	CH_2NH_2, NO
31	CH_2OH, OCH_3
32	O_2 (Air)
33	SH, CH_2F
34	H_2S
35	Cl
36	HCl
39	C_3H_3
40	$CH_2C \equiv N$, Ar(Air)
41	C_3H_5, $CH_2=C \equiv N + H$, C_2H_2NH
42	C_3H_6
43	C_3H_7, $CH_3C=O$, C_2H_5N
44	$CH_2\overset{H}{\underset{|}{C}}=O + H$, CH_3CHNH_2, CO_2, $NH_2C=O$, $(CH_3)_2N$
45	$\overset{CH_3}{\underset{|}{C}HOH}$, CH_2CH_2OH, CH_2OCH_3, $\overset{O}{\overset{\|}{C}}-OH$, $CH_3CH-O + H$

m/e	Ions
46	NO_2
47	CH_2SH, CH_3S
48	$CH_3S + H$
49	CH_2Cl
51	CHF_2
53	C_4H_5
54	$CH_2CH_2C \equiv N$
55	C_4H_7, $CH_2=CHC=O$
56	C_4H_8
57	C_4H_9, $C_2H_5C=O$
58	$CH_3-\overset{O}{\overset{\|}{C}} + H$, $C_2H_5CHNH_2$, $(CH_3)_2NCH_2$, $\underset{CH_2}{}$ $C_2H_5NHCH_2$, C_2H_2S
59	$(CH_3)_2COH$, $CH_2OC_2H_5$, $\overset{O}{\overset{\|}{C}}-OCH_3$, $NH_2\overset{O}{\underset{CH_2}{\overset{\|}{C}}}=O + H$, CH_3OCHCH_3, CH_3CHCH_2OH
60	$CH_2\overset{O}{\overset{\|}{C}} + H$, CH_2ONO
61	$\overset{O}{\overset{\|}{C}}\overset{OH}{\underset{}{-OCH_3}} + 2H$, CH_2CH_2SH, CH_2SCH_3
65	$\equiv C_5H_5$

*Ions indicated as a fragment + nH (n = 1, 2, 3, . . .) are ions that arise via rearrangement involving hydrogen transfer.

m/e	Ions

66 ⬠ (+·) ≡ C_5H_6

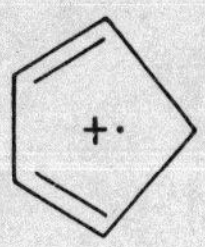

67 C_5H_7

68 $CH_2CH_2CH_2C{\equiv}N$

69 C_5H_9, CF_3, $CH_3CH{=}CHC{=}O$, $CH_2{=}C(CH_3)C{=}O$

70 C_5H_{10}

71 C_5H_{11}, $C_3H_7C{=}O$

72 $C_2H_5\overset{O}{C}+H$ (with CH_2), $C_3H_7CHNH_2$, $(CH_3)_2N{=}C{=}O$,
$C_2H_5NHCHCH_3$, and isomers

73 Homologs of 59

74 $CH_2{-}\overset{O}{C}{-}OCH_3 + H$

75 $\overset{O}{C}{-}OC_2H_5 + 2H$, $CH_2SC_2H_5$, $(CH_3)_2CSH$,
$(CH_3O)_2CH$

77 C_6H_5

78 $C_6H_5 + H$

79 $C_6H_5 + 2H$, Br

80 [pyrrole]$-CH_2$, $CH_3SS + H$

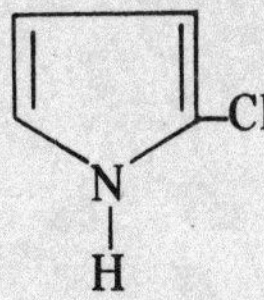

81 [furan]$-CH_2$, C_6H_9, [methylcyclohexene]

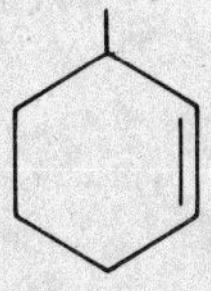

82 $CH_2CH_2CH_2CH_2C{\equiv}N$, CCl_2, C_6H_{10}

83 C_6H_{11}, $CHCl_2$, [thiophene]

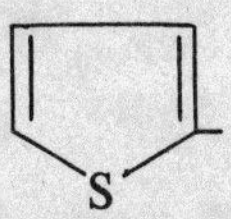

85 C_6H_{13}, $C_4H_9C{=}O$, $CClF_2$

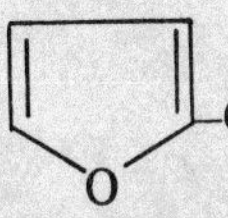

86 $C_3H_7\overset{O}{C}+H$ (with CH_2), $C_4H_9CHNH_2$, and isomers.

87 $C_3H_7\overset{O}{C}O$, homologs of 73, $CH_2CH_2\overset{O}{C}OCH_3$

88 $CH_2{-}\overset{O}{C}{-}OC_2H_5 + H$

89 $\overset{O}{C}{-}OC_3H_7 + 2H$, [benzene]$-C$

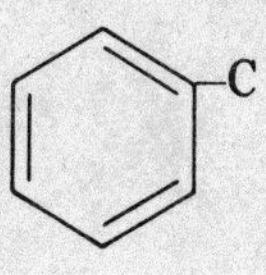

90 CH_3CHONO_2, [benzene]$-CH$

91 [benzene]$-CH_2$, [benzene]$-CH + H$, [benzene]$-C + 2H$,
$(CH_2)_4Cl$, [pyridine N]

92 [pyridine]$-CH_2$, [cyclohexadienone], [benzene]$-CH_2 + H$

93 CH_2Br, [cresol]$-OH$, C_7H_9, [pyrrole]$-C{=}O$,
[phenoxy]$-O$, C_7H_9 (terpenes)

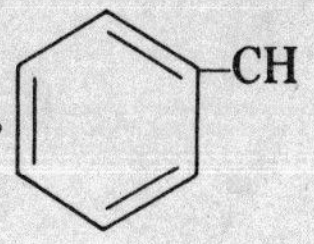

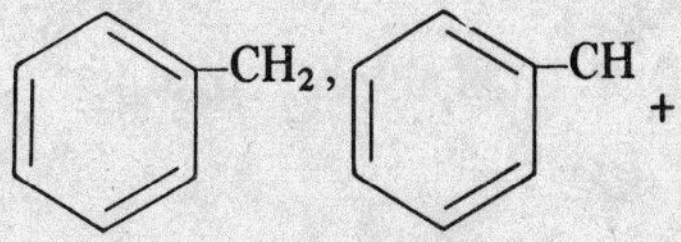

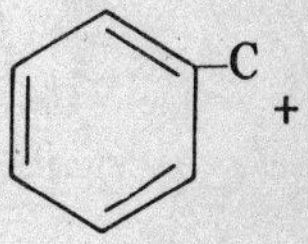

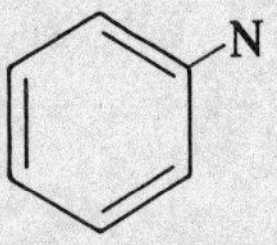

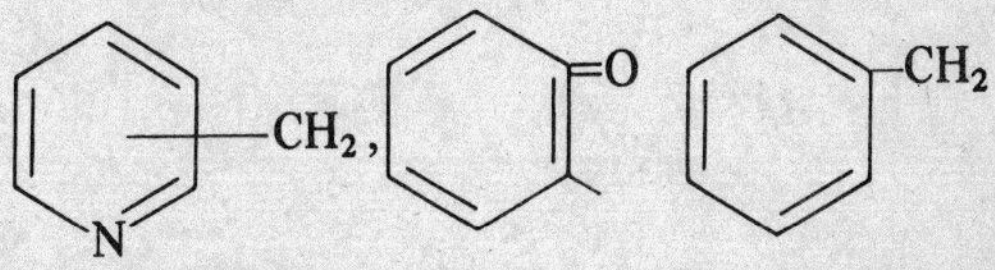

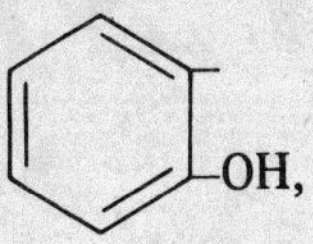

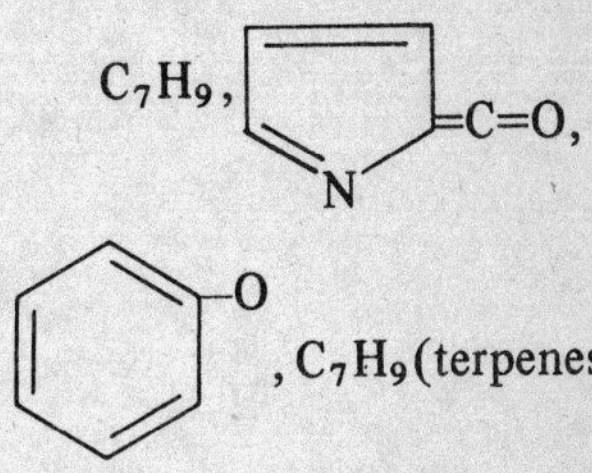

94 [phenol]$-O + H$, [pyrrole]$-C{=}O$

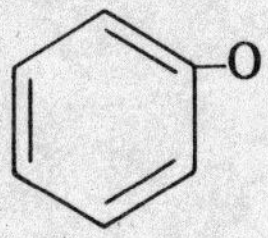

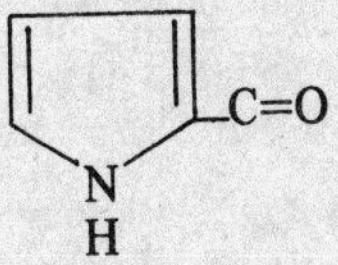

95 [furan]$-C{=}O$

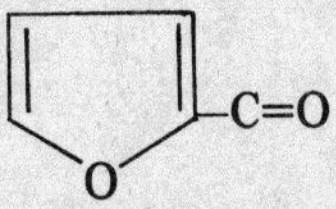

96 $CH_2CH_2CH_2CH_2CH_2C{\equiv}N$

97 C_7H_{13}, [thiophene]$-CH_2$

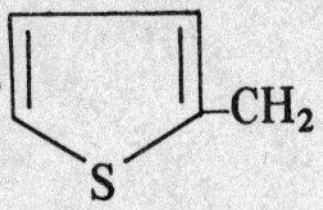

98 [furan]$-CH_2O + H$

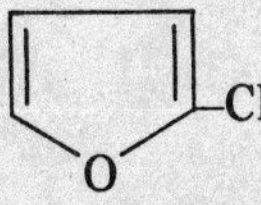

99 C_7H_{15}, $C_6H_{11}O$

100 $C_4H_9\overset{O}{C}+H$ (with CH_2), $C_5H_{11}CHNH_2$

101 $\overset{O}{C}{-}OC_4H_9$

m/e	Ions
102	$CH_2\overset{\displaystyle O}{\overset{\|}{C}}-OC_3H_7 + H$
103	$\overset{\displaystyle O}{\overset{\|}{C}}-OC_4H_9 + 2H,\ C_5H_{11}S,\ CH(OCH_2CH_3)_2$
104	$C_2H_5CHONO_2$
105	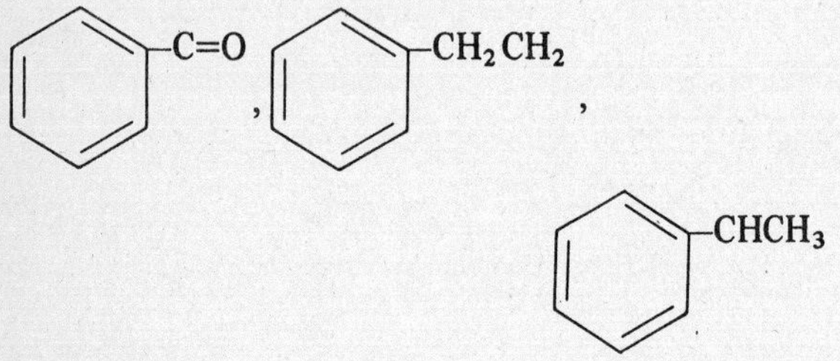 (benzene ring)–C=O , (benzene ring)–CH_2CH_2 , (benzene ring)–$CHCH_3$
106	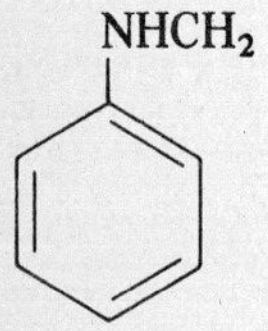 (benzene ring)–$NHCH_2$
107	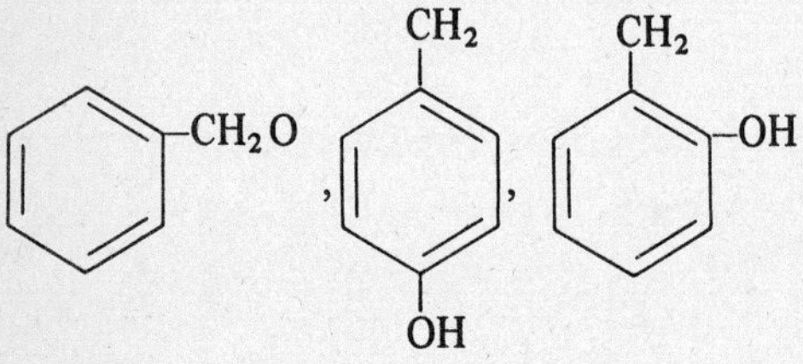 (benzene ring)–CH_2O , (ring with CH_2 and OH, para) , (ring with CH_2 and OH, ortho)
108	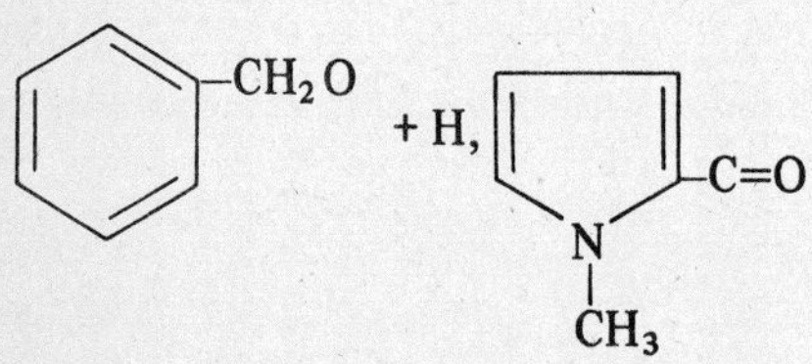 (benzene ring)–CH_2O + H , (N-methylpyrrole)–C=O
109	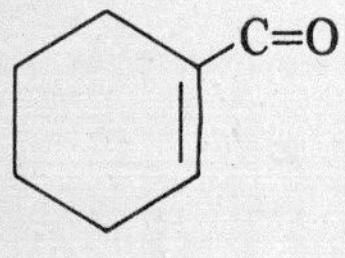 (cyclohexenyl)–C=O
111	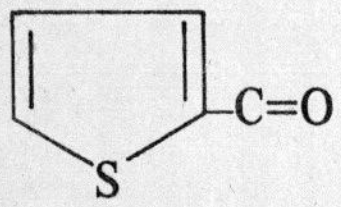 (thiophene)–C=O
119	CF_3CF_2 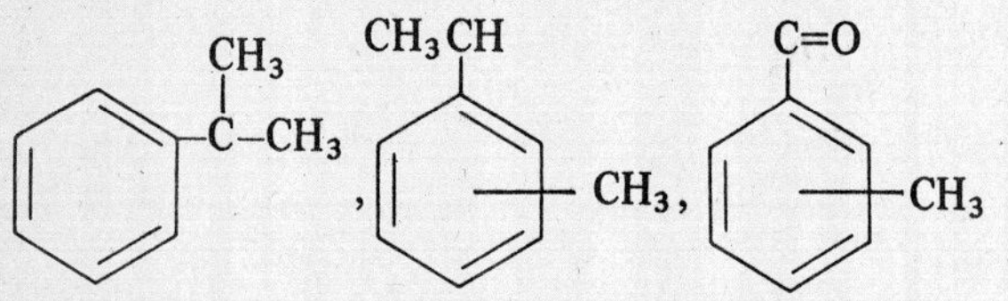 , (benzene ring)–$C(CH_3)_2$, (ring)–$CH(CH_3)$–CH_3 , (ring)–C=O with CH_3
120	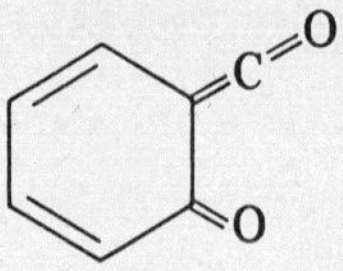 (cyclohexadiene-dione with C=O and O)
121	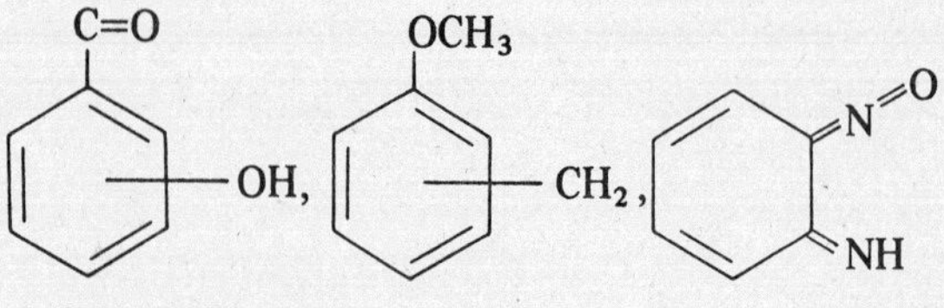 (ring)–C=O and –OH , (ring with OCH_3)–CH_2 , C_9H_{13} (terpenes) , (ring)=N–O and =NH
123	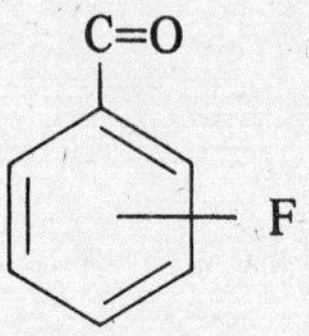 (benzene ring)–C=O , –F
125	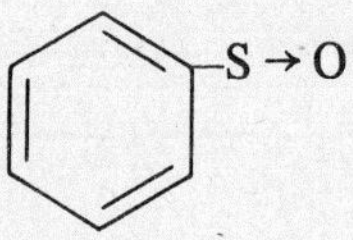 (benzene ring)–$S\rightarrow O$
127	I
131	C_3F_5 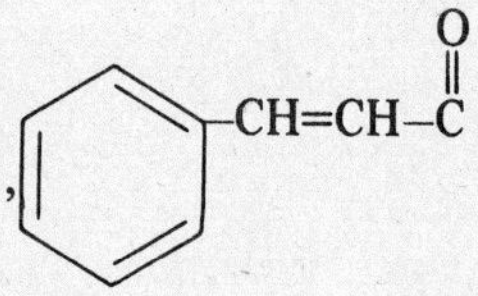 , (benzene ring)–$CH=CH$–C=O
135	$(CH_2)_4Br$
138	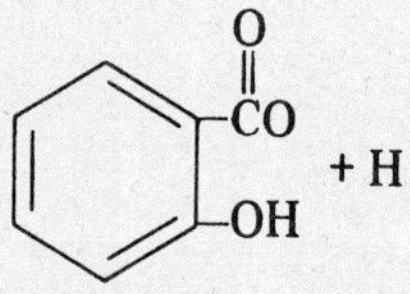 (benzene ring)–CO and –OH + H
139	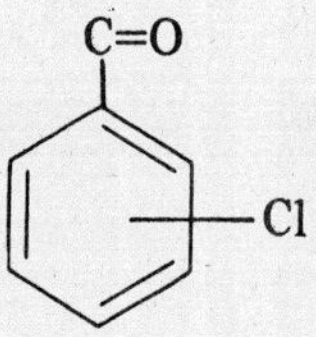 (benzene ring)–C=O and –Cl
149	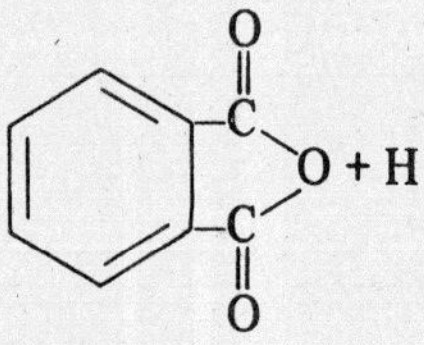 (phthalic anhydride) + H
154	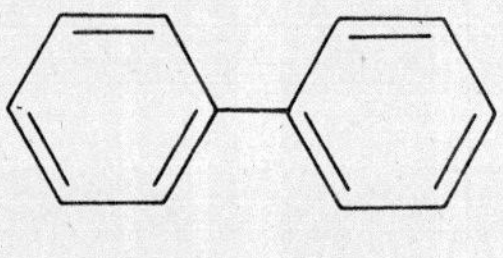 (biphenyl)

70   spectrometric identification of organic compounds

appendix c

COMMON FRAGMENTS LOST

This list is suggestive rather than comprehensive. It should be used in conjunction with Appendix B. Table 5-19 of Hamming and Foster[10f] and Table A-5 of McLafferty[10c] are recommended as supplements

Molecular Ion Minus	Fragment Lost	Molecular Ion Minus	Fragment Lost
1	H·	52	C_4H_4, C_2N_2
15	CH_3·	53	C_4H_5
17	HO·	54	CH_2=CH–CH=CH_2
18	H_2O	55	CH_2=CHCHCH$_3$
19	F·	56	CH_2=CHCH$_2$CH$_3$, CH$_3$CH=CHCH$_3$, 2CO
20	HF	57	C_4H_9·
26	CH≡CH, ·C≡N	58	·NCS, (NO + CO), CH$_3$COCH$_3$
27	CH_2=CH·, HC≡N		
28	CH_2=CH_2, CO, (HCN + H)	59	$CH_3O\overset{O}{\overset{\|}{C}}$·, $CH_3\overset{O}{\overset{\|}{C}}NH_2$, $\overset{\overset{H}{\underset{S}{}}}{\triangle}$
29	CH_3CH_2·, ·CHO	60	C_3H_7OH
30	NH_2CH_2·, CH_2O, NO		
31	·OCH_3, ·CH_2OH, CH_3NH_2	61	CH_3CH_2S·, $\overset{\overset{H}{\underset{S}{}}}{\triangle}$
32	CH_3OH, S		
33	HS·, (·CH_3 and H_2O)	62	[H_2S and CH_2=CH_2]
34	H_2S	63	·CH_2CH_2Cl
35	Cl·	64	C_5H_4, S_2, SO_2
36	HCl, $2H_2O$		CH_3
37	H_2Cl (or HCl + H)	68	CH_2=C–CH=CH_2
38	C_3H_2·, C_2N, F_2	69	CF_3·, C_5H_9·
39	C_3H_3, HC_2N	71	C_5H_{11}·
40	CH_3C≡CH		O
41	CH_2=CHCH$_2$·	73	$CH_3CH_2O\overset{O}{\overset{\|}{C}}$·
		74	C_4H_9OH
	CH_2	75	C_6H_3
42	CH_2=CHCH$_3$, CH_2=C=O, CH_2–CH_2, NCO, NCNH$_2$	76	C_6H_4, CS_2
		77	C_6H_5, CS_2H
	O	78	C_6H_6, CS_2H_2, C_5H_4N
43	C_3H_7·, $CH_3\overset{O}{\overset{\|}{C}}$·, CH_2=CH–O·, [CH_3· and CH_2=CH_2], HCNO	79	Br·, C_5H_5N
		80	HBr
44	CH_2=CHOH, CO_2, N_2O, CONH$_2$, NHCH$_2$CH$_3$	85	·CClF$_2$
		100	CF_2=CF_2
45	CH_3CHOH, CH_3CH_2O·, CO_2H, $CH_3CH_2NH_2$	119	CF_3–CF_2·
46	[H_2O and CH_2=CH_2], CH_3CH_2OH, ·NO_2	122	C_6H_5COOH
47	CH_3S·	127	I·
48	CH_3SH, SO, O_3	128	HI
49	·CH_2Cl		
51	·CHF_2		

I. INTRODUCTION

Infrared radiation refers broadly to that part of the electromagnetic spectrum between the visible and microwave regions. Of greatest practical use to the organic chemist is the limited portion between 4000 cm^{-1} and 666 cm^{-1} (2.5–15.0 μm). Recently there has been increasing interest in the near infrared region, 14,290–4000 cm^{-1} (0.7–2.5 μm) and the far infrared region, 700–200 cm^{-1} (14.3–50 μm).

From the brief theoretical discussion that follows, it is clear that even a very simple molecule can give an extremely complex spectrum. The organic chemist takes advantage of this complexity when he matches the spectrum of an unknown compound against that of an authentic sample. A peak-by-peak correlation is excellent evidence for identity. It is unlikely that any two compounds, except optical enantiomorphs, give the same infrared spectrum.

Although the infrared spectrum is characteristic of the entire molecule, it turns out that certain groups of atoms give rise to bands at or near the same frequency regardless of the structure of the rest of the molecule. It is the persistence of these characteristic bands that permits the chemist to obtain useful structural information by simple inspection and reference to generalized charts of characteristic group frequencies. We shall rely heavily upon these characteristic group frequencies.

Since we are not solely dependent upon infrared spectra for identification, a detailed analysis of the spectrum will not be required. Following our general plan, we shall present only sufficient theory to accomplish our purpose: utilization of infrared spectra in conjunction with other spectral data to determine molecular structure.

Nearly all academic and industrial laboratories make infrared spectrophotometers available as bench tools for the organic chemist. A simplified infrared spectrophotometer

73

costs about $5000. Precision instruments are available at $12,000 to $17,000. Spectrophotometers covering the range of 667–33 cm^{-1} (15–300 μm) cost approximately $35,000. Since the organic chemist very frequently obtains his own infrared spectra, we shall describe instrumentation and sample preparation in somewhat more detail than is given in the chapters on mass spectrometry and NMR spectrometry.

The increased emphasis on infrared spectrometry, as a tool of the practicing organic chemist, is readily apparent from the number of books devoted wholly or in part to discussions of applications of infrared spectrometry, which have been published. There is no lack of reference material covering all aspects of infrared spectrometry.[1–28] The text by Colthup, Daly, and Wiberly[12] presents a thorough coverage of theory, practice, and spectra-structure correlations. The compact manuals by Cross,[4] Flett,[15] Szymanski[16] and Nakanishi[13] are convenient sources of concise information. The volumes by Potts[26] and Miller[27] are valuable references for instrument operation and sample handling techniques. There are compilations of infrared spectra including indices to collections of spectra and to the literature.[29–38] The book by Conley[28d] is an introduction at an elementary level to theory, determination and interpretation. Bellamy's book and supplement include extensive literature reviews.[3]

II. THEORY

Infrared radiation of frequencies less than about 100 cm^{-1} (wavelengths longer than 100 μm) is absorbed and converted by an organic molecule into energy of molecular rotation. This absorption is quantized; thus, a molecular rotation spectrum consists of discrete lines.

Infrared radiation in the range from about 10,000–100 cm^{-1} (1–100 μm) is absorbed and converted by an organic molecule into energy of molecular vibration. This absorption is also quantized, but vibrational spectra appear as bands rather than lines because a single vibrational energy change is accompanied by a number of rotational energy changes. It is with these vibrational-rotational bands, particularly those occurring between 4000 cm^{-1} and 666 cm^{-1} (2.5–15 μm) that we shall be concerned. The frequency or wavelength of absorption depends on the relative masses of the atoms, the force constants of the bonds, and the geometry of the atoms.

Band positions in infrared spectra are presented either as wavelengths or wavenumbers. The micron ($\mu = 10^{-6}$ meters) has long been the commonly used unit for wavelength λ in infrared spectrometry, but recently the term micron has been replaced by micrometer ($\mu m = 10^{-6}$

meters). The wavenumber unit (cm^{-1}, reciprocal centimeter) is currently used since it is directly proportional to energy and since many new, sophisticated spectrometers are linear in the cm^{-1} scale. The units are related as follows:

$$\text{cm}^{-1} = \frac{1}{\mu m} \times 10^4$$

Specific absorptions and absorption ranges will be given in cm^{-1}, with the corresponding μm data adjoining (usually in parentheses). A useful table relating cm^{-1} and μm is given in Appendix E.

Notice also that the wavenumbers ($\bar{\nu}$) are often called "frequencies"; this is not rigorously correct, since wavenumbers are $1/\lambda$ and frequency (ν) is c/λ. Such "frequency" for $\bar{\nu}$ terminology is quite common and is probably not a serious error as long as one keeps in mind the missing speed of light term (c). Spectra linear in μm *and* in cm^{-1} are used in this book. A spectrum linear in wavenumbers has a very different appearance from one linear in wavelength, as shown later (Figure 7, p. 84).

Band intensities are expressed either as transmittance (T) or absorbance (A). Transmittance is the ratio of the radiant power transmitted by a sample to the radiant power incident on the sample. Absorbance is the logarithm, to the base 10, of the reciprocal of the transmittance; $A = \log_{10}(1/T)$. A concise compilation of approved spectrometry nomenclature is periodically published.[39]

There are two types of molecular vibrations: stretching and bending. A stretching vibration is a rhythmical movement along the bond axis such that the interatomic distance is increasing or decreasing. A bending vibration may consist of a change in bond angles between bonds with a common atom, or the movement of a group of atoms with respect to the remainder of the molecule without movement of the atoms in the group with respect to one another. For example, twisting, rocking, and torsional vibrations involve a change in bond angles with reference to a set of coordinates arbitrarily set up within the molecule.

Only those vibrations that result in a rhythmical change in the dipole moment of the molecule are observed in the infrared. The alternating electric field, produced by the changing charge distribution accompanying a vibration, couples the molecular vibration with the oscillating electric field of the electromagnetic radiation.

A molecule has as many degrees of freedom as the total degrees of freedom of its individual atoms. Each atom has 3 degrees of freedom corresponding to the Cartesian coordinates (X, Y, Z) necessary to describe its position relative to other atoms in the molecule. A molecule of n atoms therefore has $3n$ degrees of freedom. For nonlinear molecules, three of the degrees of freedom describe rotation and three describe translation; the remaining $3n-6$ degrees of freedom are vibrational degrees of freedom or fundamental

vibrations. Linear molecules have $3n-5$ vibrational degrees of freedom, for only two degrees of freedom are required to describe rotation.

Fundamental vibrations involve no change in the center of gravity of the molecule.

The three fundamental vibrations of the nonlinear, triatomic water molecule can be depicted as follows:

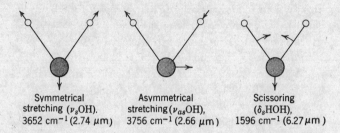

Symmetrical stretching (ν_sOH). 3652 cm^{-1} (2.74 μm)

Asymmetrical stretching (ν_{as}OH), 3756 cm^{-1} (2.66 μm)

Scissoring (δ_sHOH), 1596 cm^{-1} (6.27 μm)

Note the very close spacing of the interacting ("coupled") asymmetric and symmetric stretching above compared with the far removed scissoring mode. This will be shown to be useful later in classification of absorptions and application to structure determination.

The CO_2 molecule is linear and contains three atoms; therefore it has four fundamental vibrations $((3 \times 3) - 5)$:

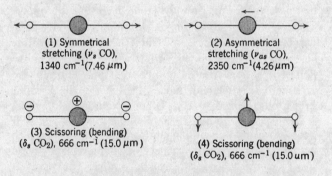

(1) Symmetrical stretching (ν_s CO), 1340 cm^{-1} (7.46 μm)

(2) Asymmetrical stretching (ν_{as} CO), 2350 cm^{-1} (4.26 μm)

(3) Scissoring (bending) (δ_s CO$_2$), 666 cm^{-1} (15.0 μm)

(4) Scissoring (bending) (δ_s CO$_2$), 666 cm^{-1} (15.0 um)

$\oplus$ and $\ominus$ indicate movement perpendicular to the plane of the page.

The symmetrical stretching vibration (1) is inactive in the infrared since it produces no change in the dipole moment of the molecule. The bending vibrations (3) and (4) are equivalent, and are the resolved components of bending motion oriented at any angle to the internuclear axis; they have the same frequency and are said to be doubly degenerate.

The various stretching and bending modes for an AX$_2$ group appearing as a portion of a molecule, e.g., the CH$_2$ group in a hydrocarbon molecule, are shown in Figure 1. The $3n-6$ rule does not apply since the CH$_2$ represents only a portion of a molecule.

The theoretical number of fundamental vibrations (absorption frequencies) will seldom be observed because overtones (multiples of a given frequency) and combination tones (sum of two other vibrations) increase the number of

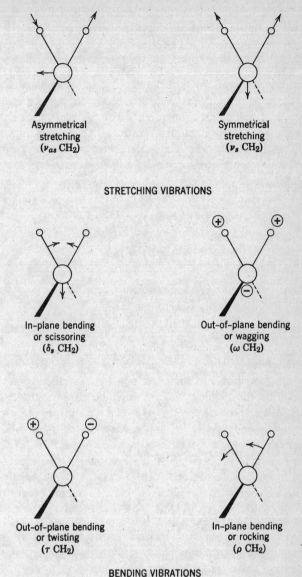

Asymmetrical stretching (ν_{as} CH$_2$)

Symmetrical stretching (ν_s CH$_2$)

STRETCHING VIBRATIONS

In-plane bending or scissoring (δ_s CH$_2$)

Out-of-plane bending or wagging (ω CH$_2$)

Out-of-plane bending or twisting (τ CH$_2$)

In-plane bending or rocking (ρ CH$_2$)

BENDING VIBRATIONS

Fig. 1.
Vibrational modes for a CH$_2$ group. ($\oplus$ and $\ominus$ indicate movement perpendicular to the plane of the page.)

bands, whereas other phenomena reduce the number of bands. The following will reduce the theoretical number of bands:

1. Fundamental frequencies that fall outside of the 2.5–15 μm region.

2. Fundamental bands that are too weak to be observed.

3. Fundamental vibrations that are so close that they coalesce.

4. The occurrence of a degenerate band from several absorptions of the same frequency in highly symmetrical molecules.

5. The failure of certain fundamental vibrations to appear in the infrared because of lack of required change in dipole character of the molecule.

Assignments for stretching frequencies can be approximated by the application of Hooke's law. In the application of the law, two atoms and their connecting bond are treated as a simple harmonic oscillator composed of two masses joined by a spring. Equation 1, derived from Hooke's law, states the relationship between frequency of oscillation, atomic masses, and the force constant of the bond.

$$v = \frac{1}{2\pi c} \left(\frac{f}{\frac{MxMy}{Mx + My}} \right)^{1/2}$$

where v = the vibrational frequency (cm^{-1})
  c = velocity of light (cm/sec)
  f = force constant of bond (dynes/cm)
Mx and My = mass of atom x and atom y, respectively (g).

The value of f is approximately 5×10^5 dynes per cm for single bonds and approximately two and three times this value for double bonds and triple bonds, respectively.
Application of the formula to C–H stretching, using 19.8 $\times 10^{-24}$ g and 1.64×10^{-24} g as mass values for C and H, respectively, places the frequency of the C–H bond vibration at 3040 cm^{-1} (3.30 μm). Actually, C–H stretching vibrations, associated with methyl and methylene groups, are generally observed in the region between 2960–2850 cm^{-1} (3.38–3.51 μm). The calculation is not precise because effects arising from the environment of the C–H within a molecule have been ignored. The frequency of infrared absorption is commonly used to calculate the force constants of bonds.

The shift in absorption frequency following deuteration is often employed in the assignment of C–H stretching frequencies. If the hydrogen in an X–H group is replaced by deuterium, equation 1 indicates that the ratio of the C–H to C–D stretching frequencies should equal $\sqrt{2}$. If the ratio of the frequencies, following deuteration, is much less than $\sqrt{2}$ we can not assume that the vibration is simply a C–H stretching vibration, but rather a mixed vibration involving interaction (coupling) with another vibration. The actual mode of vibration is a combination of both.

Calculations place the stretching frequencies of the following bonds in the general absorption regions indicated.

C–C, C–O, C–N	1300–800 cm^{-1}(7.7–12.5 μm)
C=C, C=O, C=N, N=O	1900–1500 cm^{-1}(5.3–6.7 μm)
C≡C, C≡N	2300–2000 cm^{-1}(4.4–5.0 μm)
C–H, O–H, N–H	3800–2700 cm^{-1}(2.6–3.7 μm)

To approximate the vibrational frequencies of bond stretching by Hooke's law, the relative contributions of bond strengths and atomic masses must be considered. For example, a superficial comparison of the C–H group with the F–H group, on the basis of atomic masses, might lead to the conclusion that the stretching frequency of the F–H bond should occur at a lower frequency than that for the C–H bond. However, the increase in the force constant from left to right across the first two rows of the periodic table has a greater effect than the mass increase. Thus the F–H group absorbs at a higher frequency (4138 cm^{-1}, 2.42 μm) than the C–H group (3040 cm^{-1}, 3.30 μm).

In general, functional groups that have a strong dipole give rise to strong absorptions in the infrared.

INTERACTION. When two bond oscillators share a common atom, they seldom behave as individual oscillators unless the individual oscillation frequencies are widely different. This is because there is mechanical coupling interaction between the oscillators. For example, the carbon dioxide molecule, which consists of two C=O bonds with a common carbon atom, has two fundamental stretching vibrations: an asymmetrical and a symmetrical stretching mode. The symmetrical stretching mode consists of an in-phase stretching or contracting of the C to O bonds, and absorption occurs at a wavelength longer than that observed for the carbonyl group in an aliphatic ketone. The symmetrical stretching mode produces no change in the dipole moment of the molecule and is therefore "inactive" in the infrared, but is easily observed in the Raman spectrum near 1340 cm^{-1} (7.46 μm). In the asymmetrical stretching mode, the two C to O bonds stretch out of phase; one C–O bond stretches as the other contracts. The asymmetrical stretching mode, since it produces a change in the dipole moment, is infrared active; the absorption (2350 cm^{-1}, 4.26 μm) is at a shorter wavelength (higher frequency) than observed for a carbonyl group in aliphatic ketones.

This difference in carbonyl absorption frequencies displayed by the carbon dioxide molecule results from strong mechanical coupling or interaction. In contrast, two ketonic carbonyl groups separated by one or more carbon atoms show normal carbonyl absorption near 1715 cm^{-1} (5.83 μm) because appreciable coupling is apparently prevented by the intervening carbon atom(s).

Coupling accounts for the two N–H stretching bands in the 3497–3077 cm^{-1} (2.86–3.25 μm) region in primary amine and primary amide spectra, for the two C=O stretching bands in the 1818–1720 cm^{-1} (5.50–5.81 μm) region in carboxylic anhydrides and imide spectra, and for the two C–H stretching bands in the 3000–2760 cm^{-1} (3.33–3.62 μm) region for both methylene and methyl groups.

Useful characteristic group frequency bands often involve coupled vibrations. The spectra of alcohols have a

strong band in the region between 1212 and 1000 cm^{-1} (8.25–10.00 μm) which is usually designated as the "C–O stretching band." In the spectrum of methanol this band is at 1034 cm^{-1} (9.67 μm), in the spectrum of ethanol it occurs at 1053 cm^{-1} (9.50 μm). Branching and unsaturation produce absorption characteristic of these structures[40] (see discussion of alcohols). It is evident that we are not dealing with an isolated C–O stretching vibration, but rather a coupled asymmetric vibration involving C–C–O stretching.

Vibrations resulting from bond angle bending frequently couple in a manner similar to stretching vibrations. Thus, the ring C–H out-of-plane bending frequencies of aromatic molecules depend on the number of adjacent hydrogen atoms on the ring; coupling between the hydrogen atoms is affected by the bending of the C–C bond in the ring to which the hydrogen atoms are attached.

Interaction arising from coupling of stretching and bending vibrations is illustrated by the absorption of secondary acyclic amides. Secondary acyclic amides, which exist predominantly in the *trans* conformation, show strong absorption in the 1563–1515 cm^{-1} (6.40–6.60 μm) region; this absorption involves coupling of the N–H bending and C–N stretching vibrations.

The requirements for effective coupling interaction may be summarized as follows:

1. The vibrations must be of the same symmetry species if interaction is to occur.
2. Strong coupling between stretching vibrations requires a common atom between the groups.
3. Interaction is greatest when the coupled groups absorb, individually, near the same frequency.
4. Coupling between bending and stretching vibrations can occur if the stretching bond forms one side of the changing angle.
5. A common bond is required for coupling of bending vibrations.
6. Coupling is negligible when groups are separated by one or more carbon atoms and the vibrations are mutually perpendicular.

HYDROGEN BONDING. Hydrogen bonding can occur in any system containing a proton donor group (X–H) and a proton acceptor, Y, if the s orbital of the proton can effectively overlap the p or π orbital of the acceptor group. Atoms X and Y are electronegative with Y possessing lone pair electrons. The common proton donor groups in organic molecules are carboxyl, hydroxyl, amine, or amide groups. Common proton acceptor atoms are oxygen, nitrogen, and the halogens. Unsaturated groups, such as the ethylenic linkage, can also act as proton acceptors.

The strength of the hydrogen bond is at a maximum when the proton donor group and the axis of the lone pair orbital are collinear. The strength of the bond is inversely proportional to the distance between X and Y.

Hydrogen bonding alters the force constant of both groups; thus, the frequencies of both stretching and bending vibrations are altered. The X–H stretching bands move to longer wavelengths (lower frequencies) usually with increased intensity and band widening. The stretching frequency of the acceptor group, e.g., C=O, is also reduced but to a lesser degree than the proton donor group. The H–X bending vibration usually shifts to a shorter wavelength when bonding occurs; this shift is less pronounced than that of the stretching frequency.

*Inter*molecular hydrogen bonding involves association of two or more molecules of the same or different compounds. Intermolecular bonding may result in dimer molecules (as observed for carboxylic acids) or in polymer molecules, which exist in neat samples or concentrated solutions of monohydroxy alcohols. *Intra*molecular hydrogen bonds are formed when the proton donor and acceptor are present in a single molecule under spatial conditions that allow the required overlap of orbitals, for example, the formation of a 5- or 6-membered ring. The extent of both inter- and intra-molecular bonding is temperature dependent. The effect of concentration on intermolecular and intramolecular hydrogen bonding is markedly different. The bands that result from intermolecular bonding generally disappear at low concentrations (less than about 0.01 M in nonpolar solvents). Intramolecular hydrogen bonding is an internal effect and persists at very low concentrations.

The change in frequency between "free" OH absorption and bonded OH absorption is a measure of the strength of the hydrogen bond. Ring strain, molecular geometry, and the relative acidity and basicity of the proton donor and acceptor groups affect the strength of bonding. Intramolecular bonding involving the same bonding groups is stronger when a 6-membered ring is formed than when a smaller ring results from bonding. Hydrogen bonding is strongest when the bonded structure is stabilized by resonance.

The effects of hydrogen bonding on the stretching frequencies of hydroxyl and carbonyl groups are summarized in Table I.

An important aspect of hydrogen bonding involves interaction between functional groups of solvent and solute. If the solute is polar, then it is necessary to specify the solvent used and the solute concentration.

FERMI RESONANCE. As we have seen in our discussion of interaction, coupling of two fundamental vibrational modes will produce two new modes of vibration, with frequencies higher and lower than that observed when interaction is absent. Interaction can also occur between fundamental vibrations and overtones or combination-tone vibrations. Such interaction is known as Fermi resonance. One example of Fermi resonance is afforded by the absorption pattern of carbon dioxide. In our discussion of

Table I Stretching Frequencies in Hydrogen Bonding

X–H...Y Strength	Intermolecular bonding			Intramolecular bonding		
	Frequency Reduction $(cm^{-1})^a$			Frequency Reduction $(cm^{-1})^a$		
	νOH	$\nu C=O^b$	Compound Class	νOH	$\nu C=O^b$	Compound Class
Weak	300	15	Alcohols, phenols, and intermolecular hydroxyl to carbonyl bonding.	<100	10	1,2-diols; α- and most β-hydroxy ketones; o-chloro and o-alkoxy phenols
Medium				100-300	50	1,3-diols; some β-hydroxy ketones; β-hydroxy amino compounds; β-hydroxy nitro compounds
Strong	>500	50	RCOOH dimers	>300	100	o-hydroxy aryl ketones; o-hydroxy aryl acids; o-hydroxy aryl esters; β-diketones; tropolones.

[a]Frequency shift referred to "free" stretching frequencies.
[b]Carbonyl stretching only where applicable.

interaction, we indicated that the symmetrical stretching band of CO_2 appears in the Raman spectrum near 1340 cm^{-1} (7.46 μm). Actually two bands are observed; one at 1286 cm^{-1} (7.78 μm), one at 1388 cm^{-1} (7.20 μm). The splitting results from coupling between the fundamental C=O stretching vibration, near 1340 cm^{-1} (7.46 μm) and the first overtone of the bending vibration. The fundamental bending vibration occurs near 666 cm^{-1} (15.00 μm) the first overtone near 1334 cm^{-1} (7.55 μm).

Fermi resonance is a common phenomenon in infrared and Raman spectra. It requires that the vibrational levels be of the same symmetry species and that the interacting groups be located in the molecule so that mechanical coupling is appreciable.

An example of Fermi resonance in an organic structure is the "doublet" appearance of the C=O stretch of cyclopentanone under sufficient resolution conditions. Figure 2 shows the appearance of the spectrum of cyclopentanone under the usual conditions. With adequate resolution (Figure 3), Fermi resonance with an overtone or combination band of an α-methylene group shows two absorptions in the carbonyl stretch region.[3]

III. INSTRUMENTATION

The modern double-beam infrared spectrophotometer consists of five principal sections: source (radiation), sampling area, photometer, grating (monochromator), and detector (thermocouple). A diagram of the optical system of a double-beam infrared spectrophotometer is shown in Figure 4.

RADIATION SOURCE. Infrared radiation is produced by electrically heating a source, usually a Nernst filament or a Globar, to 1000–1800°C. The Nernst filament is fabricated from a binder and oxides of zirconium, thorium, and cerium. The Globar is a small rod of silicon carbide. The image of the source must be wider than the maximum width of the slits (see Ref. 28d). The maximum radiation for the Globar occurs in the 5500–5000 cm^{-1} (1.8–2.0 μm) region and drops off by a factor of about 600 as the 600 cm^{-1} (16.7 μm) region is approached. The Nernst filament furnishes maximum radiation energy at about 7100 cm^{-1} (1.4 μm) and drops by a factor of about 1000 as the lower frequency region is approached.

The radiation from the source is divided into two beams by mirrors $M1$ and $M2$. The two beams, reference beam and sample beam, are focused into the sample area by mirrors $M3$ and $M4$.

SAMPLING AREA. Reference and sample beams enter the sampling area and pass through the reference cell and sampling cell, respectively. Opaque shutters, mounted on the source housing, permit blocking of either beam independently. The sampling area of a precision spectrophotometer accommodates a wide variety of sampling accessories varying from gas cells of 40 m effective path to microcells.

PHOTOMETER. The reference beam passes through the attenuator (see below) and is reflected by mirrors $M6$ and $M8$ to the rotating sector mirror $M7$, which alternately reflects the reference beam out of the optical system and transmits the beam to mirror $M9$. The reference beam is now an intermittent beam with a "frequency" of from 8 to

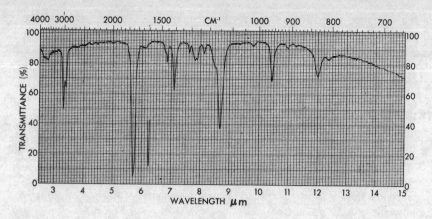

Fig. 2.
Cyclopentanone, liquid film.

WAVE NUMBER (CM⁻¹)

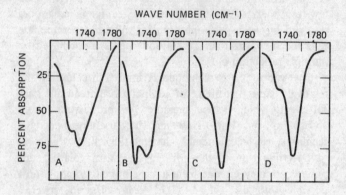

Fig. 3.
*Infrared spectrum of cyclopentanone in various media. A.
Carbon tetrachloride solution (0.015 molar). B. Carbon
disulfide solution (0.023 molar). C. Chloroform solution
(0.025 molar). D. Liquid state (thin film). (Computed spectral slit width 2 cm⁻¹.)*

13 cycles per second depending on the particular instrument. This beam is focused by mirror $M10$ on the slit $S1$.
The sample beam passes through the comb (see below) and
is reflected by mirror $M5$ to the rotating sector mirror $M7$,
which alternately transmits the beam out of the optical
system and reflects it to mirror $M9$, thence to mirror $M10$
and slit $S1$.

At any given moment, the beam focused on slit $S1$ is
either the reference beam, which was transmitted by the
rotating sector mirror $M7$, or the sample beam, which was
reflected by $M7$. In other words, the reference beam and
the sample beam have been combined into a single beam of
alternating segments; this establishes a switching frequency
at the detector equal to the speed of rotation of $M7$.

When the beams are of equal intensity, the instrument is
at an optical null. The comb in the sample beam permits
balancing the beams.[28d] The recording pen is then at 100%
transmittance when no sample is present.

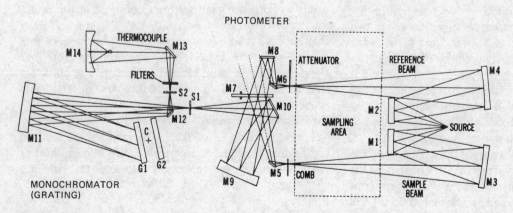

Fig. 4.
Optical system of double-beam infrared spectrophotometer.

The attenuator is driven in and out of the reference beam in response to the signal created at the detector by the sample beam. Thus, when the sample beam is absorbed by the sample, the attenuator is driven into the reference beam until its intensity matches that of the sample beam.

MONOCHROMATOR. The combined beam passes through the monochromator entrance slit $S1$ to the mirror $M11$, which reflects it to the diffraction grating $G1$. This beam is dispersed into various frequencies; this beam is reflected back to mirror $M11$, subsequently to mirror $M12$. $M12$ focuses the beam on exit slit $S2$. The width of the frequency range focused on slit $S2$ is determined by the width of entrance slit $S1$ and the dispersing power of the grating. The frequency band of one dispersed beam focused on slit $S2$ at any moment is determined by the angle of grating $G1$ at that moment. Rotation of grating $G1$ produces a scan of frequency bands at exit slit $S2$ and thus at the detector. Filters are automatically inserted into the radiation path at the exit slit in a sequence during scanning to eliminate all unwanted radiation; this unwanted radiation corresponds to harmonic multiples of the frequency to be measured.

Maximum resolution is obtained by using gratings only in the range of greatest dispersing effectiveness. Therefore, in a modern, high-resolution instrument, two or more gratings are used ($G1$ and $G2$ of Figure 4).

The narrower the slit width, the greater is the resolution. Here again, some compromise is necessary because of the decreased energy output of the source at lower frequencies (longer wavelengths). On most instruments, the slit width is programmed to increase as the emitted source energy decreases, so that constant reference beam energy enters the monochromator.

DETECTOR (THERMOCOUPLE). After leaving the exit slit of the monochromator, the beam is reflected by a flat mirror $M13$ to an ellipsoidal mirror $M14$. The foci of the ellipsoidal mirror are the exit slit $S2$ and the detector.

The detector is a device that measures radiant energy by means of its heating effect. Two common types of detectors are the thermocouple and bolometer. In the thermocouple detector, the radiant energy heats one of two bimetallic junctions, and an emf is produced between the two junctions proportional to the degree of heating. The bolometer changes its resistance upon heating. It serves as one arm of a bridge so that a change in temperature will cause an unbalanced signal across the circuit. The unbalanced signal can be amplified and recorded or used to activate a servomechanism to re-establish a balance.

Since the detector sees alternately the reference and the sample beam at a switching frequency determined by the rotation of the sector mirror, any change in the intensity of the radiation due to absorption is detected as an off-null signal.

The amplified off-null signal of the detector is used to position the optical attenuator so that the radiation from the reference and sample beam are kept at equal intensity. The amount of attenuation required is a direct measure of the absorption by the sample. The movement of the attenuator is recorded by the recording chart pen.

IV. SAMPLE HANDLING

Infrared spectra may be obtained for gases, liquids, or solids.

1. The spectra of gases or low boiling liquids may be obtained by expansion of the sample into an evacuated cell. Cells equipped with freeze-out tips are used for sample concentration and cell evacuation prior to expansion of the sample into the cell. Gas cells are available in lengths of a few centimeters to 40 meters. The sampling area of a standard infrared spectrophotometer will not accommodate cells much longer than 10 cm; long paths are achieved by multiple reflection optics.

The vapor phase technique is limited because of the relatively large percentage of compounds that do not have sufficiently high vapor pressures to produce a useful absorption spectrum. However, the usefulness of the technique can be extended by heating the cell.

Determination of the infrared spectra of volatile compounds, as they emerge from a gas chromatograph, is now possible by use of a rapid scan infrared spectrophotometer.* This instrument is reported to run a complete scan from 2.5 to 14.5 μm in 4 seconds. Spectra are obtained on 30 micrograms of volatile effluent.

2. Liquids may be examined neat or in solution. Neat liquids are examined between salt plates usually without a spacer. Pressing a liquid sample between flat plates produces a film of 0.01 mm or less in thickness, the plates being held together by capillarity. Samples of 1–10 mg are required. Thick samples of neat liquids usually absorb too strongly to produce a satisfactory spectrum. Volatile liquids are examined in sealed cells with very thin spacers. Silver chloride or KRS-5 plates may be used for samples that dissolve sodium chloride plates. Recently low-cost silver chloride windows have become available† that allow for infrared analysis via thin film, 0.025 mm, or 0.050 mm pathlengths (Figure 5).

Solutions are handled in cells of 0.1–1 mm thickness. Volumes of 0.1–1 ml of 0.05–10% solutions are required for readily available demountable cells. A compensating cell, containing pure solvent, is placed in the reference

---

* Beckman Instruments, Inc., Fullerton, Calif. Model IR-102.
† Wilks Scientific Corporation, South Norwalk, Conn.

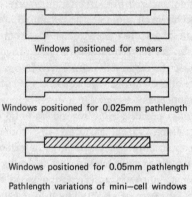

Windows positioned for smears

Windows positioned for 0.025mm pathlength

Windows positioned for 0.05mm pathlength

Pathlength variations of mini—cell windows

Fig. 5.
Equipment for thin film (neat sample) and solution IR spectral analysis.

beam. The spectrum thus obtained is that of the solute except in those regions in which the solvent absorbs strongly. For example, thick samples of carbon tetrachloride absorb strongly near 800 cm^{-1} (12.50 μm); compensation for this band is ineffective since strong absorption prevents any radiation from reaching the detector.

The solvent selected must be dry and reasonably transparent in the region of interest. When the entire spectrum is of interest, several solvents must be used. A common pair of solvents is carbon tetrachloride and carbon disulfide. Carbon tetrachloride is relatively free of absorption at frequencies above 1333 cm^{-1} (shorter than 7.50 μm) whereas carbon disulfide shows little absorption below 1333 cm^{-1} (longer than 7.50 μm). Solvent and solute combinations that react must be avoided. For example, carbon disulfide cannot be used as a solvent for primary or secondary amines. Amino alcohols react slowly with carbon disulfide and carbon tetrachloride.

When only very small samples are available, ultramicro cavity cells are used in conjunction with a beam condenser. The smallest commercially available cell has a path length of approximately 0.05 mm and a capacity of about 0.8 microliter.* A spectrum can thus be obtained on a few micrograms of sample in solution. When volatility permits, the solute can be recovered for examination by other spectrometric techniques.

The absorption pattern of selected solvents and mulling oils are presented in Appendix A.

3. Solids are usually examined as a mull, a pressed disc, or as a deposited glassy film.

Mulls are prepared by thoroughly grinding 2–5 mg of a solid in a smooth agate mortar. Grinding is continued after the addition of a drop or two of the mulling oil. The suspended particles must be less than 2 μm to avoid excessive scattering of radiation. The mull is examined as a thin film between flat salt plates. Nujol (a high-boiling petroleum oil) is commonly used as a mulling agent. When hydrocarbon bands interfere with the spectrum, Fluorolube (a completely halogenated polymer containing F and Cl), or hexachlorobutadiene may be used. The use of both Nujol and Fluorolube mulls makes possible a scan, essentially free of interfering bands, over the 4000 cm^{-1}–250 cm^{-1} (2.5–40.0 μm) region.

The pressed-disc technique depends upon the fact that dry, powdered potassium bromide (or other alkali metal halides) can be pressed under pressure *in vacuo* to form transparent discs. The sample (0.5–1.0 mg) is intimately mixed with approximately 100 mg of dry, powdered potassium bromide. Mixing can be effected by thorough grinding in a smooth agate mortar or, more efficiently, with a small vibrating ball mill, or by lyophilization. The mixture is pressed into a transparent disc, with special dies, under a pressure of 10,000–15,000 pounds per square inch. The quality of the spectrum depends upon the intimacy of mixing and the reduction of the suspended particles to 2 μm or less. Micro-discs, 0.5 to 1.5 mm in diameter, can be used with a beam condenser. The micro-disc technique permits examination of samples as small as 1 microgram. Bands near 3448 cm^{-1} and 1639 cm^{-1} (2.9 and 6.1 μm) due to moisture, frequently appear in spectra obtained by the pressed disc technique (see H_2O spectrum in Appendix A).

The use of KBr discs or pellets has often been avoided because of the demanding task of making good pellets. Such KBr techniques should be more available through the Mini-Press (Figure 6) that affords a simple procedure; the KBr-sample mixture is placed in the nut portion of the assembly with one bolt in place. The second bolt is introduced and pressure is applied by tightening the bolts.

*Barnes Engineering Co., Instrument Division, Stamford, Conn.

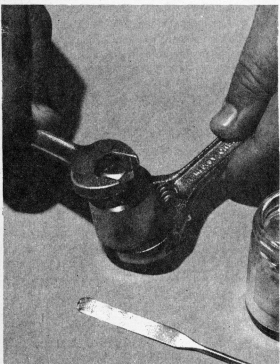

Fig. 6.
*Mini-press and operation. (a) Cell holder. (b) Nut = cell.
(c) Bolts.*

Removal of the bolts leaves a pellet in the nut that now serves as a cell.

Deposited films are useful only when the material can be deposited from solution or cooled from a melt as micro-crystals or as a glassy film. Crystalline films generally lead to excessive light scattering. Specific crystal orientation may lead to spectra differing from those observed for randomly oriented particles such as exist in a mull or halide disc. The deposited film technique is particularly useful for obtaining spectra of resins and plastics. Care must be taken to free the sample of solvent by vacuum treatment or gentle heating.

A technique known as attenuated total reflection or internal reflection spectroscopy is now available for obtaining qualitative spectra of solids regardless of thickness. The technique depends upon the fact that a beam of light that is internally reflected from the surface of a transmitting medium passes a short distance beyond the reflecting boundary and returns to the transmitting medium as a part of the process of reflection. If a material (i.e., the sample) of lower refraction index than the transmitting medium, is brought in contact with the reflecting surface, the light passes through the material to the depth of a few microns, producing an absorption spectrum. An extension of the technique provides for multiple internal reflections along the surface of the sample. The multiple internal reflection technique results in spectra with intensities comparable to transmission spectra.*

In general, a dilute solution in a nonpolar solvent furnishes the best (that is, least distorted) spectrum. Nonpolar compounds give essentially the same spectra in the condensed phase (that is, neat liquid, a mull, a KBr disc, or a film) as they give in nonpolar solvents. Polar compounds, however, often show hydrogen bonding effects in the condensed phase. Unfortunately, polar compounds are frequently insoluble in nonpolar solvents, and the spectrum must be obtained either in a condensed phase, or in a polar solvent; the latter introduces the possibility of solute-solvent hydrogen bonding.

Reasonable care must be taken in handling salt cells and plates. Moisture-free samples should be used. Fingers should not come in contact with the optical surfaces. Care should be taken to prevent contamination with silicones, which are hard to remove and have strong absorption patterns.

V. INTERPRETATION OF SPECTRA

There are no rigid rules for interpreting an infrared spectrum. Certain requirements, however, must be met before an attempt is made to interpret a spectrum.

1. The spectrum must be adequately resolved and of adequate intensity.

* Wilks Scientific Corp., South Norwalk, Conn.

2. The spectrum should be that of a reasonably pure compound.

3. The spectrophotometer should be calibrated so that the bands are observed at their proper frequencies or wavelengths. Proper calibration can be made with reliable standards such as polystyrene film.

4. The method of sample handling must be specified. If a solvent is employed, the solvent, concentration, and the cell thickness should be indicated.

A precise treatment of the vibrations of a complex molecule is not feasible; thus the infrared spectrum must be interpreted from empirical comparison of spectra, and extrapolation of studies of simpler molecules. Many questions arising in the interpretation of an infrared spectrum can be answered by data obtained from the mass, ultraviolet, and NMR spectra.

Infrared absorption of organic molecules is summarized in the chart of characteristic group frequencies in Appendix B. Many of the group frequencies vary over a wide range because the bands arise from complex interacting vibrations within the molecule. Absorption bands may, however, represent predominantly a single vibrational mode. Certain absorption bands, for example, those arising from the C—H, O—H, and C=O stretching modes, remain within fairly narrow regions of the spectrum. Important details of structure may be revealed by the exact position of an absorption band within these narrow regions. Shifts in absorption position and changes in band contours, accompanying changes in molecular environment, may also suggest important structural details.

The two important areas for a preliminary examination of a spectrum are the region 4000–1300 cm^{-1} (2.5–7.7 μm) and the 909–650 cm^{-1} (11.0–15.4 μm) region. The short-wavelength (high energy) portion of the spectrum is called the functional group region. The characteristic stretching frequencies for important functional groups such as OH, NH, and C=O occur in this portion of the spectrum. The absence of absorption in the assigned ranges for the various functional groups can usually be used as evidence for the absence of such groups in the molecule. Care must be exercised, however, in such interpretations since certain structural characteristics may cause a band to weaken and become extremely broad so that it may go unnoticed. For example, intramolecular hydrogen bonding in the enolic form of acetylacetone results in a broad, weak OH band, which may be overlooked. The absence of absorption in the 1850–1540 cm^{-1} (5.40–6.50 μm) region excludes a structure containing a carbonyl group. Weak bands in the short wavelength region, resulting from the fundamental absorption of functional groups such as S—H and C≡C, are extremely valuable in the determination of structure. Such weak bands would be of little value in the more complicated regions of the spectrum. Overtones and combination tones of longer wavelength bands frequently appear in the short wavelength region of the spectrum. Overtone- and combination-tone bands are characteristically weak except when Fermi resonance occurs. Strong skeletal bands for aromatics and heteroaromatics fall in the 1600–1300 cm^{-1} (6.25–7.70 μm) region of the spectrum.

The lack of strong absorption bands in the 909–650 cm^{-1} (11.0–15.4 μm) region generally indicates a nonaromatic structure. Aromatic and heteroaromatic compounds display strong out-of-plane C—H bending and ring bending absorption bands in this region that can frequently be correlated with the substitution pattern. Broad, moderately intense absorption in the long wavelength region suggests the presence of carboxylic acid dimers, amines or amides all of which show out-of-plane bending in this region. If the region is extended to 1000 cm^{-1} (10.0 μm) absorption bands characteristic of olefinic structures are included.

The intermediate portion of the spectrum, 1300–909 cm^{-1} (7.7–11.0 μm) is usually referred to as the "fingerprint" region. The absorption pattern in this region is frequently complex, with the bands originating in interacting vibrational modes. This portion of the spectrum is extremely valuable when examined in reference to the other regions. For example, if alcoholic or phenolic O—H stretching absorption appears in the short-wavelength region of the spectrum, the position of the C—C—O absorption band in the 1260–1000 cm^{-1} (7.93–10.00 μm) region frequently makes it possible to assign the O—H absorption to alcohols and phenols with highly specific structures. Absorption in this intermediate region is probably unique for every molecular species.

Any conclusions arrived at after examination of a particular band should be confirmed where possible by examination of other portions of the spectrum. For example, the assignment of a carbonyl band to an aldehyde should be confirmed by the appearance of a band or a pair of bands in the 2900–2695 cm^{-1} (3.45–3.71 μm) region of the spectrum, arising from the C—H stretching vibration of the aldehyde group. Similarly, the assignment of a carbonyl band to an ester should be confirmed by observation of a strong band in the C—O stretching region, 1300–1100 cm^{-1} (7.6–9.1 μm).

Finally, in a "fingerprint" comparison of spectra, or any other situation in which the *shapes* of peaks are important, one should anticipate the substantial change in the general appearance of the spectrum in changing from a spectrum that is linear in wavenumber to one that is linear in wavelength (Figure 7).

VI. CHARACTERISTIC GROUP FREQUENCIES OF ORGANIC MOLECULES

A table of characteristic group frequencies is presented as Appendix B. The ranges presented for group frequencies

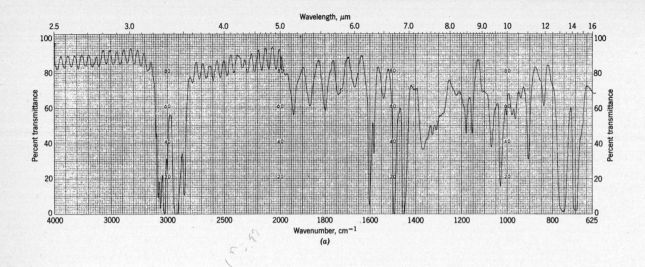

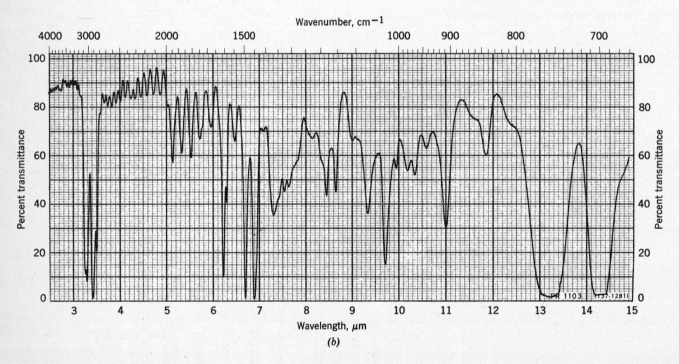

Fig. 7.
Polystyrene, same sample for both (a) and (b). (a) linear in wavenumber (cm⁻¹); (b) linear in wavelength (μm).

have been assigned following the examination of many compounds in which the groups occur. Although the ranges are quite well defined, the precise frequency or wavelength at which a specific group absorbs is dependent on its environment within the molecule and on its physical state.

This section of the chapter is concerned with a comprehensive look at these characteristic group frequencies and their relationship to molecular structure. Books that should be consulted to supplement this section are those by Bellamy,[3] Conley,[28d] and Colthup, Daly, and Wiberly.[12]

As each new type or class of molecule or functional group is introduced in the succeeding sections, an example of an infrared spectrum will be given with the important peak assignments. Reference will be made to examples of infrared spectra in Appendix A that are members of various classes of these organic molecules; the spectra in Appendix A do not have detailed absorption assignments.

NORMAL PARAFFINS (ALKANES)

The spectra of normal paraffins can be interpreted in terms of four vibrations, namely, the stretching and bending of C–H and C–C bonds. Detailed analysis of the spectra of the lower members of the alkane series has made possible detailed assignments of the spectral positions of specific vibrational modes.

Not all of the possible absorption frequencies of the paraffin molecule are of equal value in the assignment of structure. The C–C bending vibrations occur at very low frequencies (below 500 cm^{-1}, longer than 20 μm) and therefore do not appear in our spectra. The bands assigned to C–C stretching vibrations are weak, and appear in the broad region of 1200–800 cm^{-1} (8.3–12.5 μm); they are generally of little value for identification.

The most characteristic vibrations are those arising from C–H stretching and bending. Of these vibrations, those arising from methylene twisting and wagging are usually of limited diagnostic value because of their weakness and instability; this instability is a result of strong coupling to the remainder of the molecule.

The vibrational modes of paraffins are common to many organic molecules. Although the positions of C–H stretching and bending frequencies of methyl and methylene groups remain nearly constant in hydrocarbons, the attachment of CH$_3$ or CH$_2$ to atoms other than carbon, or to a carbonyl group or aromatic ring, may result in appreciable shifts of the C–H stretching and bending frequencies. These shifts are summarized in Tables I and II of Appendix C.

The spectrum of dodecane, Figure 8 is that of a typical straight-chain hydrocarbon.

C–H Stretching Vibrations

Absorption arising from C–H stretching in the alkanes occurs in the general region of 3000–2840 cm^{-1} (3.3–3.5 μm). The positions of the C–H stretching vibrations are among the most stable in the spectrum. When a spectrum is obtained with an instrument using a sodium chloride prism, the bands in this region are frequently unresolved (as in Fig. 8). Resolution is achieved through the use of a fluoride prism or a grating instrument.

METHYL GROUPS. An examination of a large number of saturated hydrocarbons containing methyl groups showed, in all cases, two distinct bands occurring at 2962 cm^{-1} (3.38 μm) and at 2872 cm^{-1} (3.48 μm). The first of these results from the asymmetrical stretching mode in which two C–H bonds of the methyl group are extending while the third one is contracting (ν_{as}CH$_3$). The second arises from symmetrical stretching (ν_sCH$_3$) in which all three of the C–H bonds extend and contract in phase. The presence of several methyl groups in a molecule results in strong absorption at these positions.

METHYLENE GROUPS. The asymmetrical stretching (ν_{as}CH$_2$) and symmetrical stretching (ν_sCH$_2$) occur, respectively, near 2926 cm^{-1} (3.43 μm) and 2853 cm^{-1} (3.51 μm). The positions of these bands do not vary more than $\pm$ 10 cm^{-1} in the aliphatic and nonstrained cyclic hydrocarbons. The frequency of methylene stretching is

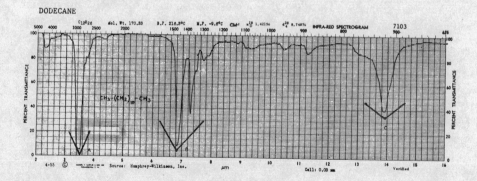

Fig. 8.

A: C–H stretch; CH$_3$ 2962 cm^{-1} (3.38 μm), ν_{as} CH$_3$
            2872 cm^{-1} (3.48 μm), ν_s CH$_3$
        CH$_2$ 2926 cm^{-1} (3.43 μm), ν_{as}
            2853 cm^{-1} (3.51 μm), ν_s

B: C–H bend; CH$_3$ 1375 cm^{-1} (7.28 μm), δ_s
            1450 cm^{-1} (6.90 μm), δ_{as}
        CH$_2$ 1465 cm^{-1} (6.83 μm), δ_s CH$_2$.

C: CH$_2$ rock; ρCH$_2$ 720 cm^{-1} (13.9 μm).

increased when the methylene group is part of a strained ring.

C–H Bending Vibrations

METHYL GROUPS. Two bending vibrations can occur within a methyl group. The first of these, the symmetrical bending vibration, involves the in-phase bending of the C–H bonds (I). The second, the asymmetrical bending vibration, involves out-of-phase bending of the C–H bonds (II).

In these diagrams the arrows tapering toward the head show the H atoms moving away from the viewer. The reverse taper shows the reverse movement moving essentially above and below the plane of the page.

The symmetrical bending vibration ($\delta_s CH_3$) occurs near 1375 cm^{-1} (7.28 μm), the asymmetrical bending vibration ($\delta_{as} CH_3$) near 1450 cm^{-1} (6.90 μm).

The asymmetrical vibration generally overlaps the scissoring vibration of the methylene group (see below). Two distinct bands are observed, however, in compounds such as diethyl ketone, in which the methylene scissoring band has been shifted to a lower frequency, 1439–1399 cm^{-1} (6.95–7.15 μm) and increased in intensity because of its proximity to the carbonyl group.

The absorption band near 1375 cm^{-1} (7.28 μm) arising from the symmetrical bending of the methyl C–H bonds, is very stable in position when the methyl group is attached to another carbon atom. The intensity of this band is greater for each methyl group in the compound than that for the asymmetrical methyl bending vibration or the methylene scissoring vibration.

METHYLENE GROUPS. The bending vibrations of the C–H bonds in the methylene group have already been shown schematically in Figure 1. The four bending vibrations are referred to as scissoring, rocking, wagging, and twisting.

The scissoring band ($\delta_s CH_2$) in the spectra of hydrocarbons occurs at a nearly constant position near 1465 cm^{-1} (6.83 μm).

The band resulting from the methylene rocking vibration (ρCH_2), in which all of the methylene groups rock in phase, appears near 720 cm^{-1} (13.9 μm) for straight chain paraffins of seven or more carbons. This band may appear as a doublet in the spectra of solid samples. In the lower members of the n-paraffin series, the band appears at somewhat higher frequencies.

Absorption of hydrocarbons, due to methylene twisting and wagging vibrations, is observed in the 1350–1150 cm^{-1} (7.4–8.7 μm) region. These bands are generally appreciably weaker than those resulting from methylene scissoring. A series of bands in this region, arising from the methylene group, is characteristic of the spectra of solid samples of long-chain acids, amides, and esters.

BRANCHED-CHAIN HYDROCARBONS

In general, the changes brought about in the spectrum of a hydrocarbon by branching result from changes in skeletal stretching vibrations and methyl bending vibrations; these occur below 1500 cm^{-1} (longer than 6.7 μm). The spectrum of Figure 9 is that of a typical branched hydrocarbon.

C–H Stretching Vibrations

TERTIARY C–H GROUPS. Absorption resulting from this vibrational mode is very weak and is usually lost in other aliphatic C–H absorption. Absorption in hydrocarbons occurs near 2890 cm^{-1} (3.46 μm).

C–H Bending Vibrations

GEM-DIMETHYL GROUPS. Configurations in which two methyl groups are attached to the same carbon atom exhibit distinctive absorption in the C–H bending region. The isopropyl group shows a strong doublet, with peaks of almost equal intensity, at 1385–1380 cm^{-1} (7.22–7.25 μm) and at 1370–1365 cm^{-1} (7.30–7.33 μm). The tertiary butyl group gives rise to two C–H bending bands, one in the 1395–1385 cm^{-1} (7.17–7.22 μm) region and one near 1370 cm^{-1} (7.30 μm). In the t-butyl doublet, the long wavelength band is more intense. When the gem-dimethyl group occurs at an internal position, a doublet is observed in essentially the same region where absorption occurs for the isopropyl and t-butyl groups. Doublets are observed for gem-dimethyl groups because of interaction between the in-phase and out-of-phase symmetrical CH$_3$ bending of the two methyl groups attached to a common carbon atom.

Weak bands result from methyl rocking vibrations in isopropyl and t-butyl groups. These vibrations are sensitive to mass and interaction with skeletal stretching modes and are generally less reliable than the C–H bending vibrations. The following assignments have been made: isopropyl

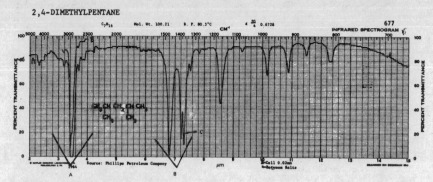

Fig. 9.

A: C–H stretch (see Figure 8).

B: C–H bend (see Figure 8).

Note gem-dimethyl (*C*) doublet 1385 cm^{-1} (7.22 μm)

1368 cm^{-1} (7.31 μm).

group, 922–919 cm^{-1} (10.85–10.88 μm), and *t*-butyl group, 932–926 cm^{-1} (10.73–10.80 μm).

CYCLOPARAFFINS (CYCLIC ALKANES)

C–H Stretching Vibrations

The methylene stretching vibrations of unstrained cyclic polymethylene structures are much the same as those observed for acyclic paraffins. Increasing ring strain moves the C–H stretching bands progressively to higher frequencies. The ring CH$_2$ and CH groups in a monoalkyl cyclopropane ring absorb in the region of 3100–2990 cm^{-1} (3.23 to 3.34 μm).

C–H Bending Vibrations

Cyclization decreases the frequency of the CH$_2$ scissoring vibration. Cyclohexane absorbs at 1452 cm^{-1} (6.89 μm), whereas *n*-hexane absorbs at 1468 cm^{-1} (6.81 μm). Cyclopentane absorbs at 1455 cm^{-1} (6.87 μm), cyclopropane at 1442 cm^{-1} (6.94 μm). This shift frequently makes it possible to observe distinct bands for methylene and methyl absorption in this region. Spectra of other saturated hydrocarbons appear in Appendix A: hexane-no. 1, nujol-no. 2 and cyclohexane-no. 3.

OLEFINIC HYDROCARBONS (ALKENES)

Olefinic structure introduces several new modes of vibration into a hydrocarbon molecule: a C=C stretching vibration, C–H stretching vibrations in which the carbon atom is present in the olefinic linkage, and in-plane and out-of-plane bending of the olefinic C–H bond. The spectrum of Figure 10 is that of a typical terminal olefin.

C=C Stretching Vibrations

UNCONJUGATED LINEAR OLEFINS. The C=C stretching mode of unconjugated olefins usually shows moderate to weak absorption at 1667–1640 cm^{-1} (6.00–6.10 μm). Monosubstituted olefins, i.e., vinyl groups, absorb near 1640 cm^{-1} (6.10 μm) with moderate intensity. Disubstituted *trans*-olefins, tri-, and tetra-alkyl substituted olefins absorb at or near 1670 cm^{-1} (5.99 μm); disubstituted *cis*-olefins and vinylidene olefins absorb near 1650 cm^{-1} (6.06 μm).

The absorption of symmetrical disubstituted *trans*-olefins or tetrasubstituted olefins may be extremely weak or absent. *Cis*-olefins, which lack the symmetry of the *trans* structure, absorb more strongly than *trans*-olefins. Internal double bonds generally absorb more weakly than terminal double bonds because of pseudosymmetry.

Abnormally high frequency absorption is observed for –CH=CF$_2$ and –CF=CF$_2$ groups. The former absorbs near 1754 cm^{-1} (5.70 μm), the latter near 1786 cm^{-1} (5.60 μm). In contrast, the absorption frequency is reduced by the attachment of chlorine, bromine, or iodine.

CYCLOOLEFINS. Absorption of the internal double bond in the unstrained cyclohexene system is essentially the same as that of a *cis* isomer in an acyclic system. The C=C stretch vibration is coupled with the C–C stretching

of the adjacent bonds. As the angle α $\left(C \overset{C}{\underset{\alpha}{\diagdown\!\!\diagup}} C \right)$ becomes

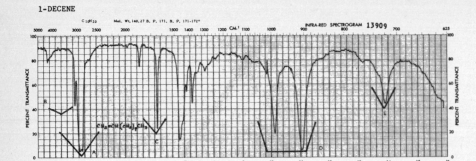

Fig. 10.

A: C–H stretch (see Figure 8)
   Note olefinic C–H stretch (B) at 3049 cm^{-1} (3.28 μm).
C: C=C stretch, 1645 cm^{-1} (6.08 μm), see Table III of Appendix C.
D: Out of plane C–H bend: 986 cm^{-1} (10.14 μm)
   (olefinic)   907 cm^{-1} (11.03 μm).
E: Methylene rock: 720 cm^{-1} (13.88 μm).

smaller the interaction becomes less until it is at a minimum at 90° in cyclobutene, (1566 cm^{-1}, 6.39 μm). In the cyclopropene structure, interaction again becomes appreciable and the absorption frequency increases (1641 cm^{-1}, 6.09 μm).

The substitution of alkyl groups for an α-hydrogen atom in strained ring systems serves to increase the frequency of C=C absorption. Cyclobutene absorbs at 1566 cm^{-1} (6.39 μm), 1-methylcyclobutene at 1641 cm^{-1} (6.09 μm).

The absorption frequency of external (exocyclic) olefinic bonds increases with decreasing ring size. Methylenecyclohexane absorbs at 1650 cm^{-1} (6.06 μm), methylenecyclopropane at 1781 cm^{-1} (5.62 μm).

CONJUGATED SYSTEMS. The olefinic bond stretching vibrations in conjugated dienes without a center of symmetry interact to produce two C=C stretching bands. The spectrum of an unsymmetrical conjugated diene, such as 1,3-pentadiene, shows absorption near 1650 cm^{-1} (6.06 μm) and near 1600 cm^{-1} (6.25 μm). The symmetrical molecule, 1,3-butadiene, shows only one band near 1600 cm^{-1} (6.25 μm), resulting from asymmetric stretching; the symmetrical stretching band is inactive in the infrared.

Conjugation of an olefinic double bond with an aromatic ring produces enhanced olefinic absorption near 1625 cm^{-1} (6.15 μm).

The absorption frequency of the olefinic bond in conjugation with a carbonyl group is lowered by about 30 cm^{-1} (ca 0.11 μm); the intensity of absorption is increased. In *s-cis* structures, the olefinic absorption may be as intense as that of the carbonyl group. *s-Trans* structures absorb more weakly than *s-cis* structures.

CUMULATED OLEFINS. A cumulated double bond system, as occurs in the allenes $\left(\diagdown C{=}C{=}CH_2 \right)$, absorbs near 2000–1900 cm^{-1} (5.00–5.26 μm). The absorption results from asymmetric C=C=C stretching. The absorption may be considered an extreme case of exocyclic C=C absorption.

Olefinic C–H Stretching Vibrations

In general, any C–H stretching bands above 3000 cm^{-1} (below 3.33 μm) result from aromatic, heteroaromatic, acetylenic, or olefinic C–H stretching. Also found in the same region, are the C–H stretching in small rings such as cyclopropane, and the C–H in halogenated alkyl groups. The frequency and intensity of olefinic C–H stretching absorption are influenced by the pattern of substitution. With proper resolution, multiple bands are observed for structures in which stretching interaction may occur. For example, the vinyl group produces three closely spaced C–H stretching bands. Two of these result from symmetrical and asymmetrical stretching of the terminal C–H groups, and the third from the stretching of the remaining single C–H.

Olefinic C–H Bending Vibrations

Olefinic C–H bonds can undergo bending either in the same plane as the C=C bond, or perpendicular to it; the bending vibrations can be either in-phase or out-of-phase with respect to each other.

Assignments have been made for a few of the more prominent and reliable in-plane bending vibrations. The vinyl group absorbs near 1416 cm^{-1} (7.06 μm), due to a scissoring vibration of the terminal methylene. The C–H

rocking vibration of a *cis*-disubstituted olefin occurs in the same general region.

The most characteristic vibrational modes of olefins are the out-of-plane C–H bending vibrations between 1000 and 650 cm^{-1} (10.0 and 15.4 μm). These bands are usually the strongest in the spectra of olefins. The most reliable bands are those of the vinyl group, the vinylidene group, and the trans disubstituted olefins. Olefinic absorption is summarized in Tables III to V of Appendix C.

In allene structures, strong absorption is observed near 850 cm^{-1} (11.76 μm), arising from =CH$_2$ wagging. The first overtone of this band may also be seen. Some spectra showing olefinic features are shown in Appendix A: trichloroethylene-no. 12 and tetrachloroethylene-no. 13.

ACETYLENIC HYDROCARBONS (ALKYNES)

The two stretching vibrations in acetylenic molecules involve C≡C and C–H stretching. Absorption due to C–H bending is characteristic of acetylene and monosubstituted acetylenes. The spectrum of Figure 11 is that of a typical terminal alkyne.

C≡C Stretching Vibrations

The weak C≡C stretching band of acetylenic molecules occurs in the region of 2260–2100 cm^{-1} (4.43–4.76 μm). Because of symmetry, no C≡C band is observed in the infrared for acetylene and symmetrically substituted acetylenes. In the infrared spectra of monosubstituted acetylenes, the band appears at 2140–2100 cm^{-1} (4.67–4.76 μm). Disubstituted acetylenes, in which the substituents are different, absorb near 2260–2190 cm^{-1} (4.43–4.57 μm). When the substituents are similar in mass, or produce similar inductive and mesomeric effects, the band may be so weak as to be unobserved in the infrared spectrum. For reasons of symmetry, a terminal C≡C produces a stronger band than an internal C≡C (pseudosymmetry). The intensity of the C≡C stretching band is increased by conjugation with a carbonyl group.

C–H Stretching Vibrations

The C–H stretching band of monosubstituted acetylenes occurs in the general region of 3333–3267 cm^{-1} (3.00–3.06 μm). This is a strong band and is narrower than the bonded OH and NH bands occurring in the same region.

C–H Bending Vibrations

The C–H bending vibration of acetylene or monosubstituted acetylenes leads to strong, broad absorption in the 700–610 cm^{-1} (14.29–16.39 μm) region. The first overtone of the C–H bending vibration appears as a weak, broad band in the 1370–1220 cm^{-1} (7.30–8.20 μm) region.

MONONUCLEAR AROMATIC HYDROCARBONS

The most prominent and most informative bands in the spectra of aromatic compounds occur in the low-frequency

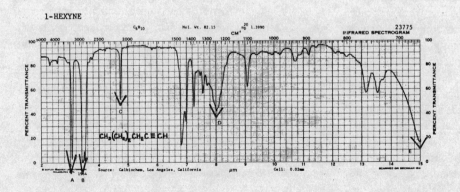

Fig. 11.
A: ≡C–H stretch, 3268 cm^{-1} (3.06 μm).
B: Normal C–H stretch (see Figure 8), 2857-2941 cm^{-1} (3.4–3.5 μm).
C: C≡C stretch, 2110 cm^{-1} (4.74 μm).
D: ≡C–H bend overtone, 1247 cm^{-1} (8.02 μm).
E: ≡C–H bend fundamental, < 667 cm^{-1} (> 15.0 μm).

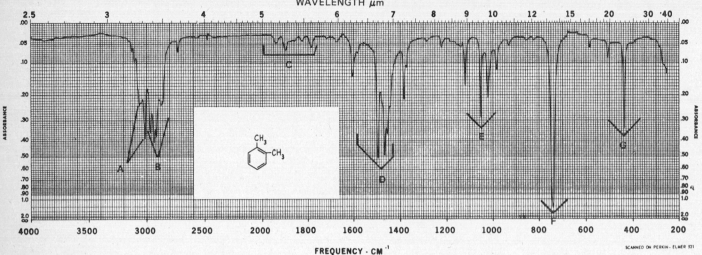

Source: The Matheson Co., Inc., E. Rutherford, N.J.

Fig. 12.

A: Aromatic C—H stretch, 3008 cm^{-1} (3.32 μm).

B: Methyl C—H stretch 2965, 2938, 2918, 2875 cm^{-1} (3.37, 3.40, 3.43, 3.48 μm), (see Figure 8).

C: Overtone or combination bands, 2000–1667 cm^{-1} (5.0–6.0 μm), (see Figure 14).

D: C$\doteq$C ring stretch 1605, 1495, 1466 cm^{-1} (6.23, 6.69, 6.82 μm).

E: In plane C—H bend 1052, 1022 cm^{-1} (9.51, 9.78 μm).

F: Out of plane $\doteq$C—H bend, 742 cm^{-1} (13.48 μm).

G: Out of plane ring C—C bend, 438 cm^{-1} (22.83 μm).

range between 900 and 675 cm^{-1} (11.11 and 14.82 μm). These strong absorption bands result from the out-of-plane bending of the ring C—H bonds. In-plane bending bands appear in the 1300–1000 cm^{-1} (7.70–10.00 μm) region. Skeletal vibrations, involving carbon to carbon stretching within the ring, absorb in the 1600–1585 cm^{-1} (6.25–6.31 μm) and in the 1500–1400 cm^{-1} (6.67–7.14 μm) regions. The skeletal bands frequently appear as doublets, depending upon the nature of the ring substituents.

Aromatic C—H stretching bands occur between 3100 and 3000 cm^{-1} (3.23–3.33 μm).

Weak combination and overtone bands appear in the 2000–1650 cm^{-1} (5.00–6.06 μm) region. The pattern of the overtone bands is characteristic of the substitution pattern of the ring. Because they are weak, the overtone and combination bands are most readily observed in spectra obtained from thick samples. The spectra of Figures 12 and 13 are those of typical aromatic (benzenoid) compounds.

Positions of substitution on benzene rings can often be determined by reference to Figure 14. This procedure is limited by the weak absorption of such overtone (or combination) bands. If there is a strong absorption in that region (such as a carbonyl stretch, see below), the weak overtone bands are obliterated.

Out-of-Plane C—H Bending Vibrations

The in-phase, out-of-plane bending of a ring hydrogen atom is strongly coupled to adjacent hydrogen atoms. The position of absorption of the out-of-plane bending bands is, therefore, characteristic of the number of adjacent hydrogen atoms on the ring. The bands are frequently intense, and may be used for the quantitative determination of the relative concentrations of isomers in mixtures.

Assignments for C—H out-of-plane bending bands in the spectra of substituted benzenes appear in the chart of characteristic group frequencies (Appendix B). These assignments are usually reliable for alkyl substituted benzenes, but caution must be observed in the interpretation of spectra when polar groups are attached directly to the ring, e.g., nitrobenzenes, aromatic acids, and esters or amides of aromatic acids.

The absorption band that frequently appears in the spectra of substituted benzenes near 710–675 cm^{-1} (14.08–14.81 μm) is attributed to out-of-plane ring bending. Some spectra showing typical aromatic absorption are shown in Appendix A: benzene-no.4, indene-no.8, diethyl phthalate-no.21, and *m*-xylene-no.6.

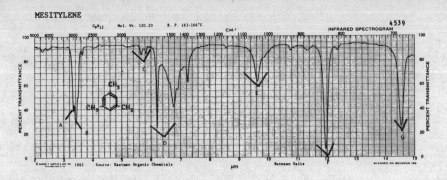

Fig. 13.
A: Aromatic C–H stretch, 3003 cm⁻¹ (3.33 μm).
B: Methyl C–H stretch, 2924 cm⁻¹ (3.42 μm)
                            2874 cm⁻¹ (3.48 μm).
C: Overtone or combination bands, 2000–1667 cm⁻¹ (5.0–6.0 μm), (see Figure 14).
D: C⁓C ring stretch, 1610, 1475 cm⁻¹ (6.21, 6.78 μm).
E: In plane C–H bend, 1038 cm⁻¹ (9.63 μm).
F: Out of plane C–H bend, 837 cm⁻¹ (11.95 μm).
G: Out of plane ring C⁓C bend, 687 cm⁻¹ (14.55 μm).

POLYNUCLEAR AROMATIC COMPOUNDS

Polynuclear aromatic compounds, like the mononuclear aromatics, show characteristic absorption in three regions of the spectrum.

The aromatic C–H stretching and the skeletal vibrations absorb in the same regions as observed for the mononuclear aromatics. The most characteristic absorption of polynuclear aromatics results from C–H out-of-plane bending in the 900–675 cm⁻¹ (11.11–14.81 μm) region. These bands can be correlated with the number of adjacent hydrogen atoms on the rings. Most β-substituted naphthalenes, for example, show three absorption bands due to out-of-plane C–H bending; these correspond to an isolated hydrogen atom and 2 adjacent hydrogen atoms on one ring and 4 adjacent hydrogen atoms on the other ring.

C–H Out-of-Plane Bending Vibrations of a β-Substituted Naphthalene

Isolated hydrogen	862–835 cm⁻¹ (11.60–11.98 μm)
2 Adjacent hydrogens	835–805 cm⁻¹ (11.98–12.42 μm)
4 Adjacent hydrogens	760–735 cm⁻¹ (13.16–13.61 μm)

In the spectra of α-substituted naphthalenes the bands for the isolated hydrogen and the two adjacent hydrogens of β-naphthalenes are replaced by a band for three adjacent hydrogens. This band is near 810–785 cm⁻¹ (12.34–12.74 μm).

Additional bands may appear due to ring bending vibrations. The position of absorption bands for more highly substituted naphthalenes and other polynuclear aromatics are summarized by Colthup[12] and by Conley.[28d]

ALCOHOLS AND PHENOLS

The characteristic bands observed in the spectra of alcohols and phenols result from O–H stretching and C–O stretching. These vibrations are sensitive to hydrogen bonding. The C–O stretching and O–H bending modes are not independent vibrational modes because they couple with the vibrations of adjacent groups. Some typical spectra of alcohols and a phenol are shown in Figures 15 to 17.

O–H Stretching Vibrations

The unbonded or "free" hydroxyl group of alcohols and phenols absorbs strongly in the 3650–3584 cm⁻¹ (2.74–2.79 μm) region. Sharp, "free" hydroxyl bands are observed only in the vapor phase or in very dilute solution in nonpolar solvents. Intermolecular hydrogen bonding increases as the concentration of the solution increases, and additional bands start to appear at lower frequencies, 3550–3200 cm⁻¹ (2.82–3.13 μm), at the expense of the "free" hydroxyl band. This effect is illustrated in Figure 18 in which the absorption bands in the O–H stretching region are shown for two different concentrations of cyclohexyl carbinol in carbon tetrachloride. For comparisons of this type, the path length of the cell must be altered with

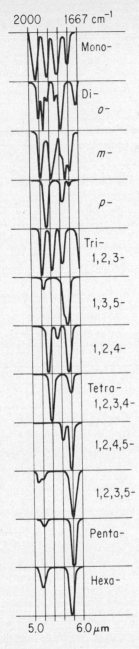

2000   1667 cm^{-1}

Mono-

Di-

o-

m-

p-

Tri-
1,2,3-

1,3,5-

1,2,4-

Tetra-
1,2,3,4-

1,2,4,5-

1,2,3,5-

Penta-

Hexa-

5.0   6.0 μm

Fig. 14.
*Schematic representation of the 5–6 μm region to be antici-
pated for benzenoid compounds of all substitution types.
(Source: J. R. Dyer,* Applications of Absorption Spec-
troscopy of Organic Compounds, *copyright 1965, p. 52.
Reprinted by permission of Prentice-Hall, Inc., Englewood
Cliffs, N.J.)*

changing concentration, so that the same number of
absorbing molecules will be present in the infrared beam at
each concentration. The band at 3623 cm^{-1} (2.76 μm)
results from the monomer, whereas the broad absorption

near 3333 cm^{-1} (3.00 μm) arises from "polymeric"
structures.

Weak intramolecular bonding frequently occurs when a
hydroxyl group is situated adjacent to a proton acceptor
group (X).

The acceptor group may be a heteroatom or a group
containing π-electrons such as a double bond. Intramolecular bonding of this type causes a slight shift of absorption
to longer wavelengths, 3600–3436 cm^{-1} (2.78–2.91 μm),
compared with "free" OH absorption. The band is generally
sharp but may undergo slight broadening. Intramolecular
bonding is essentially independent of concentration.
Studies to detect intramolecular hydrogen bonding, free
from interference from intermolecular hydrogen bonding,
are usually carried out in carbon tetrachloride at concentrations less than 0.005 molar. High resolution is necessary.

In the spectra of compounds such as methyl salicylate or
o-hydroxyacetophenone, in which strong intramolecular
bonding exists, a broad, medium to strong band is observed
near 3077 cm^{-1} (3.25 μm).

In structures such as 2,6-di-*t*-butylphenol, in which steric
hindrance prevents hydrogen bonding, no bonded O–H
band is observed even in the spectra of neat samples.

C–O Stretching Vibrations

The C–O stretching vibrations in alcohols and phenols
produce a strong band in the 1260–1000 cm^{-1}
(7.93–10.00 μm) region of the spectrum. The C–O
stretching mode is coupled with the adjacent C–C stretching vibration; thus in primary alcohols the vibration might
better be described as an asymmetric C–C–O stretching
vibration. The vibrational mode is further complicated by
branching and α, β-unsaturation. These effects are summarized as follows for a series of secondary alcohols (neat
samples):

Methylethylcarbinol	1105 cm^{-1} (9.05 μm)
Methylisopropylcarbinol	1091 cm^{-1} (9.17 μm)
Methylphenylcarbinol	1073 cm^{-1} (9.32 μm)
Methylvinylcarbinol	1058 cm^{-1} (9.45 μm)
Diphenylcarbinol	1014 cm^{-1} (9.86 μm)

The absorption ranges of the various types of alcohols
appear in Table II, p. 95. These values are for neat samples
of the alcohols.

Mulls, pellets or melts of phenols absorb at 1390–1330
cm^{-1} (7.20–7.52 μm) and 1260–1180 cm^{-1} (7.93–8.48

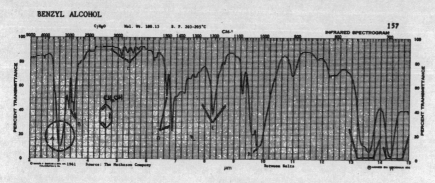

Fig. 15.

A: O—H stretch: intermolecular hydrogen bonded, 3300 cm^{-1} (3.03 μm).

B: C—H stretch: aromatic, 2985 cm^{-1} (3.35 μm)
                methylene, 2857 cm^{-1} (3.50 μm).

C: Overtone or combination bands 2000–1667 cm^{-1} (5.0–6.0 μm), (see Figure 14).

D: C$\dot{=}$C ring stretch, 1497, 1453 cm^{-1} (6.68, 6.88 μm), overlapped by CH$_2$ scissoring, ca. 1471 cm^{-1} (ca. 6.8 μm).

E: O–H bend, possibly augmented by C–H in plane bend, 1208 cm^{-1} (8.28 μm).

F: C—O stretch, primary alcohol, see Table II, 1017 cm^{-1} (9.83 μm).

G: Out of plane aromatic C—H bend, 735 cm^{-1} (13.60 μm).

H: Ring C$\dot{=}$C bend, 697 cm^{-1} (14.35 μm).

μm). These bands apparently result from interaction between O—H bending and C—O stretching. The long wavelength band is the stronger and both bands appear at longer wavelengths in spectra observed in solution. The spectrum of Figure 17 was determined on a melt, to show a high degree of association.

O–H Bending Vibrations

The O–H in-plane bending vibration occurs in the general region of 1420–1330 cm^{-1} (7.04–7.52 μm). In primary and secondary alcohols the O–H in-plane bending couples with the C–H wagging vibrations to produce two bands; the first near 1420 cm^{-1} (7.04 μm), the second near 1330 cm^{-1} (7.52 μm). These bands are of little diagnostic value. Tertiary alcohols, in which no coupling can occur, show a single band in this region, the position depending upon the degree of hydrogen bonding.

The spectra of alcohols and phenols determined in the liquid state, show a broad absorption band in the 769–650 cm^{-1} (13.00–15.40 μm) region because of out-of-plane

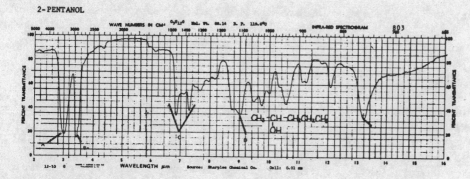

Fig. 16.

A: O—H stretch, intermolecular hydrogen bonded, 3333 cm^{-1} (3.0 μm).

B: C—H stretch (see Figure 8) 2907 cm^{-1} (3.44 μm).

C: C—H bend (see Figure 8), 1460, 1361 cm^{-1} (6.85, 7.35 μm).

D: C—O stretch (see Table II), 1101 cm^{-1} (9.08 μm).

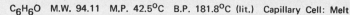

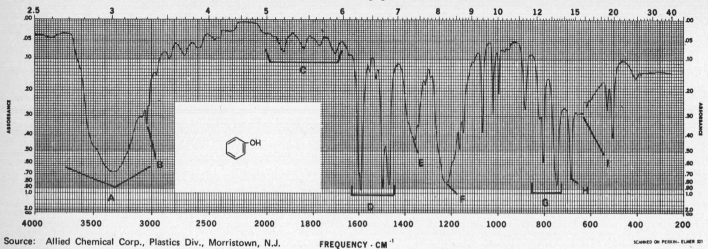

C₆H₆O M.W. 94.11 M.P. 42.5°C B.P. 181.8°C (lit.) Capillary Cell: Melt

Source: Allied Chemical Corp., Plastics Div., Morristown, N.J.   FREQUENCY - CM⁻¹

SCANNED ON PERKIN-ELMER 521

Fig. 17.

A: Broad, intermolecular hydrogen bonded, O—H stretch, 3333 cm^{-1} (3.0 μm).

B: Aromatic C—H stretch, 3045 cm^{-1} (3.28 μm).

C: Overtone or combination bands (see Figure 14), 2000–1667 cm^{-1} (5.0–6.0 μm).

D: C⋯C ring stretch, 1580, 1495, 1468 cm^{-1} (6.33, 6.69, 6.81 μm).

E: In plane O—H bend 1359 cm^{-1} (7.36 μm).

F: C—O stretch, 1223 cm^{-1} (8.18 μm).

G: Out of plane C—H bend 805, 745 cm^{-1} (12.40, 13.43 μm).

H: Out of plane ring C⋯C bend 685 cm^{-1} (14.60 μm).

I: (Broad) hydrogen bonded, out of plane O—H bend, ca. 650 cm^{-1} (ca. 15.0 μm).

bending of the bonded O—H group. Some spectra showing typical alcoholic absorptions are shown in Appendix A: ethyl alcohol-no.16 and methanol-no.15.

ETHERS, EPOXIDES, AND PEROXIDES

C–O Stretching Vibrations

The characteristic response of ethers in the infrared is associated with the stretching vibration of the C—O—C system. Since the vibrational characteristics of this system would not be expected to differ greatly from the C—C—C system, it is not surprising to find the response to C—O—C stretching in the same general region. However, since vibrations involving oxygen atoms result in greater dipole moment changes than those involving carbon atoms, more intense infrared bands are observed for ethers. The C—O—C stretching bands of ethers, as is the case with the C—O stretching bands of alcohols, involve coupling with other vibrations within the molecule. The spectrum of Figure 19 is that of a typical aryl alkyl ether. In addition, the spectra of ethyl ether-no.22 and p-dioxane (a cyclic diether, no.23) are shown in Appendix A.

In the spectra of aliphatic ethers, the most characteristic absorption is a strong band in the 1150–1085 cm^{-1} (8.70–9.23 μm) region because of asymmetrical C—O—C stretching; this band usually occurs near 1125 cm^{-1} (8.89 μm). The symmetrical stretching band is usually weak and is more readily observed in the Raman spectrum.

The C—O—C group in a six-membered ring absorbs at the same frequency as in an acyclic ether. As the ring becomes smaller, the asymmetrical C—O—C stretching vibration moves progressively to lower wavenumbers (longer wavelengths), whereas the symmetrical C—O—C stretching vibration (ring breathing frequency) moves to shorter wavelengths.

Branching on the carbon atoms adjacent to the oxygen usually leads to splitting of the C—O—C band. Isopropyl ether shows a triplet structure in the 1170–1114 cm^{-1} (8.55–8.98 μm) region, the principal band occurring at 1114 cm^{-1} (8.98 μm).

Spectra of aryl alkyl ethers display an asymmetrical C—O—C stretching band at 1275–1200 cm^{-1} (7.84–8.33 μm) with symmetrical stretching near 1075–1020 cm^{-1} (9.30–9.80 μm). Strong absorption due to asymmetrical C—O—C stretching in vinyl ethers occurs in the 1225–1200 cm^{-1} (8.16–8.33 μm) region with a strong symmetrical band at 1075–1020 cm^{-1} (9.30–9.80 μm). Resonance,

94 spectrometric identification of organic compounds

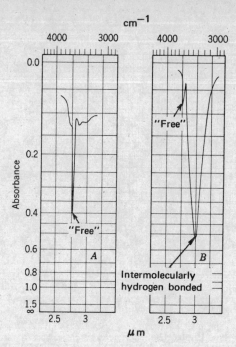

cm^{-1}

Fig. 18.
Infrared spectrum of the O—H stretching region of cyclohexyl carbinol in CCl$_4$. A) 0.03 M (0.406 mm cell); B) 1.00 M (0.014 mm cell).

band frequently appears as a doublet resulting from absorption of rotational isomers.

$$CH_2{=}C \quad \textit{trans} \text{ ca. } 1620^{-1}$$
$$(6.17\ \mu m)$$

$$\textit{cis} \text{ ca. } 1640\ cm^{-1}$$
$$(6.10\ \mu m)$$

Coplanarity in the *trans* isomer allows maximum resonance, thus more effectively reducing the doublebond character of the olefinic linkage. Steric hindrance reduces resonance in the *cis* isomer.

The two bands arising from C—H wagging in terminal olefins occur near 1000 cm^{-1} (10.00 μm) and near 909 cm^{-1} (11.00 μm). In the spectra of vinyl ethers, these bands are shifted to longer wavelengths because of resonance.

terminal CH$_2$ wag, 813 cm^{-1} (12.30 μm)
trans CH wag, 960 cm^{-1} (10.42 μm)

which results in strengthening of the C—O bond, is responsible for the shift in the asymmetric absorption band of aryl alkyl and vinyl ethers.

The C=C stretching band of vinyl ethers occurs in the 1660–1610 cm^{-1} (6.02–6.21 μm) region. This olefinic band is characterized by its higher intensity compared with the C=C stretching band in olefinic hydrocarbons. This

Alkyl and aryl peroxides display C—C—O absorption in the 1198–1176 cm^{-1} (8.35–8.50 μm) region. Acyl and aroyl peroxides display two carbonyl absorption bands in the 1818–1754 cm^{-1} (5.50–5.70 μm) region. Two bands are observed because of mechanical interaction between the stretching modes of the two carbonyl groups.

The symmetrical stretching, or ring breathing frequency,

Table II   Alcoholic C—O Absorption[28c,40]

Alcohol Type	Absorption Range
(1) Saturated tertiary (2) Secondary, highly symmetrical	1205-1124 cm^{-1} (8.30-8.90 μm)
(1) Saturated secondary (2) α-Unsaturated or cyclic tert.	1124-1087 cm^{-1} (8.90-9.20 μm)
(1) Secondary, α-unsaturated (2) Secondary, alicyclic 5- or 6-membered ring (3) Saturated primary	1085-1050 cm^{-1} (9.22-9.52 μm)
(1) Tertiary, highly α-unsaturated (2) Secondary, di-α-unsaturated (3) Secondary, α-unsaturated and α-branched (4) Secondary, alicyclic 7- or 8-membered ring (5) Primary, α-unsaturated and/or α-branched	< 1050 cm^{-1} (> 9.52 μm)

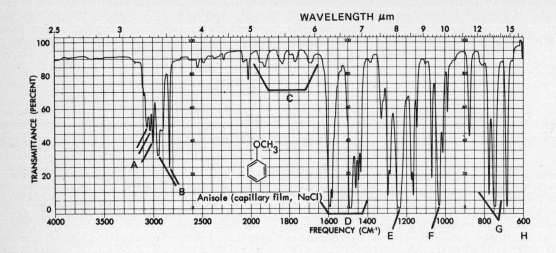

Fig. 19.
A: Aromatic C–H stretch, 3060, 3030, 3000 cm^{-1} (3.27, 3.30, 3.33 μm).
B: Methyl C–H stretch, 2950, 2835 cm^{-1} (3.39, 3.53 μm).
C: Overtone or combination region, 2000–1650 cm^{-1} (5.0–6.0 μm), (see Figure 14).
D: C⋯C ring stretch, 1590, 1480 cm^{-1} (6.29, 6.76 μm).
E: Asymmetric C–O–C stretch, 1245 cm^{-1} (8.03 μm).
F: Symmetric C–O–C stretch, 1030 cm^{-1} (9.71 μm).
G: Out of plane C–H bend, 800–740 cm^{-1} (12.6–13.5 μm).
H: Out of plane ring C⋯C bend, 680 cm^{-1} (14.7 μm).

Anisole. (Source: Daniel J. Pasto and Carl R. Johnson, Organic Structure Determination, *copyright 1969, p. 372. Reprinted by permission of Prentice-Hall, Inc., Englewood Cliffs, N.J.)*

of the epoxy ring, all ring bonds stretching and contracting in phase, occurs near 1250 cm^{-1} (8.00 μm). Another band appears in the 950–810 cm^{-1} (10.53–12.35 μm) region attributed to asymmetrical ring stretching in which the C–C bond is stretching during contraction of the C–O bond. A third band, referred to as the "12 μ band," appears in the 840–750 cm^{-1} (11.90–13.33 μm) region. The C–H stretching vibrations of epoxy rings occur in the 3050–2990 cm^{-1} (3.28–3.34 μm) region of the spectrum.

KETONES

C=O Stretching Vibrations

Ketones, aldehydes, carboxylic acids, carboxylic esters, lactones, acid halides, anhydrides, amides, and lactams show a strong C=O stretching absorption band in the region of 1870–1540 cm^{-1} (5.35–6.50 μm). Its relatively constant position, high intensity, and relative freedom from interfering bands make this one of the easiest bands to recognize in infrared spectra.

Within its given range, the position of the C=O stretching band is determined by the following factors: (1) the physical state, (2) electrical and mass effects of neighboring substituents, (3) conjugation, (4) hydrogen bonding (intermolecular and intramolecular), and (5) ring strain. Consideration of these factors leads to a considerable amount of information about the environment of the C=O group.

In a discussion of these effects, it is customary to refer to the absorption frequency of a neat sample of a saturated aliphatic ketone, 1715 cm^{-1} (5.83 μm), as "normal." Changes in the environment of the carbonyl can either lower or raise the absorption frequency from this "normal" value. A typical ketone spectrum is displayed in Figure 20.

The absorption frequency observed for a neat sample is increased when absorption is observed in non-polar solvents. Polar solvents reduce the frequency of absorption. The over-all range of solvent effects doesn't exceed 25 cm^{-1}.

Replacement of an alkyl group of a saturated aliphatic ketone by a heteroatom (X) shifts the carbonyl absorption. The direction of the shift depends on whether the inductive effect (a) or resonance effect (b) predominates:

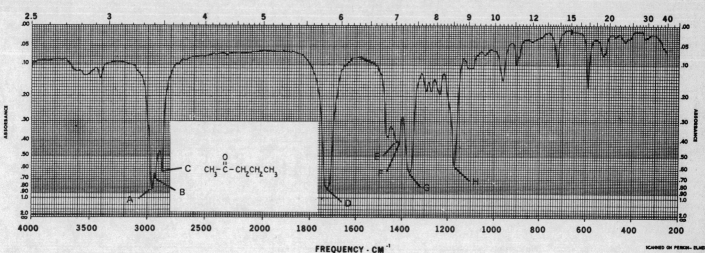

Source: The Matheson Co., Inc., E. Rutherford, N.J.

Fig. 20.

A: ν_{as}, methyl, 2955 cm^{-1} (3.39 µm).

B: ν_{as}, methylene, 2930 cm^{-1} (3.41 µm).

C: ν_s, methyl, 2866 cm^{-1} (3.49 µm).

D: Normal* C=O stretch, 1725 cm^{-1} (5.80 µm).

E: δ_{as}, CH$_3$, ca. 1430 cm^{-1} (ca. 7.0 µm), see Table I, Appendix C.

F: δ_s, CH$_2$, ca. 1430 cm^{-1} (ca. 7.0µm), see Table II, Appendix C.

G: δ_s, CH$_3$ of CH$_3$CO unit, 1370 cm^{-1} (7.30 µm), see Table I, Appendix C.

H: C–CO–C stretch and bend, 1172 cm^{-1} (8.53 µm).

*See p. 96 for a definition of normal.

The inductive effect reduces the length of the C=O bond and thus increases its force constant and the frequency of absorption. The resonance effect increases the C=O bond length and reduces the frequency of absorption.

The absorptions of several carbonyl compound classes are summarized in Table III.

Conjugation with a C=C bond results in delocalization of the π electrons of both unsaturated groups. Delocalization of the π electrons of the C=O group reduces the double bond character of the C to O bond, causing absorption at lower wavenumbers (longer wavelengths). Conjugation with an olefinic or phenyl group causes absorption in the 1685–1666 cm^{-1} (5.93–6.00 µm) region. Additional conjugation may cause a slight further reduction in frequency. This effect of conjugation is illustrated in Figure 21.

Steric effects, which reduce the coplanarity of the

Table III   The Carbonyl Absorption of Various R(CO)X Compounds

X	X Predominantly Inductive νC=O
Cl	1815-1785 cm^{-1} (5.51-5.60 µm)
F	ca. 1869 cm^{-1} (5.35 µm)
Br	1812 cm^{-1} (5.52 µm)
OH (monomer)	1760 cm^{-1} (5.68 µm)
OR	1750-1735 cm^{-1} (5.71-5.76 µm)

X	X Predominantly Resonance νC=O
NH$_2$	1695-1650 cm^{-1} (5.90-6.06 µm)
SR	1720-1690 cm^{-1} (5.82-5.92 µm)

conjugated system, reduce the effect of conjugation. In the absence of steric hindrance, a conjugated system will tend toward a planar conformation. Thus, α,β-unsaturated ketones may exist in *s-cis* and *s-trans* conformations. When

infrared spectrometry   **97**

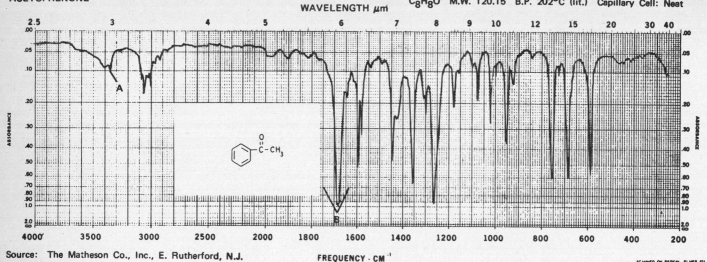

WAVELENGTH μm   C₈H₈O   M.W. 120.15   B.P. 202°C (lit.)   Capillary Cell: Neat

Source: The Matheson Co., Inc., E. Rutherford, N.J.   FREQUENCY · CM⁻¹

SCANNED ON PERKIN- ELMER 521

Fig. 21.

A: Overtone of C=O stretch, 3350 cm⁻¹ (2.98 μm); frequency twice that of C=O stretch.

B: C=O stretch, 1683 cm⁻¹ (5.94 μm), lower frequency (longer wavelength) than observed in Figure 20 because of the conjugation of the phenyl group.

---

both forms are present, absorption for each of the forms is observed. The absorption of benzalacetone in carbon disulfide serves as an example; both the *s-cis* and *s-trans* form are present at room temperature.

s-trans
1674 cm⁻¹ (5.97 μm)

s-cis
1699 cm⁻¹ (5.89 μm)

The absorption of the olefinic bond in conjugation with the carbonyl group occurs at a lower frequency than that of an isolated double bond; the intensity of the conjugated double bond absorption, when in an *s-cis* system, is greater than that of an isolated double bond.

Intermolecular hydrogen bonding between a ketone and a hydroxylic solvent such as methyl alcohol causes a slight decrease in the absorption frequency of the carbonyl group. For example, a neat sample of methyl ethyl ketone absorbs at 1715 cm⁻¹ (5.83 μm), whereas a 10% solution of the ketone in methanol absorbs at 1706 cm⁻¹ (5.86 μm).

β-Diketones usually exist as mixtures of tautomeric keto and enol forms. The enolic form does not show the normal absorption of conjugated ketones. Instead, a broad band appears in the 1640–1580 cm⁻¹ (6.10–6.33 μm) region, many times more intense than normal carbonyl absorption. The intense and displaced absorption results from intramolecular hydrogen bonding, the bonded structure being stabilized by resonance.

$$R-C=CR'-C-R'' \quad \longleftrightarrow \quad R-C-CR'=C-R''$$

Acetylacetone as a liquid at 40°C exists to the extent of 64% in the enolic form that absorbs at 1613 cm⁻¹ (6.20 μm). The keto form and a small amount of unbonded enolic form may be responsible for two bands centering near 1725 cm⁻¹ (5.80 μm). Interaction between the two carbonyl groups in the keto form has also been suggested as a cause for this doublet. The enolic O—H stretching absorption is seen as a broad shallow band at 3000–2700 cm⁻¹ (3.33–3.70 μm).

α-Diketones, in which carbonyl groups exist in formal conjugation, show a single absorption band near the frequency observed for the corresponding monoketone. Biacetyl absorbs at 1718 cm⁻¹ (5.82 μm), dibenzoyl at 1681 cm⁻¹ (5.92 μm). Conjugation is ineffective for α-diketones and the C=O groups of these diketones do not couple as do; e.g., the corresponding groups in acid anhydrides (see below).

Quinones, which have both carbonyl groups in the same ring, absorb in the 1690–1655 cm⁻¹ (5.92–6.04 μm) region. With extended conjugation, in which the carbonyl groups appear in different rings, the absorption shifts to the 1655–1635 cm⁻¹ (6.04–6.12 μm) region.

Acyclic α-chloro ketones absorb at two frequencies due to rotational isomerism. When the chlorine atom is near the oxygen, its negative field repels the nonbonding electrons of the oxygen atom, thus increasing the force constant of

98   spectrometric identification of organic compounds

the C=O bond. This conformation absorbs at a higher frequency (1745 cm^{-1}, 5.73 μm) than that in which the carbonyl oxygen and chlorine atom are widely separated (1725 cm^{-1}, 5.80 μm). In rigid molecules such as the monoketo-steroids, α-halogenation results in equatorial or axial substitution. In the equatorial orientation, the halogen atom is near the carbonyl group and the "field effect" causes an increase in the C=O stretching frequency. In the isomer in which the halogen atom is axial to the ring, and distant from the C=O, no shift is observed.

In cyclic ketones, the bond angle of the $\overset{\displaystyle C}{\underset{\displaystyle C}{\diagdown}}$C=O group influences the absorption frequency of the carbonyl group. The C=O stretching undoubtedly is affected by adjacent C–C stretching. In noncyclic ketones and in ketones with a 6-membered ring, the angle is near 120°. In strained rings in which the angle is less than 120° interaction with C–C bond stretching increases the energy required to produce C=O stretching and thus increases the stretching frequency. Cyclohexanone absorbs at 1715 cm^{-1} (5.83 μm), cyclopentanone absorbs at 1751 cm^{-1} (5.71 μm), and cyclobutanone absorbs at 1775 cm^{-1} (5.63 μm).

$$\overset{\displaystyle O}{\underset{\displaystyle \|}{C}}$$
C–C–C *Stretching and Bending Vibrations*

Ketones show moderate absorption in the 1300–1100 cm^{-1} (7.70–9.09 μm) region as a result of C–C–C stretching and bending in the C–$\overset{\overset{\displaystyle O}{\|}}{C}$–C group. The absorption may consist of multiple bands. Aliphatic ketones absorb in the 1230–1100 cm^{-1} (8.13–9.09 μm) region; aromatic ketones absorb at the higher frequency end of the general absorption region.

The spectra of 2-butanone (methyl ethyl ketone-no. 18), acetone-no.17, and cyclohexanone-no.19 in Appendix A illustrate ketonic absorptions.

ALDEHYDES

The spectrum of 2-phenylpropionaldehyde, illustrating typical aldehydic absorption characteristics, is shown in Figure 22.

C=O Stretching Vibrations

The carbonyl group of aldehydes absorb at slightly higher frequencies than that of the corresponding methyl ketones.

Aliphatic aldehydes absorb near 1740–1720 cm^{-1} (5.75–5.82 μm). Aldehydic carbonyl absorption responds to structural changes in the same manner as ketones. Electronegative substitution on the α-carbon increases the frequency of carbonyl absorption. Acetaldehyde absorbs at 1730 cm^{-1} (5.78 μm), trichloroacetaldehyde absorbs at 1768 cm^{-1} (5.65 μm). Conjugate unsaturation, as occurs in α,β-unsaturated and aryl aldehydes, reduces the frequency of carbonyl absorption. α,β-Unsaturated aldehydes and aromatic aldehydes absorb in the region of 1710–1685 cm^{-1} (5.85–5.94 μm). Internal hydrogen bonding, such as occurs in salicylaldehyde, shifts the absorption (1666 cm^{-1}, 6.00 μm, for salicylaldehyde) to lower wavenumbers (longer wavelengths). α-Dialdehydes, like the α-diketones, show only one carbonyl absorption peak with no shift from the normal absorption position of mono-aldehydic absorption.

C–H Stretching Vibrations

The majority of aldehydes show aldehydic C–H stretching absorption in the 2830–2695 cm^{-1} (3.53–3.71 μm) region. Two moderately intense bands are frequently observed in this region. The appearance of two bands is attributed to Fermi resonance between the fundamental aldehydic C–H stretch and the first overtone of the aldehydic C–H bending vibration that usually appears near 1390 cm^{-1} (7.20 μm). Only one C–H stretching band is observed for aldehydes whose C–H bending band has been shifted appreciably from 1390 cm^{-1} (7.20 μm).

Some aromatic aldehydes with strongly electronegative groups in the ortho position may absorb as high as 2900 cm^{-1} (3.45 μm).

Medium intense absorption near 2720 cm^{-1} (3.68 μm), accompanied by a carbonyl absorption band is good evidence for the presence of an aldehyde group.

The absorption frequencies of methyl and methylene groups, attached to a carbonyl, are summarized in Tables I and II of Appendix C.

CARBOXYLIC ACIDS

O–H Stretching Vibrations

In the liquid or solid state, and in carbon tetrachloride solution at concentrations much over 0.01 M, carboxylic acids exist as dimers due to strong hydrogen bonding.

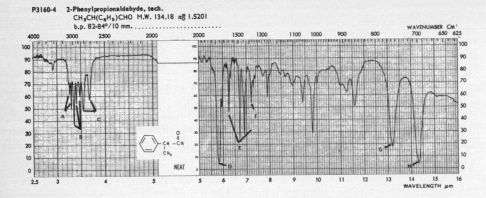

P3160-4   2-Phenylpropionaldehyde, tech.
CH$_3$CH(C$_6$H$_5$)CHO   M.W. 134.18 n_D^{20} 1.5201
b.p. 82-84°/10 mm. . . . .

Fig. 22.

*2-Phenylpropionaldehyde.**

A: Aromatic, 3077, 3040 cm^{-1} (3.25, 3.29 μm) (see Figure 12).

B: Aliphatic, 2985, 2941, 2874 cm^{-1} (3.35, 3.40, 3.48 μm) (see Figures 8 and 12).

C: Aldehydic C–H stretch, 2825, 2717 cm^{-1} (3.54, 3.68 μm). Doublet due to Fermi resonance with overtone of band at F.

D: Normal aldehydic C=O stretch, 1730 cm^{-1} (5.78 μm). Conjugated C=O stretch would be ca. 1700 cm^{-1}, 5.89 μm, for example, in C$_6$H$_5$CHO.

E: Ring C⁼C stretch, 1600, 1497, 1453 cm^{-1} (6.25, 6.68, 6.88 μm).

F: Aldehydic C–H bend, 1389 cm^{-1} (7.20 μm).

G: Out of plane C–H bend, 749 cm^{-1} (13.35 μm).

H: Out of plane C–C bend, 699 cm^{-1} (14.30 μm).

*A through C are C–H stretch absorptions.

The exceptional strength of the hydrogen bonding is explained on the basis of the large contribution of the ionic resonance structure. Because of the strong bonding, a free hydroxyl stretching vibration (near 3520 cm^{-1}, 2.84 μm) is observed only in very dilute solution in nonpolar solvents or in the vapor phase. In each case, however, there is a mixture of monomer and dimer.

Carboxylic acid dimers display very broad, intense O–H stretching absorption in the region of 3300–2500 cm^{-1} (3.03–4.00 μm). The band usually centers near 3000 cm^{-1} (3.33 μm). The weaker C–H stretching bands are generally seen superimposed upon the broad O–H band. Fine structure observed on the long wavelength side of the broad O–H band represents overtones and combination tones of fundamental bands occurring at longer wavelengths. The spectrum of a typical aliphatic carboxylic acid is displayed in Figure 23.

Other structures with strong hydrogen bonding, such as β-diketones, also absorb in the 3300–2500 cm^{-1} (3.03–4.00 μm) region, but the absorption is usually less intense. Also, the C=O stretching vibrations of structures such as β-diketones are shifted to lower frequencies than those observed for carboxylic acids.

Carboxylic acids can bond intermolecularly with ethers such as dioxane and tetrahydrofuran or with other solvents that can act as proton acceptors. Spectra determined in such solvents show bonded O–H absorption near 3100 cm^{-1} (3.23 μm).

C=O Stretching Vibrations

The C=O stretching bands of acids are considerably more intense than ketonic C=O stretching bands. The monomers of saturated aliphatic acids absorb near 1760 cm^{-1} (5.68 μm).

The carboxylic dimer has a center of symmetry; only the asymmetrical C=O stretching mode absorbs in the infrared. Hydrogen bonding and resonance weaken the C=O bond, resulting in absorption at a lower frequency than the monomer. The C=O group in dimerized saturated aliphatic acids absorbs in the region of 1720–1706 cm^{-1} (5.81–5.86 μm).

Internal hydrogen bonding reduces the frequency of the carbonyl stretching absorption to a greater degree than does intermolecular hydrogen bonding. For example, salicylic acid absorbs at 1665 cm^{-1} (6.01 μm), whereas p-hydroxybenzoic acid absorbs at 1680 cm^{-1} (5.95 μm).

Unsaturation in conjugation with the carboxylic carbonyl group decreases the frequency (increases the wavelength) of absorption of both the monomer and dimer forms only slightly. In general, α,β-unsaturated and aryl

$C_7H_{14}O_2$ M.W. 130.19 B.P. 215°C N_D^{20}1.4221 Capillary Cell: Neat

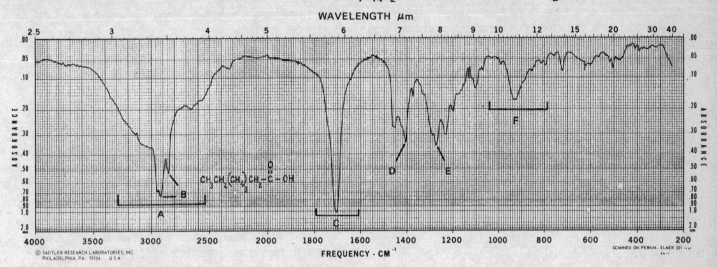

Source: Aldrich Chemical Company, Milwaukee, Wis.

Fig. 23.

A: Broad O–H stretch, 3300-2500 cm^{-1} (3.03-4.00 μm).

B: C–H stretch (see Figure 8) 2950, 2920, 2850 cm^{-1} (3.39, 3.43, 3.51 μm). (Superimposed upon O–H stretch).

C: Normal, dimeric carboxylic C=O stretch, 1715 cm^{-1} (5.83 μm).

D: C–O–H in plane bend,* 1408 cm^{-1} (7.10 μm).

E: C–O stretch,* dimer, 1280 cm^{-1} (7.81 μm).

F: O–H out of plane bend, 930 cm^{-1} (10.75 μm).

   *Bands at D and E involve C–O–H interaction.

conjugated acids show absorption for the dimer in the 1710–1680 cm^{-1} (5.85–5.95 μm) region. Extension of conjugation beyond the α,β-position results in very little additional shifting of the C=O absorption.

   Substitution in the α-position with electronegative groups, such as the halogens, brings about a slight increase in the C=O absorption frequency (10–20 cm^{-1}; 0.03–0.07 μm). The spectra of acids with halogens in the α-position, determined in the liquid state or in solution, show dual carbonyl bands due to rotational isomerism (field effect). The higher frequency (shorter wavelength) band corresponds to the conformation in which the halogen is in proximity to the carbonyl group.

C–O Stretching and O–H Bending Vibrations

Two bands arising from C–O stretching and O–H bending appear in the spectra of carboxylic acids near 1320–1210 cm^{-1} (7.58–8.26 μm) and near 1440–1395 cm^{-1} (6.94–7.17 μm) respectively. Both of these bands involve some interaction between C–O stretching and in-plane C–O–H bending. The more intense band, near 1315–1280

cm^{-1} (7.60–7.81 μm) for dimers, is generally referred to as the C–O stretching band, and usually appears as a doublet in the spectra of long-chain fatty acids. The C–O–H bending band near 1440–1395 cm^{-1} (6.94–7.17 μm) is of moderate intensity and occurs in the same region as the CH$_2$ scissoring vibration of the CH$_2$ group adjacent to the carbonyl.

   One of the characteristic bands in the spectra of dimeric carboxylic acids results from the out-of-plane bending of the bonded O–H. The band appears near 920 cm^{-1} (10.87 μm) and is characteristically broad with medium intensity.

CARBOXYLATE ANION

   The carboxylate anion has two strongly coupled carbon to oxygen bonds with bond strengths intermediate between C=O and C–O.

infrared spectrometry   **101**

The carboxylate ion gives rise to two bands: a strong asymmetrical stretching band near 1650–1550 cm^{-1} (6.06–6.45 μm), and a weaker, symmetrical stretching band near 1400 cm^{-1} (7.15 μm).

The conversion of a carboxylic acid to a salt can serve as confirmation of the acid structure. This is conveniently done by the addition of a tertiary, aliphatic amine, such as triethylamine, to a solution of the carboxylic acid in chloroform (no reaction occurs in carbon tetrachloride). The carboxylate ion, thus formed, shows the two characteristic carbonyl absorption bands in addition to an "ammonium" band in the 2700–2200 cm^{-1} (3.70–4.55 μm) region. The O–H stretching band, of course, disappears. The spectrum of ammonium benzoate, Figure 24, demonstrates most of these features.

ESTERS AND LACTONES

Esters and lactones have two characteristically strong absorption bands arising from C=O and C–O stretching. The intense C=O stretching vibration occurs at higher frequencies (shorter wavelength) than that of normal ketones. The force constant of the carbonyl bond is increased by the electron attracting nature of the adjacent oxygen atom (inductive effect). Overlapping occurs between esters in which the carbonyl frequency is lowered, and ketones in which the normal ketone frequency is raised. A distinguish-ing feature of esters and lactones, however, is the strong C–O stretching band in the region where a weaker band occurs for ketones. There is overlapping in the C=O frequency of esters or lactones and acids, but the OH stretching and bending vibrations, and the possibility of salt formation distinguish the acids.

The frequency of the ester carbonyl responds to environmental changes in the vicinity of the carbonyl group in much the same manner as ketones. The spectrum of phenyl acetate, Figure 25, illustrates most of the important absorption characteristics for esters.

C=O Stretching Vibrations

The C=O absorption band of saturated aliphatic esters (except formates) is in the 1750–1735 cm^{-1} (5.71–5.76 μm) region. The C=O absorption bands of formates, α,β-unsaturated, and aryl esters are in the region of 1730–1715 cm^{-1} (5.78–5.83 μm). Further conjugation has little or no additional effect upon the frequency of the carbonyl absorption.

In the spectra of vinyl or phenyl esters, with unsaturation adjacent to the C–O– group, a marked rise in the carbonyl frequency is observed along with a lowering of the C–O frequency. Vinyl acetate has a carbonyl band at 1776 cm^{-1} (5.63 μm), phenyl acetate absorbs at 1770 cm^{-1} (5.65 μm).

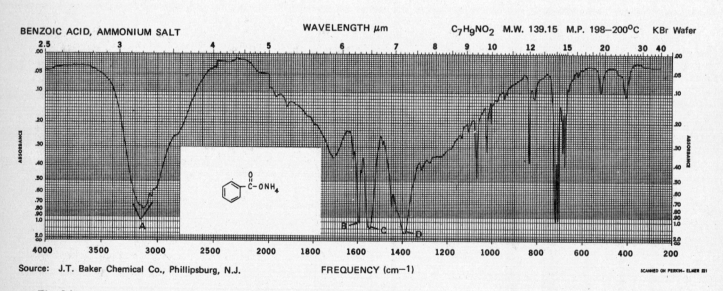

BENZOIC ACID, AMMONIUM SALT   WAVELENGTH μm   C$_7$H$_9$NO$_2$ M.W. 139.15 M.P. 198–200°C   KBr Wafer

Source: J.T. Baker Chemical Co., Phillipsburg, N.J.   FREQUENCY (cm^{-1})

Fig. 24.
A: N–H and C–H stretch, 3600–2500 cm^{-1} (2.78–4.00 μm).
B: Asymmetric carboxylate anion C($\cdots$O)$_2^-$ stretch, 1600 cm^{-1} (6.25 μm).
C: Ring C$\cdots$C stretch, 1550 cm^{-1} (6.45 μm).
D: Symmetric carboxylate anion C($\cdots$O)$_2^-$ stretch, 1385 cm^{-1} (7.22 μm) and N–H bend.

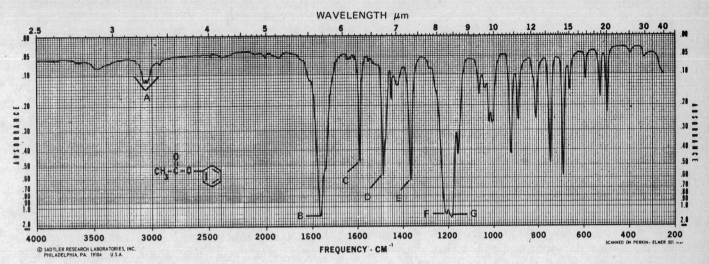

Source: The Matheson Company, Inc., East Rutherford, New Jersey

Fig. 25.

A: Aromatic C–H stretch, 3070, 3040 cm^{-1} (3.26, 3.29 μm).

B: C=O stretch, 1770 cm^{-1} (5.65 μm), this is higher frequency than that due to normal ester C=O stretch (ca. 1740 cm^{-1}, 5.75 μm, see Table III), due to phenyl conjugation with alcohol oxygen; conjugation of an aryl group or other unsaturation with the carbonyl group causes this C=O stretch to be at lower than normal frequency (e.g., benzoates absorb at ca. 1724 cm^{-1}, ca. 5.80 μm).

C: Ring C$\doteq$C stretch, 1593 cm^{-1} (6.28 μm).

D: δ_{as}, CH$_3$, 1493 cm^{-1} (6.70 μm).*

E: δ_s, CH$_3$, 1360 cm^{-1} (7.35 μm).*

F: Acetate CC(=O)–O stretch, 1205 cm^{-1} (8.30 μm).

G: O$\doteq$C$\doteq$C asym. stretch, 1183 cm^{-1} (8.45 μm).

*See Table I, Appendix C.

α-Halogen substitution results in a rise in the C=O stretching frequency. Ethyl trichloroacetate absorbs at 1770 cm^{-1} (5.65 μm).

In α-diesters and α-keto esters, as in α-diketones, there appears to be little or no interaction between the two carbonyl groups so that normal absorption occurs in the region of 1755–1740 cm^{-1} (5.70–5.75 μm). In the spectra of β-keto esters, however, where enolization can occur, a band is observed near 1650 cm^{-1} (6.06 μm) that results from bonding between the ester C=O and the enolic hydroxyl group.

The carbonyl absorption of saturated δ-lactones (six membered ring) occurs in the same region as straight-chain, unconjugated esters. Unsaturation α to the C=O reduces the C=O absorption frequency. Unsaturation α to the –O– group increases it.

α-Pyrones frequently display two carbonyl absorption bands in the 1775–1715 cm^{-1} (5.63–5.83 μm) region, probably due to Fermi resonance.

Saturated γ-lactones (five membered ring) absorb at shorter wavelengths than esters or δ-lactones: 1795–1760 cm^{-1} (5.57–5.68 μm), γ-valerolactone absorbs at 1770 cm^{-1} (5.65 μm). Unsaturation in the γ-lactone molecule affects the carbonyl absorption in the same manner as unsaturation in δ-lactones.

1800 cm^{-1} (5.56 μm)   1750 cm^{-1} (5.71 μm)

In unsaturated lactones, when the double bond is adjacent to the –O–, a strong C=C absorption is observed in the 1685–1660 cm^{-1} (5.94–6.02 μm) region.

1720 cm^{-1} (5.81 μm)   1760 cm^{-1} (5.68 μm)

infrared spectrometry **103**

C–O Stretching Vibrations

The "C–O stretching vibrations" of esters actually consist of two asymmetric coupled vibrations: C–C(=O)–O and O–C–C, the former being more important. These bands occur in the region of 1300–1000 cm^{-1} (7.70–10.00 μm). The corresponding symmetric vibrations are of little importance. The C–O stretch correlations are less reliable than the C=O stretch correlations.

The C–C(=O)–O band of saturated esters, except for acetates, absorbs strongly in the 1210–1163 cm^{-1} (8.26–8.60 μm) region. It is often broader and stronger than the C=O stretch absorption. Acetates of saturated alcohols display this band at 1240 cm^{-1} (8.07 μm). Vinyl and phenyl acetates absorb at a somewhat lower frequency, 1190–1140 cm^{-1} (8.40–8.77 μm); for example, see Figure 25. The C–C(=O)–O stretch of esters of α,β-unsaturated acids results in multiple bands in the 1300–1160 cm^{-1} (7.69–8.62 μm) region. Esters of aromatic acids absorb strongly in the 1310–1250 cm^{-1} (7.63–8.00 μm) region. The analogous type of stretch in lactones is observed in the 1250–1111 cm^{-1} (8.00–9.00 μm) region.

The O–C–C band of esters ("alcohol" carbon-oxygen stretch) of primary alcohols occurs at about 1064–1031 cm^{-1} (9.40–9.70 μm) and that of esters of secondary alcohols occurs at about 1100 cm^{-1} (9.09 μm). Aromatic esters of primary alcohols show this absorption near 1111 cm^{-1} (9.00 μm).

Methyl esters of long-chain fatty acids present a three-band pattern with bands near 1250 cm^{-1} (8.00 μm), 1205 cm^{-1} (8.30 μm), and 1175 cm^{-1} (8.51 μm). The band near 1175 cm^{-1} (8.51 μm) is the strongest.

The influence of the –C(=O)–O group upon adjacent CH$_2$ and CH$_3$ groups is summarized in the Appendix C, Tables I and II. Some spectra showing typical ester absorptions are shown in Appendix A: ethyl acetate-no.20 and diethyl phthalate-no.21.

ACID HALIDES

C=O Stretching Vibrations

Acid halides show strong absorption in the C=O stretching region. Unconjugated acid chlorides absorb in the 1815–1785 cm^{-1} (5.51–5.60 μm) region. Acetyl fluoride in the gas phase absorbs near 1869 cm^{-1} (5.35 μm). Conjugated acid halides absorb at a slightly lower frequency because resonance reduces the force constant of the C=O bond; aromatic acid chlorides absorb strongly at 1800–1770 cm^{-1} (5.56–5.65 μm). A weak band near 1750–1735 cm^{-1} (5.71–5.76 μm) appearing in the spectra of aroyl chlorides probably results from Fermi resonance

between the C=O band and the overtone of a longer wavelength band near 875 cm^{-1} (11.43 μm). The annotated spectrum of benzoyl chloride is given in Figure 26.

CARBOXYLIC ACID ANHYDRIDES

C=O Stretching Vibrations

Anhydrides display two stretching bands in the carbonyl region. The two bands result from asymmetrical and symmetrical C=O stretching modes. Saturated noncyclic anhydrides absorb near 1818 cm^{-1} (5.50 μm) and near 1750 cm^{-1} (5.71 μm). Conjugated noncyclic anhydrides show absorption near 1775 cm^{-1} (5.63 μm) and near 1720 cm^{-1} (5.81 μm); the decrease in the frequency of absorption is due to resonance. The higher frequency band is the more intense.

Cyclic anhydrides with five-membered rings show absorption at higher frequencies (lower wavelengths) than noncyclic anhydrides because of ring strain; succinic anhydride absorbs at 1865 cm^{-1} (5.37 μm) and at 1782 cm^{-1} (5.62 μm). The lower frequency (longer wavelength) C=O band is the stronger of the two carbonyl bands in five-membered-ring cyclic anhydrides.

C–O Stretching Vibrations

Other strong bands appear in the spectra of anhydrides as a result of C–C–O–C–C stretching vibrations. Unconjugated straight chain anhydrides absorb near 1047 cm^{-1} (9.55 μm). Cyclic anhydrides display bands near 952–909 cm^{-1} (10.50–11.00 μm), and near 1299–1176 cm^{-1} (7.70–8.50 μm). The C–O stretching band for acetic anhydride is at 1125 cm^{-1} (8.89 μm).

The spectrum of Figure 27 is that of a typical aliphatic anhydride.

AMIDES

All amides show a carbonyl absorption band known as the Amide I band. Its position depends upon the degree of hydrogen bonding, and thus on the physical state of the compound.

Primary amides show two N–H stretching bands resulting from symmetrical and asymmetrical N–H stretching. Secondary amides and lactams show only one N–H stretching band. As in the case of O–H stretching, the

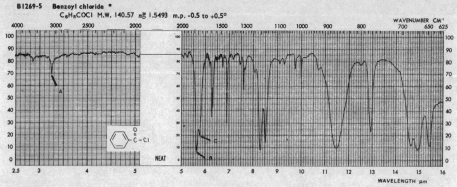

Fig. 26.

A: Aromatic C–H stretch, 3080 cm^{-1} (3.25 μm).

B: C=O stretch, 1790 cm^{-1} (5.58 μm), see Table III. (Acid chloride C=O stretch position shows very small dependence upon conjugation; aroyl chlorides identified by band such as at *C*.)

C: Fermi resonance band (of C=O stretch and overtone of 875 cm^{-1} band), 1745 cm^{-1} (5.73 μm).

frequency of the N–H stretching is reduced by hydrogen bonding, though to a lesser degree. Overlapping occurs in the observed position of N–H and O–H stretching frequencies so that an unequivocal differentiation in structure is sometimes impossible.

Primary amides and secondary amides, and a few

lactams, display a band or bands in the region of 1650–1515 cm^{-1} (6.06–6.60 μm) due primarily to NH$_2$ or NH bending: the amide II band. This absorption involves coupling between N–H bending and other fundamental vibrations and requires a *trans* configuration.

Out-of-plane NH wagging is responsible for a broad band

PROPIONIC ANHYDRIDE   C$_6$H$_{10}$O$_3$ M.W. 130.14 B.P. 165–169°C Capillary Cell: Neat

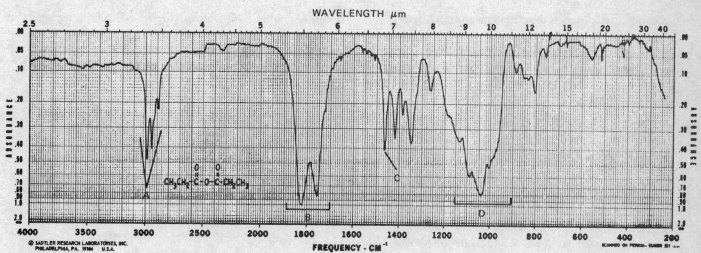

Source: The Matheson Company, Inc., East Rutherford, New Jersey

Fig. 27.

A: C–H stretch, 2990, 2950, 2880 cm^{-1} (3.34, 3.39, 3.47 μm).

B: Asymmetric and symmetric C=O coupled stretching, respectively: 1825, 1758 cm^{-1} (5.48, 5.69 μm). See Table III, p. 97.

C: δ$_s$ CH$_2$ (scissoring), 1465 cm^{-1} (6.83 μm), see Table II, Appendix C.

D: C–CO–O–CO–C stretch, 1040 cm   (9.62 μm).

infrared spectrometry   **105**

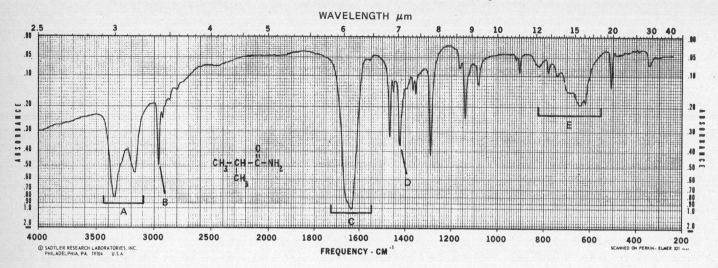

C_4H_9NO N.W. 87.12 M.P. 127–129°C KBr Wafer

Source: Aldrich Chemical Company, Milwaukee, Wis.

Fig. 28.

A: N–H stretch, coupled, primary amide, hydrogen bonded.
            Asymmetric, 3350 cm^{-1} (2.99 μm)
            Symmetric, 3170 cm^{-1} (3.15 μm).

B: Aliphatic C–H stretch, 2960 cm^{-1} (3.38 μm).

C: Overlap $\begin{cases} \text{C=O stretch, Amide I band, 1640 cm}^{-1}\text{ (6.10 μm), see Table III.} \\ \text{N–H bend, 1640 cm}^{-1}\text{ (6.10 μm), Amide II band.} \end{cases}$

D: C–N stretch, 1425 cm^{-1} (7.02 μm).

E: Broad N–H out of plane bend, 700–600 cm^{-1} (14.28–16.67 μm).

of medium intensity in the 800–666 cm^{-1} (12.5–15.0 μm) region.

The spectrum of Figure 28 is that of a typical primary amide of an aliphatic acid. The spectrum of DMF (N,N-dimethylformamide, Appendix A, no. 28) displays typical amide absorptions.

N–H Stretching Vibrations

In dilute solution in nonpolar solvents, primary amides show two moderately intense NH stretching frequencies corresponding to the asymmetrical and symmetrical NH stretching vibrations. These bands occur near 3520 cm^{-1} (2.84 μm) and 3400 cm^{-1} (2.94 μm), respectively. In the spectra of solid samples, these bands are observed near 3350 cm^{-1} (2.99 μm) and 3180 cm^{-1} (3.15 μm) because of hydrogen bonding.

In infrared spectra of secondary amides, which exist mainly in the *trans* conformation, the free NH stretching vibration observed in dilute solutions occurs near 3500–3400 cm^{-1} (2.86–2.94 μm). In more concentrated

solutions and in solid samples, the free NH band is replaced by multiple bands in the 3330–3060 cm^{-1} (3.00–3.27 μm) region. Multiple bands are observed since the amide group can bond to produce dimers, with a *cis* conformation, and polymers, with a *trans* conformation.

C=O Stretching Vibrations (Amide I Band)

The C=O absorption of amides occurs at longer wavelengths than "normal" carbonyl absorption due to the

resonance effect. (see p. 97) The position of absorption depends upon the same environmental factors as the carbonyl absorption of other compounds.

Primary amides (except acetamide, whose C=O bond absorbs at 1694 cm^{-1}, 5.90 μm), have a strong amide I band in the region of 1650 cm^{-1} (6.06 μm) when examined in the solid phase. When the amide is examined in dilute solution, the absorption is observed at a higher frequency, near 1690 cm^{-1} (5.92 μm). In more concentrated solutions, the C=O frequency is observed at some intermediate value, depending on the degree of hydrogen bonding.

Simple, open chain, secondary amides absorb near 1640 cm^{-1} (6.10 μm) when examined in the solid state. In dilute solution, the frequency of the amide I band may be raised to 1680 cm^{-1} (5.95 μm) and even to 1700 cm^{-1} (5.88 μm) in the case of the anilides. In the anilide structure there is competition between the ring and the C=O for the nonbonded electron pair of the nitrogen.

The carbonyl frequency of tertiary amides is independent of the physical state, since hydrogen bonding with another tertiary amide group is impossible. The C=O absorption occurs in the range of 1680 to 1630 cm^{-1} (5.95 to 6.13 μm). The absorption range of tertiary amides in solution is influenced by hydrogen bonding with the solvent. N,N-Diethylacetamide absorbs at 1647 cm^{-1} (6.07 μm) in dioxane and at 1615 cm^{-1} (6.20 μm) in methanol.

Electron attracting groups attached to the nitrogen increase the frequency of absorption since they effectively compete with the carbonyl oxygen for the electrons of the nitrogen, thus increasing the force constant of the C=O bond.

N–H Bending Vibrations (Amide II Band)

All primary amides show a sharp absorption band in dilute solution (amide II band) resulting from NH$_2$ bending at a somewhat lower frequency than the C=O band. This band has an intensity of one-half to one-third of the C=O absorption band. In mulls and pellets the band occurs near 1655–1620 cm^{-1} (6.04–6.17 μm) and is usually under the envelope of the Amide I band. In dilute solutions the band appears at lower frequency, 1620–1590 cm^{-1} (6.17–6.29 μm), and normally is separated from the Amide I band. Multiple bands may appear in the spectra of concentrated solutions, arising from the free and associated states. The nature of the R group (R–C(=O)–NH$_2$) has little effect upon the amide II band.

Secondary acyclic amides in the solid state display an amide II band in the region of 1570–1515 cm^{-1} (6.37–6.60 μm). In dilute solution, the band occurs in the 1550–1510 cm^{-1} (6.45–6.62 μm) region. This band results from interaction between the N–H bending and the C–N stretching of the C–N–H group. A second, weaker band near 1250 cm^{-1} (8.00 μm) also results from interaction between the N–H bending and C–N stretching.

Other Vibration Bands

The C–N stretching band of primary amides occurs near 1400 cm^{-1} (7.14 μm). A broad, medium band in the 800–666 cm^{-1} (12.5–15.0 μm) region in the spectra of primary and secondary amides results from out-of-plane N–H wagging.

LACTAMS

In lactams of medium ring size, the amide group is forced into the *cis* conformation. Solid lactams absorb strongly near 3200 cm^{-1} (3.12 μm) because of the N–H stretching vibration. This band does not shift appreciably with dilution since the *cis* form remains associated at relatively low concentrations.

C=O Stretching Vibrations

The C=O absorption of lactams with six-membered rings or larger is near 1650 cm^{-1} (6.06 μm). Five-membered ring (γ) lactams absorb in the 1750–1700 cm^{-1} (5.71–5.88 μm) region. Four-membered ring (β) lactams, unfused, absorb at 1760–1730 cm^{-1} (5.68–5.78 μm). Fusion of the lactam ring to another ring generally increases the frequency by 20–50 cm^{-1} (0.07–0.17 μm).

Most lactams do not show a band near 1550 cm^{-1} (6.45 μm) that is characteristic of *trans* non-cyclic secondary amides. The N–H out-of-plane wagging in lactams causes broad absorption in the 800–700 cm^{-1} (12.5–14.3 μm) region.

AMINES

The spectrum of a typical, primary, aliphatic amine appears in Figure 29.

N–H Stretching Vibrations

Primary amines, examined in dilute solution, display two weak absorption bands: one near 3500 cm^{-1} (2.86 μm), the other near 3400 cm^{-1} (2.94 μm). These bands

C$_8$H$_{19}$N   M.W. 129.25   B.P. 178–181°C   Capillary Cell: Neat

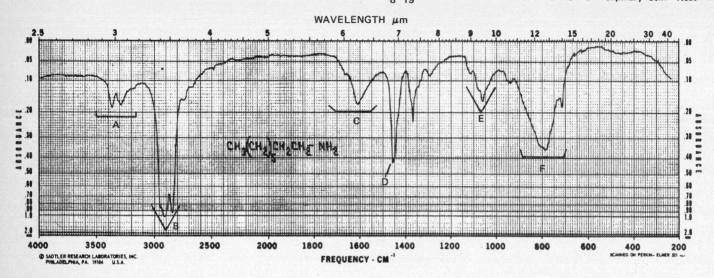

Source: The Matheson Company, Inc., East Rutherford, N.J.

Fig. 29.

A: N–H stretch, hydrogen bonded, primary amine coupled doublet:
   Asymmetric, 3365 cm^{-1} (2.97 μm)
   Symmetric, 3290 cm^{-1} (3.04 μm).
   (Shoulder at ca. 3200 cm^{-1}, ca. 3.12 μm, Fermi resonance band with overtone of band at C.

B: Aliphatic C–H stretch, 2910, 2850 cm^{-1} (3.44, 3.51 μm); ν_s, CH$_2$, 2817 cm^{-1} (3.55 μm), see Table II, Appendix C.

C: N–H bend (scissoring), 1620 cm^{-1} (6.17 μm).

D: δ_s, CH$_2$ (scissoring), 1458 cm^{-1} (6.86 μm), see Table II, Appendix C.

E: C–N stretch, 1063 cm^{-1} (9.41 μm).

F: N–H wag (neat sample), 909–666 cm^{-1} (11.00–15.00 μm).

represent, respectively, the "free" asymmetrical and symmetrical N–H stretching modes. Secondary amines show a single weak band in the 3350–3310 cm^{-1} (2.98–3.02 μm) region. These bands are shifted to longer wavelengths by hydrogen bonding. The associated N–H bands are weaker and frequently sharper than the corresponding O–H bands. Aliphatic primary amines (neat) absorb at 3400–3330 cm^{-1} (2.94–3.00 μm) and at 3330–3250 cm^{-1} (3.00–3.08 μm). Aromatic primary amines absorb at slightly higher frequencies (shorter wavelengths). In the spectra of liquid primary and secondary amines, a shoulder usually appears on the low frequency (long wavelength) side of the N–H stretching band, arising from the overtone of the NH bending band intensified by Fermi resonance. Tertiary amines do not absorb in this region of the spectrum.

N–H Bending Vibrations

The N–H bending (scissoring) vibration of primary amines is observed in the 1650–1580 cm^{-1} (6.06–6.33 μm) region

of the spectrum. The band is medium to strong in intensity and is moved to slightly higher frequencies when the compound is associated. The N–H bending band is seldom detectable in the spectra of aliphatic secondary amines, whereas secondary aromatic amines absorb near 1515 cm^{-1} (6.60 μm).

Liquid samples of primary and secondary amines display medium-to-strong broad absorption in the 909–666 cm^{-1} (11.00–15.00 μm) region of the spectrum arising from NH wagging. The position of this band depends on the degree of hydrogen bonding.

C–N Stretching Vibrations

Medium-to-weak absorption bands for the unconjugated C–N linkage in primary, secondary, and tertiary aliphatic amines appear in the region of 1250–1020 cm^{-1} (8.00–9.80 μm). The vibrations responsible for these bands involve C–N stretching coupled with the stretching of adjacent bonds in the molecule. The position of absorption in this region depends on the class of the amine and the pattern of substitution on the α-carbon.

Aromatic amines display strong C–N stretching absorption in the 1342–1266 cm⁻¹ (7.45–7.90 μm) region. The absorption appears at higher frequencies (shorter wavelengths) than the corresponding absorption of aliphatic amines because the force constant of the C–N bond is increased by resonance with the ring.

Characteristic strong C–N stretching bands in the spectra of aromatic amines have been assigned as in Table IV.

The absorption frequencies of methyl and methylene groups attached to the nitrogen atom of an amine are summarized in Tables I and II of Appendix C.

AMINE SALTS

N–H Stretching Vibrations

The ammonium ion displays strong, broad absorption in the 3300–3030 cm⁻¹ (3.03–3.30 μm) region because of N–H stretching vibrations (see Figure 24). There is also a combination band in the 2000–1709 cm⁻¹ (5.00–5.85 μm) region.

Salts of primary amines show strong, broad absorption between 3000 and 2800 cm⁻¹ (3.33–3.57 μm) arising from asymmetrical and symmetrical stretching in the NH₃⁺ group. In addition, multiple combination bands of medium intensity occur in the 2800–2000 cm⁻¹ (3.57–5.00 μm) region, the most prominent being the band near 2000 cm⁻¹ (5.00 μm). Salts of secondary amines absorb strongly in the 3000–2700 cm⁻¹ (3.33–3.70 μm) region with multiple bands extending to 2273 cm⁻¹ (4.00 μm). A medium band near 2000 cm⁻¹ (5.00 μm) may be observed. Tertiary amine salts absorb at longer wavelengths than the salts of primary and secondary amines; 2700–2250 cm⁻¹ (3.70–4.44 μm). Quaternary ammonium salts have no N–H stretching vibrations.

N–H Bending Vibrations

The ammonium ion displays a strong, broad NH₄⁺ bending band near 1429 cm⁻¹ (7.00 μm). The NH₃⁺ group of the salt of a primary amine absorbs near 1600–1575 cm⁻¹

Table IV C–N Stretch of Primary, Secondary, and Tertiary Aromatic Amines

Primary	1340–1250 cm⁻¹ (7.46–8.00 μm)
Secondary	1350–1280 cm⁻¹ (7.41–7.81 μm)
Tertiary	1360–1310 cm⁻¹ (7.35–7.63 μm)

(6.25–6.35 μm) and near 1550–1504 cm⁻¹ (6.45–6.65 μm). These bands originate in asymmetrical and symmetrical NH₃⁺ bending, analogous to the corresponding bands of the CH₃ group. Salts of secondary amines absorb near 1620–1560 cm⁻¹ (6.17–6.41 μm). The N–H bending band of the salts of tertiary amines is weak and of no practical value.

AMINO ACIDS AND SALTS OF AMINO ACIDS

Amino acids are encountered in three forms: the free amino acid (zwitterion)

$$-\underset{\underset{NH_3^+}{|}}{\overset{|}{C}}-COO^-$$

the hydrochloride (or other acid) salt,

$$-\underset{\underset{NH_3^+Cl^-}{|}}{\overset{|}{C}}-COOH$$

the sodium (or other cation) salt,

$$-\underset{\underset{NH_2}{|}}{\overset{|}{C}}-COO^-Na^+$$

Free primary amino acids are characterized by the following absorption (most of the work has been done with α-amino acids, but the relative positions of the amino and carboxyl groups seem to have little effect):

1. A broad, strong NH₃⁺ stretching band in the 3100–2600 cm⁻¹ (3.23–3.85 μm) region. Multiple combination and overtone bands extend the absorption to about 2000 cm⁻¹ (5.00 μm). This overtone region usually contains a prominent band near 2222–2000 cm⁻¹ (4.50–5.00 μm) assigned to a combination of the asymmetrical NH₃⁺ bending vibration and the torsional oscillation of the NH₃⁺ group. The torsional oscillation occurs near 500 cm⁻¹ (20.00 μm). The 2000 cm⁻¹ (5.00 μm) band is absent if the nitrogen atom of the amino acid is substituted.

2. A weak asymmetric NH₃⁺ bending band near 1660–1610 cm⁻¹ (6.03–6.21 μm); a fairly strong symmetrical bending band near 1550–1485 cm⁻¹ (6.45–6.73 μm).

3. The carboxylate ion group $\left(-C\underset{\searrow O}{\overset{\nearrow O}{}}-\right)$ absorbs strongly near 1600–1590 cm⁻¹ (6.25–6.29 μm) and more

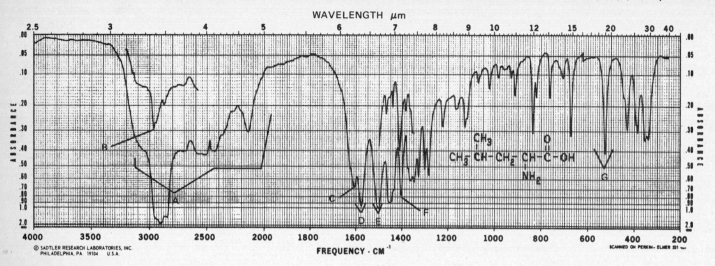

Source: The Matheson Company, Inc., E. Rutherford, N.J.

Fig. 30.

A: Broad (NH_3^+) N−H stretch, 3100−2000 cm^{-1} (3.23−5.00 μm), extended by combination band at 2140 cm^{-1} (4.67 μm), and other combination-overtone bands.

B: Aliphatic C−H stretch (superimposed on N−H stretch), 2967 cm^{-1} (3.37 μm).

C: Asymmetric (−NH_3^+) N−H bend, 1610 cm^{-1} (6.21 μm).

D: Asymmetric carboxylate (C$\cdots$O)$_2$ stretch, 1580 cm^{-1} (6.33 μm).

E: Symmetric (−NH_3^+) N−H bend, 1505 cm^{-1} (6.65 μm).

F: Symmetric carboxylate (C$\cdots$O)$_2$ stretch, 1405 cm^{-1} (7.12 μm).

G: Torsional (−NH_3^+) N−H oscillation, 525 cm^{-1} (19.0 μm).

weakly near 1400 cm^{-1} (7.14 μm). These bands result, respectively, from asymmetrical and symmetrical C−O stretching.

The spectrum of the amino acid leucine, including assignments corresponding to the preceding three categories, is shown in Figure 30.

Hydrochlorides of amino acids present the following patterns:

1. Broad strong absorption in the 3333−2380 cm^{-1} (3.00−4.20 μm) region resulting from superimposed O−H and NH_3^+ stretching bands. Absorption in this region is characterized by multiple fine structure on the long wavelength side of the band.

2. A weak, asymmetrical $\overset{+}{N}H_3$ bending band near 1610−1590 cm^{-1} (6.21−6.29 μm); a relatively strong, symmetrical, $\overset{+}{N}H_3$ bending band at 1550−1481 cm^{-1} (6.45−6.75 μm).

3. A strong band at 1220−1190 cm^{-1} (8.20−8.40 μm)

$$\overset{\overset{\text{O}}{\parallel}}{\text{arising from C−C−O stretching.}}$$

4. Strong carbonyl absorption at 1755−1730 cm^{-1} (5.70−5.78 μm) for α-amino acid hydrochlorides, and at 1730−1700 cm^{-1} (5.78−5.88 μm) for other amino acid hydrochlorides.

Sodium salts of amino acids show the normal N−H stretching vibrations at 3400−3200 cm^{-1} (2.94−3.13 μm) common to other amines. The characteristic carboxylate ion bands appear near 1600−1590 cm^{-1} (6.25−6.29 μm) and near 1400 cm^{-1} (7.14 μm).

NITRILES

The spectra of nitriles (R−C≡N) are characterized by weak to medium absorption in the triple-bond stretching region of the spectrum. Aliphatic nitriles absorb near 2260−2240 cm^{-1} (4.42−4.46 μm). Electron attracting atoms, such as oxygen or chlorine, attached to the carbon atom alpha to the C≡N group reduce the intensity of absorption. Conjugation, such as occurs in aromatic nitriles, reduces the frequency of absorption to 2240−2222 cm^{-1}

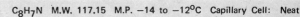

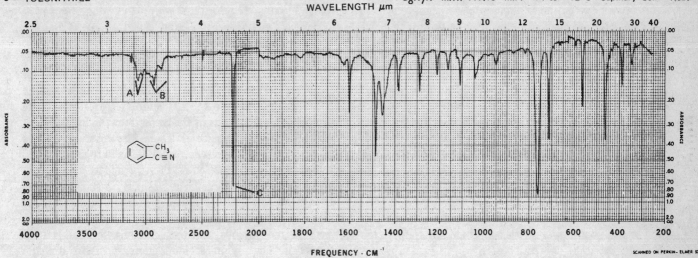

WAVELENGTH μm

FREQUENCY · CM⁻¹

SCANNED ON PERKIN- ELMER 521

Source: The Matheson Co., Inc., E. Rutherford, N.J.

Fig. 31.

A: Aromatic C–H stretch, 3070 cm⁻¹, 3025 cm⁻¹ (3.26, 3.31 μm).

B: Aliphatic C–H stretch, 2910, 2860 cm⁻¹ (3.44, 3.50 μm).

C: C≡N stretch, 2210 cm⁻¹ (4.53 μm) (intensified by aryl conjugation); aliphatic nitriles absorb at higher frequency (shorter wavelength).

(4.46 to 4.50 μm) and enhances the intensity. The spectrum of a typical nitrile, with an aryl group in conjugation with the cyano function, is shown in Figure 31.

COMPOUNDS CONTAINING C≡N, C=N, –N=C=O AND –N=C=S GROUPS

Isocyanides (isonitriles), isocyanates, thiocyanates, and isothiocyanates all show C≡N stretch or cumulated double bond (–Y=C=X; X, Y = N, S or O) stretch in the 2273–2000 cm⁻¹ (4.40–5.00 μm) region (Appendix B). Schiff's bases (RCH=NR, imines), oximes, thiazoles, iminocarbonates, guanidines, etc., show the C=N stretch in the 1689–1471 cm⁻¹ (5.92–6.80 μm) region. Although the intensity of the C=N stretch is variable, it is usually more intense than the C=C stretch.

COMPOUNDS CONTAINING – N=N– GROUP

The N=N stretching vibration of a symmetrical trans azo compound is forbidden in the infrared but absorbs in the 1576 cm⁻¹ (6.35 μm) region of the Raman spectrum. Unsymmetrical para-substituted azobenzenes in which the substituent is an electron donating group absorb near 1429

cm⁻¹ (7.00 μm). The bands are weak because of the nonpolar nature of the bond.

COVALENT COMPOUNDS CONTAINING NITROGEN-OXYGEN BONDS

Nitro compounds, nitrates, and nitramines contain an NO_2 group. Each of these classes shows absorption due to asymmetrical and symmetrical stretching of the NO_2 group. Asymmetrical absorption results in a strong band in the 1661–1499 cm⁻¹ (6.02–6.67 μm) region; symmetrical absorption occurs in the region between 1389 and 1259 cm⁻¹ (7.20 and 7.94 μm). The exact position of the bands is dependent on substitution and unsaturation in the vicinity of the NO_2 group.

NITRO COMPOUNDS. In the nitroparaffins, the bands occur near 1550 cm⁻¹ (6.45 μm) and near 1372 cm⁻¹ (7.29 μm). Conjugation lowers the frequency of both bands, resulting in absorption near 1550–1500 cm⁻¹ (6.45–6.67 μm) and near 1360–1290 cm⁻¹ (7.36–7.75 μm). Attachment of electronegative groups to the α-carbon of a nitro compound causes an increase in the frequency of the asymmetrical NO_2 band and a reduction in the frequency of the symmetrical band; chloropicrin absorbs at 1610 cm⁻¹ (6.21 μm) and at 1307 cm⁻¹ (7.65 μm).

Aromatic nitro groups absorb near the same frequencies

infrared spectrometry   **111**

as observed for conjugated aliphatic nitro compounds. Interaction between the NO_2 out-of-plane bending and ring C–H out-of-plane bending frequencies destroys the reliability of the substitution pattern observed for nitro-aromatics in the long wavelength region of the spectrum. Nitro-aromatic compounds show a C–N stretching vibration near 870 cm^{-1} (11.49 μm). The spectrum of nitrobenzene, with assignments corresponding to the preceding discussion, is shown in Figure 32.

Because of strong resonance in aromatic systems containing NO_2 groups and electron-donating groups such as the amino group ortho or para to one another, the symmetrical NO_2 vibration is shifted to lower frequencies and increases in intensity. p-Nitroaniline absorbs at 1475 cm^{-1} (6.78 μm) and 1310 cm^{-1} (7.64 μm).

The position of asymmetric and symmetric NO_2 stretching bands of nitramines $\left(\diagdown N{-}NO_2 \right)$ and the NO stretch of nitrosoamines are given in Appendix B.

NITRATES. Organic nitrates show absorption for N–O stretching vibrations of the NO_2 group and for the O–N linkage. Asymmetrical stretching in the NO_2 group results in strong absorption in the 1660–1625 cm^{-1} (6.02–6.15 μm) region; the symmetrical vibration absorbs strongly near 1300–1255 cm^{-1} (7.69–7.97 μm). Stretching of the single

bond of the N–O linkage produces absorption near 870–833 cm^{-1} (11.50–12.00 μm). Absorption observed at longer wavelengths, near 763–690 cm^{-1} (13.10–14.50 μm), likely results from NO_2 bending vibrations.

NITRITES. Nitrites display two strong N=O stretching bands. The band near 1680–1650 cm^{-1} (5.95–6.06 μm) is attributed to the *trans* isomer; the *cis* isomer absorbs in the 1625–1610 cm^{-1} (6.16–6.21 μm) region. The N–O stretching band appears in the region between 850 and 750 cm^{-1} (11.76 and 13.33 μm). The nitrite absorption bands are among the strongest observed in infrared spectra.

NITROSO COMPOUNDS. Primary and secondary aliphatic nitroso compounds are usually unstable and rearrange to oximes or dimerize. Tertiary and aromatic nitroso compounds are reasonably stable, existing as monomers in the gaseous phase or in dilute solution and as dimers in neat samples. Monomeric, tertiary, aliphatic nitroso compounds show N=O absorption in the 1585–1539 cm^{-1} (6.31–6.50 μm) region; aromatic monomers absorb between 1511 and 1495 cm^{-1} (6.62 and 6.69 μm).

The N→O stretching absorption of dimeric nitroso compounds are categorized in Appendix B as to *cis* vs. *trans* and aliphatic vs. aromatic. Nitrosoamine absorptions are given in Appendix B.

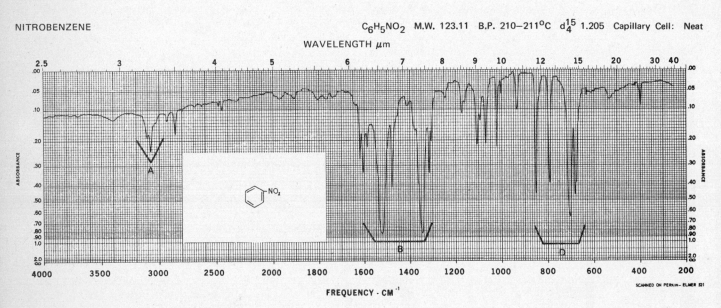

NITROBENZENE $C_6H_5NO_2$ M.W. 123.11   B.P. 210–211°C   d$_4^{15}$ 1.205   Capillary Cell: Neat

Source: J.T. Baker Chemical Co., Phillipsburg, N.J.

Fig. 32.

A: Aromatic C–H stretch, 3100, 3080 cm^{-1} (3.23, 3.25 μm).

B: Asymmetric ($ArNO_2$) N$\dot{=}$O stretch, 1520 cm^{-1} (6.58 μm)
Symmetric ($ArNO_2$) N$\dot{=}$O stretch 1345 cm^{-1} (7.44 μm).

C: $ArNO_2$ C–N stretch, 850 cm^{-1} (11.76 μm).

D: Low frequency (long wavelengths) of little use in determining the nature of ring substitution since these absorption patterns are due to interaction of NO_2 and C–H out of plane bending frequencies.

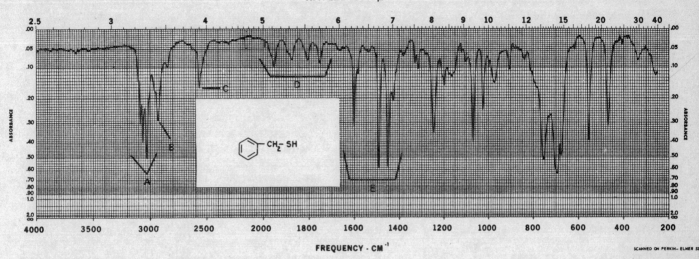

WAVELENGTH μm

FREQUENCY - CM⁻¹

SCANNED ON PERKIN- ELMER 521

Source: The Matheson Co., Inc., E. Rutherford, N.J.

Fig. 33.
A: Aromatic C—H stretch, 3085, 3060, 3030 cm⁻¹ (3.24, 3.27, 3.30 μm).
B: Aliphatic C—H stretch, 2930 cm⁻¹ (3.41 μm).
C: Moderately weak S—H stretch, 2565 cm⁻¹ (3.90 μm).
D: Overtone or combination band pattern indicative of monosubstituted aromatic (see Figure 14), 2000–1667 cm⁻¹ (5.0–6.0 μm).
E: C∷C ring stretch 1600, 1495, 1455 cm⁻¹ (6.25, 6.69, 6.87 μm).

ORGANIC SULFUR COMPOUNDS

MERCAPTANS. Aliphatic mercaptans and thiophenols, as liquids or in solution, show S—H stretching absorption in the range of 2600–2550 cm⁻¹ (3.85–3.92 μm). The S—H stretching band is characteristically weak and may go undetected in the spectra of dilute solutions or thin films. However, since few other groups show absorption in this region, it is useful in detecting S—H groups. The spectrum of benzyl mercaptan in Figure 33 is that of a mercaptan with a detectable S—H stretch band. The band may be obscured by strong carboxyl absorption in the same region. Hydrogen bonding is much weaker for S—H groups than for O—H and N—H groups.

The S—H group of thiol acids absorbs in the same region as mercaptans and thiophenols.

SULFIDES. The stretching vibrations assigned to the C—S linkage occur in the region of 700–600 cm⁻¹ (14.3–16.7 μm). The weakness of absorption and variability of position make this band of little value in structural determination.

DISULFIDES. The S—S stretching vibration is very weak and falls between 500 and 400 cm⁻¹ (20–25 μm), outside the range of sodium chloride optics. The characteristic absorptions of methyl and methylene groups attached to sulfur are summarized in Tables I and II in Appendix C.

THIOCARBONYL COMPOUNDS. Aliphatic thials or thiones exist as trimeric, cyclic sulfides. Aralkyl thiones may exist either as monomers or trimers, whereas diaryl thiones, such as thiobenzophenone, exist only as monomers. The C=S group is less polar than the C=O group and has a considerably weaker bond. In consequence, the band is not intense, and it falls at lower frequencies, where it is much more susceptible to coupling effects. Identification is therefore difficult and uncertain.

Compounds that contain a thiocarbonyl group show absorption in the 1250–1020 cm⁻¹ (8.00–9.80 μm) region. Thiobenzophenone and its derivatives absorb moderately in the 1224–1207 cm⁻¹ (8.17–8.29 μm) region. Since the absorption occurs in the same general region as C—O and C—N stretching, considerable interaction can occur between these vibrations within a single molecule.

Spectra of compounds in which the C=S group is attached to a nitrogen atom show an absorption band in the general C=S stretching region. In addition, several other bands in the broad region of 1563–700 cm⁻¹ (6.40–14.30 μm) can be attributed to vibrations involving interaction between C=S and stretching and C—N stretching.

infrared spectrometry   **113**

Thioketo compounds that can undergo enolization exist as thioketo-thioenol tautomeric systems; such systems show S–H stretching absorption. The thioenol tautomer of ethyl thiobenzoylacetate,

$$C_6H_5-\underset{\underset{SH---O}{|}}{C}=\underset{\underset{}{}}{C}-\underset{\underset{}{\|}}{C}-OC_2H_5$$

absorbs broadly at 2415 cm^{-1} (4.14 μm) due to bonded S–H stretching absorption.[41]

COMPOUNDS CONTAINING SULFUR–OXYGEN BONDS

SULFOXIDES. Alkyl and aryl sulfoxides are liquids to in solution show strong absorption in the 1070–1030 cm^{-1} (9.35–9.71 μm) region. This absorption occurs at 1050 cm^{-1} (9.52 μm) for dimethyl sulfoxide (DMSO, methyl sulfoxide) as may be seen in Appendix A, spectrum 26. Conjugation brings about a small change (10–20 cm^{-1}) in the observed frequency in contrast to the marked reduction in frequency of the C=O bond accompanying conjugation. Diallylsulfoxide absorbs at 1047 cm^{-1} (9.55 μm). Phenyl methyl sulfoxide and cyclohexyl methyl sulfoxide absorb at 1055 cm^{-1} (9.48 μm) in dilute solution in carbon tetrachloride. The sulfoxide group is susceptible to hydrogen bonding, the absorption shifting to slightly lower frequencies in methanol or chloroform solutions, or on passing from dilute solution to the liquid phase. The frequency of S=O absorption is increased by electronegative substitution.

SULFONES. Spectra of sulfones show strong absorption bands at 1350–1300 cm^{-1} (7.41–7.70 μm) and at 1160–1120 cm^{-1} (8.62–8.93 μm). These bands arise, respectively, from asymmetric and symmetric SO$_2$ stretching. Hydrogen bonding results in absorption near 1300 cm^{-1} (7.70 μm) and near 1125 cm^{-1} (8.90 μm). Splitting of the high frequency band often occurs in carbon tetrachloride solution or in the solid state.

SULFONYL CHLORIDES. Sulfonyl chlorides absorb strongly in the regions of 1410–1380 cm^{-1} (7.09–7.25 μm) and 1204–1177 cm^{-1} (8.31–8.49 μm). This increase in frequency, compared with the sulfones, results from the electronegativity of the chlorine atom.

SULFONAMIDES. Solutions of sulfonamides absorb strongly at 1370–1335 cm^{-1} (7.30–7.49 μm) and at 1170–1155 cm^{-1} (8.55–8.66 μm). In the solid phase, these frequencies are lowered by 10–20 cm^{-1}. In solid samples, the high frequency band is broadened and several submaxima usually appear.

Primary sulfonamides show strong N–H stretching bands at 3390–3330 cm^{-1} (2.95–3.00 μm) and at 3300–3247 cm^{-1} (3.03–3.08 μm) in the solid state; secondary sulfonamides absorb near 3265 cm^{-1} (3.06 μm).

SULFONATES, SULFATES, AND SULFONIC ACIDS. The asymmetric (higher frequency, shorter wavelength) and symmetric S=O stretching frequency ranges for these compounds are as follows:

Class	cm^{-1}	μm
Sulfonates (covalent)	1372–1335, 1195–1168	7.29–7.49, 8.37–8.56
Sulfates (organic)	1415–1380, 1200–1185	7.06–7.24, 8.33–8.44
Sulfonic acids	1350–1342, 1165–1150	7.41–7.45, 8.58–8.69
Sulfonate salts	ca. 1175, ca. 1055	ca. 8.5, ca. 9.5

The spectrum of a typical alkyl arenesulfonate is given in Figure 34. In virtually all sulfonates, the asymmetric stretch occurs as a doublet. Alkyl and aryl sulfonates show negligible differences; electron donating groups in the para position of arenesulfonates cause higher frequency absorption.

Sulfonic acids are listed in narrow ranges above; these apply only to anhydrous forms. Such acids hydrate readily to give bands that are probably a result of the formation of hydronium sulfonate salts, in the 1230–1120 cm^{-1} (8.13–8.93 μm) range.

ORGANIC HALOGEN COMPOUNDS

The strong absorption of halogenated hydrocarbons arises from the stretching vibrations of the carbon to halogen bond.

Aliphatic C–Cl absorption is observed in the broad region between 850 and 550 cm^{-1} (11.76–18.18 μm). When several chlorine atoms are attached to one carbon atom, the band is usually more intense and at the high frequency end of the assigned limits. Carbon tetrachloride (see Appendix A, spectrum 10) shows an intense band at 797 cm^{-1} (12.55 μm). The first overtones of the intense fundamental bands are frequently observed. Spectra of typical chlorinated hydrocarbons are shown in Appendix A: nos. 10, 11, 12, and 13. Brominated compounds absorb in the 690–515 cm^{-1} (14.49–19.42 μm) region, iodocompounds in the 600–500 cm^{-1} (16.67–20.00 μm) region. A strong CH$_2$ wagging band is observed for the

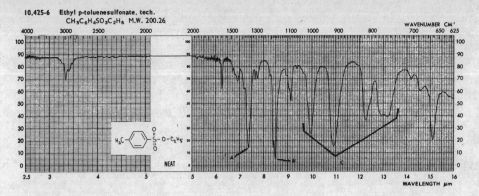

Fig. 34.
A: Asymmetric S(=O)$_2$ stretch, 1351 cm^{-1} (7.40 μm).
B: Symmetric S(=O)$_2$ stretch, 1176 cm^{-1} (8.50 μm).
C: Various strong S$-$O$-$C stretching, 1000$-$769 cm^{-1} (10.0$-$13.0 μm).

CH$_2$X (X = Cl, Br, I) group in the 1300$-$1150 cm^{-1} (7.69$-$8.70 μm) region.

Fluorine-containing compounds absorb strongly over a wide range between 1400$-$730 cm^{-1} (7.14$-$13.70 μm) due to C$-$F stretching modes. A monofluoro-alkane shows a strong band in the 1100$-$1000 cm^{-1} (9.09$-$10.00 μm) region. As the number of fluorine atoms in an aliphatic molecule increases, the band pattern becomes more complex, with multiple strong bands appearing over the broad region of C$-$F absorption. The CF$_3$ and CF$_2$ groups absorb strongly in the 1350$-$1120 cm^{-1} (7.41$-$8.93 μm) region. The spectrum of Fluorolube®, Appendix A, spectrum 14, illustrates many of the preceding absorption characteristics.

Chlorobenzenes absorb in the 1096$-$1089 cm^{-1} (9.12$-$9.18 μm) region. The position within this region depends on the substitution pattern.[12] Aryl fluorides absorb in the 1250$-$1100 cm^{-1} (8.00$-$9.10 μm) region of the spectrum. A monofluorinated-benzene ring displays a strong, narrow absorption band near 1230 cm^{-1} (8.13 μm).

SILICON COMPOUNDS

Tables of characteristic absorptions of silicon compounds are found in Appendix D. Special features related to these tables are discussed below.

Si$-$H Vibrations

Vibrations for the Si$-$H bond are listed in Table I, Appendix D. The Si$-$H stretching frequencies are increased by the attachment of an electronegative group to the silicon.

The absorption bands of methylene and methyl groups attached to silicon are summarized in Appendix C, Tables I and II.

Si$-$O Stretching Vibration

The Si$-$O stretching vibrations are summarized in Table II, Appendix D.

The OH stretching vibrations of the SiOH group absorb in the same region as the alcohols, 3700$-$3200 cm^{-1} (2.70$-$3.12 μm). As in alcohols, the absorption characteristics depend on the degree of hydrogen bonding.

Silicon-Halogen Stretching Vibrations

The absorption of groups containing Si$-$F bonds is summarized in Table III, Appendix D.

Bands resulting from Si$-$Cl stretching occur at frequencies below 666 cm^{-1} (wavelengths longer than 15 μm).

The spectrum of silicone lubricant, Appendix A no.27, illustrates some of the preceding absorptions.

PHOSPHORUS COMPOUNDS

P$-$H Stretching

The P$-$H group absorbs in the regions described in Table IV, Appendix D.

P=O Stretching Vibrations

Such absorptions are listed in Table IV, Appendix D.

P–O Stretching Vibrations

Absorption characteristics of P–O stretching vibrations are summarized in Table V, Appendix D.

The absorption of methylene and methyl groups attached to a phosphorus atom are summarized in Tables I and II of Appendix C.

HETEROAROMATIC COMPOUNDS

The spectra of heteroaromatic compounds result primarily from the same vibrational modes as observed for the aromatics.

C–H Stretching Vibrations

Heteroaromatics such as pyridines, pyrazines, pyrroles, furans, and thiophenes show C–H stretching bands in the 3077–3003 cm^{-1} (3.25–3.33 μm) region.

N–H STRETCHING FREQUENCIES

Heteroaromatics containing an N–H group show N–H stretching absorption in the region of 3500–3220 cm^{-1}

(2.86–3.11 μm). The position of absorption within this general region depends upon the degree of hydrogen bonding, and hence upon the physical state of the sample or the polarity of the solvent. Pyrrole and indole in dilute solution in nonpolar solvents show a sharp band near 3495 cm^{-1} (2.86 μm), concentrated solutions show a widened band near 3400 cm^{-1} (2.94 μm). Both bands may be seen at intermediate concentrations.

Ring-Stretching Vibrations (Skeletal Bands)

Ring stretching vibrations occur in the general region between 1600–1300 cm^{-1} (6.25–7.69 μm). The absorption involves stretching and contraction of all of the bonds in the ring and interaction between these stretching modes. The band pattern and the relative intensities depend on the substitution pattern and the nature of the substituents.

Pyridine (Figure 35) shows four bands in this region and in this respect closely resembles a monosubstituted benzene. Furans, pyrroles, and thiophenes display two to four bands in this region.

C–H Out of Plane Bending

The C–H out-of-plane bending (γCH) absorption pattern of the heteroaromatics is determined by the number of

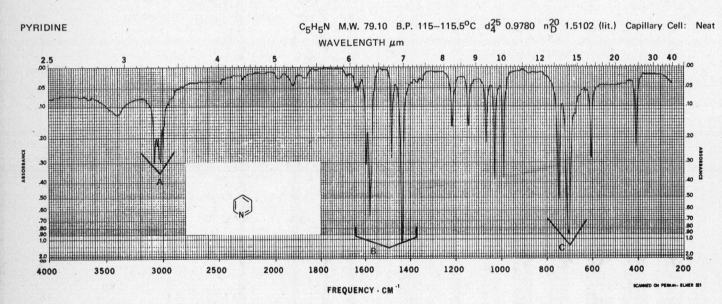

PYRIDINE   C$_5$H$_5$N M.W. 79.10 B.P. 115–115.5°C d$_4^{25}$ 0.9780 n$_D^{20}$ 1.5102 (lit.) Capillary Cell: Neat

Source: The Matheson Co., Inc. E. Rutherford, N.J.

Fig. 35.
A: Aromatic C–H stretch, 3080–3010 cm^{-1} (3.25–3.32 μm).
B: C–C, C–N ring stretching (skeletal bands), 1600–1430 cm^{-1} (6.25–7.00 μm).
C: C–H out of plane bending, 748, 703 cm^{-1} (13.37, 14.22 μm).
    See Table VI, Appendix D, for patterns in region *C* for substituted pyridines.

adjacent hydrogen atoms bending in phase. The C—H out-of-plane and ring bending (β ring) absorption of the alkyl pyridines are summarized in Table VI, Appendix D.

Absorption data for the out-of-plane C—H bending (γCH), and ring bending (β ring) modes of three common five-membered heteroaromatic rings are presented in Table VII, Appendix D. The spectrum of pyridine is given in Figure 35.

The ranges in Table VII (p. 151) include polar as well as nonpolar substituents on the ring.

REFERENCES

1. Jones, R. N., and C. Sandorfy, "The Application of Infrared and Raman Spectrometry to the Elucidation of Molecular Structure," Chapter IV, Vol. IX, in A. Weissberger, *Technique of Organic Chemistry*, Interscience, New York, 1956, pp. 247–580.   UI

2. Miller, F. A., "Applications of Infrared and Ultraviolet Spectra to Organic Chemistry," Vol. III, in H. Gilman, *Organic Chemistry, An Advance Treatise*, Wiley, New York, 1953, pp. 122–77.

3. Bellamy, L. J., *The Infra-red Spectra of Complex Organic Molecules*, Wiley, 2nd ed., New York, 1958; *Advances in Infrared Group Frequencies*, Methuen, London, 1968.

4. Cross, A. D.: "*Introduction to Practical Infra-red Spectroscopy*," 3rd. ed., Butterworth, London, 1969.

5. Randall, H. M., R. G. Fowler, N. Fuson, and J. R. Dangl, *Infrared Determination of Organic Structures*, D. Van Nostrand, New York, 1949.

6. Williams, V. Z., "Infrared Instrumentation and Techniques," *Rev. Sci. Instruments, 19,* 135–78 (1948).

7. Herzberg, G., *Infrared and Raman Spectra of Polyatomic Molecules*, D. Van Nostrand, New York, 1945. Theory.

8. Barnes, R. B., R. C. Gore, U. Liddel, and V. Z. Williams, *Infrared Spectroscopy*, Reinhold, New York, 1944.

9. Sutherland, G. B. G. M., *Infrared and Raman spectra*, Methuen, London, 1935. Theory.

10. Bauman, R. P., *Absorption Spectroscopy*, Wiley, New York, 1962.

11. Randall, H. M., R. G. Fowler, N. Fuson, and J. R. Dangl, *Infrared Determination of Organic Structures*, D. Van Nostrand, New York, 1943.

12. Colthup, N. B., L. H. Daly, and S. E. Wiberley, *Introduction to Infrared and Raman Spectroscopy*, Academic Press, New York and London, 1964.

13. Nakanishi, Koji, *Infrared Absorption Spectroscopy—Practical*, Holden-Day, San Francisco and Nankodo Company Ltd. Tokyo, 1962.

14. Kendall, D. N., Ed., *Applied Infrared Spectroscopy*, Reinhold, New York, 1966.

15. Flett, M. St. C., *Characteristic Frequencies of Chemical Groups in the Infra-Red*, Elsevier, Amsterdam, 1963.

16. Szymanski, H. A., *Correlation of Infrared and Raman Spectra of Organic Compounds*, Hertillon Press, 1969.

17. Szymanski, H. A., ed., *Raman Spectroscopy: Theory and Practice*, Plenum, New York, 1961, vol. 1, and 1970, vol. 2.

18. Strouts, C. N. R., H. N. Wilson, and R. T. Parry-Jones, *Chemical Analysis—The Working Tools*, Vol. II., Clarendon Press, Oxford, 1962.

19. Dyer, J. R., *Applications of Absorption Spectroscopy of Organic Compounds*, Prentice-Hall, Englewood Cliffs, N.J., 1965.

20. Katritzky, A. R., ed., *Physical Methods in Heterocyclic Chemistry*, Vol. II, Academic Press, New York and London, 1963.

21. Szymanski, H. A., *IR Theory and Practice of Infrared Spectroscopy*, Plenum Press, New York, 1964.

22. Szymanski, H. A., *Interpreted Infrared Spectra*, Vol. I, II, III, Plenum Press, New York, 1964–1967.

23. Davies, Mansel, ed., *Infrared Spectroscopy and Molecular Structure—An Outline of Principles*, Elsevier, Amsterdam and London, 1963.

24. Rao, C. N. R., *Chemical Applications of Infrared Spectroscopy*, Academic Press, New York and London, 1963.

25. Phillips, J. P., *Spectra-Structure Correlation*, Academic Press, New York and London, 1964.

26. Potts, W. J., Jr., *Chemical Infrared Spectroscopy, Vol. I, Techniques*, Wiley, New York, 1963.

27. Miller, R. G. J., ed., *Laboratory Methods in Infrared Spectroscopy*, Heyden, London, 1965.

28. Meloan, C. E., *Elementary Infrared Spectroscopy*, Macmillan, New York, 1963.

28a. Cole, A. R. H., "Applications of Infrared Spectroscopy," Vol. XI (Part I), pp. 133–173, in *Technique of Organic Chemistry*, A. Weissberger, Ed., Interscience, New York, 1963.

28b. Tichy, M. "The Determination of Intramolecular Hydrogen Bonding by Infrared Spectroscopy and Its Applications in Stereochemistry," Vol. 5, pp. 115–299, in *Advances in Organic Chemistry: Methods and Results*, R. R. Raphael, Ed., Interscience, New York, 1964.

28c. Stewart, J. E., "Far Infrared Spectroscopy" in *Interpretive Spectroscopy*. Freeman, S. K., Ed., New York, Reinhold, 1965, p. 131.

28d. Conley, R. T., *Infrared Spectroscopy*, Allyn and

Bacon, Boston, 2nd Ed., 1972.

28e. Loader, E. J., *Basic Laser Raman Spectroscopy*, Heyden-Sadtler, London, New York, 1970.

28f. Van der Maas, J. M., *Basic Infrared Spectroscopy*, Heyden, London, 1969.

28g. Finch, A., P. N. Gates, K. Radcliffe, F. N. Dickson, and F. F. Bentley, *Chemical Applications of Far Infrared Spectroscopy*, Academic Press, New York, 1970.

28h. Jones, R. N., J. B. DiGiorgio, J. J. Elliott, and G. A. Nonnenmacher, *J. Org. Chem., 30*, 1822(1965), Raman Spectroscopy.

28i. Bentley, F. F., L. D. Smithson, and A. L. Rozek, "*Infrared Spectra and Characteristic Frequencies ~ 700 to 300 cm^{-1}, A Collection of Spectra, Interpretation and Bibliography*," Interscience, Division of Wiley, New York, 1968.

29. *Catalog of Infrared Spectrograms*, Sadtler Research Laboratories, 3314–20 Spring Garden St., Philadelphia, Penn., Spectra indexed by name and by major bands in each micron interval. Trade name: Spec-Finder.

30. *Catalog of Infrared Spectrograms*, American Petroleum Institute Research Project 44, Carnegie Institute of Technology, Pittsburgh, Pa.

31. *Catalog of Infrared Spectral Data*, Manufacturing Chemists Association Research Project, Chemical and Petroleum Research Laboratories, Carnegie Institute of Technology, Pittsburgh, Pa., to June 30, 1960; Chemical Thermodynamics Properties Center, Agriculture and Mechanical College of Texas, College Station, Tex., from July 1, 1960.

32. *ASTM-Wyandotte Index, Molecular Formula List of Compounds, Names and References to Published Infrared Spectra.* ASTM Special Technical Publications 131 (1962) and 131-A (1963), ASTM, 1916 Race St., Philadelphia 3, Pa. Lists about 57,000 compounds. Covers infrared, near infrared, and far infrared spectra.

33. Hershenson, H. H., *Infrared Absorption Spectra Index*, Academic Press, New York and London, 1959 and 1964. Two volumes cover 1945–1962.

34. *An Index of Published Infra-Red Spectra*, M. B. B. Thomas, Ed., Vols. I and II, British Information Service, New York, 1960. Complete to 1957. Gives literature references to spectra, lists state, absorption range, and optics.

35. *Documentation of Molecular Spectroscopy (DMS)*, Butterworths Scientific Publications, London; and Verlag Chemie GMBH, Weinheim/Bergstrasse, West Germany, in cooperation with the infrared Absorption Data Joint Committee, London, and the Institut für Spectrochemie und Angewandte Spectroskopie, Dortmund. Spectra are presented on coded cards.

Coded cards containing abstracts of articles relating to infrared spectrometry are also issued.

36. Szymanski, H. A., *Infrared Band Handbook*, Plenum, New York, 1963.
Infrared Band Handbook, Supplements 1 and 2, Plenum Press, New York, 1964, 259 pp.
Infrared Band Handbook, Supplements 3 and 4, Plenum Press, New York, 1965.
Infrared Band Handbook, Second Revised and Enlarged Edition, IFI/Plenum, New York, N.Y., 1970, 2 vol.

37. IRDC Cards: *Infrared Data Committee of Japan* (S. Mizushima). Handled by Nankodo Co., Haruki-cho, Tokyo.

38a. *The NRC-NBS (Creitz) File of Spectrograms*, issued by National Research Council-National Bureau of Standards Committee on Spectral Absorption Data, National Bureau of Standards, Washington 25, D.C. Spectra presented on edge-punched cards.

38b. *Sadtler Reference Spectra–Commonly Abused Drugs IR & UV Spectra.* 1972, 600 IR Spectra, 300 UV Spectra, Sadtler Research Laboratories, Inc.

38c. *Sadtler Reference Spectra–Gases & Vapors High Resolution Infrared.* 1972, 150 Spectra, Sadtler Research Laboratories, Inc.

38d. *Sadtler Reference Spectra–Inorganics IR Grating.* 1967, 1300 spectra, Sadtler Research Laboratories, Inc.

38e. *Sadtler Reference Spectra–Organometallics IR Grating.* 1966, 400 spectra Sadtler Research Laboratories, Inc.

38f. *Sadtler Standard Infrared Grating Spectra.* 1972, 26,000 spectra in 26 volumes, Sadtler Research Laboratories, Inc.

38g. *Sadtler Standard Infrared Prism Spectra.* 1972, 43,000 spectra in 43 volumes, Sadtler Research Laboratories, Inc.

38h. Bentley, F. F., L. D. Smithson, and A. L. Rozek, *Infrared Spectra and Characteristic Frequencies*, Wiley-Interscience Publishers, 1968.

38i. Flett, M. St. C., *Characteristic Frequencies of Chemical Groups in the Infrared*, American Elsevier, New York, 1969.

38j. Ministry of Aviation Technical Information and Library Services, Ed., *An Index of Published Infra-Red Spectra*, Vol. 1 and 2. London, Her Majesty's Stationery Office, 1960.

38k. Brown, C. R., M. W. Ayton, T. C. Goodwin, and T. J. Derby, *Infrared–A Bibliography.* Washington, D.C., Library of Congress, Technical Information Division, 1954.

38l. Dobriner, K., E. R. Katzenellenbogen, and R. N. Jones, *Infrared Absorption Spectra of Steroids–An Atlas*, Vol. 1. New York, Interscience, 1953.

38m. Lawson, K. E., *Infrared Absorption of Inorganic Substances.* New York, Reinhold, 1961.

39. "Spectrometry Nomenclature," *Anal. Chem.*, December Issue, every year.

40. Zeiss, H. H., and Minoru Tsutsui, *J. Am. Chem. Soc.*, *75*, 897 (1953).

41. Reyes, Z., and R. M. Silverstein, *J. Am. Chem. Soc.*, *80*, 6367 (1958).

appendix a

CHART AND SPECTRAL PRESENTATIONS OF ORGANIC SOLVENTS, MULLING OILS, AND OTHER COMMON LABORATORY SUBSTANCES

Transparent regions of solvents and mulling oils.

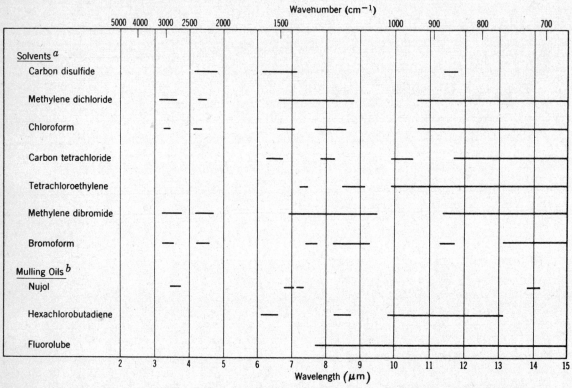

[a]The open regions are those in which the solvent transmits more than 25% of the incident light at 1 mm thickness.
[b]The open regions for mulling oils indicate transparency of thin films.

Alphabetical listing of spectra shown on succeeding pages (pp. 121–134).
Spectra courtesy of Sadtler Laboratories and Aldrich Chemical Co.

	Spectrum No.		Spectrum No.
Acetone	17	Fluorolube®	14
Benzene	4	Hexane	1
2-Butanone	18	Indene	8
Carbon disulfide	25	Methanol	15
Carbon tetrachloride	10	Methyl sulfoxide (DMSO)	26
Chloroform	9	Nujol®	2
Cyclohexane	3	Petroleum ether	24
Cyclohexanone	19	Phthalic acid, diethyl ester	21
1,2-Dichloroethane	11	Polystyrene	7
N,N-Dimethylformamide (DMF)	28	Silicone lubricant	27
p-Dioxane	23	Tetrachloroethylene	13
Ethyl Acetate	20	Toluene	5
Ethyl alcohol	16	Trichloroethylene	12
Ethyl ether	22	*m*-Xylene	6

HEXANE

C$_6$H$_{14}$ Mol. Wt. 86.18   B. P. 68-69°C
Source: Phillips Petroleum Co.
Capillary Cell

No. 1

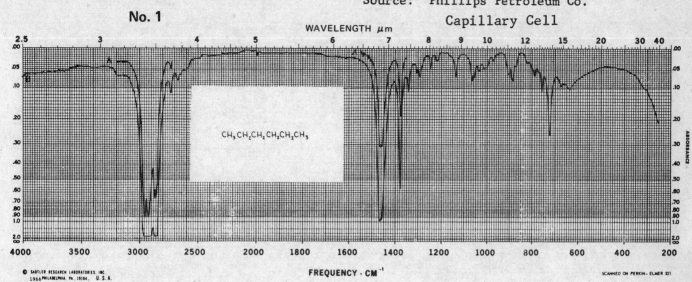

CH$_3$CH$_2$CH$_2$CH$_2$CH$_2$CH$_3$

© SADTLER RESEARCH LABORATORIES, INC.
1966 PHILADELPHIA, PA. 19104, U.S.A.

SCANNED ON PERKIN- ELMER 521

NUJOL®

Source: Plough, Inc.
Capillary Cell

No. 2

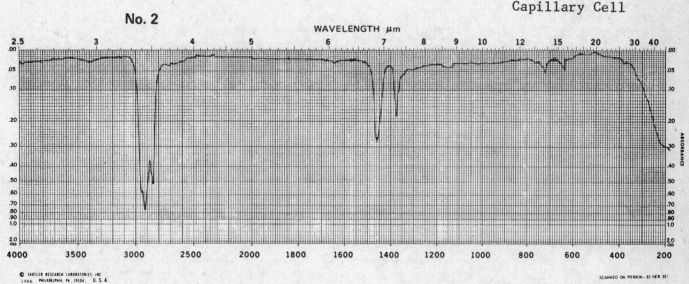

© SADTLER RESEARCH LABORATORIES, INC.
1966 PHILADELPHIA, PA. 19104, U.S.A.

SCANNED ON PERKIN- ELMER 52

appendix a   **121**

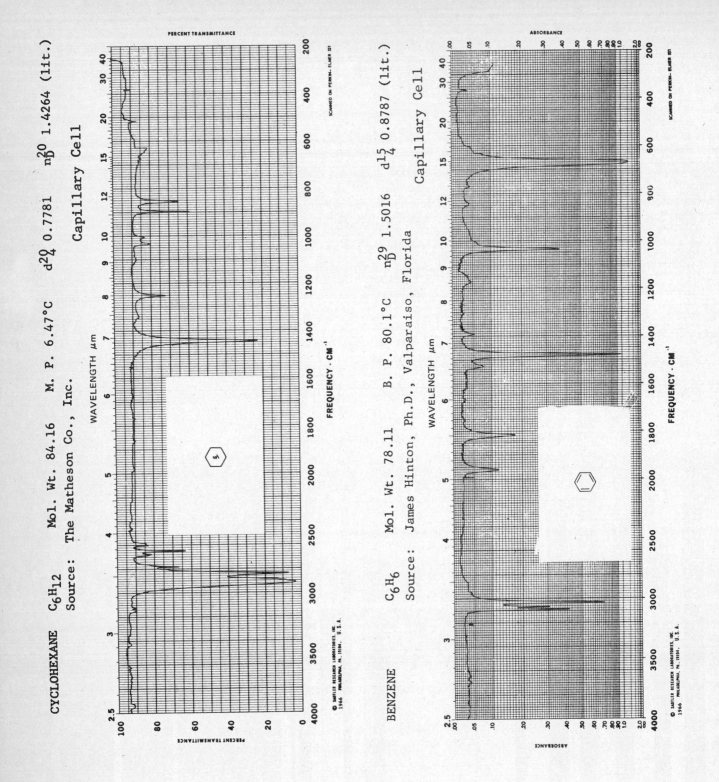

CYCLOHEXANE C6H12 Mol. Wt. 84.16 M. P. 6.47°C d_4^{20} 0.7781 n_D^{20} 1.4264 (lit.)
Source: The Matheson Co., Inc. Capillary Cell

No. 3

BENZENE C6H6 Mol. Wt. 78.11 B. P. 80.1°C n_D^{29} 1.5016 d_4^{15} 0.8787 (lit.)
Source: James Hinton, Ph.D., Valparaiso, Florida Capillary Cell

No. 4

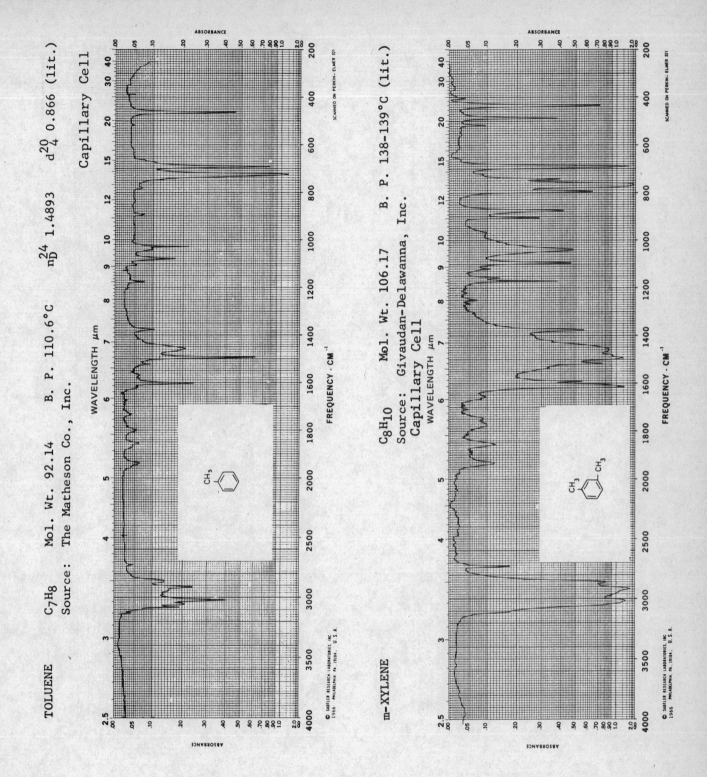

TOLUENE C_7H_8 Mol. Wt. 92.14   B. P. 110.6°C n_D^{24} 1.4893 d_4^{20} 0.866 (lit.)

Source: The Matheson Co., Inc.   Capillary Cell

m-XYLENE C_8H_{10} Mol. Wt. 106.17   B. P. 138–139°C (lit.)

Source: Givaudan-Delawanna, Inc.   Capillary Cell

No. 5

No. 6

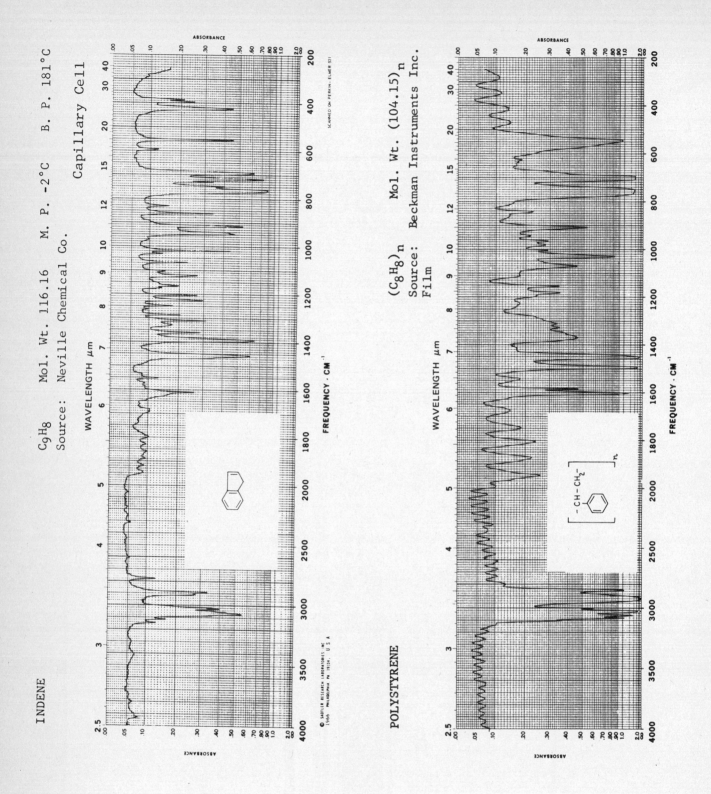

INDENE

C_9H_8 Mol. Wt. 116.16   M. P. -2°C   B. P. 181°C
Source: Neville Chemical Co.

Capillary Cell

No. 8

POLYSTYRENE

$(C_8H_8)_n$ Mol. Wt. (104.15)$_n$
Source: Beckman Instruments Inc.
Film

No. 7

124   spectrometric identification of organic compounds

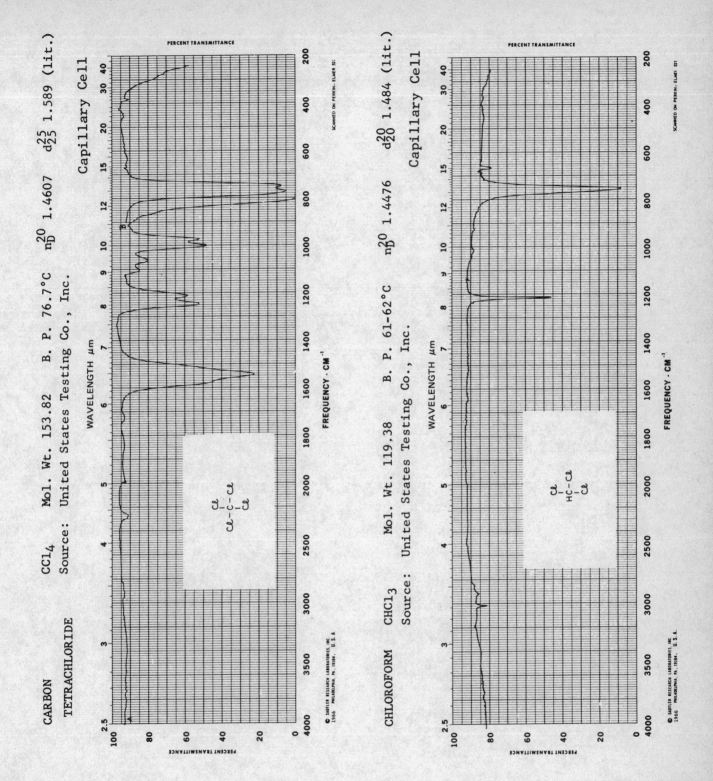

CARBON TETRACHLORIDE   CCl₄   Mol. Wt. 153.82   B. P. 76.7°C n_D^{20} 1.4607 d_{25}^{25} 1.589 (lit.)

Source: United States Testing Co., Inc.

Capillary Cell

No. 10

CHLOROFORM   CHCl₃   Mol. Wt. 119.38   B. P. 61-62°C n_D^{20} 1.4476 d_{20}^{20} 1.484 (lit.)

Source: United States Testing Co., Inc.

Capillary Cell

No. 9

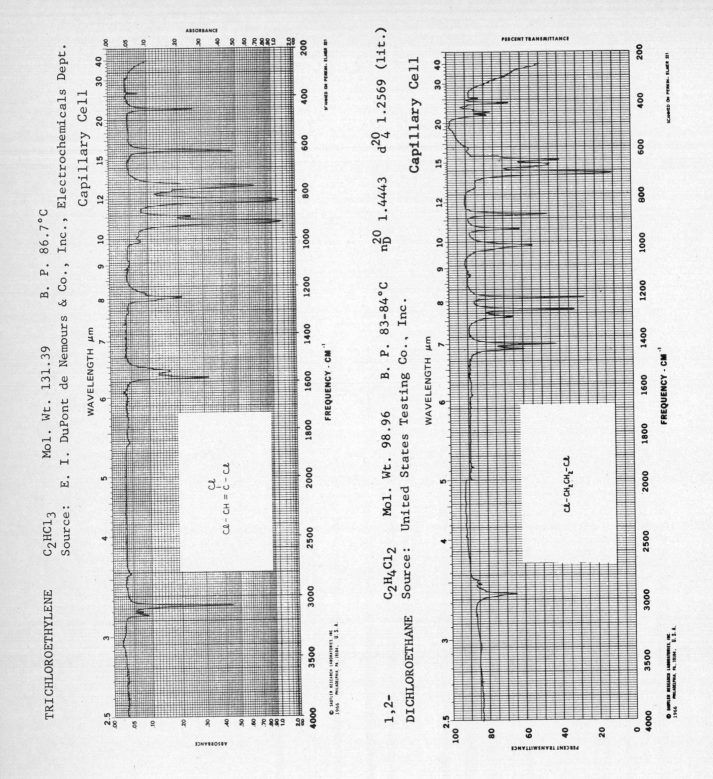

TRICHLOROETHYLENE   C$_2$HCl$_3$ Mol. Wt. 131.39   B. P. 86.7°C
Source: E. I. DuPont de Nemours & Co., Inc., Electrochemicals Dept.
Capillary Cell

Cℓ–CH=C–Cℓ
|
Cℓ

ABSORBANCE

WAVELENGTH μm

FREQUENCY - CM^{-1}

SCANNED ON PERKIN- ELMER 521

© SADTLER RESEARCH LABORATORIES, INC.
1966   PHILADELPHIA, PA. 19104 .   U. S. A.

No. 12

1,2-DICHLOROETHANE   C$_2$H$_4$Cl$_2$ Mol. Wt. 98.96   B. P. 83-84°C   n$_D^{20}$ 1.4443   d$_4^{20}$ 1.2569 (lit.)
Source: United States Testing Co., Inc.
Capillary Cell

Cℓ–CH$_2$CH$_2$–Cℓ

PERCENT TRANSMITTANCE

WAVELENGTH μm

FREQUENCY - CM^{-1}

SCANNED ON PERKIN- ELMER 521

© SADTLER RESEARCH LABORATORIES, INC.
1966   PHILADELPHIA, PA. 19104 ,   U. S. A.

No. 11

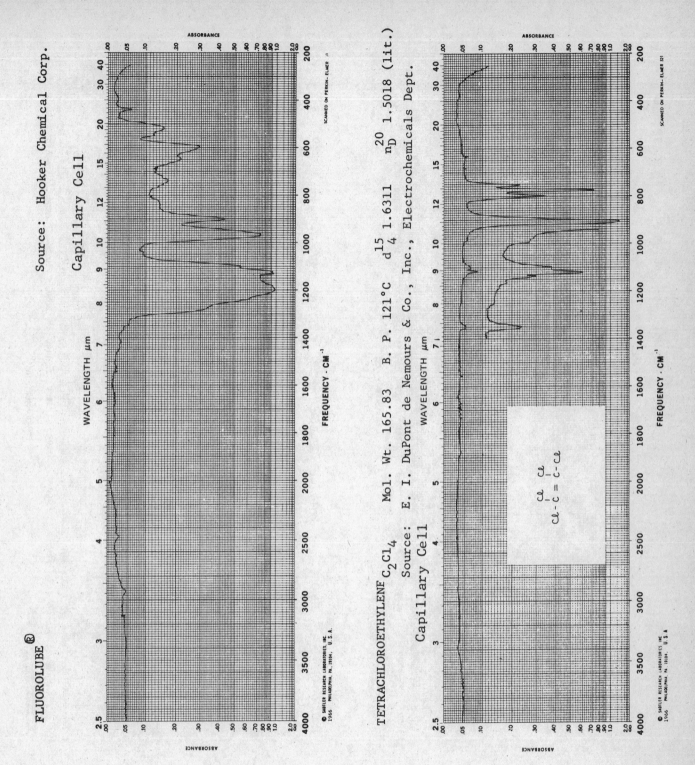

FLUOROLUBE ®

Source: Hooker Chemical Corp.

Capillary Cell

No. 14

TETRACHLOROETHYLENE C_2Cl_4 Mol. Wt. 165.83   B. P. 121°C d_4^{15} 1.6311 n_D^{20} 1.5018 (lit.)

Source: E. I. DuPont de Nemours & Co., Inc., Electrochemicals Dept.

Capillary Cell

No. 13

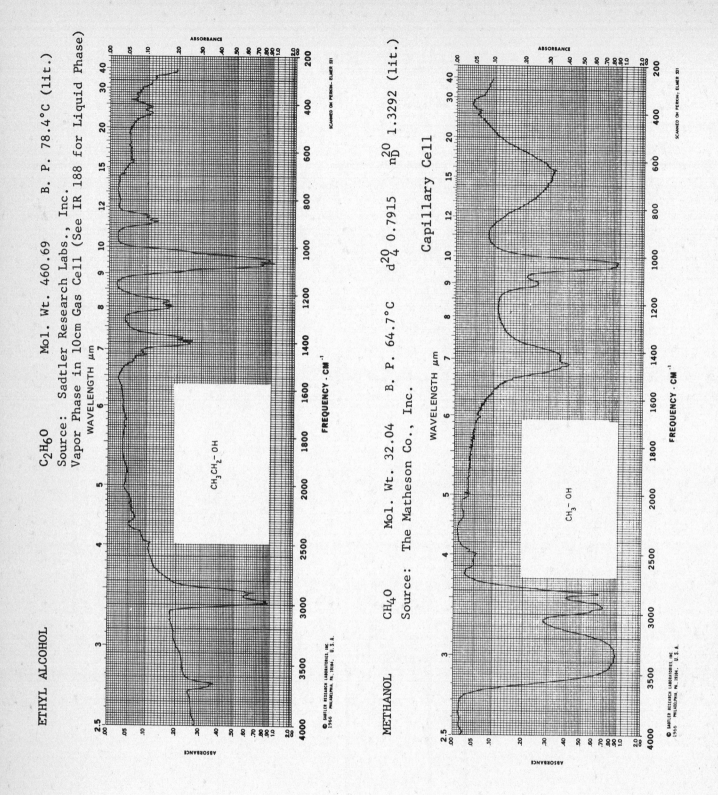

ETHYL ALCOHOL

C₂H₆O   Mol. Wt. 460.69   B. P. 78.4°C (lit.)
Source: Sadtler Research Labs., Inc.
Vapor Phase in 10cm Gas Cell (See IR 188 for Liquid Phase)

CH₃CH₂—OH

No. 16

METHANOL

CH₄O   Mol. Wt. 32.04   B. P. 64.7°C   d²⁰₄ 0.7915   n²⁰_D 1.3292 (lit.)
Source: The Matheson Co., Inc.   Capillary Cell

CH₃—OH

No. 15

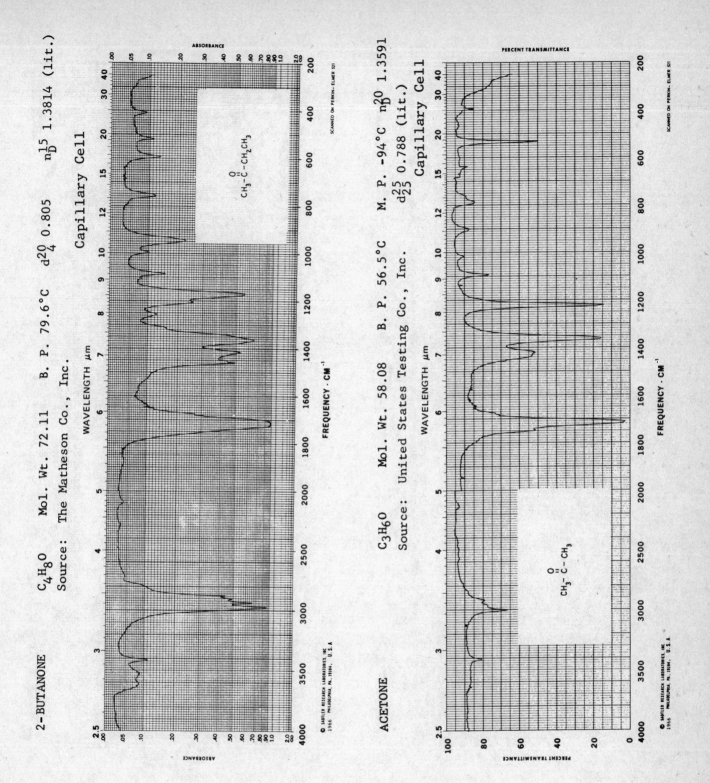

2-BUTANONE C_4H_8O Mol. Wt. 72.11   B. P. 79.6°C d_4^{20} 0.805 n_D^{15} 1.3814 (lit.)
Source: The Matheson Co., Inc.   Capillary Cell

$$CH_3-\overset{O}{\overset{\|}{C}}-CH_2CH_3$$

No. 18

ACETONE C_3H_6O Mol. Wt. 58.08   B. P. 56.5°C   M. P. -94°C n_D^{20} 1.3591 d_{25}^{25} 0.788 (lit.)
Source: United States Testing Co., Inc.   Capillary Cell

$$CH_3-\overset{O}{\overset{\|}{C}}-CH_3$$

No. 17

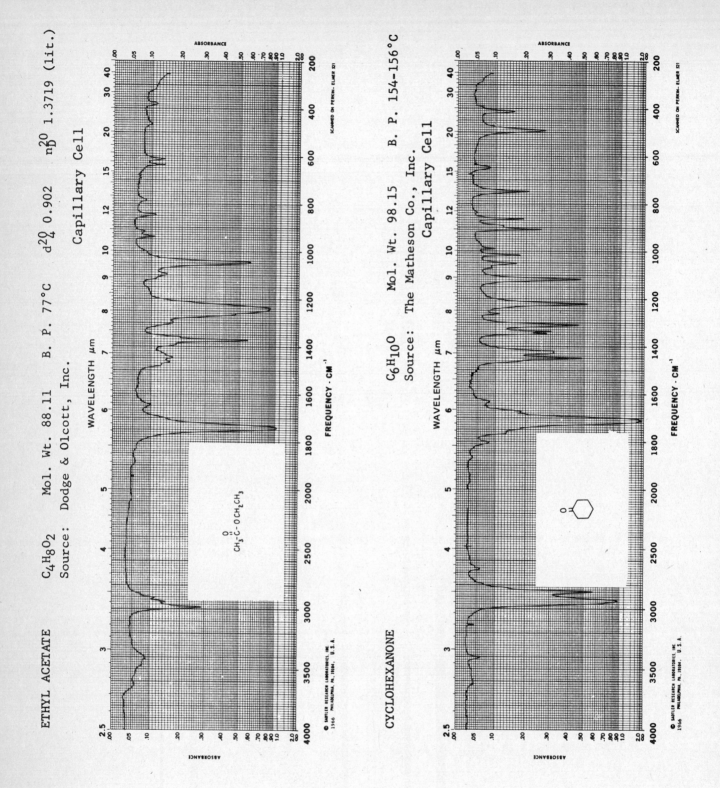

ETHYL ACETATE

$C_4H_8O_2$ Mol. Wt. 88.11   B. P. 77°C d_4^{20} 0.902 n_D^{20} 1.3719 (lit.)

Source: Dodge & Olcott, Inc.

Capillary Cell

No. 20

CYCLOHEXANONE

$C_6H_{10}O$ Mol. Wt. 98.15   B. P. 154–156°C

Source: The Matheson Co., Inc.

Capillary Cell

No. 19

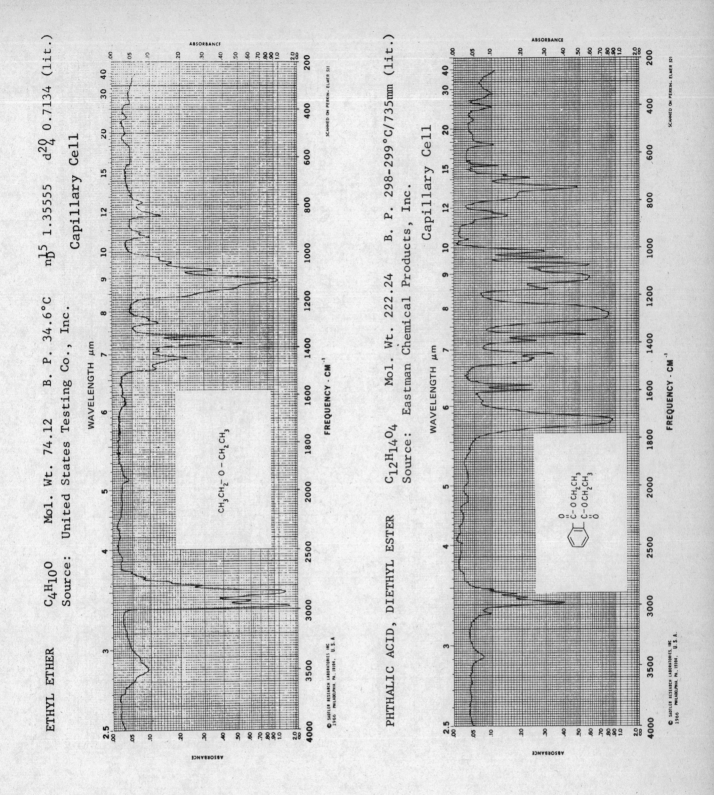

ETHYL ETHER   C₄H₁₀O   Mol. Wt. 74.12   B. P. 34.6°C   n$_D^{15}$ 1.35555   d$_4^{20}$ 0.7134 (lit.)
Source: United States Testing Co., Inc.   Capillary Cell

$CH_3CH_2 - O - CH_2CH_3$

No. 22

PHTHALIC ACID, DIETHYL ESTER   C₁₂H₁₄O₄   Mol. Wt. 222.24   B. P. 298-299°C/735mm (lit.)
Source: Eastman Chemical Products, Inc.   Capillary Cell

No. 2I

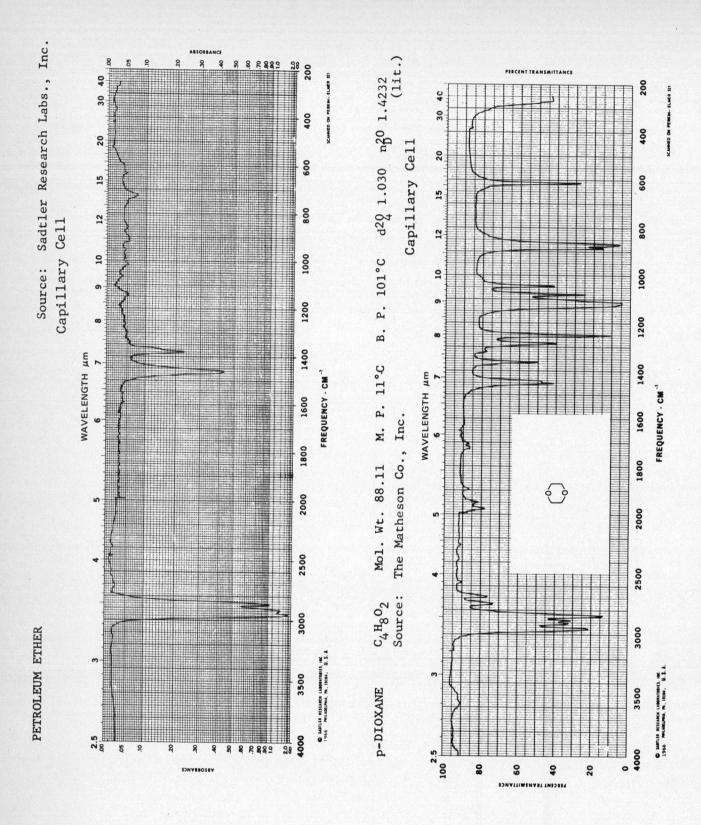

PETROLEUM ETHER

Source: Sadtler Research Labs., Inc.
Capillary Cell

No. 24

p-DIOXANE

$C_4H_8O_2$ Mol. Wt. 88.11   M. P. 11°C   B. P. 101°C d_4^{20} 1.030 n_D^{20} 1.4232 (lit.)

Source: The Matheson Co., Inc.

Capillary Cell

No. 23

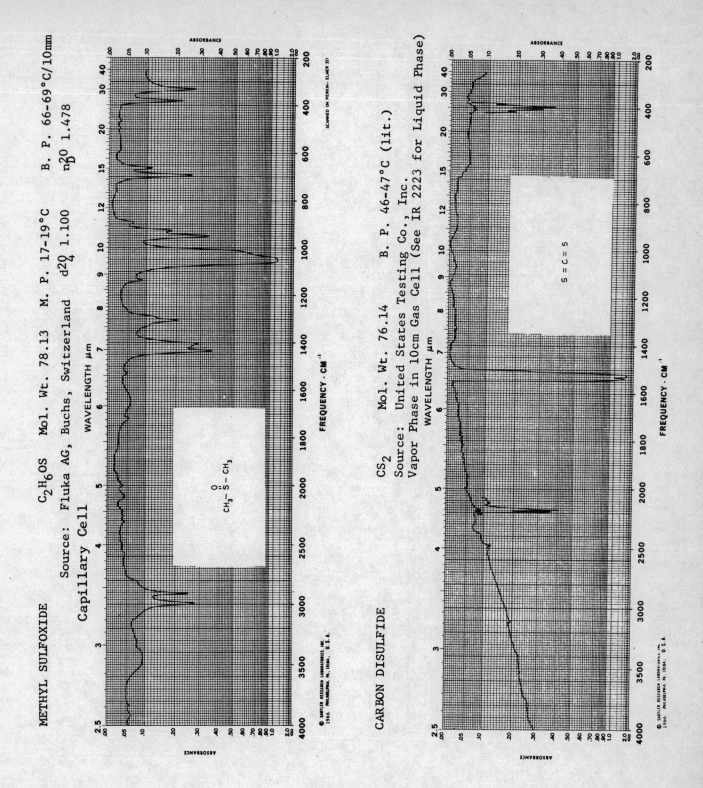

METHYL SULFOXIDE C_2H_6OS Mol. Wt. 78.13   M. P. 17-19°C   B. P. 66-69°C/10mm

Source: Fluka AG, Buchs, Switzerland d_4^{20} 1.100 n_D^{20} 1.478

Capillary Cell

$$CH_3 - \overset{O}{\underset{\|}{S}} - CH_3$$

SCANNED ON PERKIN-ELMER 521

© SADTLER RESEARCH LABORATORIES, INC.
1966   PHILADELPHIA, PA. 19104, U.S.A.

No. 26

CARBON DISULFIDE CS_2 Mol. Wt. 76.14   B. P. 46-47°C (lit.)

Source: United States Testing Co., Inc.
Vapor Phase in 10cm Gas Cell (See IR 2223 for Liquid Phase)

$$S = C = S$$

© SADTLER RESEARCH LABORATORIES, INC.
1966   PHILADELPHIA, PA. 19104, U.S.A.

No. 25

appendix a   **133**

Source: Dow Corning Corp.
Capillary Cell

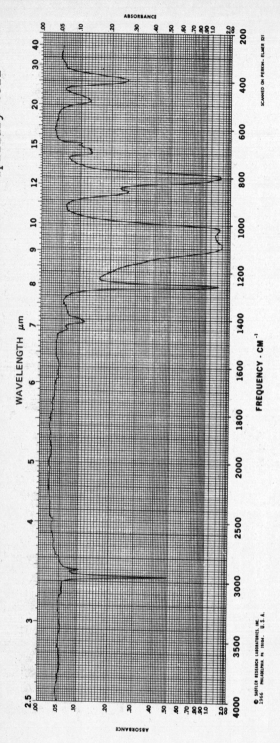

ABSORBANCE

WAVELENGTH μm

FREQUENCY · CM⁻¹

ABSORBANCE

© SADTLER RESEARCH LABORATORIES, INC.
1966 PHILADELPHIA PA 19104 · U. S. A.

SCANNED ON PERKIN - ELMER 521

No. 27

D15,855-0 N,N-Dimethylformamide
HCON(CH₃)₂ M.W. 73.10 n_D^{25} 1.4290   NEAT

WAVENUMBER CM¹

WAVELENGTH μm

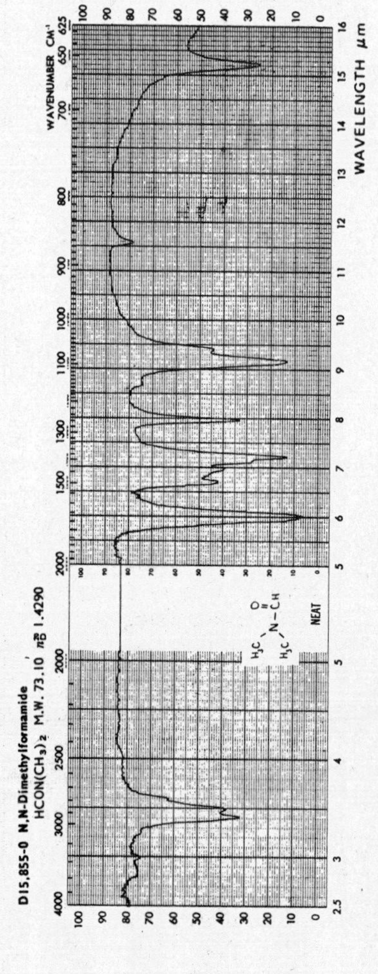

No. 28

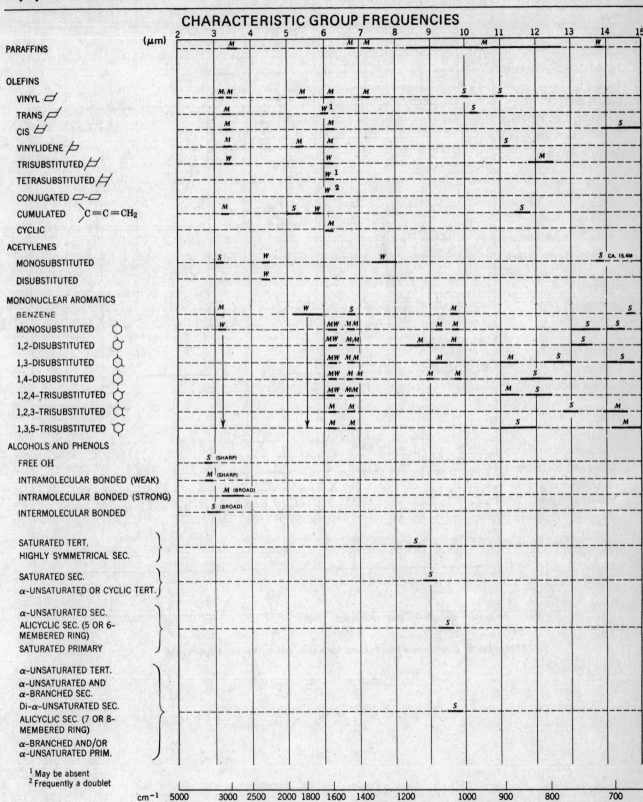

CHARACTERISTIC GROUP FREQUENCIES

(µm)

PARAFFINS

OLEFINS
  VINYL
  TRANS
  CIS
  VINYLIDENE
  TRISUBSTITUTED
  TETRASUBSTITUTED
  CONJUGATED
  CUMULATED $\ce{C=C=CH2}$
  CYCLIC

ACETYLENES
  MONOSUBSTITUTED
  DISUBSTITUTED

MONONUCLEAR AROMATICS
  BENZENE
  MONOSUBSTITUTED
  1,2-DISUBSTITUTED
  1,3-DISUBSTITUTED
  1,4-DISUBSTITUTED
  1,2,4-TRISUBSTITUTED
  1,2,3-TRISUBSTITUTED
  1,3,5-TRISUBSTITUTED

ALCOHOLS AND PHENOLS
  FREE OH
  INTRAMOLECULAR BONDED (WEAK)
  INTRAMOLECULAR BONDED (STRONG)
  INTERMOLECULAR BONDED

SATURATED TERT.
HIGHLY SYMMETRICAL SEC.

SATURATED SEC.
α-UNSATURATED OR CYCLIC TERT.

α-UNSATURATED SEC.
ALICYCLIC SEC. (5 OR 6-MEMBERED RING)
SATURATED PRIMARY

α-UNSATURATED TERT.
α-UNSATURATED AND α-BRANCHED SEC.
Di-α-UNSATURATED SEC.
ALICYCLIC SEC. (7 OR 8-MEMBERED RING)
α-BRANCHED AND/OR α-UNSATURATED PRIM.

[1] May be absent
[2] Frequently a doublet

cm⁻¹

*All absorption ranges are designated by bars. The intensities of bands within the range are denoted by the letters: s = strong, м = medium, w = weak. For example, a monosubstituted mononuclear aromatic has two bands between 6.2 and 6.4 µm, the shorter wavelength band of medium intensity, the longer wavelength band of weak intensity.

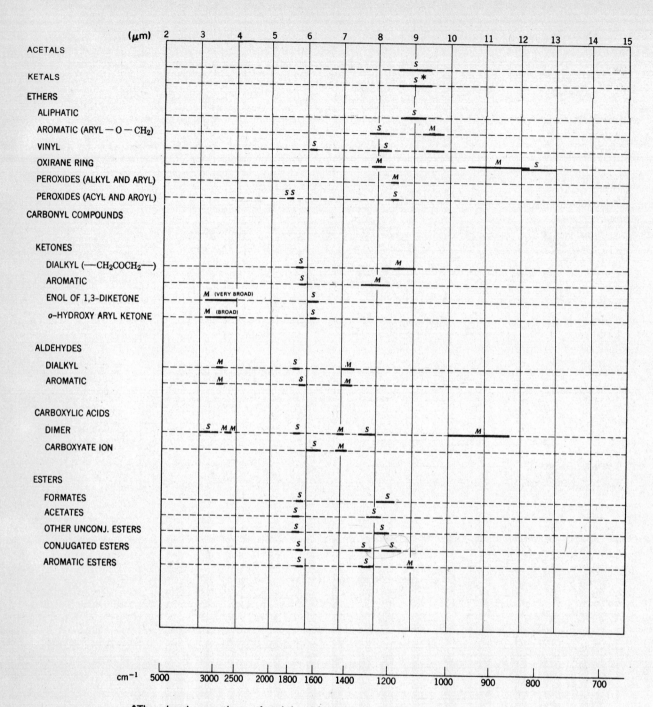

ACETALS

KETALS

ETHERS
  ALIPHATIC
  AROMATIC (ARYL — O — CH₂)
  VINYL
  OXIRANE RING
  PEROXIDES (ALKYL AND ARYL)
  PEROXIDES (ACYL AND AROYL)

CARBONYL COMPOUNDS

  KETONES
    DIALKYL (—CH₂COCH₂—)
    AROMATIC
    ENOL OF 1,3-DIKETONE
    o-HYDROXY ARYL KETONE

  ALDEHYDES
    DIALKYL
    AROMATIC

  CARBOXYLIC ACIDS
    DIMER
    CARBOXYATE ION

  ESTERS
    FORMATES
    ACETATES
    OTHER UNCONJ. ESTERS
    CONJUGATED ESTERS
    AROMATIC ESTERS

*Three bands, sometimes a fourth band for ketals and a fifth band for acetals.

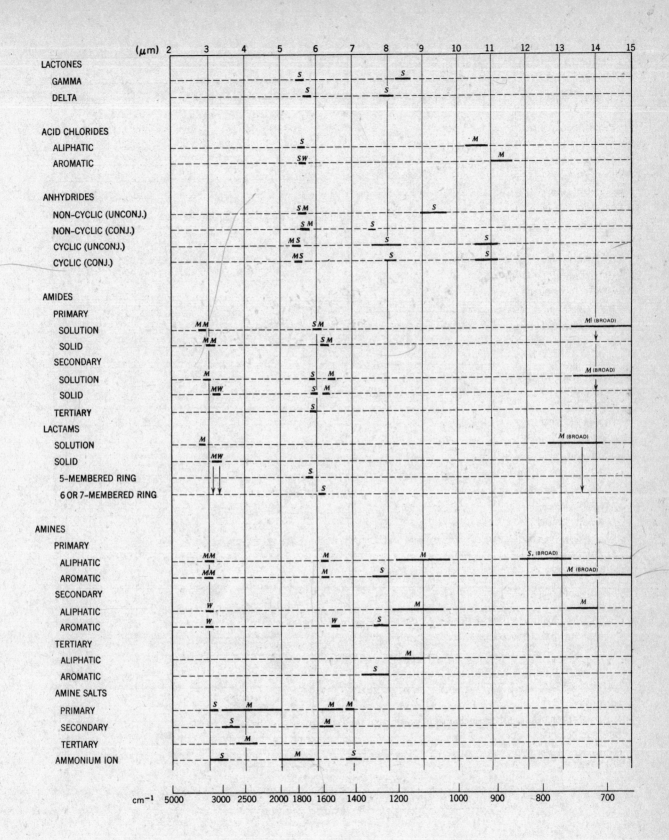

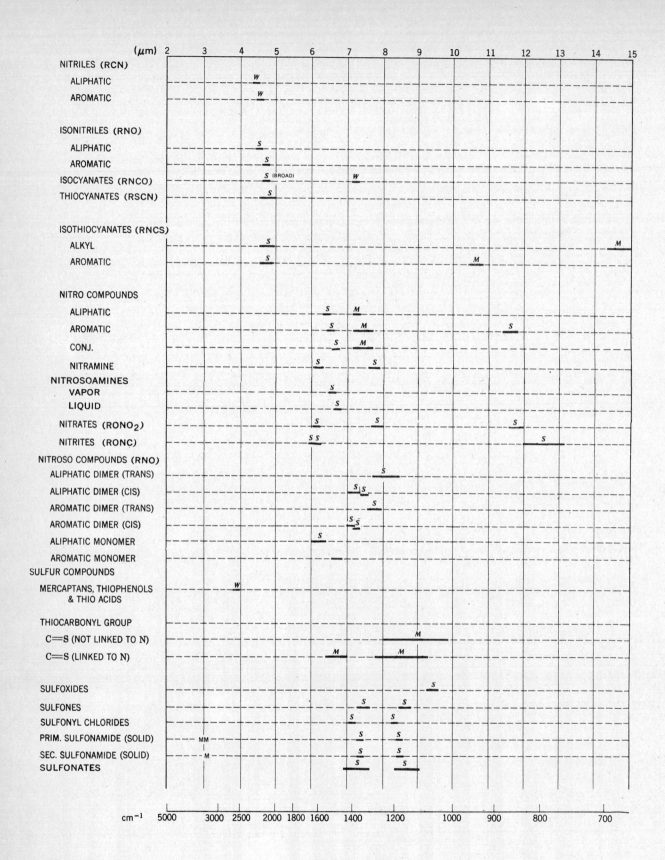

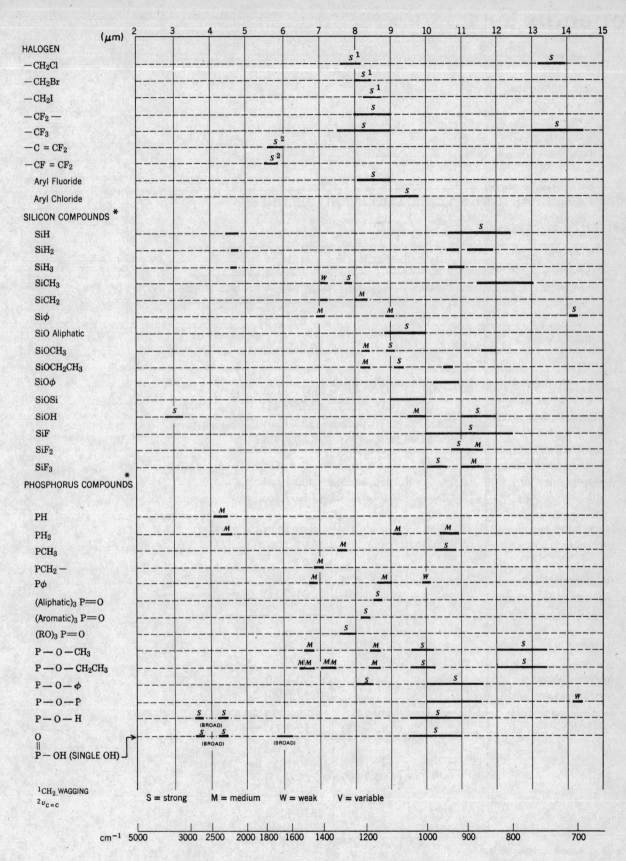

HALOGEN
—CH₂Cl
—CH₂Br
—CH₂I
—CF₂—
—CF₃
—C = CF₂
—CF = CF₂
Aryl Fluoride
Aryl Chloride
SILICON COMPOUNDS *
SiH
SiH₂
SiH₃
SiCH₃
SiCH₂
Siϕ
SiO Aliphatic
SiOCH₃
SiOCH₂CH₃
SiOϕ
SiOSi
SiOH
SiF
SiF₂
SiF₃
PHOSPHORUS COMPOUNDS *

PH
PH₂
PCH₃
PCH₂ —
Pϕ
(Aliphatic)₃ P=O
(Aromatic)₃ P=O
(RO)₃ P=O
P — O — CH₃
P — O — CH₂CH₃
P — O — ϕ
P — O — P
P — O — H
O
‖
P — OH (SINGLE OH)

¹CH₂ WAGGING
²$v_{C=C}$

S = strong   M = medium   W = weak   V = variable

*Tables relating to these absorptions are found in Appendix D.

appendix c

Table I CH₃ Group Absorption

Structure	$\nu_s CH_3$ cm^{-1} (μm)	$\delta_{as} CH_3$ cm^{-1} (μm)	$\delta_s CH_3$ cm^{-1} (μm)
CH₃—C	2872 (3.48)	1450 (6.90)	1375 (7.28) See Table IA
CH₃—C—(ketone) (C=O)	3000-2900 [a]w	1450-1400 (6.90-7.15)	1375-1350 (7.28-7.41)
CH₃—O—R	2832-2815[b] (3.53-3.55)	1470-1440 (6.80-6.95)	1470-1440 (6.80-6.95) See Table IB
CH₃—C—O— (C=O)	—	1450-1400 (6.90-7.15)	1400-1340 (7.15-7.46)
CH₃—N (amines and imines)	2820-2760 (3.55-3.62)	—	1440-1390 (6.95-7.20) See Table IC
CH₃—NC—(amides) (C—O)	—	1500-1450 (6.67-6.90)	1420-1405 (7.04-7.11)
CH₃—O—CR (C=O)	—	ca. 1440 (ca. 6.95)	ca. 1360 (ca. 7.35)
CH₃—X (X = halogen)	—	—	1500-1250 (6.67-8.00)
CH₃—S—	—	1440-1415 (6.95-7.07)	1330-1290 (7.52-7.75) See Table ID
CH₃—Si—	—	1429-1389 (7.00-7.20)	1280-1255 (7.81-7.97) See Table IE
CH₃—P	—	—	1330-1290 (7.52-7.81)

[a]w = weak; m = medium; s = strong.
[b]Overtone of CH₃ bending band.

Supplements to Table I Spectral Bands for the CH$_3$ Group Given in cm^{-1} (μm)[12]

IA	(aliphatic)—CH$_3$	2972–2952 strong (asymmetric stretch) (3.36–3.39) 2882–2862 weaker than 2960 (symmetric stretch) (3.47–3.50) (3.38) 1475–1450 medium (asymmetric deformation) (6.78–6.90) 1383–1377 medium (symmetric deformation) (7.23–7.26)
	(aromatic)—CH$_3$	2930–2920 medium (3.41–3.43) 2870–2860 medium (3.48–3.50)
	CH$_2$ | —CH$_2$—CH—CH$_2$—	1159–1151 medium (8.63–8.69)
	CH$_3$ (R)CH CH$_3$ (R = hydrocarbon group)	1389–1381 medium (7.20–7.24) 1372–1368 equal to the 1385 intensity (7.29–7.31) 1171–1168 medium (8.54–8.56) 1150–1130 variable (8.70–8.85)
	CH$_3$ | (R)—C—(R) | CH$_3$	1391–1381 medium (7.19–7.24) 1368–1366 intensity 5/4 1385 (7.31–7.32) (7.22) 1221–1206 weak (8.19–8.29) 1191–1185 medium (8.40–8.44)
	CH$_3$ | (R)—C—CH$_3$ | CH$_3$	1401–1393 medium (7.14–7.18) 1374–1366 about twice the 1390 intensity (7.28–7.32) 1253–1235 medium (7.98–8.10) 1208–1163 medium (8.28–8.60)
IB	(RO)CH$_3$	2992–2955 strong (asymmetric stretch) (3.34–3.38) 2897–2867 strong (symmetric stretch) (3.45–3.49) 2832–2815 variable (deformation overtone) (3.53–3.55) 1470–1440 medium (asymmetric and symmetric deformation (6.80–6.94)
IC	(aliphatic)—N—CH$_3$ (amine) (aromatic)—N—CH$_3$ (aliphatic)—N—(CH$_3$)$_2$ (aromatic)—N—(CH$_3$)$_2$	2805–2780 strong (3.46–3.60) 2820–2810 (3.55–3.56) 2825–2810 2775–2765 (3.54–3.56) (3.60–3.62) 2810–2790 (3.56–3.58)
ID	(RS)CH$_3$ (R—S—S)CH$_3$	2992–2955 medium (asymmetric stretch) (3.34–3.38) 2897–2867 medium (symmetric stretch) (3.45–3.49) 1440–1415 medium (asymmetric deformation) (6.94–7.06) 1330–1290 weaker (symmetric deformation) (7.52–7.75)
IE	Si—CH$_3$	1440–1410 weak (asymmetric deformation) (6.95–7.09) 1270–1255 strong (symmetric deformation) (7.88–7.97)

Table II CH₂ Group Absorption

Structure	$\nu_s CH_2$ cm⁻¹ (μm)	$\delta_s CH_2$ (scissoring) cm⁻¹ (μm)	ωCH_2 (wagging) cm⁻¹ (μm)	
$-CH_2-R$	—	—	—	see Table IIA
$-CH_2-\overset{\displaystyle O}{\overset{\|}{C}}-$ (acyclic)	—	1435–1405 (6.97–7.11)	—	see Table IIB
$-CH_2-\overset{\displaystyle O}{\overset{\|}{C}}-$ (small ring)	—	1475–1425[a] (6.78–7.02)	—	
$-CH_2-O-$ (acyclic)	—	1470–1435 (6.80–6.97)	—	see Table IIC
$-CH_2-O-$ (small ring)	—	1500–1470[a] (6.67–6.80)	—	
$-CH_2-N-$ (sec. and tert. amines)	2820–2760 (3.55–3.62)	1475–1445 (6.78–6.92)	—	
$-CH_2-\overset{\displaystyle O}{\overset{\|}{NC}}-$ (amides)	—	1450–1405 (6.89–7.11)	—	
$-CH_2-O-\overset{\displaystyle O}{\overset{\|}{C}}$	—	1475–1460 (6.78–6.85)	—	
$-CH_2-S-$	—	1440–1415 (6.94–7.06)	—	see Table IID
$-CH_2-halogen$	—	1460–1430 (6.85–7.00)	1275–1170 (7.85–8.55)	see Table IIE
$-CH_2 NO_2$	—	1425–1415 (7.02–7.06)	—	see Table IIB
$-CH_2-CN$	—	1425 (7.02)	—	see Table IIB
$-CH_2-N=C=S$	—	—	1351–1316 (7.40–7.60)	
$-Si-CH_2-$	—	near 1410 (7.09)	1250–1200 (8.00–8.33)	
$-CH_2-P\diagdown^{\diagup}$	—	1445–1405 (6.92–7.22)	—	

[a]Multiple bands.

IIA	(R)–CH$_2$–(R) (R = hydrocarbon group)	2936–2916 strong (asymmetric stretch) (3.41–3.43) 2863–2843 weaker than 2930 (symmetric stretch) (3.49–3.52) (3.41) 1475–1450 medium (deformation) (6.78–6.90)
	–(CH$_2$)$_6$–(CH$_3$) –(CH$_2$)$_5$–(CH$_3$) –(CH$_2$)$_4$–(CH$_3$) –(CH$_2$)$_3$–(CH$_3$) –(CH$_2$)$_2$–(CH$_3$) (4th carbon branched) –CH$_2$–(CH$_3$) (3rd carbon branched)	724– 722 weak (CH$_2$ rock) (13.81–13.85) 724– 723 (13.81–13.84) 726– 724 (13.77–13.81) 729– 726 (13.72–13.77) 743– 734 (13.46–13.63) 785– 770 (12.75–12.98)
	(R)–CH$_2$–(CH=CH$_2$) (R)–CH$_2$–(C≡CH)	2936–2916 strong (3.41–3.43) 2863–2843 weaker than 2930 (3.49–3.52) (3.41) 1455–1435 medium (6.87–6.97)
IIB	(R)–CH$_2$–(C=O) (R)–CH$_2$–(C≡N) (R)–CH$_2$–(NO$_2$)	3000–2900 medium (stretch) (3.33–3.45) 1445–1405 stronger than in hydrocarbons (deformation) (6.92–7.11)
IIC	(R)–CH$_2$–(O–R) (R)–CH$_2$–(OH) (R)–CH$_2$–(NH$_2$)	2955–2922 strong (3.38–3.42) 2878–2835 as strong as 2930 (3.47–3.53) 1475–1445 medium (6.78–6.92)
	(R)–CH$_2$–(NHR) (R)–CH$_2$–(NR$_2$)	2960–2920 strong (3.38–3.42) 2820–2760 as strong as 2930 (3.55–3.62) (3.41) 1475–1445 medium (6.78–6.92)
IID	(R)–CH$_2$–(SH) (R)–CH$_2$–(S–C) (R)–CH$_2$–(S–S)	2948–2922 strong (3.39–3.42) 2878–2846 weaker than 2930 (3.47–3.51) (3.41) 1440–1415 medium (deformation) (6.95–7.07) 1270–1220 strong (CH$_2$ wag) (7.88–8.20)
IIE	(R)–CH$_2$(Cl)	3000–2950 weak (3.33–3.39) 1460–1430 medium (deformation) (6.85–7.00) 1300–1240 strong (CH$_2$ wag) (7.69–8.06)
IIF	CH$_2$ in cyclopropanes	3100–3072 (asymmetric stretch) (3.23–3.26) 3033–2995 (symmetric stretch) (3.30–3.34) 1050–1000 (9.52–10.0)
	CH$_2$ in cyclobutanes	2999–2977 (asymmetric stretch) (3.33–3.36) 2924–2875 (symmetric stretch) (3.42–3.48)
	CH$_2$ in cyclopentanes	2959–2952 (asymmetric stretch) (3.38–3.39) 2866–2853 (symmetric stretch) (3.49–3.50)
	CH$_2$ in cyclohexanes	ca. 2927 (asymmetric stretch) (3.42) ca. 2854 (symmetric stretch) (3.50)
	Epoxy	3958–3029 (3.27–3.30) 3004–2990 (usually one band only) (3.33–3.34)

Table III Olefinic Absorption

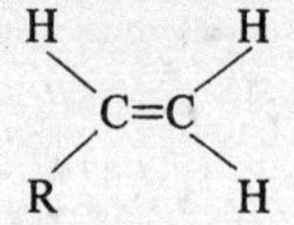

vinyl

1648–1638 cm^{-1} (6.07–6.10 μm)
995–985 cm^{-1} (10.05–10.15 μm)(s)[a]
915–905 cm^{-1} (10.93–11.05 μm)(s)

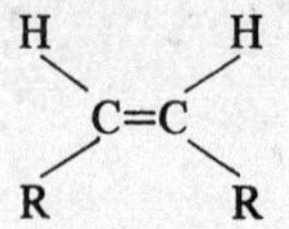

cis

1662–1626 cm^{-1} (6.02–6.15 μm)(v)
730–665 cm^{-1} (13.70–15.04 μm)(s)

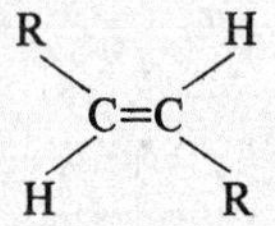

trans

1678–1668 cm^{-1} (5.96–6.00 μm)(v)
980–960 cm^{-1} (10.20–10.42 μm)(s)[b]

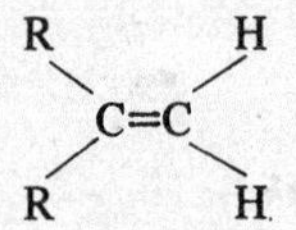

vinylidine

1658–1648 cm^{-1} (6.03–6.07 μm)(m)
895–885 cm^{-1} (11.17–11.30 μm)(s)

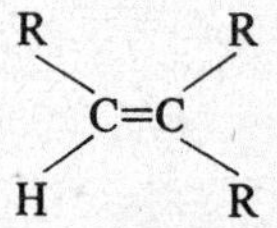

trisubstituted

1675–1665 cm^{-1} (5.97–6.01 μm)(w)
840–790 cm^{-1} (11.91–12.66 μm)(m)

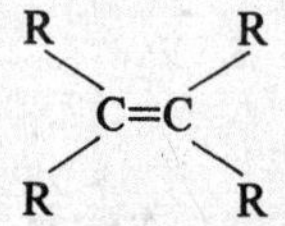

tetrasubstituted

1675–1665 cm^{-1} (5.97–6.01 μm very weak or absent.

[a]This band also shows a strong overtone band.
[b]This band occurs near 1000 cm^{-1} (10.00 μm) in conjugated trans-trans systems such as the esters of sorbic acid.
s = strong; m = medium; w = weak; v = variable.

Table IV C=C Frequencies in Open-Chain Systems. Effects of Substitution[3]

R_1	$CH_2{=}CHR_1$	$CH_2{=}C(R_1)_2$	$R_1{-}CH{=}CH{-}R_1$	$CHCH_3{=}C(R_1)_2$
H	1623[a] (6.16)	1623[a] (6.16)	1623[a] (6.16)	1648 (6.07)
D	—	—	—	—
F	1650 (6.06)	1728 (5.79)	—	—
Cl	1610 (6.21)	1620 (6.17)	cis 1590 (5.79)	—
			trans 1653 (6.05)	
Br	1605 (6.23)	1593 (6.28)	—	—
I	1593 (6.28)	—	—	—
CH_3	1648 (6.07)	1661 (6.07)	cis 1661 (6.02)	1681 (5.95)
			trans 1676 (5.97)	
CF_3	—	—	1700 (cis) (5.88)	—
OCH_3	1618 (6.18)	1639 (6.10)	—	1675 (5.97)
SCH_3	1586 (6.30)	—	—	1605 (6.23)

[a]Raman value.

Table V $=CH_2$ Wagging Frequencies[12]

R–O–CH–CH=CH$_2$	961 cm^{-1} (10.41 μm)
N≡C–CH–CH$_2$	960 (10.42)
R–CH=CH$_2$	910 (10.99)
Cl–CH=CH$_2$	894 (11.11
R–CO–O–CH=CH$_2$	870 (11.50)
R–CO–N–CH=CH$_2$	840 (11.90)
R–O–CH=CH$_2$	813 (12.30)
(N≡C)(N≡C)C=CH$_2$	985 (10.15)
(R)(R–O–CO)C=CH$_2$	939 (10.65)
(N≡C)(Cl)C=CH$_2$	916 (10.92)
(R)(R)C=CH$_2$	890 (11.23)
(Cl)(Cl)C=CH$_2$	867 (11.53)
(R)(R–O)C=CH$_2$	795 (12.58)
(R–O)(R–O)C=CH$_2$	711 (14.06)

Table VI C=C Stretching Frequencies and Ring Strain cm^{-1} (μm)[3]

Ring or Chain	$\underset{C}{\overset{H}{}}C=C\underset{C}{\overset{H}{}}$	$\underset{C}{\overset{H}{}}C=C\underset{C}{\overset{CH_3}{}}$	$\underset{C}{\overset{CH_3}{}}C=C\underset{C}{\overset{CH_3}{}}$	$\overset{C}{\underset{C}{}}C=CH_2$
Chain *cis*	1661 (6.02)	1681 (5.95)	1672 (5.98)	1661 (6.02)
Chain *trans*	1676 (5.97)			
Three-membered ring	1641 (6.10)		1890 (5.29)	1780 (5.62)
Four-membered ring	1566 (6.33)		1685 (5.94)	1678 (5.96)
Five-membered ring	1611 (6.21)	1658	1686 (5.93)	1657 (6.03)
Six-membered ring	1649 (6.06)	1678	1685 (5.93)	1651 (6.06)
Seven-membered ring	1651 (6.06	1673	—	
Eight-membered ring	1653 (6.05)	—	—	

Table VII Carbonyl Frequencies (cm^{-1})[3]

$$\underset{X-C-Y}{\overset{\displaystyle O \atop \displaystyle \|}{}}$$

Y	CH$_3$	CH$_2$Cl	CHCl$_2$	CCl$_3$	CH$_2$Br	CF$_3$	F	Cl	CHF$_2$	OCH$_3$
X = Y :	1719	1746 1730	1774 1764	1830 1751		1825	1928V	1813		1832
H	1730	1748	1742	1768		1784V	1837V			
CH$_3$	1719	1752 1726	1743 1724C			1765	1872V	1806		
C$_6$H$_5$	1692	1716 1692	1715 1692	1718	1709 1688		1812	1773 1736		
OCH$_3$	1748	1773 1749	1775 1755	1775	1764 1749			1786		
OC$_2$H$_5$	1742	1767 1742	1772 1750	1769	1761 1743	1789		1779	1780 1766	1780
NH Butyl	1688	1684	1705	1726	1681	1736			1718	
NBu$_2$	1647	1656	1684 1656			1692			1686 1667	
Cl	1806	1821 1785	1810 1779	1803	1802 1773L	1810V	1868V	1813		1806
F	1872V					1901V	1928V	1868V	1874V	
SButyl	1695	1699 1671	1703 1682	1699	1691 1669	1710				
SC$_6$H$_5$	1711	1725 1691	1736 1700	1711	1710 1695	1722				
OH	1769m 1715d	1791m 1735d	1784m 1743d	1788m 1754d	1772m 1726d	1810m 1787d			1756d	
ONa		1600 1420S				1687 1457S				
OCH$_2$CH=CH$_2$	1746	1768 1746	1773 1752	1770						
NH$_2$	1678 1702	1695S$_1$	1716S$_1$ (double)	1732S$_1$		1750S$_1$				

To save space, data on this and the following page are presented in cm^{-1} only; the corresponding μm value may be obtained from Appendix E.

More extensive tables of this type may be found in the books by Bellamy.[3] See notation footnote on p. 147.

O
‖
Y–C–X

Saturated and Unsaturated Ketones, Aldehydes and Esters[3]

X

Y	H	CH₃	C₂H₅	C₆H₅	OCH₃	OC₂H₅	OH	Y
C_6H_5	1710	1692	1693	1667	1727	1720	1695	1667
$CH_2{=}CH$	1703	1706	1707		1734	1730		
		1686	1690					
$HC{\equiv}C$					1719			
SButyl	1675	1695	1691	1685				
SC_6H_5	1693	1711	1710	1685				
OCH_3	1735	1748	1746	1727	1758	1756		1758
OC_6H_5		1749	1767	1743	1755	1757		1786
OOH	1780V	1760V	1760V	1732				

O
‖
Y–C–X

Amides, Urethanes, Anhydrides etc.[3]

X

Y	H	CH₃	C₆H₅	OCH₃	OC₂H₅	CF₃	Y	C₃H₅
NH_2	1722	1714m / 1690d	1768S₁	1733 / 1700S	1725S₁	1750	1686 / 1630S	1687S₁
$NHCH_3$	1666L	1700 / (1653L)	1660S₁		1731		1695	
$N(CH_3)_2$	1687 / (1670L)	1653			1684		1640	
NHC_6H_5		1688S₁	1679S₁	1707	1725 / 1701S	–	1632S	
$N(C_6H_5)_2$		1679						
NHC_2H_5		1686m / 1647d	1651S₂					
$N(C_2H_5)_2$	1684	1652	1640S₁					
NHOH	1650S	1665S	1660					
CH_2COCH_3		1727 / 1717 / 1616e k	1724k / 1600e		1733 / 1709 / 1645e k			
C_2H_5	1738	1722	1693	1745	1742	1761		
$OCH{=}CH_2$		1763				1800		
$NHCOCH_3$		1714 / 1690			1754 / 1704S₁			

All in carbon tetrachloride except: V = Vapour; L = Liquid; S = Solid; S₁ = chloroform solution; S₂ = acetonitrile solution; S₃ = CS₂; m = monomer; d = dimer; t = *trans*; c = *cis*; k = keto; e = enol.

Table VIII Carbonyl Frequencies (aromatics) (cm^{-1})[3]

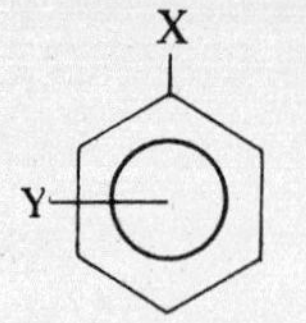

Y	COOH (monomer)	COOCH$_3$	COCH$_3$	NHCOCH$_3$	OCOCH$_3$	CHO
m-N(CH$_3$)$_2$	1740	1727				
m-OCH$_3$	1741	1728	1695	1694c	1773	
m-CH$_3$	1742			1692c	1770	
H	1744	1730				1708
m-F	1748	1733				
m-Cl	1748	1735	1696	1695c	1770	
m-Br	1748	1734			1770	
m-NO$_2$	1752	1738	1701	1706d	1779	
m-COCH$_3$				1709	1773	
p-OH			1686	1686c		1698
p-OCH$_3$	1737	1723	1684	1684c	1770	1699
p-CH$_3$	1740		1690	1691c		
p-F	1745	1732				
p-Cl	1745	1731	1694			1708
p-Br	1746	1734	1695	1698c		
p-NO$_2$	1752	1737	1700	1712d	1779	1713
p-CN				1706c		
p-COCH$_3$				1715	1773	
p-CHO				1715	1773	
p-C$_6$H$_5$			1692	1689c	1776	
o-OCH$_3$	1760, 1751	1736, 1718				
o-F	1755, 1739	1741, 1726				
o-Cl	1756, 1738	1744, 1727				
o-Br	1757, 1738	1744, 1727				
o-I	1753, 1736	1740, 1727				
o-NO$_2$	1760	1747				
o-OCOCH$_3$	1747	1733				
o-CH$_3$	1742	1728				

c, Chloroform solution; d, Dioxan solution.

To save space, data on this page are presented in cm^{-1} only; the corresponding μm value may be obtained from Appendix E.

More extensive tables of this type may be found in the books by Bellamy.[3]

Table IX Carbonyl Frequencies in Selected Heterocyclic Compounds. νCO Values in cm^{-1} (μm) for Carbon Tetrachloride Solution[3]

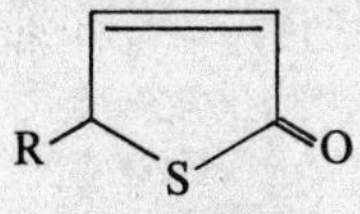

$\sim$1675
(5.97)

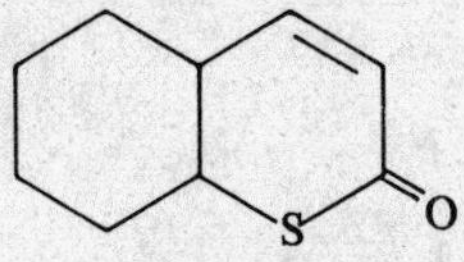

1683
(5.97)

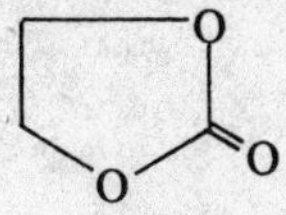

1819
(5.50)

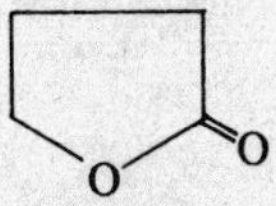

1774
(5.64)

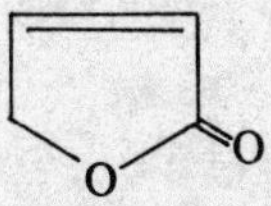

1784, 1742
(5.61, 5.74)

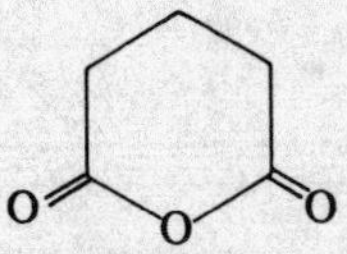

1822, 1780
(5.49, 5.62)

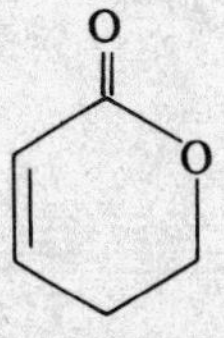

1743(CHCl$_3$)
(5.74)

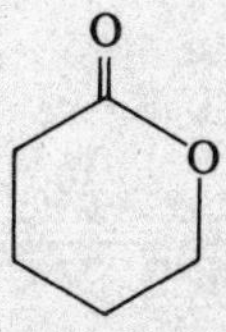

1750(CHCl$_3$)
(5.71)

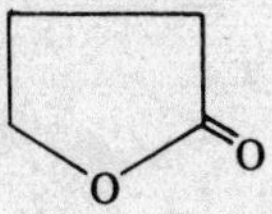

1876, 1799
(5.33, 5.56)

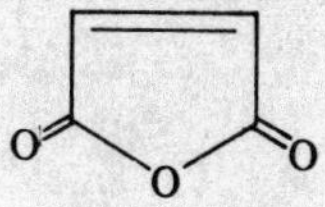

1855, 1784
(5.39, 5.61)

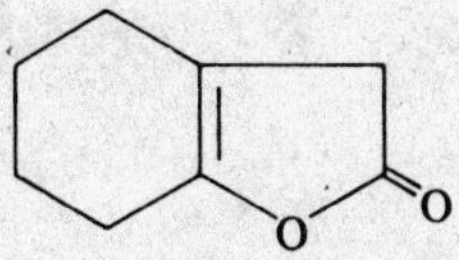

1819
(5.50)

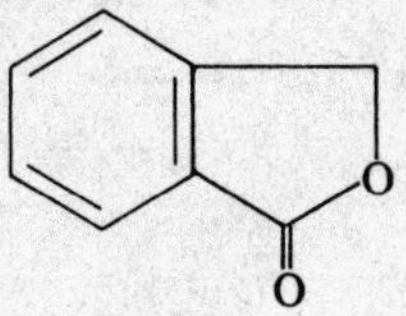

1778

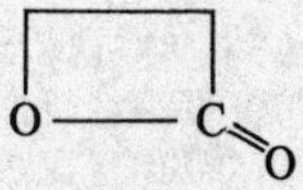

1841
(5.43)

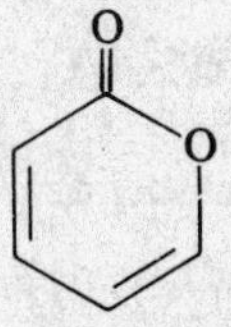

1752, 1731
(5.71, 5.78)

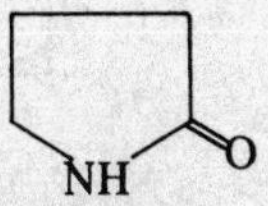

1706
(5.86)

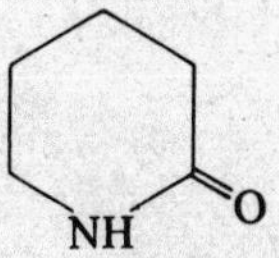

1690
(5.92)

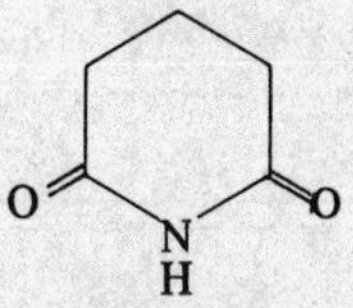

1727, 1672
(solid)
(5.79, 5.98)

appendix d

Table I   Si–H IR Absorptions

Mode	Position cm^{-1} (μm) intensity
Si–H stretch	2250–2100 (4.44–4.76) m-s
SiH$_2$ bend (scissoring)	942–923 (10.62–10.83)
SiH$_3$, two bands	
sym. and asym. bend	959–900 (10.43–11.11)
SiH$_2$ wag	900–840 (11.11–11.90)
R$_3$SiH bend	950–800 (10.53–12.50)

Table II   Si–O Stretching Vibrations

Group	ν_{Si-O} Bands
Si–O–R (aliphatic)	1110–1000 cm^{-1} (9.01–10.00 μm)s
Si–O–Si	1110–1000 cm^{-1} (9.01–10.00 μm)s
Si–O–R (aromatic)	970–920 cm^{-1} (10.31–10.87 μm)s
Si–OH	910–830 cm^{-1} (11.00–12.05 μm)s

s = strong.

Table III   Absorption of Si–F Bonds

Group	ν_{Si-F} Bands	
$\diagdown$ –SiF $\diagup$	1000–800 cm^{-1} (10.00–12.50 μm)	
$\diagdown$ SiF$_2$ $\diagup$	943–910 cm^{-1} (10.60–11.00 μm)s	910–870 cm^{-1} (11.00–11.50 μm)m
–SiF$_3$	980–945 cm^{-1} (10.20–10.58 μm)s	910–860 cm^{-1} (11.00–11.63 μm)m

s = strong; m = medium.

Table IV PH, PO IR Absorptions

Mode	Position cm^{-1} (μm) intensity
P-H stretch	2440–2275 (4.10–4.40)m
alkyl, aryl, P-H stretch	2326–2275 (4.30–4.40)
PH$_2$ bend	
scissoring	1090–1080 (9.19–9.26)m
wagging	940–909 (10.64–11.00)
P=O stretch	
phosphine oxides	
aliphatic	ca. 1150 (8.70)
aromatic	ca. 1190 (8.40)
phosphate esters[a]	1299–1250 (7.70–8.00)

[a]The increase in P = O stretching frequency of the ester, relative to the oxides, results from the electronegativity of the attached alkoxy groups.

Table V P–O Stretching Vibrations

Group	ν_{P-O} Bands	
P–OH	1040–910 cm^{-1} (9.62–11.00 μ)s	
P–O–P	1000–870 cm^{-1} (10.00–11.50 μm)s	ca. 700 cm^{-1} (14.30 μm)w
P–O–C (aliph.)	1050–970 cm^{-1} (9.52–10.30 μm)s[a]	830–740 cm^{-1} (12.05–13.51 μm)s[b]
P–O–C (arom.)	1260–1160 cm^{-1} (7.94–8.62 μm)s	994–855 cm^{-1} (10.06–11.70 μm)s

[a]May be a doublet.
[b]May be absent.
 s = strong; w = weak.

Table VI γCH and Ring Bending Bands (β Ring) of Pyridines[20]

Substitution	Number Adjacent H's	γCH	β Ring
2-	4	781–740 cm^{-1} (12.80–13.50 μm)	752–746 cm^{-1} (13.30–13.40 μm)
3-	3	810–789 cm^{-1} (12.35–12.67 μm)	715–712 cm^{-1} (13.99–14.04 μm)
4-	2	820–794 cm^{-1} (12.19–12.60 μm)	775–709 cm^{-1} (12.90–14.10 μm)

Table VII Characteristic γCH or β Ring Bands of Furans, Thiophenes, and Pyrroles[20]

Nucleus	Position of Substitution	Phase	γCH or β Ring Modes			
			cm^{-1} (μm)	cm^{-1} (μm)	cm^{-1} (μm)	cm^{-1} (μm)
Furan	2-	CHCl$_3$	ca. 925 (10.8)	ca. 884 (11.3)	835–780 (12.0–12.8)	
	2-	liq.	960–915 (10.4–10.9)	890–875 (11.2–11.4)		780–725 (12.8–13.8)
	2-	Solid	955–906 (10.5–11.0)	887–860 (11.3–11.6)	821–793 (12.2–12.6)	750–723 (13.3–13.8)
	3-	liq.		885–870 (11.3–11.5)	741 (13.5)	
Thiophene	2-	CHCl$_3$	ca. 925 (10.8)	ca. 853 (11.7)	843–803 (11.9–12.5)	
	3-	liq.				755 (13.2)
Pyrrole	2-Acyl	Solid			774–740 (12.9–13.5)	ca. 755 (13.2)

appendix e

WAVELENGTH–WAVE NUMBER CONVERSION TABLE

wavelength (μm)	Wave number (cm^{-1})									
	0	1	2	3	4	5	6	7	8	9
2.0	5000	4975	4950	4926	4902	4878	4854	4831	4808	4785
2.1	4762	4739	4717	4695	4673	4651	4630	4608	4587	4566
2.2	4545	4525	4505	4484	4464	4444	4425	4405	4386	4367
2.3	4348	4329	4310	4292	4274	4255	4237	4219	4202	4184
2.4	4167	4149	4232	4115	4098	4082	4065	4049	4032	4016
2.5	4000	3984	3968	4953	3937	3922	3006	3891	3876	3861
2.6	3846	3831	3817	3802	3788	3774	3759	3745	3731	3717
2.7	3704	3690	3676	3663	3650	3636	3623	3610	3597	3584
2.8	3571	3559	3546	3534	3521	3509	3497	3484	3472	3460
2.9	3448	3436	3425	3413	3401	3390	3378	3367	3356	3344
3.0	3333	3322	3311	3300	3289	3279	3268	3257	3247	3236
3.1	3226	3215	3205	3195	3185	3175	3165	3155	3145	3135
3.2	3125	3115	3106	3096	3086	3077	3067	3058	3049	3040
3.3	3030	3021	3012	3003	2994	2985	2976	2967	2959	2950
3.4	2941	2933	2924	2915	2907	2899	2890	2882	2874	2865
3.5	2857	2849	2841	2833	2825	2817	2809	2801	2793	2786
3.6	2778	2770	2762	2755	2747	2740	2732	2725	2717	2710
3.7	2703	2695	2688	2681	2674	2667	2660	2653	2646	2639
3.8	2632	2625	2618	2611	2604	2597	2591	2584	2577	2571
3.9	2654	2558	2551	2545	2538	2532	2525	2519	2513	2506
4.0	2500	2494	2488	2481	2475	2469	2463	2457	2451	2445
4.1	2439	2433	2427	2421	2415	2410	2404	2398	2387	2387
4.2	2381	2375	2370	2364	2358	2353	2347	2342	2336	2331
4.3	2326	2320	2315	2309	2304	2299	2294	2288	2283	2278
4.4	2273	2268	2262	2257	2252	2247	2242	2237	2232	2227
4.5	2222	2217	2212	2208	2203	2198	2193	2188	2183	2179
4.6	2174	2169	2165	2160	2155	2151	2146	2141	2137	2132
4.7	2128	2123	2119	2114	2110	2105	2101	2096	2092	2088
4.8	2083	2079	2075	2070	2066	2062	2058	2053	2049	2045
4.9	2041	2037	2033	2028	2024	2020	2016	2012	2008	2004
5.0	2000	1996	1992	1988	1984	1980	1976	1972	1969	1965
5.1	1961	1957	1953	1949	1946	1942	1938	1934	1931	1927
5.2	1923	1919	1916	1912	1908	1905	1901	1898	1894	1890
5.3	1887	1883	1880	1876	1873	1869	1866	1862	1859	1855
	0	1	2	3	4	5	6	7	8	9

Wave number (cm^{-1})										
	0	1	2	3	4	5	6	7	8	9
5.4	1852	1848	1845	1842	1838	1835	1832	1828	1825	1821
5.5	1818	1815	1812	1808	1805	1802	1799	1795	1792	1788
5.6	1786	1783	1779	1776	1773	1770	1767	1764	1761	1757
5.7	1754	1751	1748	1745	1742	1739	1736	1733	1730	1727
5.8	1724	1721	1718	1715	1712	1709	1706	1704	1701	1698
5.9	1695	1692	1689	1686	1684	1681	1678	1675	1672	1669
6.0	1667	1664	1661	1658	1656	1653	1650	1647	1645	1642
6.1	1639	1637	1634	1631	1629	1626	1623	1621	1618	1616
6.2	1613	1610	1608	1605	1603	1600	1597	1595	1592	1590
6.3	1587	1585	1582	1580	1577	1575	1572	1570	1567	1565
6.4	1563	1560	1558	1555	1553	1550	1548	1546	1543	1541
6.5	1538	1536	1534	1531	1529	1527	1524	1522	1520	1517
6.6	1515	1513	1511	1508	1506	1504	1502	1499	1497	1495
6.7	1493	1490	1488	1486	1484	1481	1479	1477	1475	1473
6.8	1471	1468	1466	1464	1462	1460	1458	1456	1453	1451
6.9	1449	1447	1445	1443	1441	1439	1437	1435	1433	1431
7.0	1429	1427	1425	1422	1420	1418	1416	1414	1412	1410
7.1	1408	1406	1404	1403	1401	1399	1397	1395	1393	1391
7.2	1389	1387	1385	1383	1381	1379	1377	1376	1374	1372
7.3	1370	1368	1366	1364	1362	1361	1359	1357	1355	1353
7.4	1351	1350	1348	1346	1344	1342	1340	1339	1337	1335
7.5	1333	1332	1330	1328	1326	1325	1323	1321	1319	1318
7.6	1316	1314	1312	1311	1309	1307	1305	1304	1302	1300
7.7	1299	1297	1295	1294	1292	1290	1289	1287	1285	1284
7.8	1282	1280	1279	1277	1276	1274	1272	1271	1269	1267
7.9	1266	1264	1263	1261	1259	1258	1256	1255	1253	1252
8.0	1250	1248	1247	1245	1244	1242	1241	1239	1238	1236
8.1	1235	1233	1232	1230	1229	1227	1225	1224	1222	1221
8.2	1220	1218	1217	1215	1214	1212	1211	1209	1208	1206
8.3	1205	1203	1202	1200	1199	1198	1196	1195	1193	1192
8.4	1190	1189	1188	1186	1185	1183	1182	1181	1179	1178
8.5	1176	1175	1174	1172	1171	1170	1168	1167	1166	1164
8.6	1163	1161	1160	1159	1157	1156	1155	1153	1152	1151
8.7	1149	1148	1147	1145	1144	1143	1142	1140	1139	1138
8.8	1136	1135	1134	1133	1131	1130	1129	1127	1126	1125
8.9	1124	1122	1121	1120	1119	1117	1116	1115	1114	1112
9.0	1111	1110	1109	1107	1106	1105	1104	1103	1101	1100
9.1	1099	1098	1096	1095	1094	1093	1092	1091	1089	1088
9.2	1087	1086	1085	1083	1082	1081	1080	1079	1078	1076
9.3	1075	1074	1073	1072	1071	1070	1068	1067	1066	1065
9.4	1064	1063	1062	1060	1059	1058	1057	1056	1055	1054
9.5	1053	1052	1050	1049	1048	1047	1046	1045	1044	1043
9.6	1042	1041	1040	1038	1037	1036	1035	1034	1033	1032
9.7	1031	1030	1029	1028	1027	1026	1025	1024	1022	1021
9.8	1020	1019	1018	1017	1016	1015	1014	1013	1012	1011
9.9	1010	1009	1008	1007	1006	1005	1004	1003	1002	1001
10.0	1000	999	998	997	996	995	994	993	992	991
	0	1	2	3	4	5	6	7	8	9

Wavelength (μm)

Wavelength (μm)	0	1	2	3	4	5	6	7	8	9
10.1	990	989	988	987	986	985	984	983	982	981
10.2	980	979	978	978	977	976	975	974	973	972
10.3	971	970	969	968	967	966	965	964	963	962
10.4	962	961	960	959	958	957	956	955	954	953
10.5	952	951	951	950	949	948	947	946	945	944
10.6	943	943	942	941	940	939	938	937	936	935
10.7	935	934	933	932	931	930	929	929	928	927
10.8	926	925	924	923	923	922	921	920	919	918
10.9	917	917	916	915	914	913	912	912	911	910
11.0	909	908	907	907	906	905	904	903	903	902
11.1	901	900	899	898	898	897	896	895	894	894
11.2	893	892	891	890	890	889	888	887	887	886
11.3	885	884	883	883	882	881	880	880	879	878
11.4	877	876	876	875	874	873	873	872	871	870
11.5	870	869	868	867	867	866	865	864	864	863
11.6	862	861	861	860	859	858	858	857	856	855
11.7	855	854	853	853	852	851	850	850	849	848
11.8	847	847	846	845	845	844	843	842	842	841
11.9	840	840	839	838	838	837	836	835	835	834
12.0	833	833	832	831	831	830	829	829	828	827
12.1	826	826	825	824	824	823	822	822	821	820
12.2	820	819	818	818	817	816	816	815	814	814
12.3	813	812	812	811	810	810	809	808	808	807
12.4	806	806	805	805	804	803	803	802	801	801
12.5	800	799	799	798	797	797	796	796	795	794
12.6	794	793	792	792	791	791	790	789	789	788
12.7	787	787	786	786	785	784	784	783	782	782
12.8	781	781	780	779	779	778	778	777	776	776
12.9	775	775	774	773	773	772	772	771	770	770
13.0	769	769	768	767	767	766	766	765	765	764
13.1	763	763	762	762	761	760	760	759	759	758
13.2	758	757	756	756	755	755	754	754	753	752
13.3	752	751	751	750	750	749	749	748	747	747
13.4	746	746	745	745	744	743	743	742	742	741
13.5	741	740	740	739	739	738	737	737	736	736
13.6	735	735	734	734	733	733	732	732	731	730
13.7	730	729	729	728	728	727	727	726	726	725
13.8	725	724	724	723	723	722	722	721	720	720
13.9	719	719	718	718	717	717	716	716	715	715
14.0	714	714	713	713	712	712	711	711	710	710
14.1	709	709	708	708	707	707	706	706	705	705
14.2	704	704	703	703	702	702	701	701	700	700
14.3	699	699	698	698	697	697	696	696	695	695
14.4	694	694	693	693	693	692	692	691	691	690
14.5	690	689	689	688	688	687	687	686	686	685
14.6	685	684	684	684	683	683	682	682	681	681
14.7	680	680	679	679	678	678	678	677	677	676
14.8	676	675	675	674	674	673	673	672	672	672
14.9	671	671	670	670	669	669	668	668	668	667
	0	1	2	3	4	5	6	7	8	9

Wave number (cm⁻¹) → $\text{Wave number (cm}^{-1})$

Wavelength (μm)	0	1	2	3	4	5	6	7	8	9
15.0	666.7	666.2	665.8	665.3	664.9	664.5	664.0	663.6	663.1	662.7
15.1	662.3	661.8	661.4	660.9	660.5	660.1	659.6	659.2	658.8	658.3
15.2	657.9	657.5	657.0	656.6	656.2	655.7	655.2	654.9	654.5	654.0
15.3	653.6	653.2	652.7	652.3	651.9	651.5	651.0	650.6	650.2	649.8
15.4	649.4	648.9	648.5	648.1	647.7	647.2	646.8	646.4	646.0	645.6
15.5	645.2	644.7	644.3	643.9	643.5	643.1	642.7	642.3	641.8	641.4
15.6	641.0	640.6	640.2	639.8	639.4	639.0	638.6	638.2	637.8	637.3
15.7	636.9	636.5	636.1	635.7	635.3	634.9	634.5	634.1	633.7	633.3
15.8	632.9	632.5	632.1	631.7	631.3	630.9	630.5	630.1	629.7	629.3
15.9	628.9	628.5	628.1	627.7	627.4	627.0	626.6	626.2	625.8	625.4
16.0	625.0	624.6	624.2	623.8	623.4	623.1	622.7	622.3	621.9	621.5
16.1	621.1	620.7	620.3	620.0	619.6	619.2	618.8	618.4	618.0	617.7
16.2	617.3	616.9	616.5	616.1	615.8	615.4	615.0	614.6	614.3	613.9
16.3	613.5	613.1	612.7	612.4	612.0	611.6	611.2	610.9	610.5	610.1
16.4	609.8	609.4	609.0	608.6	608.3	607.9	607.5	607.2	606.8	606.4
16.5	606.1	605.7	605.3	605.0	604.6	604.2	603.9	603.5	603.1	602.8
16.6	602.4	602.0	601.7	601.3	601.0	600.6	600.2	599.9	599.5	599.2
16.7	598.8	598.4	598.1	597.7	597.4	597.0	596.7	596.3	595.9	595.6
16.8	595.2	594.9	594.5	594.2	593.8	593.5	593.1	592.8	592.4	592.1
16.9	591.7	591.4	591.0	590.7	590.3	590.0	589.6	589.3	588.9	588.6
17.0	588.2	587.9	587.5	587.2	586.9	586.5	586.2	585.8	585.5	585.1
17.1	584.8	584.5	584.1	583.8	583.4	583.1	582.8	582.4	582.1	581.7
17.2	581.4	581.1	580.7	580.4	580.0	579.7	579.4	579.0	578.7	578.4
17.3	578.0	577.7	577.4	577.0	576.7	576.4	576.0	575.7	575.4	575.0
17.4	574.7	574.4	574.1	573.7	573.4	573.1	572.7	572.4	572.1	571.8
17.5	571.4	571.1	570.8	570.5	570.1	569.8	569.5	569.2	568.8	568.5
17.6	568.2	567.9	567.5	567.2	566.9	566.6	566.3	565.9	565.6	565.3
17.7	565.0	564.7	564.3	564.0	563.7	563.4	563.1	562.7	562.4	562.1
17.8	561.8	561.5	561.2	560.9	560.5	560.2	559.9	559.6	559.3	559.0
17.9	558.7	558.3	558.0	557.7	557.4	557.1	556.8	556.5	556.2	555.9
18.0	555.6	555.2	554.9	554.6	554.3	554.0	553.7	553.4	553.1	552.8
18.1	552.5	552.2	551.9	551.6	551.3	551.0	550.7	550.4	550.1	549.8
18.2	549.5	549.1	548.8	548.5	548.2	547.9	547.6	547.3	547.0	546.7
18.3	546.4	546.1	545.9	545.6	545.3	545.0	544.7	544.4	544.1	543.8
18.4	543.5	543.2	542.9	542.6	542.3	542.0	541.7	541.4	541.1	540.8
18.5	540.5	540.2	540.0	539.7	539.4	539.1	538.8	538.5	538.2	537.9
18.6	537.6	537.3	537.1	536.8	536.5	536.2	535.9	535.6	535.3	535.0
18.7	534.8	534.5	543.2	533.9	533.6	533.3	533.0	532.8	532.5	532.2
18.8	531.9	531.6	531.3	531.1	530.8	530.5	530.2	529.9	529.7	529.4
18.9	529.1	528.8	528.5	528.3	528.0	527.7	527.4	527.1	526.9	526.6
19.0	526.3	526.0	525.8	525.5	525.2	524.9	524.7	524.4	524.1	538.8
19.1	523.6	523.3	523.0	522.7	522.5	522.2	521.9	521.6	521.4	521.1
19.2	520.8	520.6	520.3	520.0	519.8	519.5	519.2	518.9	518.7	518.4
19.3	518.1	517.9	517.6	517.3	517.1	516.8	516.5	516.3	516.0	515.7
19.4	515.4	515.2	514.9	514.7	514.4	514.1	513.9	513.6	513.3	513.1
19.5	512.8	512.6	512.3	512.0	511.8	511.5	511.2	511.0	510.7	510.5
19.6	510.2	509.9	509.7	509.4	509.2	508.9	508.6	508.4	508.1	507.9
19.7	507.6	507.4	507.1	506.8	506.6	506.3	506.1	505.8	505.6	505.3
19.8	505.1	504.8	504.5	504.3	504.0	503.8	503.5	503.3	503.0	502.8
19.9	502.5	502.3	502.0	501.8	501.5	501.3	501.0	500.8	500.5	500.3
	0	1	2	3	4	5	6	7	8	9

Wavelength (μm)	0	1	2	3	4	5	6	7	8	9
20.0	500.0	499.8	499.5	499.3	499.0	498.8	498.5	498.3	498.0	497.8
20.1	497.5	497.3	497.0	496.8	496.5	496.3	496.0	495.8	495.5	495.3
20.2	*495.0*	*494.8*	*494.6*	*494.3*	*494.1*	*493.8*	*493.6*	*493.3*	*498.1*	*492.9*
20.3	492.6	492.4	492.1	491.9	491.6	491.4	491.2	490.9	490.7	490.4
20.4	490.2	490.0	489.7	489.5	489.2	489.0	488.8	488.5	488.3	488.0
20.5	487.8	487.6	487.3	487.1	486.9	486.6	486.4	486.1	485.9	485.7
20.6	485.4	485.2	485.0	484.7	484.5	484.3	484.0	483.8	483.6	483.3
20.7	*483.1*	*482.9*	*482.6*	*482.4*	*482.2*	*481.9*	*481.7*	*481.5*	*481.2*	*481.0*
20.8	480.8	480.5	480.3	480.1	479.8	479.6	479.4	479.2	478.9	478.7
20.9	478.5	478.2	478.0	477.8	477.6	477.3	477.1	476.9	476.6	476.4
21.0	476.2	476.0	475.7	475.5	475.3	475.1	474.8	474.6	474.4	474.2
21.1	473.9	473.7	473.5	473.3	473.0	472.8	472.6	472.4	472.1	471.9
21.2	*471.7*	*471.5*	*471.8*	*471.0*	*470.8*	*470.6*	*470.4*	*470.1*	*469.9*	*469.7*
21.3	469.5	469.3	469.0	468.8	468.6	468.4	468.2	467.9	467.7	467.5
21.4	467.3	467.1	466.9	466.6	466.4	466.2	466.0	465.8	465.5	465.3
21.5	465.1	464.9	464.7	464.5	464.3	464.0	463.8	463.6	463.4	463.2
21.6	463.0	462.7	462.5	462.3	462.1	461.9	461.7	461.5	461.3	461.0
21.7	*460.8*	*460.6*	*460.4*	*460.2*	*460.0*	*459.8*	*459.6*	*459.3*	*459.1*	*458.9*
21.8	458.7	458.5	458.3	458.1	457.9	457.7	457.5	457.2	457.0	456.8
21.9	456.6	456.4	456.2	456.0	455.8	455.6	455.4	455.2	455.0	454.8
22.0	454.5	454.3	454.1	453.9	453.7	453.5	453.3	453.1	452.9	452.7
22.1	452.5	452.3	452.1	451.9	451.7	451.5	451.3	451.1	450.9	450.7
22.2	*450.5*	*450.2*	*450.0*	*449.8*	*449.6*	*449.4*	*449.2*	*449.0*	*448.8*	*448.6*
22.3	448.4	448.2	448.0	447.8	447.6	447.4	447.2	447.0	446.8	446.6
22.4	446.4	446.2	446.0	445.8	445.6	445.4	445.2	445.0	444.8	444.6
22.5	444.4	444.2	444.0	443.9	443.7	443.5	443.3	443.1	442.9	442.7
22.6	442.5	442.3	442.1	441.9	441.7	441.5	441.3	441.1	440.9	440.7
22.7	*440.5*	*440.3*	*440.1*	*439.9*	*439.8*	*439.6*	*439.4*	*439.2*	*439.0*	*438.8*
22.8	438.6	438.4	438.2	438.0	437.8	437.6	437.4	437.3	437.1	436.9
22.9	436.7	436.5	436.3	436.1	435.9	435.7	435.5	435.4	435.2	435.0
23.0	434.8	434.6	434.4	434.2	434.0	433.8	433.7	433.5	433.3	433.1
23.1	432.9	432.7	432.5	432.3	432.2	432.0	431.8	431.6	431.4	431.2
23.2	*431.0*	*430.8*	*430.7*	*430.5*	*430.3*	*430.1*	*429.9*	*429.7*	*429.6*	*429.4*
23.3	429.2	429.0	428.8	428.6	428.4	428.3	428.1	427.9	427.7	427.5
23.4	427.4	427.2	427.0	426.8	426.6	426.4	426.3	426.1	425.9	425.7
23.5	425.5	425.4	425.2	425.0	424.8	424.6	424.4	424.3	424.1	423.9
23.6	423.7	423.5	423.4	423.2	423.0	422.8	422.7	422.5	422.3	422.1
23.7	*421.9*	*421.8*	*421.6*	*421.4*	*421.2*	*421.1*	*420.9*	*420.7*	*420.5*	*420.3*
23.8	420.2	420.0	419.8	419.6	419.5	419.3	419.1	418.9	418.8	418.6
23.9	418.4	418.2	418.1	417.9	417.7	417.5	417.4	417.2	417.0	416.8
24.0	416.7	416.5	416.3	416.1	416.0	415.8	415.6	415.5	415.3	415.1
24.1	414.9	414.8	414.6	414.4	414.3	414.1	413.9	413.7	413.6	413.4
24.2	*413.2*	*413.1*	*412.9*	*412.7*	*412.5*	*412.4*	*412.2*	*412.0*	*411.9*	*411.7*
24.3	411.5	411.4	411.2	411.0	410.8	410.7	410.5	410.3	410.2	410.0
24.4	409.8	409.7	409.5	409.3	409.2	409.0	408.8	408.7	408.5	408.3
24.5	408.2	408.0	407.8	407.7	407.5	407.3	407.2	407.0	406.8	406.7
24.6	406.5	406.3	406.2	406.0	405.8	405.7	405.5	405.4	405.2	405.0
24.7	*404.9*	*404.7*	*404.5*	*404.4*	*404.2*	*404.0*	*403.9*	*403.7*	*403.6*	*403.4*
24.8	403.2	403.1	402.9	402.7	402.6	402.4	402.3	402.1	401.9	401.8
24.9	401.6	401.4	401.3	401.1	401.0	400.8	400.6	400.5	400.3	400.2
	0	1	2	3	4	5	6	7	8	9

four

magnetic resonance spectrometry

I. INTRODUCTION AND THEORY

Nuclear magnetic resonance (NMR) spectrometry is basically another form of absorption spectrometry, akin to infrared or ultraviolet spectrometry. Under appropriate conditions, a sample can absorb electromagnetic radiation in the radio-frequency region at frequencies governed by the characteristics of the sample. Absorption is a function of certain nuclei in the molecule. A plot of the frequencies of the absorption peaks versus peak intensities constitutes an NMR spectrum. This discussion will emphasize proton magnetic resonance (pmr) spectra.

With some mastery of basic theory, interpretation of NMR spectra merely by inspection is usually feasible in greater detail than is the case for infrared or ultraviolet spectra. To supplement the material presented here, several lucid nonmathematical introductions to nuclear magnetic resonance are recommended.[1–11,11a,11b] For greater depth, the classic treatise by Pople, Schneider, and Bernstein,[12] the work by Emsley, Feeney, and Sutcliff[12a] and the book by Carrington and McLachlan[11f] are available. Two volumes of NMR spectra have been published by Varian Associates.[13] These volumes are indexed by a unique code that shows the type of proton, nearest neighbor functional group, and next nearest neighbor. The well-known Sadtler IR and UV are supplemented by NMR spectra.[14]
Sets of spectra are also issued by the American Petroleum Institute.[15] Indexes to the NMR literature have been published.[16,16a,16b] Jackman and Sternhell's book[4] contains references to over 2600 papers.

The present account will suffice for the immediate limited objective: identification of organic compounds in conjunction with other spectrometric information.

We begin by describing some magnetic properties of nuclei. All nuclei carry a charge. In some nuclei this charge "spins" on the nuclear axis, and this circulation of nuclear charge generates a magnetic dipole along the axis (Figure

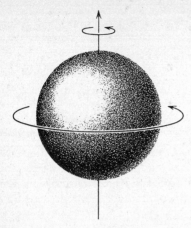

Fig. 1.
Spinning charge in proton generates magnetic dipole.

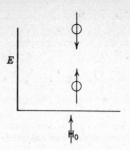

Fig. 2.
Energy levels of a proton.

1). The angular momentum of the spinning charge can be described in terms of spin numbers I; these numbers have values of 0, 1/2, 1, 3/2, and so forth ($I = 0$ denotes no spin). The intrinsic magnitude of the generated dipole is expressed in terms of nuclear magnetic moment, μ.

Each proton and neutron has its own spin, and I is a resultant of these spins. If the sum of protons and neutrons is even, I is zero or integral (0, 1, 2, . . .); if the sum is odd, I is half-integral (1/2, 3/2, 5/2, . . .); if both protons and neutrons are even-numbered, I is zero. Both ^{12}C and ^{16}O fall in the latter category and give no NMR signal.

Several nuclei (^1H, ^{19}F, ^{13}C, and ^{31}P) have a spin number I of 1/2 and a uniform spherical charge distribution (Figure 1). Nuclei with a spin number I of 1 or higher have a nonspherical charge distribution. This asymmetry is described by an electrical quadrupole moment which, as we shall see later, affects the relaxation time and, consequently, the coupling with neighboring nuclei. ^{14}N and ^2H have a spin number I of 1. ^{11}B, ^{35}Cl, ^{37}Cl, ^{79}Br, and ^{81}Br are examples of nuclei with $I = 3/2$.

The spin number I determines the number of orientations a nucleus may assume in an external uniform magnetic field in accordance with the formula $2I + 1$. We shall be concerned primarily with the proton whose spin number I is 1/2. Thus the proton has two orientations in an applied uniform magnetic field: parallel with the applied field (aligned with the field) or antiparallel (aligned against the field). The former is the low-energy (stable) state, the latter, the high-energy (unstable) state (Figure 2). The energy levels are a function of the magnitude of the nuclear magnetic moment μ, and the strength of the applied external magnetic field H_0.

Two energy levels for the proton having been established, it should now be possible to introduce quanta of energy $h\nu$ (h is Planck's constant; ν is the frequency of electromagnetic radiation) such that the parallel orientation (low-energy state) can be flipped to the antiparallel orientation (high-energy state) in a magnetic field of given strength H_0. The fundamental NMR equation correlating electromagnetic frequency with magnetic field strength is

$$\nu = \frac{\gamma H_0}{2\pi}$$

The constant γ is called the magnetogyric (or more commonly but less properly, gyromagnetic) ratio and is a fundamental nuclear constant; it is the proportionality constant between the magnetic moment μ and the spin number I

$$\gamma = \frac{2\pi\mu}{hI}$$

where h is Planck's constant.

The bald statement made earlier that nuclear magnetic resonance spectrometry is akin to other forms of absorption spectrometry may now seem somewhat more plausible. The problem is how to inject electromagnetic energy into protons aligned in a magnetic field so as to flip the proton spin into a higher energy level, and how to measure the energy thus absorbed. Before we describe the instrumentation, we have to consider one peculiarity of a small magnet spinning in an external magnetic field: The axis of the small magnet (the proton) will precess about the axis of the external magnetic field in the same manner in which a spinning gyroscope precesses under the influence of gravity (Figure 3). The precessional angular velocity, ω_0 is equal to the product of the magnetogyric ratio, γ and the strength of the applied magnetic field H_0.

$$\omega_0 = \gamma H_0$$

We recall from the fundamental NMR equation that

$$\gamma H_0 = 2\pi\nu$$

Therefore,

$$\omega_0 = 2\pi\nu$$

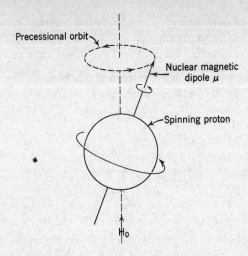

Fig. 3.
Proton precessing in a magnetic field H_0.

The precessional angular velocity is quantized, so that the difference between angular velocities in the ground state and excited state corresponds to a precise frequency (i.e., energy). Thus, if the proper frequency of electromagnetic radiation is introduced, the energy can be absorbed resulting in excitation of the spinning precessing nucleus to the next highest energy level. In pictorial terms, the precessing nucleus has "jumped" or "flipped" to a different orientation. The energy involved in this transition is about 10^{-6} kcal/mole. A frequency of 60 megaherz (MHz) is needed at a magnetic field H_0 of 14,092 gauss for the proton (or any other desired combination in the same ratio). When a proton on a magnetic field is subjected to electromagnetic radiation, a field strength of 14,092 gauss can be expressed as its equivalent, 60 MHz.

We are now in a position to arrange the geometry for a nuclear magnetic resonance experiment. We subject the protons to a powerful uniform magnetic field. The protons are now aligned with and against the field and are precessing about the axis of the applied magnetic field. Because of thermal disorder, actually only a small fraction of the total population of protons is properly aligned, but this fraction is sufficient. The electromagnetic frequency is applied in such a way that its magnetic component H_1 is at right angles to the main magnetic field H_0 and is rotating with the precessing proton. An oscillator coil whose axis is at right angles to the axis of the main magnetic field H_0 will generate a linear oscillating magnetic field H_1 along the direction of the coil axis as shown in Figure 4. A linear oscillating magnetic field can be resolved into two components rotating in opposite directions. One of these components is rotating in the same direction as the precessional orbit of the nuclear magnetic dipole (the proton); the oppositely rotating component of H_1 is disregarded. If H_0 is held constant and the oscillator frequency is varied, the angular velocity of the component of rotating magnetic field H_1 will vary until it is equal to (in resonance with) the angular velocity ω_0 of the precessing proton. At this point, energy is absorbed, the nucleus flips to its higher energy level, and the recorder shows a peak. Frequently, in practice, the oscillator frequency is held constant, and H_0 is swept over a narrow range.

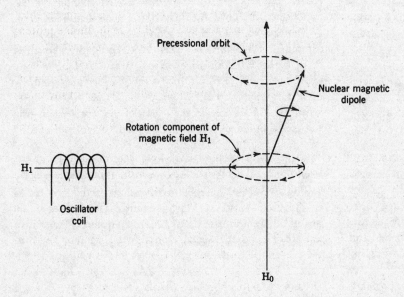

Fig. 4.
Oscillator generates rotating component of magnetic field H_1.

The experiment described and the explanation given represent a nonrigorous classical interpretation of quantum-mechanical phenomena. However, it is sufficient for our purposes and a more rigorous discussion can be found in references cited previously.

Now that we have briefly discussed how a nucleus is excited to a higher energy state by the absorption of energy, we need to account for the return of the nucleus to the ground state. In the absence of such a mechanism, all of the small excess population of nuclei in the lower-energy state will be raised to the higher-energy state, and no more energy will be absorbed. Fortunately, there exists a mechanism whereby the nucleus in the higher energy state can lose energy to its environment and thus return to its lower energy state. The mechanism is called a spin-lattice or longitudinal relaxation process and involves transfer of energy from the nucleus in its high-energy state to the molecular lattice. Its efficiency is described as the time T_1 taken for the transfer. An efficient relaxation process involves a short time T_1 and results in broadening of the absorption peak. The line width is inversely proportional to the lifetime of the excited state. In neat liquids, solutions, and gases, the time T_1 is of the proper duration to produce a peak of usable width. In solids, this mechanism is not effective; T_1 is therefore very long, and in the absence of any other effects, a crystalline solid would show extremely narrow lines. There is another effect, called spin-spin or transverse relaxation, that is especially important in solids. This involves transfer of energy from one high-energy nucleus to another. There is no net loss of energy, but the spread of energy among the nuclei concerned results in line broadening. In fact, this latter mechanism causes line broadening of such magnitude as to render NMR spectra of solids of little interest to the organic chemist.

We can now consider very briefly the instrumentation necessary to obtain a nuclear magnetic resonance spectrum.

II. APPARATUS AND SAMPLE HANDLING

A high-resolution (defined in Section III) nuclear magnetic resonance spectrometer is commercially available from several sources. Instruments fitted with permanent magnets with field strengths of about 14,000 gauss give proton spectra at 60 MHz. Electromagnets permit greater flexibility so that spectra can be obtained at various frequencies up to 100 MHz (23,500 gauss). Newer superconducting magnets allow for frequencies up to 220 and 300 MHz, but are not commonly available. In addition, ^{19}F, ^{11}B, ^{13}C, ^{2}H, ^{15}N, and ^{31}P nuclei can be examined at appropriate combinations of frequencies and magnetic field strengths.

A schematic diagram of an NMR spectrometer is shown in Figure 5. The instrument can be described in terms of the following components:

1. A strong magnet whose homogeneous field can be varied continuously and precisely over a relatively narrow range. This is accomplished by means of the sweep generator.
2. A radio-frequency oscillator.
3. A radio-frequency receiver.
4. A recorder, calibrator, and integrator.
5. A sample holder that positions the sample relative to the main magnetic field, the transmitter coil and the receiver coil. The sample holder also spins the sample to increase the apparent homogeneity of the magnetic field. A variable-temperature sample holder is also available.

Operational details will not be described. Suffice it to say, the desired frequency and field strength are selected, and the sweep generator sweeps the magnetic field over a narrow range near the field strength selected. Proton spectra are usually run at 60 MHz (megaherz) or 100 MHz and the usual range of field sweep is equivalent to about 1000 Hz (1 hertz, Hz is 1 cycle per second, cps) at 60 MHz or about 1700 Hz at 100 MHz. By measuring frequency shifts from a reference marker, an accuracy of ± 1 Hz can be achieved. The recording is presented as a series of peaks, the peak areas being proportional to the number of protons they represent. Peak areas are measured by an electronic integrator that traces a series of steps whose heights are proportional to the peak areas. The steps can be superimposed on the peaks. As will be obvious in Chapter 6, proton counting with the integrator is extremely useful. Peaks hidden under other peaks can thus be detected. Proton counting is often useful for determining sample purity and, of course, for quantitative analytical work.

The sample, a liquid or a solution in a suitable solvent, is contained in a 5-mm O.D. glass tube. Ordinarily about 0.4 ml of a neat liquid or somewhere between 10 to 50 mg of a liquid or a solid dissolved in 0.4 ml of a solvent is used. By using inert plugs to cut down on dead space—that is, by restricting the sample to the region of the receiver coil—the volume of liquid or solution can be reduced to about 0.025 ml.

A microtube (Nuclear Magnetic Resonance Specialities, New Kensington, Pa.) consisting of a thick-wall capillary leading to a 25-μl spherical cavity allows usable spectra to be obtained on about 1 mg of sample. To overcome the lack of inherent sensitivity of NMR spectrometry, repetitive scans can be used. Signals can thus be accumulated and noise averaged out. A small computer attachment (computer of average transients, CAT) is available as an accessory for this purpose. Usable spectra can be obtained on samples of the order of 150 μg.

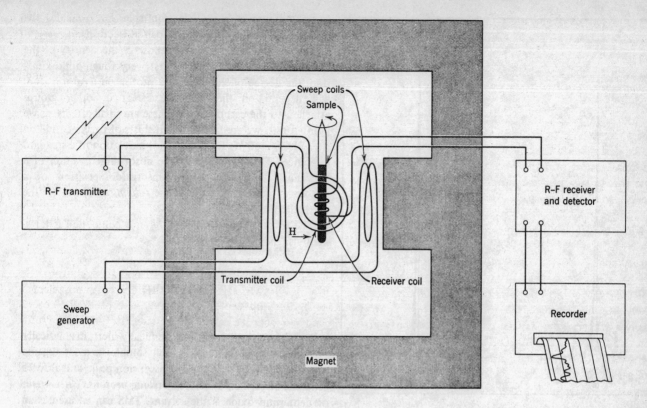

Fig. 5.
Schematic diagram of an NMR Spectrometer. (Courtesy of Varian Associates, Palo Alto, Calif.)

An important recent development for detection of weak signals is Fourier Transform (FT) NMR.[11e] A series of short rf pulses is used instead of a continuous signal. The rf pulse encompasses the entire frequency range in which the nuclei absorb energy. All the absorbing nuclei are excited. After the pulse of energy has passed, the nuclei return to the ground state by relaxation processes. The energy lost by the relaxing nuclei causes the Fourier transform signal. The output signal does not look like a conventional NMR signal. A pulse takes about a millisecond, and several hundred pulses (at suitable intervals) take only a few minutes. The repetitive output signals are stored by a computer. Then the signals are converted into a conventional NMR spectrum by a second computer. Usable spectra can be obtained on samples of the order of 10 to 50 μg.

The ideal solvent should contain no protons in its structure, be inexpensive, low-boiling, nonpolar, and inert. Carbon tetrachloride is ideal when the sample is sufficiently soluble in it. The most widely used solvent is deuterated chloroform ($CDCl_3$). The small sharp proton peak from the $CHCl_3$ impurity present (see Appendix A) rarely interferes seriously. Carbon disulfide is also a useful solvent. Almost all of the common solvents are available in the deuterated form with an isotopic purity (atom %D) of 98–99.8%. A list of these solvents and the positions of their proton impurity peaks are given in Appendix A.

Small "spinning side bands" (see Figure 6) are often seen, symmetrically disposed on both sides of a strong absorption peak; these result from inhomogeneities in the magnetic field and in the spinning tube. They are readily recognized because of their symmetrical appearance and because their separation from the absorption peak is proportional to the rate of spinning. The oscillations often seen at the high-field end of a strong sharp peak are called "ringing" (Figure 7). These are "beat" frequencies resulting from "fast" (normal operation) passage through the absorption peak.

Traces of ferromagnetic impurities cause severe broadening of absorption peaks. Common sources are tap water, steel wool, Raney nickel, and particles from metal spatulas or fittings (Figure 8). These impurities can be removed by dipping a thin bar magnet into the NMR tube, by filtration, or centrifugation.

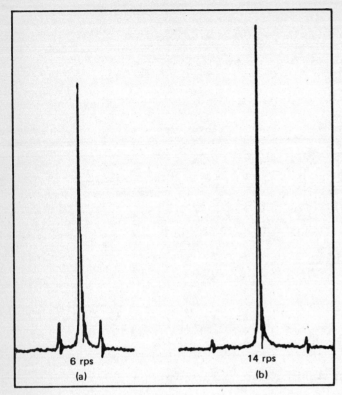

Fig. 6.
Signal of neat chloroform with spinning side bands produced by spinning rate of (a) 6 rps and (b) 14 rps. (From F. A. Bovey, NMR Spectrometry, copyright 1969, Academic Press.)

III. CHEMICAL SHIFT

Thus far, we have obtained a single peak from the interaction of radio frequency and a strong magnetic field on a proton in accordance with the basic NMR equation in which γ, the magnetogyric ratio, is an intrinsic property of the nucleus. The peak area (measured by the integrator) is proportional to the number of protons it represents. Fortunately, the situation is not quite so simple. The nucleus is shielded to a small extent by its electron cloud whose density varies with the environment. This variation gives rise to different absorption positions within the range of about 1000 Hz or so in a magnetic field corresponding to 60 MHz or about 1700 Hz in a field corresponding to 100 MHz. The ability to discriminate among the individual absorptions describes high resolution NMR spectrometry.

Electrons under the influence of a magnetic field will circulate, and, in circulating, will generate their own magnetic field opposing the applied field; hence, the shielding effect (Figure 9). This effect accounts for the diamagnetism exhibited by all organic materials. In the case of materials with an unpaired electron, the paramagnetism

associated with the net electron spin far overrides the diamagnetism of the circulating, paired electrons.

The degree of shielding depends on the density of the circulating electrons, and as a first, very rough approximation, the degree of shielding of a proton on a carbon atom will depend on the inductive effect of other groups attached to the carbon atom. These are small effects; as we pointed out, we are talking about shifts of parts per million (i.e., Hz in a 60- or 100-MHz field) in relation to a standard reference. The difference in the absorption position of a particular proton from the absorption position of a reference proton is called the *chemical shift* of the particular proton.

The most generally useful reference compound is tetramethylsilane (TMS).

$$CH_3 - \overset{\overset{\displaystyle CH_3}{|}}{\underset{\underset{\displaystyle CH_3}{|}}{Si}} - CH_3$$

It has several advantages: it is chemically inert, magnetically isotropic, volatile (b.p. 27°), and soluble in most organic solvents; it gives a single sharp absorption peak, and absorbs at higher field than almost all organic protons. When water or deuterium oxide is the solvent, TMS can be used as an "external reference," i.e., sealed in a capillary immersed in the solution. The methyl protons of sodium 2,2-dimethyl-2-silapentane-5-sulfonate (DSS):

$$(CH_3)_3 SiCH_2 CH_2 CH_2 SO_3 Na$$

are sometimes used as an internal reference in aqueous solution. Since only enough is used to give a small methyl peak, the diffuse pattern of the CH_2 peaks barely shows on the base line. Unless hydrogen-bonding effects are involved, a proton peak referenced to TMS in deuterochloroform will be within 0.01 to 0.03 ppm of the same peak referenced to DSS in water or deuterium oxide. Acetonitrile and dioxane are also used as references in aqueous solution.

Let us set up an NMR scale (Figure 10) and set the TMS peak at 0 Hz at the right-hand edge. The magnetic field increases toward the right. When chemical shifts are given in Hz (designated ν), the applied frequency must be specified. Chemical shifts can be expressed in dimensionless units (δ), independent of the applied frequency, by dividing ν by the applied frequency and multiplying by 10^6. Thus a peak at 60 Hz (ν 60) from TMS at an applied frequency of 60 MHz would be at δ 1.00 or 1.00 ppm.

$$\delta \text{ or ppm} = \frac{60}{60 \times 10^6} \times 10^6 = 1.00$$

Since δ units are expressed in parts per million, the expression ppm is often used. The same peak at an applied frequency of 100 MHz would be at ν 100, but would still be at δ 1.00.

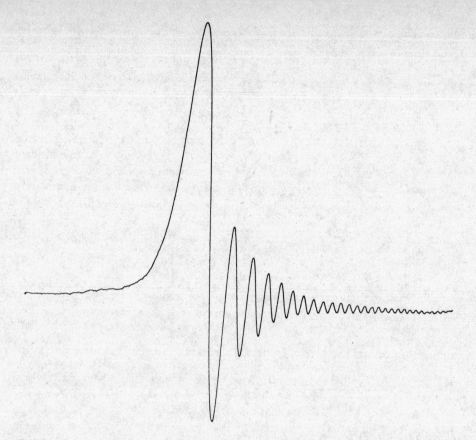

Fig. 7.
Ringing (or "wiggles") seen after passage through resonance. Direction of scan from left to right. (From E. C. Becker, High Resolution NMR, *copyright 1969, Academic Press.)*

$$\delta \text{ or ppm} = \frac{100}{100 \times 10^6} \times 10^6 = 1.00$$

This scheme has been criticized because δ values increase in the downfield direction; the rejoinder is that these are really negative numbers. The other commonly used system assigns a value of 10.00 for tetramethylsilane, and describes chemical shifts in terms of τ values.

$$\tau = 10.00 - \delta$$

It should be noted that δ is treated as a positive number. We shall make our assignments in both δ and τ values. Shifts at higher field than TMS (δ 0.00, τ 10.00) will be encountered very rarely; δ values are then shown with a negative sign, and τ values merely increase numerically.

It is important to realize that the chemical shift in Hz is directly proportional to the strength of the applied field H_0, and therefore to the applied frequency. This is understandable because the chemical shift is dependent on the diamagnetic shielding induced by H_0. The strongest magnetic field consistent with field homogeneity should be used to spread out the chemical shifts. This is made clear in

Figure 11 in which increased applied magnetic field in the NMR spectrum of acrylonitrile means increased separation of signals.

The concept of electronegativity is a dependable guide, up to a point, to chemical shifts. It tells us that the electron density around the protons of TMS is high (silicon is electropositive relative to carbon), and these protons will therefore be highly shielded and their peak will be found at high field. We could make a number of good estimates as to chemical shifts, using concepts of electronegativity and proton acidity. For example, the following values are reasonable on these grounds:

	δ	τ
$(CH_3)_2O$	3.27	6.73
CH_3F	4.30	5.70
RCOOH	10.8 (approx.)	-0.8

But finding the protons of acetylene at δ 2.35, τ 7.65, that is more shielded than ethylene protons (δ 4.60, τ 5.40), is

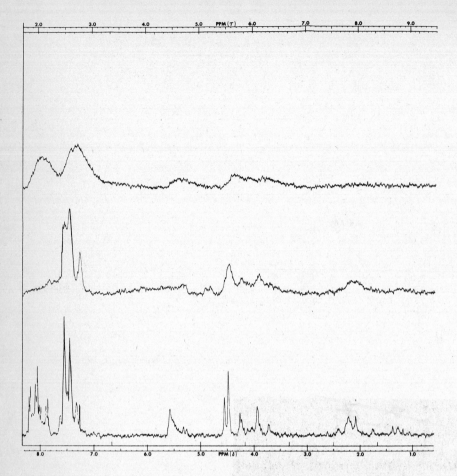

Fig. 8.
The effect of a tiny ferromagnetic particle on the proton resonance spectrum of a benzoylated sugar. The top and middle curves are repeat runs with the particle present; the bottom curve is the spectrum with the particle removed. (From E. C. Becker, High Resolution NMR, copyright 1969, Academic Press.)

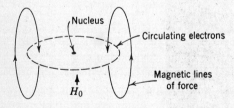

Fig. 9.
Diamagnetic shielding of nucleus by circulating electrons.

unsettling. And finding the aldehydic proton of acetaldehyde at δ 9.97, τ 0.03 definitely calls for some augmentation of the electronegativity concept. We shall use diamagnetic anisotropy to explain these and other apparent anomalies, such as the unexpectedly large deshielding effect of the benzene ring (benzene protons δ 7.27, τ 2.73).

Let us begin with acetylene. The molecule is linear, and the triple bond is symmetrical about the axis. If this axis is aligned with the applied magnetic field, the π-electrons of the bond can circulate at right angles to the applied field, thus inducing their own magnetic field opposing the applied field. Since the protons lie along the magnetic axis, the magnetic lines of force induced by the circulating electrons act to shield the protons (Figure 12) and the NMR peak is found further upfield than electronegativity would predict. Of course, only a small number of the rapidly tumbling molecules are aligned with the magnetic field, but the overall average shift is affected by the aligned molecules.

This effect depends upon diamagnetic anisotropy, which means that shielding and deshielding depend on the orientation of the molecule with respect to the applied magnetic field. Similar arguments can be adduced to rationalize the unexpected low field position of the aldehydic proton. In this case, the effect of the applied

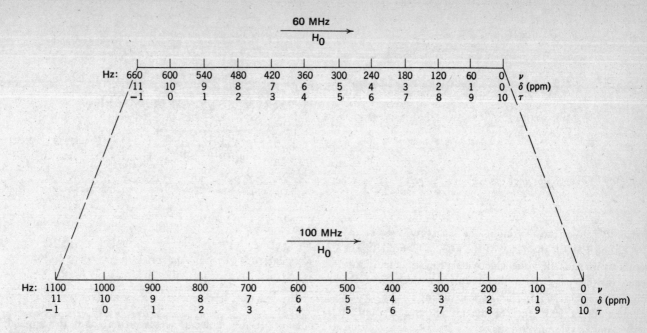

Fig. 10.
NMR Scale at 60 MHz and 100 MHz.

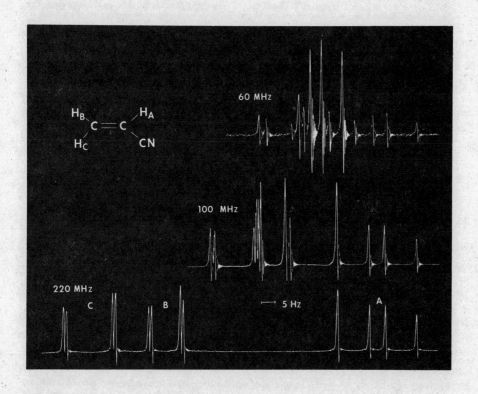

Fig. 11.
60-, 100-, and 220-MHz spectra of acrylonitrile. (Reprinted from Anal. Chem. *Copyright 1971 by the American Chemical Society. Reprinted by permission of the copyright owner.)*

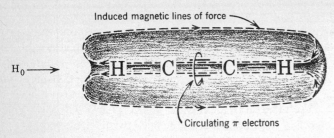

Fig. 12.
Shielding of acetylenic protons.

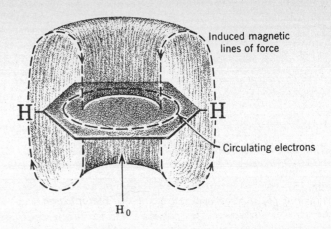

Fig. 14.
Ring current effects in benzene.

magnetic field is greatest along the transverse axis of the C=O bond (i.e., in the plane of the page in Figure 13). The geometry is such that the aldehydic proton, which lies in front of the page, is in the deshielding portion of the induced magnetic field. The same argument can be used to account for at least part of the rather large amount of deshielding of olefinic protons.

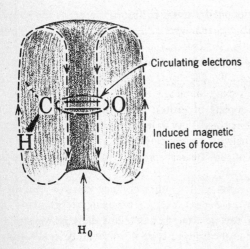

Fig. 13.
Deshielding of aldehydic protons.

The so-called "ring-current effect" is another example of diamagnetic anisotropy and accounts for the large deshielding of benzene ring protons. Figure 14 shows this effect. It also indicates that a proton held directly above or below the ring should be shielded. This has actually been found to be the case for some of the methylene protons in 1,4-polymethylenebenzenes.

All the ring protons of acetophenone are found downfield because of the ring current effect. Moreover, the ortho protons are shifted slightly further downfield (meta, para δ ~ 7.40, τ 2.60; ortho δ ~ 7.85, τ 2.15) because of the additional deshielding effect of the carbonyl group. In Figure 15 the carbonyl bond and the benzene ring are coplanar. If the molecule is oriented so that the applied magnetic field H_0 is perpendicular to the plane of the molecule, the circulating π electrons of the C=O bond shield the conical zones above and below them, and deshield the lateral zones in which the ortho proton is located. Both ortho protons are equally deshielded since another, equally populated, conformation can be written in which the "lefthand" ortho proton is more proximate to the anisotropy cone.

A spectacular example of shielding and deshielding by ring currents is furnished by some of the annulenes.[17] The protons outside the ring of [18] annulene are strongly deshielded (δ 8.9, τ 1.1), and those inside are strongly shielded (δ − 1.8, τ 11.8).

[18] Annulene

Demonstration of such a ring current is probably the best evidence available for aromaticity.

In contrast with the striking anisotropic effects of circulating π electrons, the σ electrons of a C−C bond produce a small effect. The axis of the C−C bond is the axis of the deshielding cone (Figure 16).

This figure accounts for the deshielding effect of successive alkyl substituents on a proton attached to a carbon atom. Thus the protons are found progressively downfield in the sequence RCH_3, R_2CH_2, and R_3CH. The

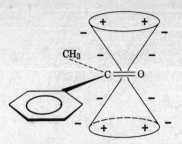

Fig. 15.
Shielding (+) and deshielding (−) zones of acetophenone.

observation that an equatorial proton is consistently found further downfield by 0.1–0.7 ppm than the axial proton on the same carbon atom in a rigid six-membered ring can also be rationalized (Figure 17). The axial and equatorial proton on C_1 are oriented similarly with respect to C_1–C_2 and C_1–C_6, but the equatorial proton is within the deshielding cone of the C_2–C_3 bond (and C_5–C_6).

Chemical shifts of protons bound to carbon and near a single functional group are shown in chart form in Appendix B. Values assigned (δ and τ) are rough averages designed to indicate a region rather than an exact number. Except where otherwise specified, the shifts shown are for

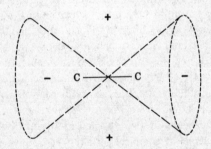

Fig. 16.
Shielding (+) and deshielding (−) zones of C—C.

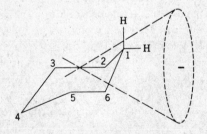

Fig. 17.
Deshielding of equatorial proton of a rigid six-membered ring.

aliphatic compounds in deuterochloroform or carbon tetrachloride. Variations due to changes in concentration are usually small in the absence of hydrogen bonding effects. Solvent changes often cause appreciable changes in shift position. A separate value is given for methyl, methylene, and methine protons. Inspection of these charts gives the very useful general impression that chemical shifts of protons in organic compounds fall roughly into eight regions as summarized in Figure 18.

Appendix C gives shift positions for methylene protons carrying two functional groups. The calculated values were obtained from the set of Shoolery's constants,[18] also presented in Appendix C. Appendix D lists shift ranges for protons subject to hydrogen bonding effects, and Appendix E has data useful for predicting olefinic proton chemical shifts.

IV. SIMPLE SPIN-SPIN COUPLING

We have obtained a series of absorption peaks representing protons in different chemical environments: each absorption area is proportional to the number of protons it represents. This achievement alone furnishes considerable information. We have now to consider one further refinement, spin-spin coupling. This can be described as the indirect coupling of proton spins through the intervening bonding electrons. Very briefly, it occurs because there is some tendency for a bonding electron to pair its spin with the spin of the nearest proton; the spin of a bonding electron, having been thus influenced, the electron will affect the spin of the other bonding electron and so on through to the next proton. Coupling is ordinarily not important beyond three bonds unless there is ring strain as in small rings or bridged systems, or bond delocalization as in aromatic or unsaturated systems.

Suppose two protons are in very different chemical environments from one another as in the compound
$$\text{RO–CH–CH–CR}_3$$
with OR and CR_3 substituents. Each proton will give rise to an absorption, and the absorptions will be quite widely separated. But the spin of each proton is affected slightly by the two orientations of the other proton through the intervening electrons so that each absorption appears as a doublet (Figure 19). The distance between the component peaks of a doublet is proportional to the effectiveness of the coupling, and is denoted by a coupling constant J, which is independent of the applied magnetic field H_0. Whereas chemical shifts can range over about 1700 Hz at 100 MHz, coupling constants between protons rarely exceed 20 Hz (see Appendix F). So long as the chemical shift difference in Hz is much larger than the coupling

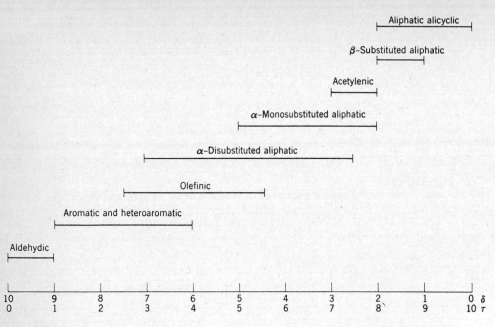

Fig. 18.
General regions of chemical shifts.

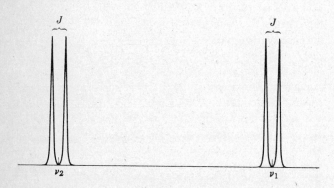

Fig. 19.
Spin-spin coupling between two protons with very different chemical shifts.

constant ($\Delta\nu/J$ is greater than about 10), the simple pattern of two doublets appears. As $\Delta\nu/J$ becomes smaller, the doublets approach one another, the inner two peaks increase in intensity, and the outer two peaks decrease (Figure 20). The shift position of each proton is no longer midway between its two peaks as was the case in Figure 19 but is at the "center of gravity" (Figure 21); it can be estimated with fair accuracy by inspection, or determined precisely by the following formula in which the peak positions (1, 2, 3, and 4 from left to right) are given in Hz from the reference.

$$(1-3) = (2-4) = \sqrt{(\Delta\nu)^2 + J^2}$$

The shift position of each proton is $\Delta\nu/2$ from the mid-point of the pattern. When $\Delta\nu = J\sqrt{3}$, the two pairs resemble a quartet resulting from splitting by three equivalent vicinal protons. Failure to note the small outer peaks (i.e., 1 and 4) may lead to mistaking the two large inner peaks for a doublet. When the chemical shift difference becomes zero, the middle peaks coalesce to give a single peak and the end peaks vanish—that is, the protons are equivalent. (Equivalent protons do spin-spin couple with one another, but splitting is not observed). A further point to be noted is the obvious one that the spacing between the peaks of two coupled multiplets is the same. The dependence of chemical shift on the applied magnetic field and the independence of the spin-spin coupling afford a method of distinguishing between them. The spectrum is merely run at two different applied magnetic fields. Chemical shifts are also solvent dependent, but J values are usually only slightly affected by change of solvent, at least to a lesser degree than are chemical shifts. The chemical shift of the methyl and acetylenic protons of methylacetylene are (fortuitously) coincident (δ 1.80, γ 8.20) when the spectrum is obtained in a $CDCl_3$ solvent; the spectrum of a neat sample of this alkyne shows the acetylenic proton at δ 1.80 (γ 8.20) and the methyl protons at δ 1.76 (γ 8.24). Fig. 22 illustrates the chemical shift dependence of the protons of biacetyl on solvent. The change from a chlorinated solvent (e.g. $CDCl_3$) to an aromatic solvent (e.g. C_6D_6) often drastically influences the position and appearance of NMR signals.[1]

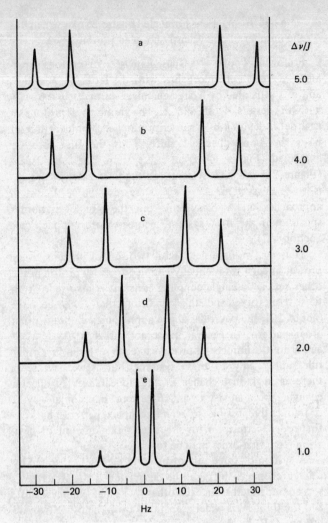

Fig. 20.
*Change in an AX system spin coupling with a decreasing
difference in chemical shifts and a large J value (10 Hz); the
AX notation is explained in the text.*

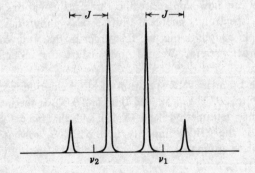

Fig. 21.
*"Center of gravity," instead of linear midpoints, for shift
position location (due to "low" $\Delta v/J$ ratio).*

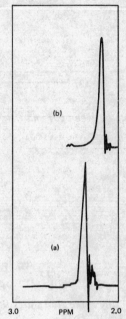

Fig. 22.
*The NMR spectrum of biacetyl (2,3-butanedione): (a)
$CDCl_3$; (b) C_6D_6.*

Look at the next stage in complexity of spin-spin
coupling (Figure 23). Consider the system $-HC-CH_2-$ in
the compound $RO-\overset{\displaystyle OR}{\underset{\displaystyle |}{CH}}-CH_2-CR_3$ in which the single
methine proton is in a very different chemical environment
from the two methylene protons. As before, we see two
sets of absorptions widely separated, and now the absorption areas are in the ratio of 1:2. The methine proton
couples with the methylene protons and splits the methylene proton absorption into a symmetrical doublet, as
explained above. The two methylene protons split the
methine proton absorption into a triplet because three
combinations of proton spins exist in the two methylene
protons (*a* and *b*) of Figure 24. Since there are two
equivalent combinations of spin (pairs 2 and 3) that do not
produce any net opposing or concerted field relative to the
applied field, there is an absorption of relative intensity two
at the center of the multiplet. Since there are single pairs (1
and 4) respectively opposed and in concert with the applied
field, there are equally spaced (*J*) lines of relative intensity
one upfield and one downfield from the center line. In
summary, the intensities of the peaks in the triplet are in
the ratio 1:2:1.

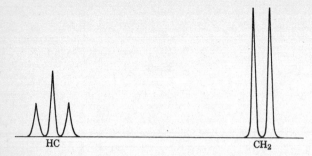

Fig. 23.
Spin-spin coupling between CH and CH₂ with very different chemical shifts.

$$\begin{array}{c}
\rightarrow b \\
\rightarrow a \\
\underset{2}{\overset{\rightarrow}{H_0}} \quad \overset{\leftarrow b}{\underset{\rightarrow a}{}} \quad^4 \quad \overset{\rightarrow b}{\underset{3}{\leftarrow a}} \\
\underset{1}{\overset{\leftarrow b}{\leftarrow a}}
\end{array}$$

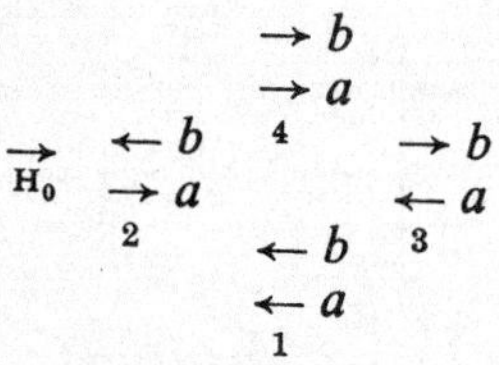

Fig. 24.
Energy levels for the three spin states of a methylene group (protons a and b).

When the methine and methylene protons in the system $-CH-CH_2-$ are in similar environments (i.e., $\Delta \nu/J$ is small), the simple doublet-triplet pattern degenerates to a complex pattern of from seven to nine lines as a result of second-order splitting; analysis by inspection is no longer possible, since the peak spacings may not correspond to the coupling constants.

Simple splitting patterns that are produced by the coupling of protons that have very different chemical shifts ($\Delta \nu/J$ is greater than about 10 or so) are called *first order* splitting patterns. These can usually be interpreted by using two rules:

1. Splitting of a proton absorption is done by neighboring protons, and the multiplicity of the split is determined by the number of these protons. Thus, one proton causes a doublet, and two equally coupled protons cause a triplet. The multiplicity then is $n + 1$, n being the number of neighboring equally coupled protons. The general formula, which covers all nuclei is $2nI + 1$, I being the spin number.

2. The relative intensities of the peaks of a multiplet also depend on n. We have seen that doublet ($n = 1$) peaks are in the ratio 1:1, and triplet peaks are in the ratio 1:2:1. Quartets are in the ratio 1:3:3:1. The general formula is $(a + b)^n$; when this is expanded to the desired value of n, the coefficients give the relative intensities. The multiplicity and relative intensities may be easily

obtained from Pascal's triangle (Figure 25), in which n is the number of equivalently coupled protons.

Following Pople,[12] we designate sets of protons separated by a small chemical shift with the letters A, B, and C, and sets separated by a large chemical shift ($\Delta \nu/J > 6$ or 7) with the letters A, M, and X. The number of protons in each set is denoted by a subscript number. Members of a set have the same chemical shift. Thus the first case we examined (Figure 19) is an AX system. The second case (Figure 21) is an AB system, and the third case (Figure 23) is an A_2X system. As $\Delta \nu/J$ decreases, the A_2X system approaches an A_2B system, and the simple first-order splitting of the A_2X system becomes more complex (see Section VII).

Thus far, we have dealt with two sets of protons; every proton in each set is equally coupled to every proton in the other set, i.e. a single coupling constant is involved. Given these conditions and the condition that $\Delta \nu/J$ be large (about 10), the two rules above apply, and we obtain a first order pattern. In general, these are the A_aX_x systems (a and x are the number of protons in each set); the first order rules apply only to these systems, but as we have seen, there is a gradual change in the appearances of spectra changing from an AX to an AB pattern. In a similar way, it is frequently possible to relate complex patterns back to first-order patterns. With practice, a fair amount of deviation from first-order may be tolerated.

A system of three sets of protons, each set separated by a large chemical shift, can be designated $A_aM_mX_x$. If two sets are separated from each other by a small chemical shift, and the third set is widely separated from the other two, we use an $A_aB_bX_x$ designation. And if all shift positions are close, the system is $A_aB_bC_c$. Both end sets may be coupled to the middle set with the same or different coupling constants, whereas the end sets may or may not be coupled to one another. AMX systems are first-order; ABX systems approximate first-order, but ABC systems cannot be analyzed by inspection. These more complex patterns are treated in Section VII.

We can now appreciate the three main features of an NMR spectrum: chemical shifts, peak intensities, and spin-spin couplings that are first-order or that approximate first-order patterns. We can now analyze first-order NMR spectra.

The 60 MHz NMR spectrum of ethyl chloride is shown in Figure 26. The peak at δ 0.00 (τ 10.00) is the internal reference tetramethylsilane (TMS) and that at δ 7.25 (τ 2.75) is the $CHCl_3$ impurity in the $CDCl_3$ solvent. The methylene group (α)

$$\overset{\beta}{C}H_3 - \overset{\alpha}{C}H_2 - Cl$$

(δ 3.57, τ 6.43) is more deshielded by the chlorine than is the methyl group (δ 1.48, τ 8.52, see Appendix B) and thus

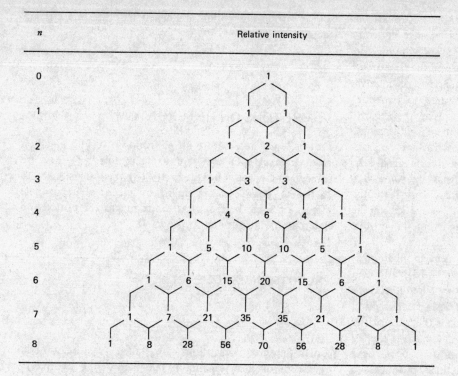

n	Relative intensity

Fig. 25. Pascal's triangle.
Relative intensities of first-order multiplets from coupling with n nuclei of spin 1/2 (e.g., protons).

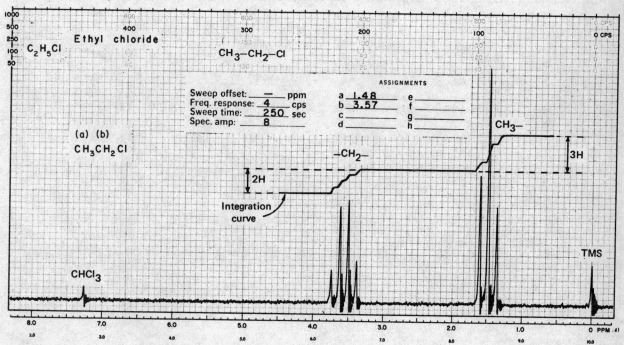

Fig. 26.
Ethyl chloride in $CDCl_3$ at 60 MHz. (Courtesy of Varian Associates, Palo Alto, Calif.)

the methylene group is downfield (lower τ, higher δ) from the methyl by 2.09 ppm. or about 125 Hz. Since the coupling is about 9 Hz, $\Delta \nu / J$ is about 14, a large enough ratio for first order analysis. The system is A_3X_2 and the first order rules correctly predict a triplet and a quartet with relative intensities (see integration curves) of three to two corresponding to the number of protons causing the absorptions. Note that even at $\Delta \nu / J = 14$ that there is a "leaning" of the two interacting signals toward each other; i.e., the intensity of the upfield lines of the methylene signal and the downfield lines of the methyl signal are somewhat larger than they would be if these signals were perfectly symmetrical. This fact, together with the same spacing in both multiplets, is valuable in identifying interacting proton signals in more complex spectra.

Consider the NMR spectrum of cumene in Figure 27. The five aromatic protons, δ 7.25, (τ 2.75), although actually not chemical shift equivalent (see Section VII) are coincidentally equivalent and occur as a single absorption downfield from the remaining absorptions (because of the benzene ring current, Figure 14). The side chain is treated as an A_6X system. The six methyl protons occur as a doublet at δ 1.25 (τ 8.75), the methine proton as a 1:6:15:20:15:6:1 septet at δ 2.90 (τ 7.10). Note that this signal is completely seen only when the sample is run at high gain (upper lines). Outer lines of complex multiplets may be overlooked, especially when these lines are part of a

single proton absorption and when base line noise is substantial.

V. PROTONS ON HETEROATOMS

Protons on a heteroatom differ from protons on a carbon atom in that: (1) they are exchangeable, (2) they are subject to hydrogen-bonding, and (3) they are subject to partial or complete decoupling by electrical quadrupole effects of some heteroatoms. Shift ranges for protons on heteroatoms are given in Appendix D.

PROTONS ON OXYGEN

Alcohols

Unless special precautions are taken (see below) the spectrum of neat ethanol usually shows the hydroxylic proton as a sharp peak at δ 5.35, τ 4.65. At the commonly used concentration of about 5 to 20% in a nonpolar solvent (e.g., carbon tetrachloride or deuterochloroform), the hydroxylic peak is found between δ 2, τ 8, and δ 4, τ 6. On

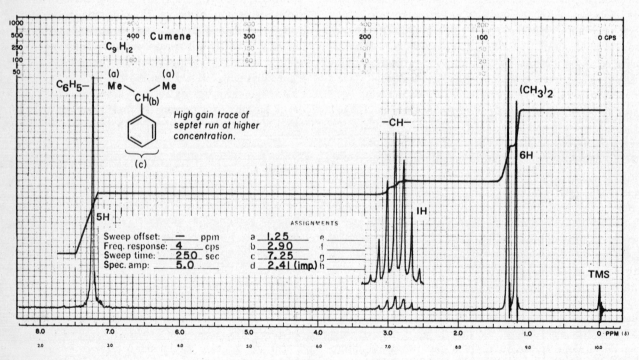

Fig. 27.
Cumene in CDCl$_3$ at 60 MHz. (Courtesy of Varian Associates, Palo Alto, Calif.)

extrapolation to infinite dilution or in the vapor phase, the peak is near δ 0.5, τ 9.5. A change in solvent or temperature will also shift the hydroxylic peak.

Hydrogen bonding explains why the shift position of the hydroxylic proton depends on concentration, temperature, and solvent. Hydrogen bonding decreases the electron density around the proton and thus moves the proton absorption to lower field. The extent of intermolecular hydrogen bonding is decreased by dilution with a nonpolar solvent and with increased temperature. Polar solvents introduce the additional complication of hydrogen bonding between the hydroxylic proton and the solvent. Intramolecular hydrogen bonds are less affected by their environment than intermolecular bonds; in fact the enolic hydroxylic absorption of β-diketones, for example, is hardly affected by change of concentration or solvent, though it can be shifted upfield somewhat by warming. NMR spectrometry is a powerful tool for studying such effects as hydrogen bonding and keto-enol tautomerism.

Exchangeability explains why the hydroxylic peak of ethanol is usually seen as a sharp singlet. Under ordinary conditions, enough acidic impurities are present in solution to catalyze rapid exchange of the hydroxylic proton. The proton is not on the oxygen atom long enough for it to "see" the three states of the methylene protons, and there is no coupling. The rate of exchange can be decreased by treating the solution or the solvent with anhydrous sodium carbonate, alumina, or molecular sieves immediately before obtaining the spectrum.[19,20] Purified deuterated dimethyl sulfoxide or acetone, in addition to allowing a slower rate of exchange, shifts the hydroxylic proton to lower field, even in dilute solution, by hydrogen-bonding between solute and solvent.[21,22] Since the hydroxylic proton can now "see" the protons on the α-carbon, a primary alcohol will show a triplet, a secondary alcohol a doublet, and a tertiary alcohol a singlet. This is illustrated in Figure 28[11d] and a list of successful applications is given in Table I. Exceptions have been reported;[23] these may be due to the sensitive concentration dependence of this phenomenon.[23a] At intermediate rates of exchange, the multiplet merges into a broad absorption band; at this point, the exchange rate in Hz is equal to $\pi J/\sqrt{2}$.

A dihydroxy alcohol may show separate absorption peaks for each hydroxylic proton; in this case, the rate of exchange in Hz is much less than the difference in Hz between the separate absorptions. As the rate increases, the two absorption peaks broaden, then merge to form a single broad peak; at this point, the exchange rate in Hz is equal to the original separation in Hz. The relative position of each peak depends on the extent of hydrogen bonding of each hydroxylic proton; steric hindrance to hydrogen bonding frequently accounts for a relative upfield absorption.

Table I[22] Hydroxyl Proton Resonances in Dimethyl Sulfoxide

Compound[a]	Chemical Shift, τ	Multi-plicity[b]
Methanol	5.92	q
Ethanol	5.65	t
Isopropyl alcohol	5.65	d
t-Butyl alcohol	5.84	s
t-Amyl alcohol	6.01	s
Propylene glycol, 1-OH	5.55	t
2-OH	5.62	d
Cyclohexanol	5.62	d
cis-4-t-Butylcyclohexyl alcohol	5.89	d
trans-4-t-Butylcyclohexyl alcohol	5.55	d
Benzyl alcohol	4.84	t
Phenol	0.75	s
β-L-Arabopyranose, O—C—OH[c]	4.02	d
α-D-Glucopyranose, O—C—OH[c]	3.84	d
α-D-Fructopyranose, O—C—OH[c]	4.88	s

[a]All spectra were taken of dimethyl sulfoxide solutions with concentrations 10 mole % or less.
[b]q = quartet, t = triplet, d = doublet, s = singlet. All splittings fall within the range 3.5–5.0 Hz.
[c]Other hydroxyl protons (τ 5.0–5.9) show splitting but overlap to such an extent that specific assignments are not possible.

The spectrum of a compound containing exchangeable protons can be simplified, and the exchangeable proton absorption removed, simply by shaking the solution with excess deuterium oxide or by obtaining a spectrum in deuterium oxide solution if the compound is soluble. A peak due to HOD will appear, generally between δ 5, τ 5, and δ 4.5, τ 5.5 in nonpolar solvents, and near δ 3.3, τ 6.7 in dimethyl sulfoxide (see Appendix D).

Acetylation or benzoylation of a hydroxyl group moves the α protons of a primary alcohol downfield about 0.5 ppm, and that of a secondary alcohol about 1.0–1.2 ppm.

Phenols

The behavior of a phenolic proton resembles that of an alcoholic proton. The phenolic proton peak is usually a sharp singlet (rapid exchange, no coupling) and its range, depending on concentration, solvent and temperature, is generally downfield (δ ∼7.5, τ 2.5 to δ ∼4.0, τ 6.0) compared with the alcoholic proton. This is illustrated in Figure 29. Note the concentration dependence of the OH peak. A carbonyl group in the ortho position shifts the phenolic proton absorption downfield to the range of about δ 12.0, τ −2.0 to δ 10.0, τ 0.0 because of intramolecular hydrogen bonding. Thus o-hydroxyacetophenone shows a

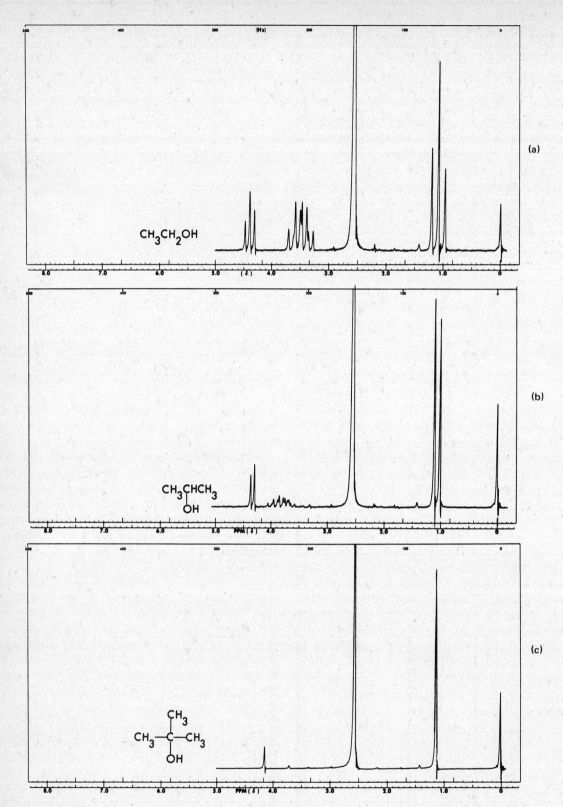

Fig. 28.
NMR spectra of typical primary, secondary, and tertiary alcohols run in dimethyl sulfoxide. Absorption at δ 2.6 is due to dimethyl sulfoxide.   (Source: Daniel J. Pasto and Carl R. Johnson, Organic Structure Determination, © 1969, pp. 358-359. Reprinted by permission of Prentice-Hall, Inc., Englewood Cliffs, N.J.)

spectrometric identification of organic compounds

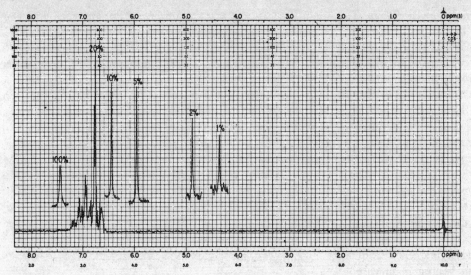

Fig. 29.

Phenol, in CCl₄, at various w/v %, at 60 MHz. Complete sweep is at 20%; single absorptions represent the OH proton at the indicated w/v %. (From J. R. Dyer, Applications of Absorption Spectroscopy of Organic Compounds, *copyright 1965, p. 90. Reprinted by permission of Prentice-Hall, Inc., Englewood Cliffs, N.J.)*

peak at about δ 12.05, τ − 2.05 almost completely invariant with concentration. The much weaker intramolecular hydrogen bonding in *o*-chlorophenol explains its shift range ($\delta \sim 6.3$, τ 3.7 at 1 molar concentration to $\delta \sim 5.6$, τ 4.4 at infinite dilution) which is broad compared with that of *o*-hydroxyacetophenone, but narrow compared with that of phenol.

Enols

Enols are usually stabilized by intramolecular hydrogen-bonding, which varies from very strong in aliphatic β-diketones to weak in cyclic α-diketones. The enolic proton is downfield relative to alcohol protons and, in the case of the enolic form of some β-diketones, may be found as far downfield as δ 16.6, τ − 6.6 (the enolic proton of acetylacetone absorbs at δ 15.0, τ − 5.0, and that of dibenzoylmethane at δ 16.6, τ − 6.6). The enolic proton peak is frequently broad at room temperature because of slow exchange. Furthermore, the keto-enol conversion is slow enough so that absorption peaks of both forms can be observed, and the equilibrium measured.

When strong intramolecular bonding is not involved, the enolic proton absorbs in about the same range as the phenolic proton.

Carboxylic Acids

Carboxylic acids exist as stable hydrogen-bonded dimers in nonpolar solvents even at high dilution. The carboxylic proton therefore absorbs in a characteristically narrow range, $\delta \sim 13.2$, τ − 3.2 to $\delta \sim 10.0$, τ 0.0, and is affected only slightly by concentration. Polar solvents partially disrupt the dimer and shift the peak accordingly.

The peak width at room temperature ranges from sharp to broad, depending on the exchange rate of the particular acid. The carboxylic proton exchanges quite rapidly with protons of water and alcohols (or hydroxy groups of hydroxyacids) to give a single peak whose position depends on concentration. Sulfhydryl or enolic protons do not exchange rapidly with carboxylic protons, and individual peaks are observed.

PROTONS ON NITROGEN

The ^{14}N nucleus has a spin number $I = 1$ and, in accordance with the formula $2I + 1$, should cause a proton attached to it and a proton on an adjacent carbon atom to show three equally intense peaks. There are two factors, however, that complicate the picture: the rate of exchange of the proton on the nitrogen atom, and the electrical quadrupole moment of the ^{14}N nucleus.

The proton on a nitrogen atom may undergo rapid, intermediate, or slow exchange. If the exchange is rapid, the NH proton(s) is decoupled from the N atom and from protons on adjacent carbon atoms. The NH peak is therefore a sharp singlet, and the adjacent CH protons are not split by NH. Such is the case for most aliphatic amines.* At an intermediate rate of exchange, the NH proton is

---

* H–C–N–H coupling in several amines has been observed[24a] following rigorous removal (with Na-K alloy) of traces of water. This effectively stops proton exchange on the NMR time scale.

partially decoupled, and a broad NH peak results. The adjacent CH protons are not split by the NH proton. Such is the case for N-methyl-*p*-nitroaniline. If the NH exchange rate is slow, the NH peak is still broad because the electrical quadrupole moment of the nitrogen nucleus induces a moderately efficient spin relaxation and thus an intermediate lifetime for the spin states of the nitrogen nucleus. The proton thus sees three spin states of the nitrogen nucleus (spin number = 1) which are changing at a moderate rate, and the proton responds by giving a broad peak. In this case, coupling of the NH proton to the adjacent protons is observed. Such is the case for pyrroles, indoles, secondary and primary amides and carbamates (Figure 30). Note that H$-$N$-$C$-$H coupling takes place through the C$-$H, C$-$N, and N$-$H bonds, but coupling between nitrogen and protons on adjacent carbons is negligible. The coupling is observed in the signal due to hydrogen on carbon; the N$-$H proton signal is severely broadened by the quadrupolar interaction with nitrogen. In the spectrum of ethyl N-methyl carbamate (Figure 30) $CH_3NHCOCH_2CH_3$,

the NH proton shows a broad absorption centered about δ 5.16, τ 4.84, and the N$-CH_3$ absorption at δ 2.78, τ 7.22 is split into a doublet ($J \sim 5$ Hz) by the NH proton. The ethoxy protons are represented by the triplet at δ 1.23, τ

8.77, and the quartet at δ 4.14, τ 5.86. The small peak at δ 2.67, τ 7.33 is an impurity.

Aliphatic and cyclic amine NH protons absorb from $\delta \sim 3.0$, τ 7.0 to $\delta \sim 0.5$, τ 9.5; aromatic amines absorb from $\delta \sim 5.0$, τ 5.0 to $\delta \sim 3.0$, τ 7.0. Because amines are subject to hydrogen bonding, the shift position depends on concentration, solvent, and temperature. Amides, pyrroles, and indoles absorb from $\delta \sim 8.5$, τ 1.5 to $\delta \sim 5.0$, τ 5.0; the effect on the absorption position of concentration, solvent, and temperature is generally smaller than in the case of amines. The nonequivalence of the protons on the nitrogen atom of a primary amide and of the methyl groups of N,N-dimethylamides is caused by hindered rotation around the C$-$N bond because of the contribution of the canonical form

$$C=N^+$$

Protons on the nitrogen atom of an amine salt exchange at a slow rate; they are seen as a broad peak downfield ($\delta \sim 8.5$, τ 1.5 to $\delta \sim 6.0$, τ 4.0), and they are coupled to protons on adjacent carbon atoms ($J \sim 7$ Hz); the α-protons are recognized by their downfield position in the salt compared with that in the free amine. The use of trifluoroacetic acid as both a protonating agent and a solvent frequently allows classification of amines as pri-

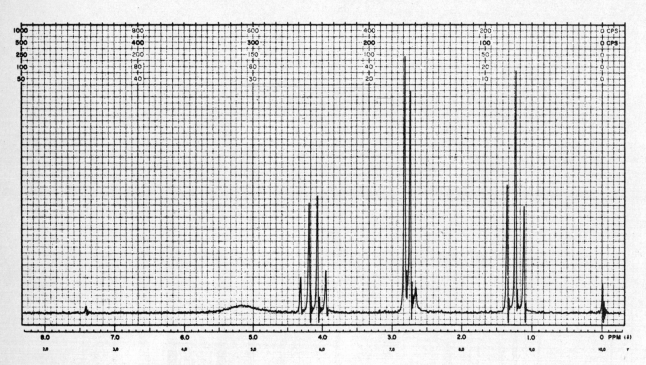

Fig. 30.
Ethyl N-methyl carbamate $CH_3NHCOCH_2CH_3$, 60 MHz.

Table II Classification of Amines by NMR of Ammonium Salts in Trifluoroacetic Acid[24]

Amine Precursor Class	Ammonium Salt Structure	Multiplicity of Methylene Unit
Primary	$C_6H_5CH_2NH_3^+$	Quartet (Figure 31)
Secondary	$C_6H_5CH_2NH_2R^+$	Triplet
Tertiary	$C_6H_5CH_2NHR_2^+$	Doublet

mary, secondary, or tertiary.[24] This is illustrated in Table II in which the number of protons on nitrogen determines the multiplicity of the methylene unit in the salt (Figure 31). Sometimes the broad $^+$NH, $^+$NH$_2$, or $^+$NH$_3$ absorption can be seen to consist of three broad humps. These humps represent splitting by the nitrogen nucleus ($J \sim 50$ Hz). With good resolution, it is sometimes possible to observe splitting of each of the humps by the protons on adjacent carbon atoms ($J \sim 7$ Hz).

PROTONS ON SULFUR

Sulfhydryl protons usually exchange at a slow rate so that at room temperature they are coupled to protons on adjacent carbons ($J \sim 8$ Hz). Nor do they exchange rapidly with hydroxylic, carboxylic, or enolic protons on the same or on other molecules; thus separate peaks are seen.

$C_6H_5CH_2NH_3^+$

δ 4.4
τ 5.6

20 Hz

Fig. 31.
NMR spectrum of α methylene unit of a primary amine in CF_3CO_2H; corresponds to Table II, first line.

However, exchange is rapid enough so that shaking for a few minutes with deuterium oxide replaces sulfhydryl protons with deuterium. The absorption range for aliphatic sulfhydryl protons is $\delta \sim 1.6$, τ 8.4, to $\delta \sim 1.2$, τ 8.8; for aromatic sulfhydryl protons, $\delta \sim 3.6$, τ 6.4 to $\delta \sim 2.8$, τ 7.2. Concentration, solvent, and temperature affect the position within these ranges.

PROTONS ON HALOGENS

Chlorine, bromine, and iodine nuclei are completely decoupled from protons directly attached, or on adjacent carbon atoms, because of strong electrical quadrupole moments. The absorption positions for protons of halogen acids vary over a wide range as a function of concentration, solvent, and temperature (in the vapor phase, for example HF absorbs at δ 2.7, τ 7.3, and HI absorbs at $\delta -13$, τ 23).

The ^{19}F atom has a spin number of 1/2 and couples strongly with protons (see Appendix G). The rules for coupling of protons with fluorine are the same as proton-proton coupling; in general, the proton-fluorine coupling constants are somewhat larger, and long-range effects are frequently found. The ^{19}F nucleus can be observed at 56.4 MHz at 14,092 gauss. Of course, its spin is split by proton and fluorine spins, and the multiplicity rules are the same as those observed in proton spectra.

VI. COUPLING OF PROTONS TO OTHER NUCLEI

The organic chemist may encounter proton coupling with such other nuclei (besides ^{1}H, ^{19}F, and ^{14}N) as ^{31}P, ^{13}C, ^{2}H, and ^{29}Si. Three factors must be considered: natural abundance, spin number, and electrical quadrupole moment; the nuclear magnetic moment and relative sensitivity are important when a spectrum of the particular nucleus is considered. These properties are listed for a number of nuclei in Appendices G and H.

The magnetogyric ratio, γ, describes the combined effect of the magnetic moment and spin number of a given type of nucleus. The ratio of the magnetogyric ratios of one type of nucleus to that of another type of nucleus is a measure of the relative coupling constants of those two nuclei to a given reference nucleus[11e]. Consider the relative magnitudes of coupling of hydrogen and of deuterium to a particular nucleus, X: Since

$$\frac{J_{HX}}{J_{DX}} = \sim \frac{\gamma_H}{\gamma_D} \qquad D = {}^2H$$

from the magnetogyric ratios in Appendix H, one calculates

$$\frac{J_{HX}}{J_{DX}} = \sim \frac{26.753}{4.107} = 6.51$$

Thus one anticipates that deuterium coupling is lower than the corresponding hydrogen coupling by a factor of about 6.5.

^{31}P has a natural abundance of 100% and a spin number of 1/2 (therefore no electrical quadrupole moment). The multiplicity rules for proton-phosphorus splitting are the same as those for proton-proton splitting. The coupling constants are large ($J_{H-P} \sim 200$ to 700 Hz and J_{HC-P} is $0.5 - 20$ Hz) and are observable through at least four bonds. The ^{31}P nucleus can be observed at the appropriate frequency and magnetic field.

^{13}C has a natural abundance (relative to ^{12}C = 100%) of 1.1%, and a spin number of 1/2. Typically, absorption due to coupling of a proton with a ^{13}C nucleus is seen as weak "^{13}C satellite peaks" on both sides of a very strong proton peak. Coupling constants for H–^{13}C have been correlated with hybridization of the ^{13}C atom: $J_{sp^3} \sim 120$ Hz, $J_{sp^2} \sim 170$ Hz, and $J_{sp} \sim 250$ Hz. The proton attached to the ^{13}C is also split by protons on adjacent ^{12}C atoms with the usual H–C–C–H coupling constant of about 7 Hz. Thus the configuration, $^{13}CH_3 - ^{12}CH_2-$, shows a triplet on both sides of an amplified $^{12}CH_3$ absorption from the ^{13}C–H and the $^{13}CH_3 - ^{12}CH_2$ couplings. Other molecules in the magnetic field will have the configuration $^{12}CH_3 - ^{13}CH_2-$; these molecules also show a triplet on each side of the amplified $^{12}CH_3$ absorption, but since $J_{CH_3 - ^{13}C}$ is only ~ 4 to 7 Hz, these peaks are buried on the sides of the $^{12}CH_3$ absorption. It is statistically unlikely that a molecule would have adjacent ^{13}C atoms. The ^{13}C satellite peaks can be distinguished from spinning side bands by their invariance with the rate of spinning of the sample tube. Parenthetically, note that each spinning side band has the same multiplicity as its "parent," but the satellite peaks do not necessarily have the same multiplicity as the parent peak (see page 141 of Jackman and Sternhell[4]).

Deuterium (^{2}H or D) is introduced into a molecule usually to detect a group or to simplify a spectrum. ^{2}H has a spin number of 1, a small coupling constant with protons, and a small electrical quadrupole moment. A proton-deuterium coupling constant is approximately 1/6.5 of the corresponding proton-proton constant (see the section above on ratio of the magnetogyric ratios). Suppose the protons on the α-carbon atom of a ketone

$$X-\overset{\gamma}{CH_2}-\overset{\beta}{CH_2}-\overset{\alpha}{CH_2}-\overset{\overset{\textstyle O}{\|}}{C}-$$

were replaced by deuterium:

$$X-\overset{\gamma}{CH_2}-\overset{\beta}{CH_2}-\overset{\alpha}{CD_2}-\overset{\overset{\textstyle O}{\|}}{C}-$$

The original spectrum would consist of a triplet for the α protons, a quintet (assuming equal coupling) for the β-protons, and a triplet for the γ protons. In the deuterated compound, the α-proton absorption would be absent, the β-protons would appear as a slightly broadened triplet, and the γ protons would be unaffected. Actually, each peak of the β-proton triplet is a very closely spaced triplet ($J_{CH-CD} \sim 1$ Hz or less), but the effect under ordinary resolution is peak broadening. Even this modest broadening due to the deuterium coupling can be removed by double resonance experiments (see below) involving irradiation at the deuterium resonance; this technique has been used to obtain a more exact measurement of the remaining proton-proton coupling. Most deuterated solvents have residual protonated impurities; the CHD_2 group in deuterated acetone or dimethyl sulfoxide, for example, is frequently encountered as a closely spaced quintet ($J \sim 2$ Hz, intensities 1:2:3:2:1).

^{29}Si has a natural abundance of 5.1% (based on ^{28}Si = 100%), and a spin number of 1/2. $J_{^{29}Si-CH}$ is about 6 Hz. The small doublet caused by the ^{29}Si–CH_3 coupling can often be seen straddling (± 3 Hz) an amplified peak of tetramethylsilane; the ^{13}C–H_3 doublet can also be seen at ± 59 Hz.[25]

$J_{^{13}CH_3} \sim 120$ Hz

$J_{^{13}CH_3 - ^{12}CH_2} \sim 7$ Hz $^{12}CH_3$ $J_{^{13}CH_3 - ^{12}CH_2} \sim 7$ Hz

VII. MORE COMPLEX SPIN-SPIN COUPLING

As was pointed out in Section IV, A_aX_x and $A_aM_mX_x$ systems are first order. The deviations from a first order spectrum as an AX pattern moves through an AB to an A_2 pattern were traced; the AB pattern could still be analyzed by inspection. But more complex changes occur as an A_2X becomes an A_2B pattern, and a complete analysis by inspection is no longer possible; the ratios of peak intensities are not those predicted by first order calculation, additional splitting occurs, and the spacings are not necessarily equal to the coupling constants. Most spectra encountered are not first order spectra. On the other hand, the resemblance of many spectra to first order spectra can be recognized because there is a gradual transition from A_aX_x to A_aB_b types; Wiberg's collection of calculated spectra[25a] can be used for matching fairly complex splitting patterns.

DEFINITIONS

To develop more complex cases, let us list the pertinent definitions here (some of which have been alluded to in Section IV).

1. A *spin system* consists of sets of nuclei that "interact (spin couple) among each other but do not interact with any nuclei outside the spin system. It is not necessary for all nuclei within a spin system to be coupled to all the other nuclei"[4] in the spin system. The spin system definition requires that spin systems be "insulated" from one another: e.g., the ethyl protons in ethyl isopropyl ether constitute one system, and the isopropyl protons another.
2. A *set of nuclei* consists of chemical shift equivalent nuclei.
3. If nuclei are interchangeable by a symmetry operation or a rapid mechanism, they are *chemical shift equivalent;* that is, they have exactly the same chemical shift under all achiral conditions. The interchange by a symmetry operation may occur in any reasonable conformation of the molecule. A rapid mechanism means one that occurs faster than once in about 10^{-3} seconds.
4. Nuclei are *magnetically equivalent* if they couple to all other nuclei in the spin system in exactly the same way. Chemical shift equivalence is presupposed.

CHEMICAL SHIFT EQUIVALENCE

The symmetry operations are: rotation about a symmetry axis (C_n), reflection at a center of symmetry (**i**), reflection at a plane of symmetry (σ), or higher orders of rotation about an axis followed by reflection in a plane normal to

this axis (S_n). The symmetry element (axis, center, or plane) must be a symmetry element for the entire molecule. The term "interchange" will be clarified by the examples below. Chemical shift equivalent protons are given the same letter of the alphabet in the Pople notation (i.e., placed in a set as described above) and if magnetically nonequivalent are distinguished by primes such as $A, A', A'', \ldots$, etc.

Protons a and b in *trans*-1,2-dichlorocyclopropane are chemical shift equivalent, as are the protons c and d (Figure 32). The molecule has an axis of symmetry passing through C_1 and bisecting the C_1-C_2 bond. Rotation of the molecule by 180° around the axis of symmetry interchanges proton H_a for H_b and H_c for H_d. If the protons were not labeled, it would not be possible to tell if the symmetry operation had been performed merely by inspecting the molecule before and after the operation.

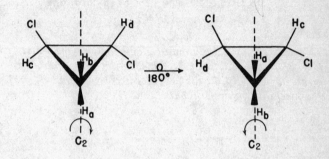

Fig. 32.
Trans-*1,2-Dichlorocyclopropane showing axis of symmetry and effect of rotation around the axis.*

An interesting situation occurs in a molecule such as 1,3-dibromo-1,3-diphenylpropane which has a methylene group between two centers of optical activity (Figure 33). There are two sets of protons in the molecule, H_a, H_b and H_c, H_d. In (1**R**,3**R**)-1,3-dibromo-1,3-diphenylpropane (one of a *dl* pair), H_a and H_b are chemical shift equivalent and so are H_c and H_d, because of an axis of symmetry (C_2) in the molecule. Rotation around that axis interchanges H_a with H_b and H_c with H_d. On the other hand, in (1**S**,3**R**)-1,3-dibromo-1,3-diphenylpropane (a *meso* compound), H_c and H_d are not chemical shift equivalent: they cannot be interchanged by a symmetry operation. (1**S**,3**R**)-1,3-dibromo-1,3-diphenylpropane does have a plane of symmetry (σ), but both H_c and H_d are in that plane. H_a and H_b are chemical shift equivalent because they can be interchanged by reflection through the plane of symmetry shown in Figure 33.

Although it is of no consequence unless a chiral solvent is used, note that only those protons that are interchangeable through an axis of rotation are completely identical

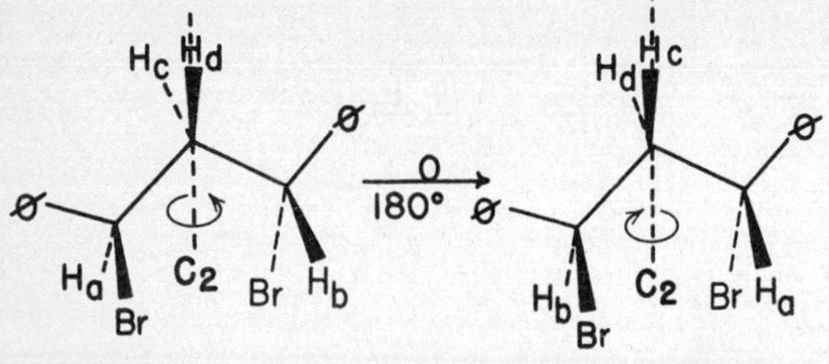

(1R,3R)-1,3-dibromo-1,3-diphenylpropane
(dl)

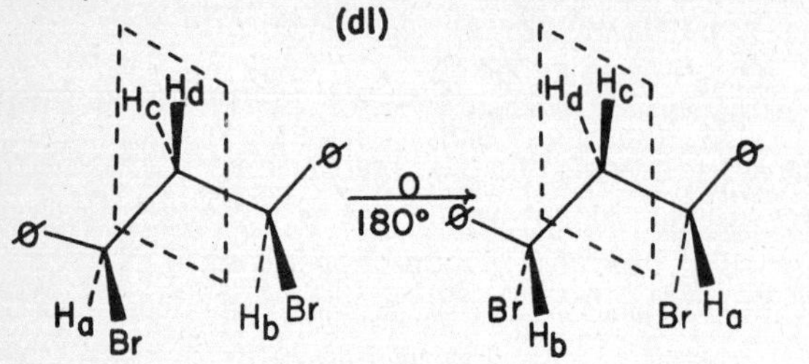

(1S,3R)-1,3-dibromo-1,3-diphenylpropane
(meso)

Fig. 33.
Two isomers of 1,3-dibromo-1,3-diphenylpropane. In the (1R, 3R) isomer, H_a and H_b are chemical shift equivalent, as are H_c and H_d. In the (1S, 3R) isomer, H_a and H_b are chemical shift equivalent, but H_c and H_d are not.

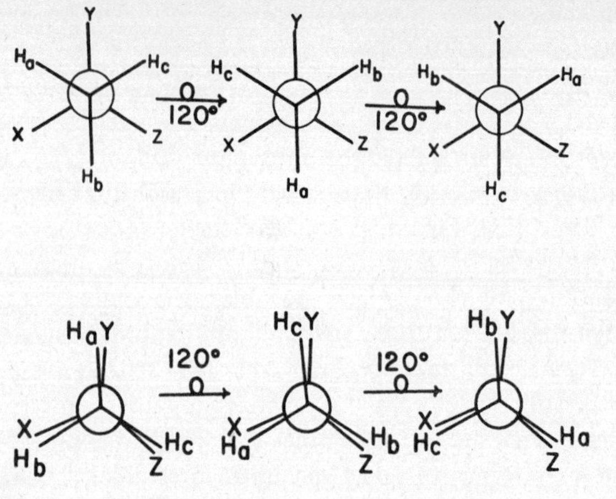

Fig. 34.
Newman projections of staggered and eclipsed conformers of a molecule with a methyl group attached to an optically active sp^3 carbon atom.

protons, whereas those that are interchangeable through other symmetry operations are enantiotopic, i.e., nonsuperimposable mirror images. Noninterchangeable protons are termed diastereotopic. Identical protons are chemical shift equivalent in any environment, chiral or achiral. Enantiotopic protons are chemical shift nonequivalent only in a chiral solvent. Diastereotopic protons are not chemical shift equivalent in any environment, although they may fortuitously absorb at the same position.

The concept of exchange through a rapid mechanism can be illustrated by the rapidly exchanging protons on some heteroatoms and by protons in molecules that are rapidly changing conformations. If the exchange is rapid enough, a single peak will result from, say, the carboxylic acid proton and the hydroxylic proton of a hydroxycarboxylic acid. Chemical shift equivalence of protons on a CH_3 group results from rapid rotation around a carbon-carbon single bond even in the absence of a symmetry element. Figure 34 shows Newman projections of the three staggered and three eclipsed conformers of a molecule containing a methyl group attached to another sp³ carbon atom having three different substituents, i.e., an optically active center. In any single conformation, protons $H_a H_b H_c$ are not chemical shift equivalent because the protons cannot be interchanged by a symmetry operation. However, the protons are rapidly changing position. The time spent in any one conformation

is short ($\sim 10^{-6}$ second), because the energy barrier for rotation around a C–C single bond is small. The chemical shift of the methyl group is an average of the shifts of each of the three protons. In other words, each proton can be interchanged with the others by a rapid rotational operation. Thus without the *a,b,c* labels the various staggered and eclipsed conformations, within each type, are indistinguishable.

Consider, in contrast with the methyl group, a methylene group next to a center of optical activity, as in 1-bromo-1,2-dichloroethane (Figure 35). Protons H_a and H_b are not chemical shift equivalent, since they cannot be interchanged by a symmetry operation in any conformation. In fact the molecule has no axis, plane, center, or

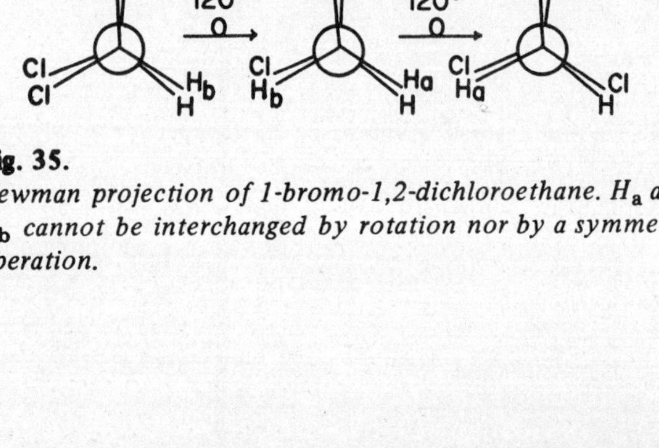

Fig. 35.
Newman projection of 1-bromo-1,2-dichloroethane. H_a and H_b cannot be interchanged by rotation nor by a symmetry operation.

reflection axis of symmetry. Although there is a rapid rotation around the carbon-carbon single bond, the protons are not interchangeable by a rotational operation. An observer can detect the difference before and after rotating the methylene group; the molecules are not superimposable.

The chemical shift of cyclohexane protons is an average of the shifts of the axial and the equatorial protons. The chemical shift equivalence of the cyclohexane protons results from rapid interchange of axial and equatorial protons as the molecule flips between chair forms. Since the planar conformation of cyclohexane is not a reasonable one, the chemical equivalence of its protons is not a consequence of achieving a planar conformation.

MAGNETIC EQUIVALENCE

Magnetic equivalence presupposes chemical shift equivalence. To determine whether chemical shift equivalent nuclei are magnetically equivalent, one determines whether they are coupled equally to each nucleus (probe nucleus) in every other set in the spin system. This is done by examining geometrical relationships. If the bond distances and angles from each nucleus in relation to the probe nucleus are identical, the nuclei in question are magnetically equivalent. Magnetically nonequivalent nuclei in a set are designated by primes (e.g., AA').

Consider the protons a,b and c,d in p-fluoronitrobenzene (Figure 36). Protons a and b are chemical shift equivalent by interchange through an axis or plane of symmetry. Protons a and b are coupled to nucleus F through the same bond distances and angles. However, protons a and b are coupled to proton c (or d) with different geometry. Thus, protons a and b fail the test for magnetic equivalence. Protons c and d are treated in the same way and are also found not to be magnetically equivalent. The system is $AA'BB'X$.

We propose a somewhat more practical definition and test for magnetic equivalence: two chemical shift equivalent protons are magnetically equivalent if they are symmetrically disposed with respect to each nucleus (probe) in any other set in the spin system. This means that the two protons under consideration can be interchanged through a reflection plane passing through the probe nucleus and perpendicular to a line joining the chemical shift equivalent protons. Note first that this plane is not necessarily a molecular plane of symmetry. Note also that the test is valid in any reasonable molecular conformation.

The three isomeric difluoroethylenes furnish additional examples of chemical shift equivalent nuclei that are not magnetically equivalent.

$$
\begin{array}{ccc}
\underset{F_a}{\overset{F_b}{\diagdown}}C=C\underset{H_a}{\overset{H_b}{\diagup}} & \underset{H_a}{\overset{F_a}{\diagdown}}C=C\underset{H_b}{\overset{F_b}{\diagup}} & \underset{H_a}{\overset{F_a}{\diagdown}}C=C\underset{F_b}{\overset{H_b}{\diagup}}
\end{array}
$$

In each case, the protons H_a and H_b and fluorines F_a and F_b comprise sets (of chemical shift equivalent nuclei) that are not magnetically equivalent.

Note that in the 1R,3R-compound of Figure 33, H_a and H_b are not magnetically equivalent (since they do not identically couple to H_c); H_c and H_d also are not magnetically equivalent since $J_{cb} \neq J_{db}$ and $J_{da} \neq J_{ca}$. Observe also that in the 1S,3R compound of Figure 33, $J_{ad} = J_{bd}$ and $J_{ac} = J_{bc}$; thus, in this molecule $H_a H_b$ *are* magnetically equivalent.

The question of magnetic equivalence in freely rotating methylene groups of aliphatic compounds becomes complex. The following assignments illustrate the nuances involved (if substituents Y and Z cause large chemical shift differences, the AMX designation is used instead of ABC):

CH_3CH_2Y	A_3B_2
ZCH_2CH_2Y	$AA'BB'$
$YCH_2CH_2CH_2Y$	$AA'BB'AA'$
$CH_3CH_2CH_2Y$	$A_3BB'CC'$
$ZCH_2CH_2CH_2Y$	$AA'BB'CC'$
$YCH_2CH_2CH_2CH_2Y$	$AA'BB'B''B'''A''A'''$
$ZCH_2CH_2CH_2CH_2Y$	$AA'BB'CC'DD'$

Methyl protons are magnetically equivalent by rotational averaging of the couplings to an adjacent methylene group whose protons are also magnetically equivalent if there is no other coupling involved. Averaging by rotation is valid since the rotational conformers are equivalent, thus equally populated. In ZCH_2CH_2Y, however, the population of the *anti* conformer is probably different from those of the enantiometic *gauche* conformers, and rotational averaging is not valid. As pointed out by Ault[11g]: "True A_2X_2 systems are quite rare (examples include difluoromethane, 1,1-difluoroallene, and 1,1,3,3-tetrachloropropane), and most

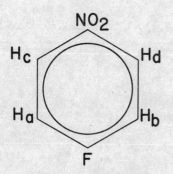

Fig. 36.
p-Fluoronitrobenzene.

systems which are described as A_2X_2 systems should really be classified as $AA'XX'$ systems."

$AA'XX'$ systems often give deceptively simple A_2X_2 spectra; the spectrum of 2-dimethylaminoethyl acetate (Figure 37a) is a case in point. Figure 37 shows the progressive distortions as $AA'XX' \rightarrow AA'BB'$ (i.e., $\Delta\nu/J$ decreases) in compounds of type ZCH_2CH_2Y. As the absorptions move closer together, the inner peaks increase in intensity, additional splitting occurs, and some of the outer peaks disappear in the baseline noise. The general appearance of symmetry throughout aids recognition of the type of spin system involved. At the extreme, the two methylene groups become chemical shift equivalent and a single A_4 peak results.

Although 1,3-dichloropropane is properly described as an $XX'AA'XX'$ system, it also presents a deceptively simple spectrum (Figure 40) that resembles an A_2X_4 system (triplet and quartet). A discussion of conditions leading to these types of deceptively simple spectra is given by Ault[11g].

In p-chloronitrobenzene,

the protons ortho to the nitro group (H_A and $H_{A'}$) are chemical shift equivalent to each other, and the protons ortho to the chlorine group (H_X and $H_{X'}$) are chemical shift equivalent to each other. J_{AX} and $J_{A'X'}$ are the same, approximately 7 to 10 Hz, and $J_{A'X}$ and $J_{AX'}$ are also the same but much smaller, approximately 0 to 1 Hz. Since H_A and $H_{A'}$ couple differently to another specific proton, they are not magnetically equivalent, and first-order rules do not apply. (Similarly, H_X is not magnetically equivalent to $H_{X'}$). In fact for calculations, H_A and $H_{A'}$, and H_X and

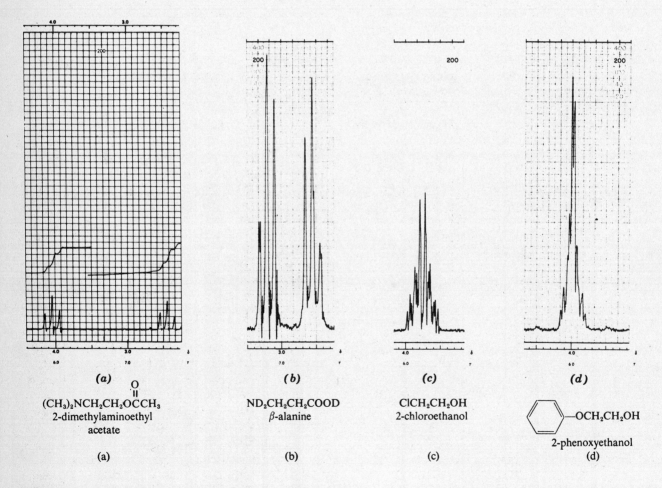

(a)

O
‖
$(CH_3)_2NCH_2CH_2OCCH_3$
2-dimethylaminoethyl
acetate

(a)

(b)

$ND_2CH_2CH_2COOD$
β-alanine

(b)

(c)

$ClCH_2CH_2OH$
2-chloroethanol

(c)

(d)

—OCH_2CH_2OH

2-phenoxyethanol
(d)

Fig. 37.
Progressive distortions as $AA'XX' \rightarrow AA'BB'$ in the configuration $Z-CH_2-CH_2Y$. 60 MHz.

$H_{X'}$ are also coupled ($J \sim 3$ Hz). The system is described as $AA'XX'$, and the spectrum is actually very complex. Fortunately, the pattern is readily recognized because of its symmetry and *apparent* simplicity; under ordinary resolution it resembles an AB pattern of two distorted doublets. (Closer inspection reveals many additional splittings, Figure 38.) As the para substituents become more similar to each other (in their shielding properties), the system tends toward $AA'BB'$; even these absorptions resemble AB patterns until they overlap.

The aromatic protons of symmetrically *o*-disubstituted benzenes also give $AA'BB'$ spectra. An example is *o*-dichlorobenzene (Figure 39).

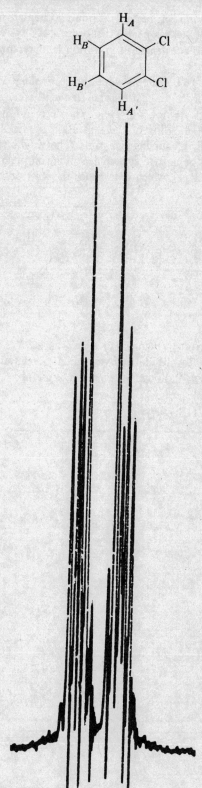

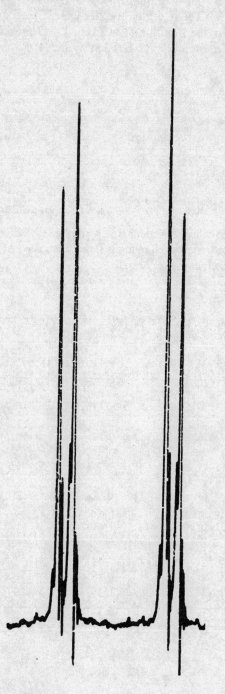

Fig. 38.
p-Chloronitrobenzene, 100 MHz CCl.

Fig. 39.
o-Dichlorobenzene, 100 MHz CCl$_4$.

The spectrum of methyl-2-furoate (Chapter 6, compound 4) represents a nearly first-order *AMX* system with three coupling constants. Each proton is represented by a pair of doublets.

Although *ABX* systems with three coupling constants are not first-order, the patterns are frequently recognized if the distortions are not too severe. The degree of distortion depends on the ratios of the separation of the *A* and *B* protons to their coupling constants. The vinylic structure gives an *ABX* or an *ABC* system depending on the nature of the substituent *Y*, which determines the shift positions of the protons.

p-Chlorostyrene (Figure 41) shows an *ABX* spectrum that can readily be related to an *AMX* pattern. The following analysis, though not rigorous, is useful.

Protons *A* and *B* are not chemically equivalent. Proton *A* ($\delta \sim 5.70$, τ 4.30) is deshielded about 25 Hz, compared with proton *B*, because of its relative proximity to the ring.[4] Proton *X* ($\delta \sim 6.70$, τ 3.30) is strongly deshielded by the ring, and is split by proton *A* ($J \sim 18$ Hz) and by proton *B* ($J \sim 11$ Hz). The *A* proton is split by the *X* proton ($J \sim 18$ Hz) and by the *B* proton ($J \sim 2$ Hz). The *B* proton is split by the *X* proton ($J \sim 11$ Hz) and by the *A* proton ($J \sim 2$ Hz).

The coupling constants for a vinylic system are characteristic: the *trans* coupling is larger than the *cis*, and the geminal coupling is very small.

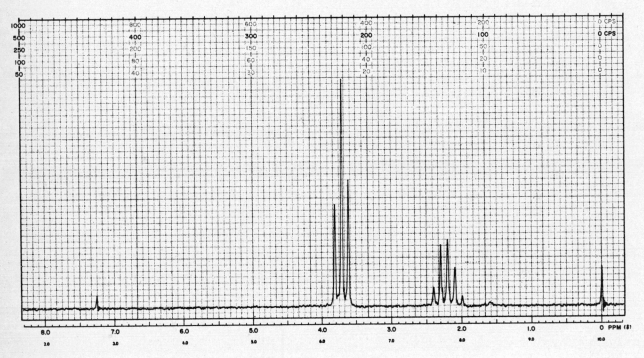

Fig. 40.
1,3-Dichloropropane in CDCl₃. 60 MHZ. XX′AA′XX′ system.

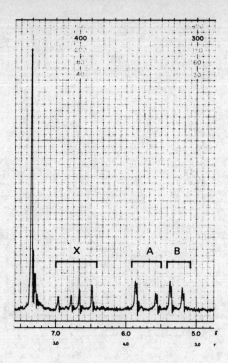

Fig. 41.
p-Chlorostyrene, 60 MHz.

The preceding analysis lacks rigor in several details. The splittings in proton X do not correspond exactly to J_{BX} and J_{AX}, although this is a good approximatión when $\Delta\nu$ is greater than about 10 for protons AB. The only exact information obtainable from the X pattern is that the spread between the outside peaks is equal to $J_{AX} + J_{BX}$. The distortions in peak intensities from an AMX pattern are obvious. The pattern of two closely spaced pairs should not be confused with the quartet resulting from splitting by three equivalent protons.

As the shift positions of protons A and B approach each other so that $\Delta\nu$ becomes much smaller than 10, the deviations from a first-order spectrum become severe; the A and B patterns overlap, and the middle peaks of the X pattern merge. In the extreme case that $\nu_A = \nu_B$ and $J_{AX} = J_{BX}$, protons A and B are magnetically equivalent, the spectrum is simply A_2X and the X proton absorption is a triplet. Note that a very similar spectrum would arise if the A pair were chemically, but not magnetically, equivalent: i.e., an $AA'X$ system. As the shift position of the X proton approaches the A and B absorptions, the spectrum degenerates to a very complex ABC pattern.

An equatorial and an axial proton on the same carbon of a fused 6-membered ring may form the AB part of an ABX or an ABC pattern, depending on the nature of Y on an adjacent carbon. Typically, H_A (equatorial) is deshielded relative to H_B (axial) by about 0.1 to 0.7 ppm (Section III), J_{AB} is about 12 to 15 Hz, J_{BX} *(ax.-ax.)* is about 5-10 Hz, and J_{AX} (eq.-eq.) is about 2 to 3 Hz (Section X).

VIRTUAL COUPLING

The spectrum at 60 MHz of 1-nitropropane (Figure 42), an $A_3MM'XX'$ system, consists of an upfield and a downfield triplet and a slightly broadened sextet in between. Such a spectrum is neatly, though not rigorously, rationalized as consisting of an A_3M portion and an M_2X_2 portion with $J_{AM} \cong J_{MX}$.

Why is it that one can use this kind of oversimplified analysis for 1-nitropropane, but one runs into trouble when one predicts a triplet for the methyl group (Figure 43a) of heptaldehyde (A_3B_2 portion of the system), or a doublet for the methyl group (Figure 43b) of β-methylglutaric acid, $HOOC-CH_2-CH-CH_2COOH$ (A_2B portion of the system)?
        |
       CH_3

The answer is that we have weakly coupled protons in 1-nitropropane, but strongly coupled protons in the latter two compounds. Or, to put it another way, the spins of the methyl protons in 1-nitropropane are affected only (or almost entirely) by the adjacent methylene protons. But in heptaldehyde, the methyl proton spins are affected by all of the adjoining methylene groups that have practically the same chemical shift; i.e., $\Delta\nu/J$ for four of the methylene protons is almost zero, and they are considered to be strongly coupled. A similar situation exists in β-methylglutaric acid, in which the methylene groups and the methine group are strongly coupled (the system is $A_3BCC'DD'$). To explain the difficulties arising from attempting to segregate portions of a strongly coupled system, the concept of virtual coupling is invoked. Even though the coupling constant of the methyl group to the methylene group once removed in heptaldehyde or the methine in β-methylglutaric acid is almost zero, the methyl group is considered to be "virtually coupled" to those once removed methylene or methine groups.

The concept of virtual coupling, then, is convenient but only necessary when one incorrectly considers a portion of a strongly coupled spin system as a separate entity and attemps to apply first order rules to that portion.

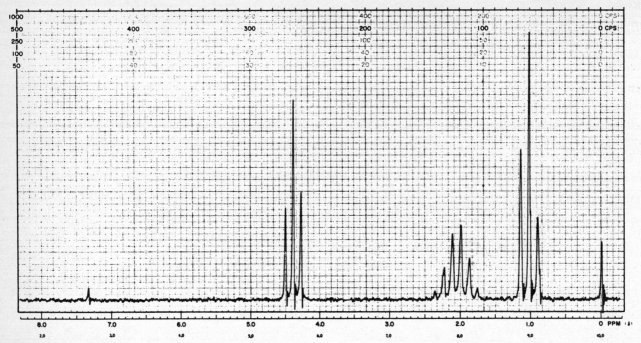

Fig. 42.
1-Nitropropane in CDCl$_3$. 60 MHz.
CH$_3$-CH$_2$-CH$_2$-NO$_2$
A$_3$MM'XX'

The modus operandi of the organic chemist in interpreting an NMR spectrum is to look initially for "first order" absorptions, and then for patterns that he can recognize as distortions of first order absorptions. He is not usually very interested in precise solutions of multispin systems and, in fact, he frequently thinks in terms of portions of spin systems. The above considerations should be of help in setting limits for this convenient, although not rigorous, approach to the interpretation of NMR spectra.

VIII. DECEPTIVELY SIMPLE SPECTRA

A "deceptively simple absorption"[4] is furnished by the methine proton (X) absorption of α-bromosuccinic acid:

$$HO_2C-\underset{\underset{H_a}{|}}{\overset{\overset{H_b}{|}}{C}}-\underset{\underset{Br}{|}}{\overset{\overset{H}{|}}{C}}-CO_2H$$

The methlene protons are not shift equivalent since the adjacent carbon atom is a chiral center (Section IX); the low field methine proton (X), however, is an apparent triplet. Such spectra are called "deceptively simple"[4] since they resemble first order spectra. For the di-acid, deceptive simplicity means that an apparent triplet, rather than a greater number of lines (see ABX systems of p-chlorostyrene

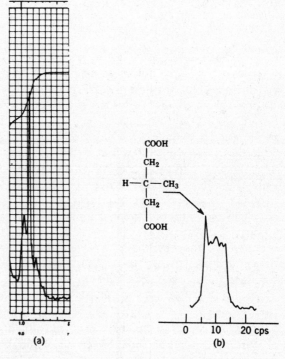

Fig. 43.
(a) Methyl group of heptaldehyde, 60 MHz. (b) Methyl group of β-methylglutaric acid, 60 MHz, expanded.

above), is observed. This "deceptively simple triplet" appears because $\Delta v/J$ is small for the AB pair, and we observed spacings in the "triplet" equal to $1/2\ (J_{AX}+J_{BX})$.[4,11b] Thus one should not leap to the conclusion that one has an A_2X system just because a triplet is observed. If such a problem is recognized, the solvent should be changed, the spectrum run at higher spectometer frequency, or shift reagents used (Section XIII).

IX. EFFECTS OF AN ASYMMETRIC CENTER

The protons of a methylene group near an asymmetric center are not chemical shift equivalent as explained in Section VII. They couple with each other, and each may have a different coupling to a vicinal proton. Even assuming fast rotation around the C–C bond, the methylene protons

of a compound
$$\begin{array}{c} H_a\ M \\ | \quad | \\ R-C-C-L \\ | \quad | \\ H_b\ S \end{array}$$
may show an AB pattern;

i.e., they are diastereotopic since they cannot be operationally interchanged (Section VII).

However, a large part of the nonequivalence may result from unequal populations of conformers even at fast rotation. The spectrum may be further complicated by slow rotation caused by low temperature or bulky substituents. Nonequivalence of the methylene protons persists in a

compound such as
$$\begin{array}{c} H_a\ M\ H_a \\ | \quad | \quad | \\ R-C-C-C-R \\ | \quad | \quad | \\ H_b\ S\ H_b \end{array}$$
even though the middle

carbon atom is not an asymmetric center.

The methylene group may display nonequivalence even though it is once removed from the asymmetric center.

Examples of the type
$$\begin{array}{c} H_a \quad\ \ M \\ | \qquad | \\ R-C-O-C-L \\ | \qquad | \\ H_b \quad\ \ S \end{array}$$
have been reported.

Nor need the asymmetric center be a carbon atom. The phenomenon has been reported for the methylene protons in quarternary ammonium salts, sulfites, sulfoxides, diethyl sulfide-borane complexes, and thiophosphonates.[26]

Chemical shift nonequivalence of the methyl groups of an isopropyl moiety near an asymmetric center is frequently observed; the effect has been measured through as many as seven bonds between the asymmetric center and the methyl protons.[26b]

A single asymmetric center in a terpene alcohol[27] effected nonequivalence of the protons of the adjacent methylene groups and of the methyl groups (Figure 44). The shift positions of the methyl groups are δ 0.92, τ 9.08,

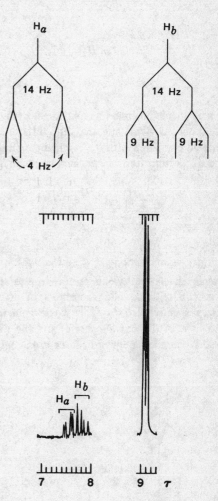

2-Methyl-6-methylene-7-octen-4-ol

and δ 0.90, τ 9.10; each absorption is split by the vicinal CH, giving rise to two overlapping doublets (J = 7 Hz). The nonequivalent protons of the CH_2 (a and b) between the CH_2=C and CHOH groups absorb, respectively, at δ 2.47, τ 7.53 and δ 2.18, τ 7.82. Each proton is split by the other (J_{gem} = 14 Hz) and unequally by the neighboring proton (J_{vic} = 9 Hz and 4 Hz). The protons of the other CH_2 groups are also nonequivalent; additional splitting by the adjacent methine proton results in a partially resolved multiplet.

Fig. 44.
5-Methylene and methyl protons of 2-methyl-6 methylene-7-octen-4-ol, 100 MHz.

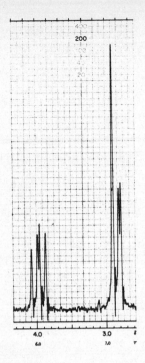

Fig. 45.
Asparic acid in D₂O, 60 MHz.

The methylene protons in aspartic acid (Figure 45) in D_2O are nonequivalent, and the shift difference between them is small compared with the geminal coupling constant. Thus the AB pattern from the geminal coupling

$$DO-\overset{O}{\underset{\parallel}{C}}-\overset{ND_2}{\underset{\mid}{C}}-\overset{H}{\underset{\mid}{C}}-\overset{O}{\underset{\parallel}{C}}-OD$$
$$\underset{H}{\mid}\quad\underset{H}{\mid}$$

is quite distorted; the inner peaks are strong, the outer peaks weak. Each methylene proton is also split by the vicinal proton with coupling constants of ~ 7.5 Hz and ~ 5.5 Hz, respectively. Two of the peaks coincide, the end peaks are lost in the baseline noise, and the net result is three peaks. The methine proton absorption consists of two pairs.

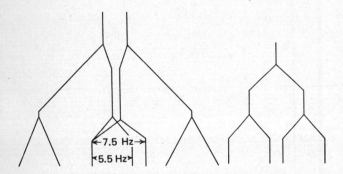

X. VICINAL AND GEMINAL COUPLING IN RIGID SYSTEMS

Coupling between protons on vicinal carbon atoms in rigid systems depends primarily on the dihedral angle ϕ between the $H-C-C'$ and the $C-C'-H'$ planes. This angle can be visualized by an end-on view of the bond between the

vicinal carbon atoms and by the perspective in Figure 46 in which the calculated relationship between dihedral angle and vicinal coupling constant[28] is graphed. Karplus emphasized[29] that his calculations are approximations and do not take into account such factors as electronegative substituents, the bond angles θ ($\angle H-C-C'$ and $\angle C-C-H'$), and bond lengths. Deductions of dihedral angles from measured coupling constants are safely made only by comparison with closely related compounds. The correlation has been very useful in cyclopentanes, cyclohexanes, carbohydrates, and bridged polycyclic systems. In cyclopentanes, the observed values of about 8 Hz for vicinal *cis* protons and about 0 Hz for vicinal *trans* protons are in accord with the corresponding angles of about 0° and about 90°, respectively. In substituted or fused cyclohexane rings, the following relations obtain:

	Calc. J	Observed J (Hz)
axial-axial	9	8–14 (usually 8–10)
axial-equatorial	1.8	1–7 (usually 2–3)
equatorial-equatorial	1.8	1–7 (usually 2–3)

A modified Karplus equation[30] can be applied to vicinal coupling in olefins. The prediction of a larger *trans* coupling ($\varphi = 180°$) than *cis* coupling ($\varphi = 0°$) is borne out. The *cis* coupling in unsaturated rings decreases with (increasing bond angle θ) decreasing ring size as follows: cyclohexenes $J = 8.8$ to 10.5, cyclopentenes $J = 5.1$ to 7.0, cyclobutenes $J = 2.5$ to 4.0, and cyclopropenes $J = 0.5$ to 2.0.[31]

The calculated relationship[32] between the $H-C-H$ angle of geminal protons is shown in Figure 47. This relationship is quite susceptible to other influences and should be used with due caution. However, it is useful for characterizing methylene groups in a fused cyclohexane ring (approximately tetrahedral, $J \sim 12$ to 18), methylene groups of a cyclopropane ring ($J \sim 5$), or a terminal methylene group ($J \sim 0$ to 3). Geminal coupling constants are actually negative numbers, but this can be ignored except for calculations.

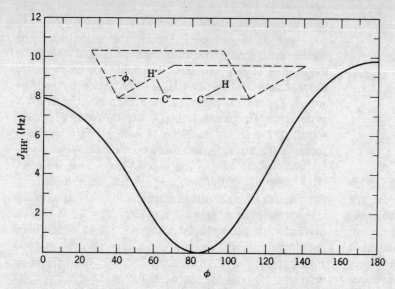

Fig. 46.
The vicinal Karplus correlation. Relationship between dihedral angle and coupling constant for vicinal protons.

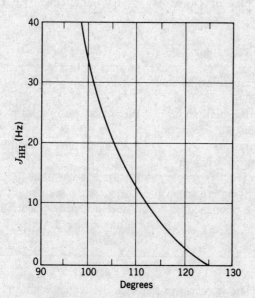

Fig. 47.
The geminal Karplus correlation. J_{HH} for CH_2 groups as function of $\angle H-C-H$.

In view of the many factors other than angle dependence that influence coupling constants, it is not surprising that there have been "uses and abuses" of the Karplus correlation.[4] Direct "reading off" of the angle from the magnitude of the J value is a very dangerous practice. The safest application of the relationships is to structure determinations in which molecular geometries have provided the extrema of the high and low expected J values and for which the 0° and 90° or 90° and 180° structures are known for a given system. A more detailed description of the limitations of the Karplus correlations has been provided in Jackman and Sternhell.[4]

XI. LONG-RANGE COUPLING

Proton-proton coupling beyond three bonds may occur in olefins, acetylenes, aromatics, and heteroaromatics, and in strained ring systems (small or bridged rings). Allylic (H–C–C=C–H) coupling constants are about 0 to 3 Hz. Homoallylic (H–C–C=C–CH) couplings are usually negligible but may be as much as 1.6 Hz. Coupling through conjugated polyacetylenic chains may occur through as many as nine bonds. Meta coupling in a benzene ring is 1 to 3 Hz, and para, 0 to 1 Hz. In five-membered heteroaromatic rings, coupling between the 2, 4 protons is 0 to 2 Hz.

J_{AB} in the bicyclo [2.1.1] hexane system is about 7 Hz. This unusually high long-range coupling constant is attri-

buted to the "W-conformation"[4] of the four sigma bonds between H_A and H_B:

XII. SPIN-SPIN DECOUPLING

Spin-spin decoupling (double irradiation or double resonance) is a powerful tool for simplifying a spectrum, for determining the relative positions of protons in a molecule, or for locating a buried absorption.

Spin-spin decoupling is simply a technique for irradiating a nucleus with a strong radiofrequency signal at its resonance frequency, while scanning other nuclei to detect which ones are affected by decoupling from the irradiated nucleus. The early experiments in spin-spin decoupling were limited to irradiation of other nuclei such as ^{14}N, ^{31}P, or ^{19}F because of the very large frequency difference from the proton resonance. For example, decoupling has been used on the NH proton absorption of a pyrrole or an amide, which can be so broad as to be merely a slight bulge on the baseline. As previously pointed out (Section V), this is a result of partial decoupling by the nitrogen electrical quadrupole moment. Irradiating the N nucleus at its resonance frequency results in complete decoupling and a sharp peak for the proton (or peaks if the NH proton is coupled to vicinal protons). Thus, irradiation of the nitrogen atom of an amine salt changes the broad ^+NH absorption to a sharp absorption whose multiplicity depends on the number of α protons.

Protons can be decoupled provided they are more than about 20 Hz apart at 100 MHz. The utility of proton-proton decoupling is shown in the 100 MHz partial spectrum of methyl-2,3,4-tri-O-benzoyl-β-L-lyxopyranoside[33] (Figure 48).

The integration (not shown) gives the following ratios in Figure 48 a from high field to low: 3:1:1:1:1:2. The sharp peak at δ 3.53, τ 6.47 is the OCH_3 group. Decoupling the two-proton multiplet at δ 5.75, τ 4.25 causes the multiplet at δ 5.45, τ 4.55 to collapse to four peaks, and the doublet at δ 5.00, τ 5.00 to a sharp singlet (Figure 48b). Decoupling the multiplet at δ 5.45, τ 4.55 partially collapses the

multiplet at δ 5.75, τ 4.25, and collapses the two upfield pairs of doublets (at δ 4.45, τ 5.55 and δ 3.77, τ 6.23) to two doublets (Figure 48c). The H_5 absorption should be upfield since these two protons are deshielded only by a single ether oxygen, whereas H_1 is deshielded by two ether oxygens, and H_2, H_3, and H_4 are deshielded by benzoyl groups. The two H_5 protons are the AM portion of an AMX pattern; the H_4 proton is the X portion (with additional splitting). The pair of doublets at δ 4.45, τ 5.55 is one (H_5') proton at C_5 strongly deshielded by the benzoyl group on C_4, and the pair of doublets at δ 3.77, τ 6.23 is the other (H_5) absorption. Further confirmation is provided by the collapse of each pair of doublets to doublets with the characteristic large geminal coupling (J = 12.5 Hz) on irradiation of the multiplet at δ 5.45, τ 4.55, which must therefore be the H_4 absorption (i.e., the X proton that was further split). The multiplet at δ 5.75, τ 4.25 must therefore represent H_2 and H_3 since irradiation of this multiplet collapsed the multiplet that we identified as H_4 (this now appears as the X portion of the AMX pattern), and also the doublet at δ 5.00, τ 5.00 which must be the H_1 absorption.

In addition to complete spin-spin decoupling there are four types of partial spin-spin decoupling. These are listed in the order of descending energy requirements at the decoupling frequency:

Designation	Effect on Spectra
1. Complete spin-spin decoupling	Total collapse of multiplets
2. Selective spin-spin decoupling	Partial collapse of multiplets
3. Spin tickling	Splitting of lines in multiplets
4. Nuclear Overhauser effect (NOE)	Change in absorption area
5. INDOR (internuclear double resonance)	Change in the more sensitive nucleus signal while irradiating the less sensitive nucleus

Complete vs. selective spin-spin decoupling is illustrated by the complete analysis of the spin system in the following tetrahydropyridine[4]:

The multiplicity of the signal because of the absorption of the protons at position 2 is examined while irradiating at

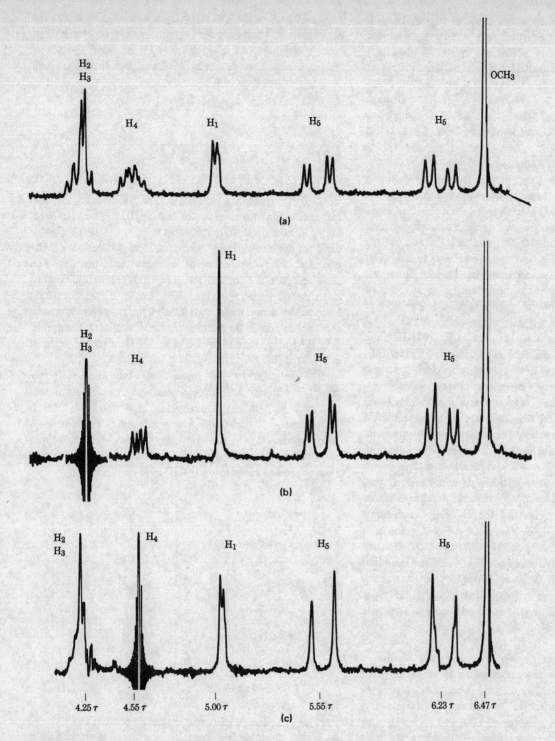

Fig. 48.
(a) Partial spectrum of methyl-2,3,4-tri-O-benzoyl-β-L-lyxopyranoside, 100 MHz CDCl₃. (b) H₂ and H₃ decoupled. (c) H₄ decoupled. Note that there are two H₅ protons.

positions corresponding to other protons in the system. Examination of the spectrum while irradiating (double resonance) at the position of the absorption of protons 3 and 4 results in only a partial collapse of the protons at position 2 (to a triplet); the remaining multiplicity might be expected to be due to the coupling J_{25}. This is confirmed by a triple resonance experiment in which simultaneous irradiation at positions corresponding to both the 3–4 and 5 proton groups results in total collapse of the absorption of the protons at position 2 to a singlet.

Spin tickling involves irradiating a nucleus with a much less intense radiation than is necessary for complete or selective decoupling. The effect is to increase the number of lines in the coupled absorption. The main advantage is that the nuclei that have a smaller chemical shift difference than that suitable for complete decoupling can be examined. The entire spectrum also is less perturbed by this technique. But since the experiment is much more demanding in terms of adjustment of instrumental parameters and stability of irradiating frequency, it is less often used.[11e]

If even less intense radiation is used than in spin tickling, a phenomenon known as a nuclear Overhauser effect (NOE) appears in the spectrum of some rather specially arranged nuclei. The low intensity energy used perturbs the Boltzman distribution of the energy states of the nucleus being irradiated. In the special case in which nuclei are coupled by dipole-dipole interaction (coupling through space instead of through bonding electrons), the Boltzman distribution of the coupled nuclei will also be effected. This causes a change in the intensity of the resonances line of that nucleus. The technique is valuable because the dipole-dipole interaction varies inversely with the cube of the distance between nuclei. Thus, if the change in the absorption area of the coupled nucleus is accurately measured, the distance between nuclei can be determined. The technique is limited to those cases in which nuclei are widely separated in terms of the number of bonds between them, but are spatially close. At the present time it is not used widely in structure determination, because of the very demanding technique required.[11e]

INDOR (internuclear double resonance) is a technique related to other double resonance methods. It is used when one is examining coupling between two nuclei of very different relative sensitivities. For example, ^{1}H and ^{15}N differ in sensitivity by a factor of 100, and it can be very difficult to obtain the spectrum of the less sensitive nucleus. But if the nuclei are coupled, the spectrum of the less sensitive nucleus can be obtained indirectly. This is accomplished by observing the height of a peak in the spectrum of the high-sensitivity nucleus while irradiating the low sensitivity nucleus with a narrow band frequency. Irradiation of the low-sensitivity peak results in a proportionate decrease in the intensity of the high-sensitivity peak. A plot of these changes vs. sweep frequency in Hz

gives an image of the low intensity spectrum. The technique is difficult and time-consuming, and simple spectra are not always obtained. As more interest in the NMR spectra of nuclei other than protons develops, the technique will probably be used more frequently.[11e]

XIII. SHIFT REAGENTS

Shift reagents, introduced by Hinckley in 1969,[35] provide a method for spreading out NMR absorption patterns without increasing the strength of the applied magnetic field; i.e., addition of shift reagents to appropriately functionalized samples results in substantial magnification of the chemical shift differences of nonequivalent protons. The shift reagents are ions in the rare earth (lanthanide) series coordinated to organic ligands. Although it had been long known that some metal ions caused shifts, it was not until the application of the more recent shift reagents, Eu(dpm)$_3$ and Eu(fod)$_3$, that these shifts could be effected without substantial line broadening.

The notation for the more commonly used shift reagents, Eu(dpm)$_3$ and Eu(fod)$_3$, come from their names *tris*-(*di*pivalomethanato) europium and *tris*-1,1,1,2,2,3,3-heptafluoro-7,7-dimethyl-3,5-*o*ctane*d*ionato europium. A more systematic name for the former is *tris*-2,2,6,6-*t*etramethyl-3,5-*h*eptane-4,6-*d*ionato europium and thus the (less common) substitute Eu(thd)$_3$ for Eu(dpm)$_3$.

Eu(dpm)$_3$ Eu(fod)$_3$

The use of such shift reagents is illustrated in Figure 49 in which the NMR spectrum of 1-heptanol is simplified by the addition of Eu(dpm)$_3$.[36] As is the usual case for such an alcohol in the absence of shift reagents, the only interpretable signal (Figure 49a) is that of the methylene adjacent to OH (proton set A, δ 3.8, triplet) and the terminal methyl (distorted set G, triplet, δ 0.9). Upon addition of the shift reagent, Eu(dpm)$_3$, the signals of the methylene groups closer to the OH group are moved downfield so that a separate signal is available for each of

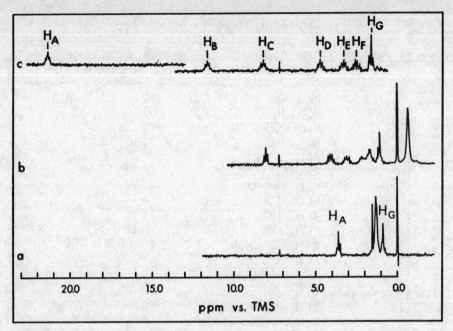

Fig. 49.
60 MHz proton NMR spectra of 0.40 ml of CDCl₃ solution containing 0.300M n-heptanol at various mole ratios [moles of Eu(dpm)₃ per mole of 1-heptanol]: (a) 0.00, (b) 0.19, (c) 0.78. Temperature, 30°C. From Anal. Chem., Vol. 43, p. 1599, Copyright 1971 by the American Chemical Society. Reprinted by permission of the copyright owner.

the methylene units (Figure 49c):

$$CH_3-CH_2-CH_2-CH_2-CH_2-CH_2-CH_2-OH$$
$$G \quad F \quad E \quad D \quad C \quad B \quad A$$

The closer the group to the functional group, the greater is the shift per increment of shift reagent (Figure 50).[36] Not all plots analogous to Figure 50 show linearity.

There are two major applications of the shift reagent to structure determination: (1) to simplify the spectrum and (2) to assign the protons from data on (or from) the response curves as in Figure 50. The former is straight-forward and is subject only to the limit of the dependence of the coupling constant on shift reagent concentration.[37] The latter is less certain because of the many parameters on which these slopes depend (see below).[38]

It is necessary to report the mole ratio (as well as solvent, concentration, etc.) of shift reagent to substrate for a given shift reagent experiment (see legend to Figure 49). Another way to report results is to give the slope of graphs such as Figure 50: ppm per mole of shift reagent per mole of substrate. Since these plots often display curvature, the method described by Kelsey[38a] could be used for obtaining more quantitative measure of the effect of shift reagents upon protons in a molecule.

Table III shows the general correlation between basicity of the functional group of the substrate and shift magnitude as induced by Eu(dpm)₃. Eu(fod)₃ is normally a "stronger" shift reagent since it is a stronger Lewis acid; it has the additional advantage of much greater solubility. One should, however, treat quantitative assessments such as Table III with caution, since such data were assembled before it was realized that performance of shift reagents is susceptible to many parameters including the purity[39,40] of the shift reagent, the method of data handling,[38a] and the presence of water. The adverse effect of a very small amount of water on shift reagent performance is displayed in Figure 51. Water does not always completely destroy the effectiveness of the shift reagent (as in Figure 51), but the reagents are much more effective when anhydrous than when damp, or hydrated.

Experimental limitations are set by the solvents in which the reagents can be used and by the absorption of the protons of the organic ligands on the reagent in the NMR spectrum. Shift reagents are most effective in "noncompetitive" solvents, i.e., those that do not coordinate strongly to the shift reagent. The most effective solvent is carbon tetrachloride; benzene and deuterochloroform are also useful, being respectively 90% and 75% as efficient as CCl₄.[38] Part of such solvent limitation is the solubility of

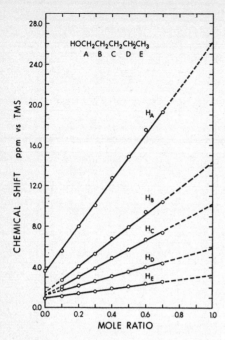

Fig. 50.

Variation in chemical shift for carbon-bonded protons of n-pentanol in CDCl₃ (0.40 ml of 0.300M solution) with increasing concentration of Eu(dpm)₃. Temperature, 30°C. Straight lines obtained by least squares calculations From Anal. Chem., Vol. 43, p. 1599. Copyright by the American Chemical Society (1971) Printed by permission of the copyright owner.

Table III[38] Variation of the Magnitude of Induced Shift with Functionality

Functional Group	Ppm per Mol of Eu(dpm)₃ per Mol of Substrate in CCl₄
$RCH_2\underline{N}H_2$	~150
$RCH_2O\underline{H}$	~100
$R_2C=NO\underline{H}$	~40
$RC\underline{H}_2NH_2$	30-40
$RC\underline{H}_2OH$	20-25
$R\underline{H}C=NOH$	14-30
$RC\underline{H}_2RC=NOH$	14-10
$RC\underline{H}_2COR$	10-17
$RC\underline{H}_2CHO$	11-19
$RC\underline{H}_2SOR$	9-11
$RC\underline{H}_2-O-C\underline{H}_2R$	10 (17-28 in CDCl₃[c])
$RC\underline{H}_2CO_2Me$	7
$RCH_2CO_2C\underline{H}_3$	6-5
$RC\underline{H}_2CN$	3-7
$RC\underline{H}_2NO_2$	~0
halides, indoles, alkenes	0
RCO_2H and phenols	decompose reagent

the shift reagent alone in the NMR solvent; Eu(dpm)₃ dissolves to the extent of about 100 mg/ml in benzene and about 200–300 mg/ml in the chlorinated solvents. Effectiveness of the shift reagent analysis may be increased by the use of low temperature NMR. The substrates under study may be recovered by the use of tlc or glc. The typical resonances of the ligands in the shift reagents in the presence of substrate are: Eu(fod)₃, δ 0.4 to 2.0; Eu(dpm)₃, δ −1.0 to −2.0; Pr(dpm)₃, δ 3.0 to 5.0. Deuterated reagents are available to alleviate this interference.

Most explanations for the effect of shift reagents are based on a pseudocontact mechanism. The dependence of the magnitude of the shift on the distance of the protons of interest from the functional group is only an approximation in the pseudocontact shift equation; this dependence breaks down when the shift reagent coordination with the substrate is not a simple, linear situation, i.e., when unique molecular geometry causes an acute proton-functional group-rare earth atom angle.[38, 41] The equilibrium for the shift reagent with the substrate rapidly establishes and thus

only a single set of time-averaged peaks is observed. Upfield shifts can be effected by the use of diamagnetic shift reagents, e.g., complexes of Pr.[38] The use of the magnitude of J values for molecules in the presence of shift reagents (e.g., for dihedral angle deductions) should be approached with caution in view of the reported J dependence upon shift reagent.[37] Shift reagents may be used to carry out highly specialized tasks, e.g., determination of optical purity of a substrate by using shift reagent with optically active ligands. An excellent review of many pertinent references is provided by Campbell.[38]

XIV. ¹³C NMR SPECTROMETRY

¹³C has a nuclear spin of 1/2 and can be observed by NMR at a frequency of 10.705 MHz at a field of 10 kilogauss. The analysis is limited by the following: the relative abundance of ¹³C is only 1.1% (compared to ¹²C), the ¹³C resonance has only 1.6% the sensitivity of the ¹H resonance, and the relaxation time for ¹³C is longer than for ¹H. These limitations can be overcome by using relatively large samples and pulsed (FT) instrumentation (Section II). Since the spin number for ¹³C is the same as for ¹H, the same rules apply for predicting the multiplicity of these absorptions. The coupling constants for ¹³C–¹H are large (100–250 Hz) and thus interpretation of ¹³C spectra can be difficult because of overlapping ¹³C–¹H multiplets. To

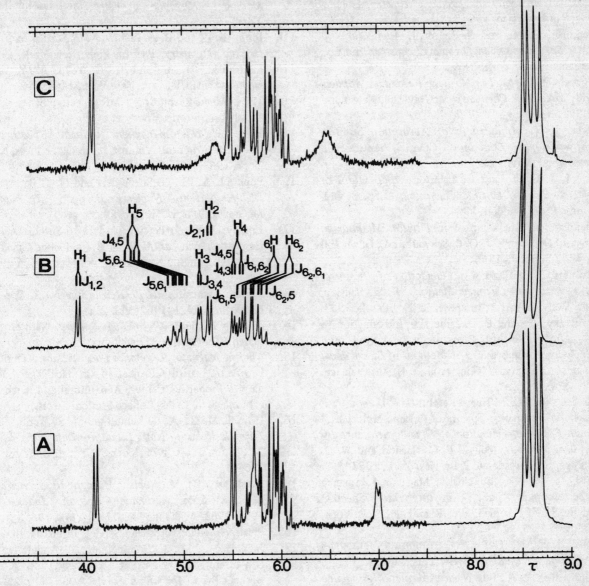

Fig. 51.
^{1}H *Nuclear magnetic resonance spectra (100 MHz) of 1,2:5,6-di-O-isopropylidene-α-D-glucofuranose (0.0842g) in deuterochloroform (0.5 ml.). A The normal spectrum; B The spectrum after the addition of (tris (dipi-valoylmethanato) europium) (9.84 × 10⁻² molar equivalents); C As for B but with the further addition of water (0.01 ml: 1.55 molar equivalents per mole of Eu (dpm)₃). A diagrammatic representation of the first-order assignment is given above the spectrum shown in B.*[38]

simplify the spectrum, ^{13}C nuclei are usually completely decoupled from all of the ^{1}H nuclei by use of double resonance. The spectrum is then simply a series of singlets corresponding to each variety of carbon atom present. A textbook of ^{13}C spectrometry[42] and a catalog of ^{13}C spectra[43] are available. Recently, a book oriented toward organic chemists has been published.[44]

REFERENCES

1. Bovey, F. A., *NMR Spectrometry*, Academic Press, New York, 1969.

2. Bible, R. H., Jr., *Interpretation of NMR Spectra*, Plenum Press, New York, 1965.

3. Bhacca, N. S., and D. H. Williams, *Applications of*

NMR Spectroscopy in Organic Chemistry, Holden-Day, San Francisco, 1964.

4. Jackman, L. M., and S. Sternhell, *Applications of NMR Spectroscopy in Organic Chemistry,* 2nd Ed., Pergamon, New York, 1969.

5. Roberts, J. D., *Nuclear Magnetic Resonance Applications to Organic Chemistry,* McGraw-Hill, New York, 1959.

6. Dyer, J. R., *Applications of Absorption Spectroscopy of Organic Compounds,* Chap. 4, Prentice-Hall, Englewood Cliffs, N.J., 1965.

7. Brand, J. C. D., and G. Eglinton, *Applications of Spectroscopy to Organic Chemistry,* Chap. 3, Oldbourne Press, London, 1964.

8. Jackman, L. M., Chap. 5 in *Physical Methods in Organic Chemistry,* J. C. P. Schwarz, Ed., Oliver and Boyd, Edinburgh, 1964.

9. Stothers, J. B., Chap. 4 in *Elucidation of Structures by Physical and Chemical Methods,* K. W. Bentley, Ed., Vol. XI, Part 1, Interscience, New York, 1963.

10. Jardetzky, O., and C. D. Jardetzky, Introduction to Magnetic Resonance Spectroscopy. Methods and Biochemical Applications, in *Methods of Biochemical Analysis,* Vol. IX, D. Glick et al., Eds., Interscience, New York, 1962.

11. Phillips, W. D., "High Resolution ^{1}H and ^{19}F Magnetic Resonance Spectra of Organic Molecules," Chap. 6, in *Determination of Organic Structure by Physical Methods,* Vol. 2, F. C. Nachod and W. D. Phillips, Eds., Academic Press, New York, 1962.

11a. Bose, A. K., "Proton Nuclear Magnetic Resonance Spectroscopy," Chap. 5, in *Interpretive Spectroscopy,* S. K. Freeman, Ed., Rheinhold, New York, 1965.

11b. Bible, R. H., Jr., *Guide to the Empirical Method: A Workbook,* Plenum Press, New York, 1967.

11c. Mathieson, D. W., Ed., *Nuclear Magnetic Resonance for Organic Chemists,* Academic Press, New York, 1967.

11d. Pasto, D. J., and C. R. Johnson, *Organic Structure Determination,* Prentice-Hall, Englewood Cliffs, N.J., 1969.

11e. Becker, E. D., *High Resolution NMR,* Academic Press, New York, 1969.

11f. Carrington, A. and A. D. McLachlan, *Introduction to Magnetic Resonance,* Harper and Row, New York, 1967.

11g. Ault, A., *J. Chem. Educ., 47,* 812 (1970).

11h. Meinwald, J., et al., *J. Amer. Chem. Soc., 89,* 68 (1967).

11i. Bovey, F. A., *NMR Data Tables for Organic Compounds,* Vol. I, Wiley-Interscience, New York, 1967.

11k. Williams, D. H., and I. Fleming, *Spectroscopic Methods in Organic Chemistry,* McGraw-Hill, New York, 1966.

11l. Ionin, B. I., and B. A. Ershov, *NMR Spectroscopy in Organic Chemistry,* Plenum Press, New York, 1970.

11m. Szymanski, H. A., and R. E. Yelin, *NMR Band Handbook,* IFI/Plenum, New York, 1968.

11n. JEOL High Resolution NMR Spectra, Sadtler Research Laboratories, Inc.

11o. Brugel, W., *Nuclear Magnetic Resonance Spectra and Chemical Structure,* Volume 1, Academic Press, New York, 1967.

12. Pople, J. A., W. G. Schneider, and H. J. Bernstein, *High-Resolution Nuclear Magnetic Resonance,* McGraw-Hill, New York, 1959.

12a. Emsley, J. W., J. Feeney, and L. H. Sutcliffe, *High Resolution Nuclear Magnetical Resonance Spectroscopy,* Pergamon, New York, Vol. 1, 1965; Vol. 2, 1966.

13. Varian Associates, *High Resolution NMR Spectra Catalogue,* Vol. 1, 1962; Vol. 2, 1963.

14. *Nuclear Magnetic Resonance Spectra.* Sadtler Research Laboratories, Philadelphia, Pa.

15. *Nuclear Magnetic Resonance Spectra Data,* American Petroleum Institute, Project 44, Chemical Thermodynamics Properties Center, Agricultural and Mechanical College of Texas, College Station, Texas.

16. Howell, M. G., A. S. Kende, and J. S. Webb, Eds., *Formula Index to NMR Literature Data,* Vols. 1 and 2, Plenum Press, New York, Vol. 1, 1964; Vol. 2, 1966.

16a. Hershenson, H. M., *Nuclear Magnetic Resonance and Electron Spin Resonance Spectra. Index for 1958–1963.* Academic Press, New York, 1965.

16b. Preston Technical Abstract Service, Evanston, Ill.

17. Jackman, L. M., F. Sondheimer, Y. Amiel, D. A. Ben-Efraim, Y. Gaoni, R. Wolovsky, and A. A. Bothner-By, *J. Am. Chem. Soc., 84,* 4307 (1962).

18. Shoolery, J. N., *Technical Information Bulletin, 2,* No. 3, Varian Associates, Palo Alto, Calif. B. P. Dailey and J. N. Shoolery, *J. Am. Chem. Soc., 77,* 3977 (1955).

19. Bruce, J. M., and P. Knowles, *Proc. Chem. Soc.,* 1964, 294.

20. Moniz, W. B., C. F. Poranski, Jr., and T. N. Hall, *J. Am. Chem. Soc., 88,* 190 (1966).

21. McGreer, D. E., and M. M. Mocek, *J. Chem. Ed., 40,* 358 (1963).

22. Chapman, O. L., and R. W. King, *J. Am. Chem. Soc., 86,* 1256 (1964).

23. Traynham, J. G., and G. A. Knesel, *J. Am. Chem. Soc., 87,* 4220 (1965).

23a. Slocum, D. W., and C. A. Jennings, Tetrahedron Letters, *34,* 3543, 3547 (1972).

24. Anderson, W. R., Jr., and R. M. Silverstein, *Anal. Chem.*, *37*, 1417 (1965).

24a. Henold, K. L., *Chem. Comm.*, 1340 (1970).

25. Ref. 2, p. 64, Fig. 3–7.

25a. Wiberg, K. B., and B. J. Nist, *The Interpretation of NMR Spectra*, W. A. Benjamin, New York, 1962.

26. For earlier references, see (a) E. I. Snyder, *J. Am. Chem. Soc.*, *85*, 2624 (1963), and (b) G. M. Whitesides, D. Holtz, and J. D. Roberts, ibid., *86*, 2628 (1964).

27. Silverstein, R. M., J. O. Rodin, D. L. Wood, and L. E. Browne, *Tetrahedron*, *22*, 1929 (1966).

28. Karplus, M., *J. Chem. Phys.*, *30*, 11 (1959).

29. Karplus, M., *J. Am. Chem. Soc.*, *85*, 2870 (1963).

30. Bothner-By, A. A., and C. Naar-Colin, *J. Am. Chem. Soc.*, *83*, 231 (1961).

31. Ref. 3, p. 54.

32. Gutowsky, H. S., M. Karplus, and D. M. Grant, *J. Chem. Phys.*, *31*, 1278 (1959).

33. Reist, E. J., L. V. Fisher, D. E. Gueffroy, and L. Goodman, *J. Org. Chem.*, *31*, 1506 (1966).

34. Freeman, R., and W. A. Anderson, *J. Chem. Phys.*, *37*, 2053 (1962). Varian Technical Information Bulletin, Summer 1965, p. 6.

35. Hinckley, C. C., *J. Amer. Chem. Soc.*, *91*, 5160 (1969).

36. Rabenstein, D. L., *Anal. Chem.*, *43*, 1599 (1971).

37. Shapiro, B. L., et. al., *J. Amer. Chem. Soc.*, *94*, 4382 (1972).

38. Campbell, J. R., *Aldrichimica Acta* (Aldrich Chemical Co.), *4*, 55 (1971).

38a. Kelsey, Donald R., *J. Amer. Chem. Soc.*, *94*, 1766 (1972).

39. Eisentraut, K. J., and R. E. Sievers, *J. Amer. Chem. Soc.*, *87*, 5254 (1965).

40. Sievers, R. E., et al., *Inorg. Chem.*, *6*, 1105 (1967).

41. Shapiro, B. L., et al., *J. Amer. Chem. Soc.*, *93*, 3281 (1971).

42. Stothers, J. B., *Carbon-13 NMR Spectroscopy*, Academic Press, New York, 1972.

43. Johnson, L. F., and W. C. Jankowski, *Catalog of Carbon-13 NMR Spectra*, Wiley, New York, 1972.

44. Levy, G. C., and G. L. Nelson, *Organic Chem. Carbon-13 Nuclear Magnetic Resonance for Organic Chemists*, Wiley, New York, 1972.

appendix a

SHIFT POSITIONS OF RESIDUAL PROTONS IN COMMERCIALLY AVAILABLE DEUTERATED SOLVENTS. DATA FURNISHED BY MERCK SHARP AND DOHME OF CANADA, LTD.

Appendix A   Shift Positions of Residual Protons in Commercially Available Deuterated Solvents. Data Furnished by Merck Sharp and Dohme of Canada, Ltd.

Solvent	Isotopic Purity Atom % D	Positions of Residual Protons (τ values)					
		Group	τ	Group	τ	Group	τ
Acetic Acid-d_4	99.5	methyl	7.95	hydroxyl	-1.53^a		
Acetone-d_6	99.5	methyl	7.95				
Acetonitrile-d_3	98	methyl	8.05				
Benzene-d_6	99.5	methine	2.80				
Chloroform-d	99.8	methine	2.75				
Cyclohexane-d_{12}	99	methylene	8.60				
Deuterium Oxide	99.8	hydroxyl	5.25^a				
1,2-Dichloroethane-d_4	99	methylene	6.31				
Diethyl-d_{10} Ether	98	methyl	8.84	methylene	6.64		
Dimethylformamide-d_7	98	methyl	7.24	methyl	7.06	formyl	1.95
Dimethyl-d_6 Sulfoxide	99.5	methyl	7.50				
p-Dioxane-d_8	98	methylene	6.45				
Ethyl Alcohol-d_6 (anh.)	98	methyl	8.83	methylene	6.41	hydroxyl	7.40^a
Hexafluoroacetone Deuterate	99.5	hydroxyl	1.0^a				
Methyl Alcohol-d_4	99	methyl	6.65	hydroxyl	5.16^a		
Methylcyclohexane-d_{14}	99	methyl	9.08	methylene	8.46	methine	8.35
Methylene-d_2 Chloride	99	methylene	4.65				
Pyridine-d_5	99	alpha	1.30	beta	2.80	gamma	2.42
Silanar^c-C (CDCl$_3$ + 1% TMS)	99.8	methyl	10.00^b	methine	2.75		
Tetrahydrofuran-d_8	98	α-methylene	6.40	β-methylene	8.25		
Tetramethylene-d_8 Sulfone	98	α-methylene	7.08	β-methylene	7.84		

[a] This value may vary considerably, depending upon the solute.
[b] By definition.
[c] Trademark.

NMR Spectra of Common Solvents and Common Impurities (Source: Sadtler Research Laboratories, Inc.)
Index to Spectra Found on Succeeding Pages.

	Spectrum Number		Spectrum Number
Acetone	1	Ethylene glycol	14
Antifoam A®	2	Isopropyl Alcohol	15
Apiezon® L Grease	3	Lubriseal® (Stopcock Grease)	16
Benzene	4	Methanol (+ Acid)	17
Chloroform	5	Methanol	18
Cyclohexane	6	Methyl ethyl ketone	19
Difluoroacetic acid	7	Phthalic acid, bis(2-ethylhexyl) ester	20
N,N-Dimethylformamide	8	Pyridine-d_5	21
Dimethyl Sulfoxide-d_6	9	Toluene	22
p-Dioxane	10	3-(Trimethyl silyl)-propane sulfonic Acid, sodium salt (DSS)	23
Dow Corning Stopcock Grease	11	Vertex® flakes (soap)	24
Ethanol	12		
Ethyl Acetate	13		

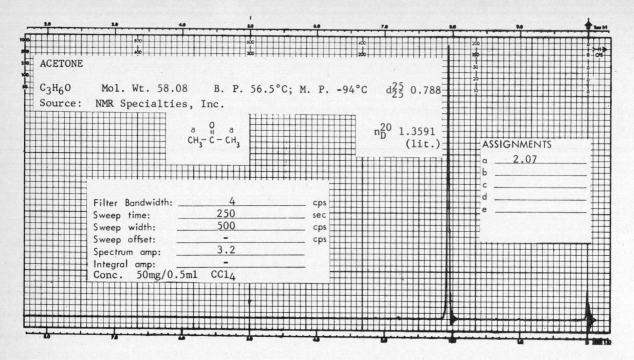

ACETONE

C_3H_6O Mol. Wt. 58.08   B. P. 56.5°C; M. P. -94°C d_{25}^{25} 0.788
Source: NMR Specialties, Inc.

$$\begin{array}{ccc} a & O & a \\ CH_3 & -\overset{\|}{C}- & CH_3 \end{array}$$

n_D^{20} 1.3591
(lit.)

ASSIGNMENTS

a   2.07
b
c
d
e

Filter Bandwidth: _____ 4 _____ cps
Sweep time: _____ 250 _____ sec
Sweep width: _____ 500 _____ cps
Sweep offset: _____ - _____ cps
Spectrum amp: _____ 3.2 _____
Integral amp: _____ - _____
Conc. 50mg/0.5ml CCl_4

No. 2

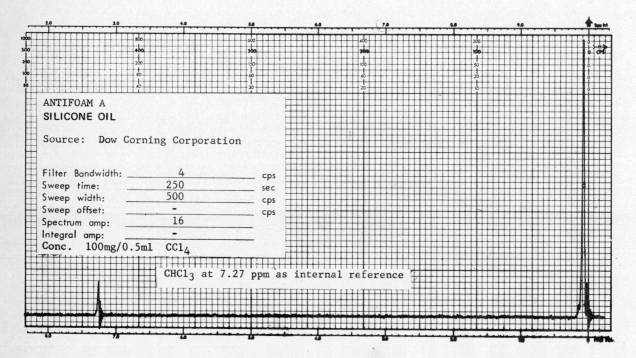

ANTIFOAM A
SILICONE OIL

Source: Dow Corning Corporation

Filter Bandwidth: _____ 4 _____ cps
Sweep time: _____ 250 _____ sec
Sweep width: _____ 500 _____ cps
Sweep offset: _____ - _____ cps
Spectrum amp: _____ 16 _____
Integral amp: _____ - _____
Conc. 100mg/0.5ml CCl_4

$CHCl_3$ at 7.27 ppm as internal reference

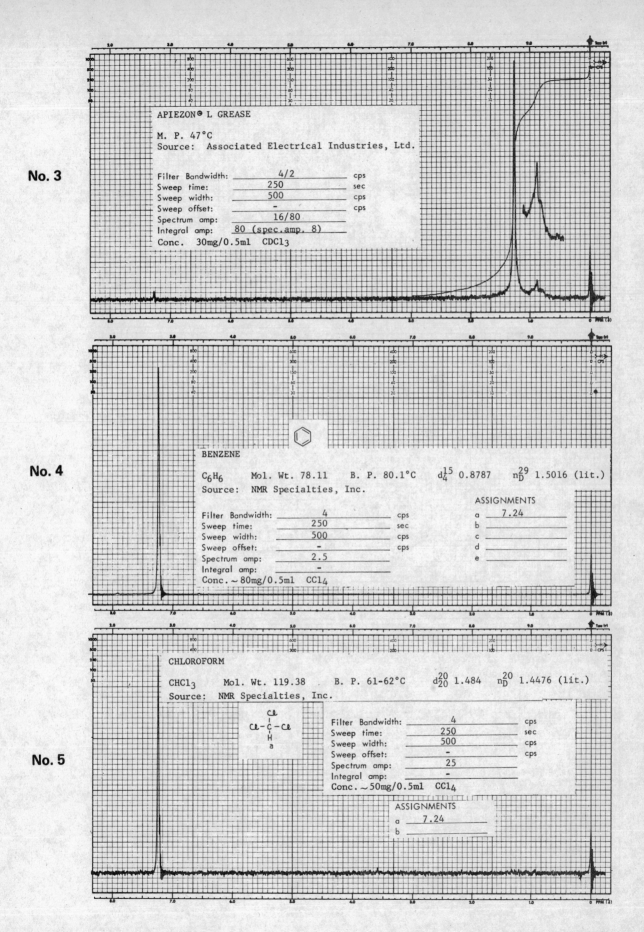

No. 3

APIEZON® L GREASE

M. P. 47°C
Source: Associated Electrical Industries, Ltd.

Filter Bandwidth: _____4/2_____ cps
Sweep time: _____250_____ sec
Sweep width: _____500_____ cps
Sweep offset: _____-_____ cps
Spectrum amp: _____16/80_____
Integral amp: __80 (spec.amp. 8)__
Conc. 30mg/0.5ml CDCl$_3$

No. 4

BENZENE

C_6H_6 Mol. Wt. 78.11   B. P. 80.1°C d_4^{15} 0.8787 n_D^{29} 1.5016 (lit.)
Source: NMR Specialties, Inc.

Filter Bandwidth: _____4_____ cps
Sweep time: _____250_____ sec
Sweep width: _____500_____ cps
Sweep offset: _____-_____ cps
Spectrum amp: _____2.5_____
Integral amp: _____-_____
Conc. ~80mg/0.5ml CCl$_4$

ASSIGNMENTS
a _____7.24_____
b _____
c _____
d _____
e _____

No. 5

CHLOROFORM

CHCl$_3$ Mol. Wt. 119.38   B. P. 61-62°C d_{20}^{20} 1.484 n_D^{20} 1.4476 (lit.)
Source: NMR Specialties, Inc.

Cl
|
Cl – C – Cl
|
H
a

Filter Bandwidth: _____4_____ cps
Sweep time: _____250_____ sec
Sweep width: _____500_____ cps
Sweep offset: _____-_____ cps
Spectrum amp: _____25_____
Integral amp: _____-_____
Conc. ~50mg/0.5ml CCl$_4$

ASSIGNMENTS
a _____7.24_____
b _____

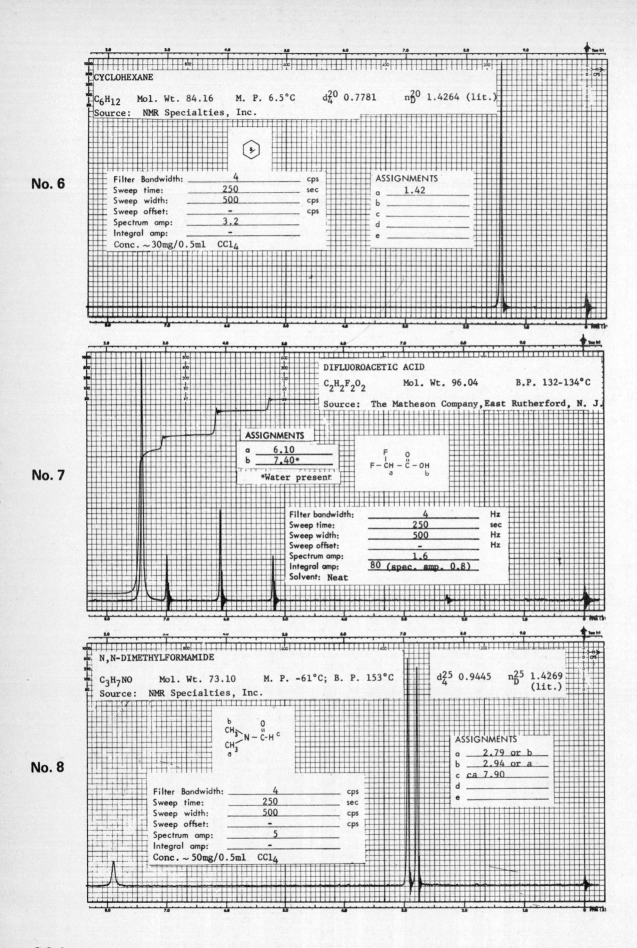

CYCLOHEXANE

C_6H_{12} Mol. Wt. 84.16   M. P. 6.5°C d_4^{20} 0.7781 n_D^{20} 1.4264 (lit.)
Source: NMR Specialties, Inc.

No. 6

Filter Bandwidth:	4	cps
Sweep time:	250	sec
Sweep width:	500	cps
Sweep offset:	-	cps
Spectrum amp:	3.2	
Integral amp:	-	

Conc. ~30mg/0.5ml CCl_4

ASSIGNMENTS
a   1.42
b
c
d
e

DIFLUOROACETIC ACID

$C_2H_2F_2O_2$ Mol. Wt. 96.04   B.P. 132-134°C

Source: The Matheson Company, East Rutherford, N. J.

No. 7

ASSIGNMENTS
a   6.10
b   7.40*

*Water present.

F   O
| ||
F – CH – C – OH
      a   b

Filter bandwidth:	4	Hz
Sweep time:	250	sec
Sweep width:	500	Hz
Sweep offset:	-	Hz
Spectrum amp:	1.6	
Integral amp:	80 (spec. amp. 0.8)	

Solvent: Neat

N,N-DIMETHYLFORMAMIDE

C_3H_7NO Mol. Wt. 73.10   M. P. -61°C; B. P. 153°C d_4^{25} 0.9445 n_D^{25} 1.4269 (lit.)
Source: NMR Specialties, Inc.

b
CH_3 O
         \ ||
          N – C–H ^{c}
         /
CH_3
a

No. 8

Filter Bandwidth:	4	cps
Sweep time:	250	sec
Sweep width:	500	cps
Sweep offset:	-	cps
Spectrum amp:	5	
Integral amp:	-	

Conc. ~50mg/0.5ml CCl_4

ASSIGNMENTS
a   2.79 or b
b   2.94 or a
c   ca 7.90
d
e

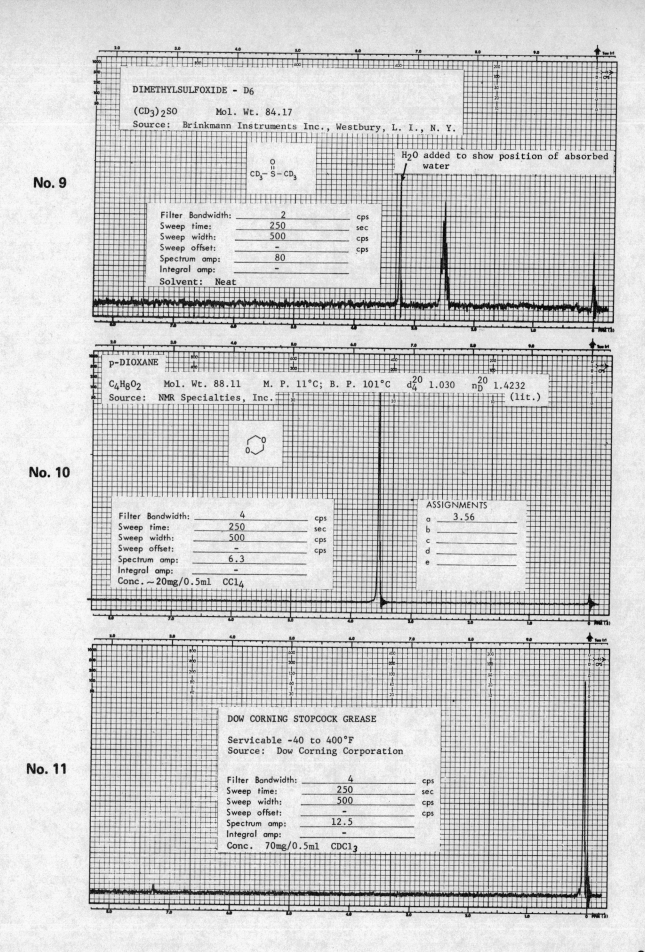

No. 9

DIMETHYLSULFOXIDE - D$_6$

(CD$_3$)$_2$SO   Mol. Wt. 84.17

Source: Brinkmann Instruments Inc., Westbury, L. I., N. Y.

$$CD_3 - \overset{\overset{\textstyle O}{\textstyle \|}}{S} - CD_3$$

H$_2$O added to show position of absorbed water

Filter Bandwidth:	2	cps
Sweep time:	250	sec
Sweep width:	500	cps
Sweep offset:	-	cps
Spectrum amp:	80	
Integral amp:	-	
Solvent: Neat		

No. 10

p-DIOXANE

C$_4$H$_8$O$_2$ Mol. Wt. 88.11   M. P. 11°C; B. P. 101°C   d$_4^{20}$ 1.030   n$_D^{20}$ 1.4232

Source: NMR Specialties, Inc.   (lit.)

Filter Bandwidth:	4	cps
Sweep time:	250	sec
Sweep width:	500	cps
Sweep offset:	-	cps
Spectrum amp:	6.3	
Integral amp:	-	
Conc. ~20mg/0.5ml CCl$_4$		

ASSIGNMENTS

a	3.56
b	
c	
d	
e	

No. 11

DOW CORNING STOPCOCK GREASE

Servicable -40 to 400°F
Source: Dow Corning Corporation

Filter Bandwidth:	4	cps
Sweep time:	250	sec
Sweep width:	500	cps
Sweep offset:	-	cps
Spectrum amp:	12.5	
Integral amp:	-	
Conc. 70mg/0.5ml CDCl$_3$		

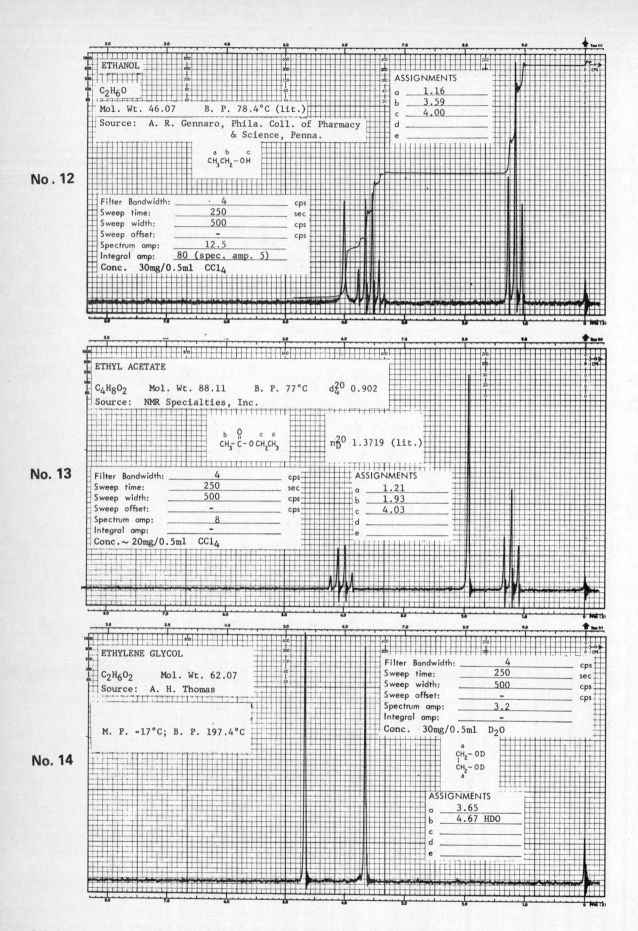

ETHANOL

C_2H_6O

Mol. Wt. 46.07   B. P. 78.4°C (lit.)

Source: A. R. Gennaro, Phila. Coll. of Pharmacy
& Science, Penna.

a   b   c
CH_3CH_2-OH

No. 12

ASSIGNMENTS	
a	1.16
b	3.59
c	4.00
d	
e	

Filter Bandwidth: ___ 4 ___ cps
Sweep time: ___ 250 ___ sec
Sweep width: ___ 500 ___ cps
Sweep offset: ___ - ___ cps
Spectrum amp: ___ 12.5 ___
Integral amp: ___ 80 (spec. amp. 5) ___
Conc. 30mg/0.5ml CCl_4

ETHYL ACETATE

$C_4H_8O_2$ Mol. Wt. 88.11   B. P. 77°C d_4^{20} 0.902

Source: NMR Specialties, Inc.

b   O   c   a
$CH_3-C-O CH_2CH_3$

n_D^{20} 1.3719 (lit.)

No. 13

Filter Bandwidth: ___ 4 ___ cps
Sweep time: ___ 250 ___ sec
Sweep width: ___ 500 ___ cps
Sweep offset: ___ - ___ cps
Spectrum amp: ___ 8 ___
Integral amp: ___ - ___
Conc.~ 20mg/0.5ml CCl_4

ASSIGNMENTS	
a	1.21
b	1.93
c	4.03
d	
e	

ETHYLENE GLYCOL

$C_2H_6O_2$ Mol. Wt. 62.07

Source: A. H. Thomas

M. P. -17°C; B. P. 197.4°C

Filter Bandwidth: ___ 4 ___ cps
Sweep time: ___ 250 ___ sec
Sweep width: ___ 500 ___ cps
Sweep offset: ___ ___ cps
Spectrum amp: ___ 3.2 ___
Integral amp: ___ - ___
Conc. 30mg/0.5ml D_2O

a
CH_2-OD
CH_2-OD
a

No. 14

ASSIGNMENTS	
a	3.65
b	4.67 HDO
c	
d	
e	

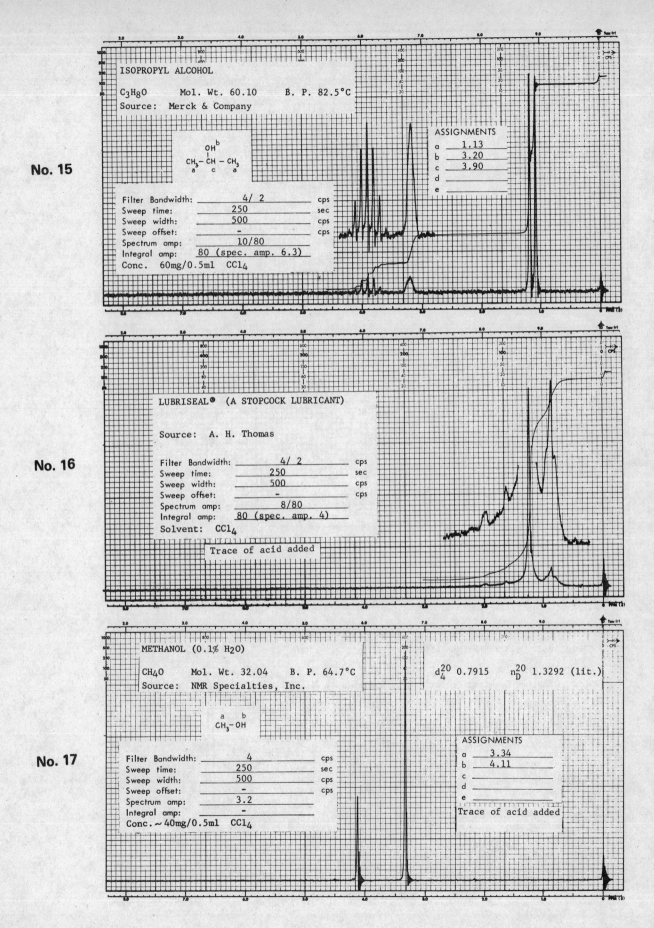

No. 15

ISOPROPYL ALCOHOL

C₃H₈O   Mol. Wt. 60.10   B. P. 82.5°C

Source: Merck & Company

$$OH$$
$$CH_3 - CH - CH_3$$

ASSIGNMENTS

a	1.13
b	3.20
c	3.90
d	
e	

Filter Bandwidth: ___4/ 2___ cps
Sweep time: ___250___ sec
Sweep width: ___500___ cps
Sweep offset: ___-___ cps
Spectrum amp: ___10/80___
Integral amp: ___80 (spec. amp. 6.3)___
Conc. 60mg/0.5ml CCl₄

No. 16

LUBRISEAL® (A STOPCOCK LUBRICANT)

Source: A. H. Thomas

Filter Bandwidth: ___4/ 2___ cps
Sweep time: ___250___ sec
Sweep width: ___500___ cps
Sweep offset: ___-___ cps
Spectrum amp: ___8/80___
Integral amp: ___80 (spec. amp. 4)___
Solvent: CCl₄

Trace of acid added

No. 17

METHANOL (0.1% H₂O)

CH₄O   Mol. Wt. 32.04   B. P. 64.7°C d_4^{20} 0.7915 n_D^{20} 1.3292 (lit.)
Source: NMR Specialties, Inc.

$$CH_3 - OH$$

ASSIGNMENTS

a	3.34
b	4.11
c	
d	
e	

Filter Bandwidth: ___4___ cps
Sweep time: ___250___ sec
Sweep width: ___500___ cps
Sweep offset: ___-___ cps
Spectrum amp: ___3.2___
Integral amp: ___-___
Conc. ~ 40mg/0.5ml CCl₄

Trace of acid added

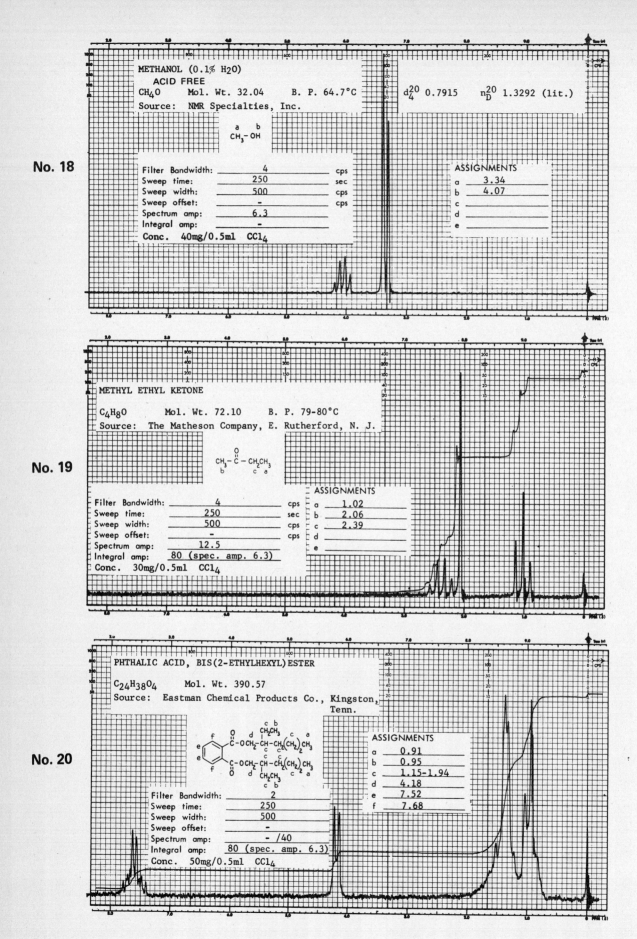

No. 18

METHANOL (0.1% H2O)
ACID FREE
CH_4O Mol. Wt. 32.04   B. P. 64.7°C d_4^{20} 0.7915 n_D^{20} 1.3292 (lit.)
Source: NMR Specialties, Inc.

a   b
$CH_3 - OH$

Filter Bandwidth:	4	cps
Sweep time:	250	sec
Sweep width:	500	cps
Sweep offset:	-	cps
Spectrum amp:	6.3	
Integral amp:	-	

Conc.   40mg/0.5ml   CCl_4

ASSIGNMENTS
a   3.34
b   4.07
c
d
e

No. 19

METHYL ETHYL KETONE

C_4H_8O Mol. Wt. 72.10   B. P. 79-80°C
Source: The Matheson Company, E. Rutherford, N. J.

$CH_3 - \overset{O}{\overset{||}{C}} - CH_2CH_3$
b   c   a

Filter Bandwidth:	4	cps
Sweep time:	250	sec
Sweep width:	500	cps
Sweep offset:	-	cps
Spectrum amp:	12.5	
Integral amp:	80 (spec. amp. 6.3)	

Conc.   30mg/0.5ml   CCl_4

ASSIGNMENTS
a   1.02
b   2.06
c   2.39
d
e

No. 20

PHTHALIC ACID, BIS(2-ETHYLHEXYL)ESTER

$C_{24}H_{38}O_4$ Mol. Wt. 390.57
Source: Eastman Chemical Products Co., Kingston, Tenn.

Filter Bandwidth:	2	
Sweep time:	250	
Sweep width:	500	
Sweep offset:	-	
Spectrum amp:	- /40	
Integral amp:	80 (spec. amp. 6.3)	

Conc.   50mg/0.5ml   CCl_4

ASSIGNMENTS
a   0.91
b   0.95
c   1.15-1.94
d   4.18
e   7.52
f   7.68

208   spectrometric identification of organic compounds

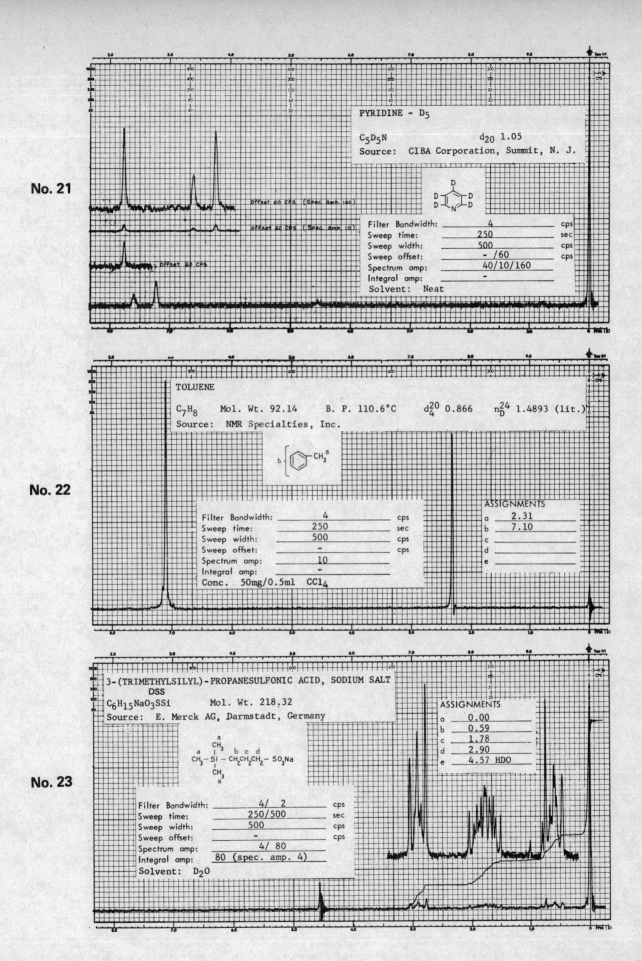

No. 21

PYRIDINE - D₅

C_5D_5N d_{20} 1.05
Source: CIBA Corporation, Summit, N. J.

Offset 60 CPS (Spec. amp. 160)
Offset 60 CPS (Spec. amp. 10)
Offset 60 CPS

Filter Bandwidth: _____ 4 _____ cps
Sweep time: _____ 250 _____ sec
Sweep width: _____ 500 _____ cps
Sweep offset: _____ - /60 _____ cps
Spectrum amp: _____ 40/10/160 _____
Integral amp: _____ - _____
Solvent: Neat

No. 22

TOLUENE

C_7H_8 Mol. Wt. 92.14   B. P. 110.6°C d_4^{20} 0.866 n_D^{24} 1.4893 (lit.)
Source: NMR Specialties, Inc.

Filter Bandwidth: _____ 4 _____ cps
Sweep time: _____ 250 _____ sec
Sweep width: _____ 500 _____ cps
Sweep offset: _____ - _____ cps
Spectrum amp: _____ 10 _____
Integral amp: _____ - _____
Conc. 50mg/0.5ml CCl_4

ASSIGNMENTS
a _____ 2.31
b _____ 7.10
c _____
d _____
e _____

No. 23

3-(TRIMETHYLSILYL)-PROPANESULFONIC ACID, SODIUM SALT
      DSS
$C_6H_{15}NaO_3SSi$ Mol. Wt. 218,32
Source: E. Merck AG, Darmstadt, Germany

ASSIGNMENTS
a _____ 0.00
b _____ 0.59
c _____ 1.78
d _____ 2.90
e _____ 4.57 HDO

Filter Bandwidth: _____ 4/ 2 _____ cps
Sweep time: _____ 250/500 _____ sec
Sweep width: _____ 500 _____ cps
Sweep offset: _____ - _____ cps
Spectrum amp: _____ 4/ 80 _____
Integral amp: 80 (spec. amp. 4)
Solvent: D_2O

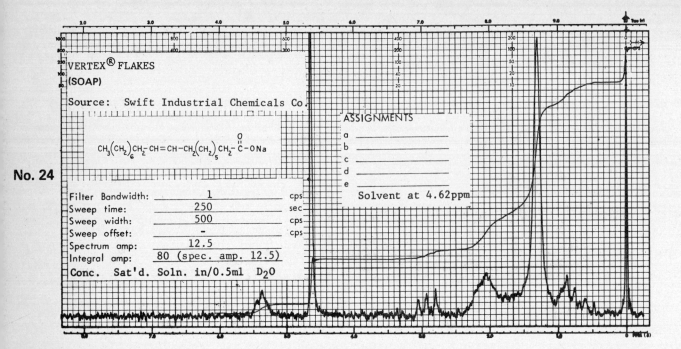

VERTEX® FLAKES
(SOAP)

Source: Swift Industrial Chemicals Co.

$$CH_3(CH_2)_6CH_2-CH=CH-CH_2(CH_2)_5CH_2-\overset{\overset{O}{\|}}{C}-ONa$$

No. 24

Filter Bandwidth: _____1_____ cps
Sweep time: _____250_____ sec
Sweep width: _____500_____ cps
Sweep offset: _____-_____ cps
Spectrum amp: _____12.5_____
Integral amp: 80 (spec. amp. 12.5)
Conc. Sat'd. Soln. in/0.5ml D$_2$O

ASSIGNMENTS

a _____
b _____
c _____
d _____
e _____
Solvent at 4.62ppm

appendix b

SHIFT POSITIONS OF PROTONS BOUND TO CARBON, AND NEAR A SINGLE FUNCTIONAL GROUP

These are two thought processes involved in examining an NMR spectrum with the object of identifying an organic compound. These are exemplified by the following questions:

1. What can we expect to find in the vicinity of δ 3.0, τ 7.0?
2. Where would we expect to find the peak of a proton on a carbon atom to which a chlorine atom is attached?

The following charts are designed to answer such questions. The shift values shown are average values for ranges that are usually less than about 0.5 ppm wide. The exact shift position will vary with changes in solvent or concentration, but, in general, these variations are small in the absence of hydrogen bonding.

A separate value is given for methyl, methylene, and methine protons. In a hydrocarbon, these protons are found, respectively, at about δ 0.9, τ 9.1; δ 1.25, τ 8.75; and δ 1.5, τ 8.5. The first line of the second chart, for example, shows that the protons of CH_3Br absorb near δ 2.7, τ 7.3; the α-protons of RCH_2Br near δ 3.4, τ 6.6; and the α-proton of R_2CHBr near δ 4.1, τ 5.9.

The shift values assigned have been gathered from a variety of sources. Two privately circulated collections are very useful: NMR Summary prepared by G. V. D. Tiers, Minnesota Mining and Manufacturing Company, St. Paul 6, Minn.; and Nuclear Magnetic Resonance by N. F. Chamberlain et al., Esso Research and Engineering Company, P. O. Box 4255, Baytown, Texas. A large number of values have been published by K. Nukada et al., *Anal. Chem., 35,* 1892 (1963). Also see Jackman and Sternhell[4] for an extensive list of shift positions.

An alternative to use of the first two tables of Appendix B is the application of the data from the table on p. 215 for calculation of chemical shifts of protons with various α and β substituents.

The use of the table on p. 215 is illustrated by the following examples:

$$HOCH_2CHCH_2OH \quad\quad -\overset{|}{C}H- \quad\quad \delta 1.55 \text{(unsubstituted)}$$

HOCH₂CHCH₂OH	$-\overset{\vert}{C}H-$	δ1.55(unsubstituted)
$\vert$	α–OH,	+ 2.20($\pm$ 0.10)
OH	2(β-CH),	+ 0.00
		δ3.75($\pm$ 0.10)(calculated)
		δ3.69(observed)

$$-CH_2- \quad \delta 1.20\text{(unsubstituted)}$$
$$\alpha-OH, + 2.30(\pm 0.03)$$
$$\beta-OH, + \underline{0.13(\pm 0.03)}$$
$$\delta 3.63(\pm 0.06)\text{(calculated)}$$
$$\delta 3.61\text{(observed)}$$

In a similar manner, the following shifts are calculated for β,β'-dibromoethyl ether:

$$(Br-CH_2-CH_2-)_2O$$
$$\delta 3.53(\pm 0.11), \delta 4.15(\pm 0.18)$$

Compare the calculated results to those calculated by another method in Compound Number 6 − 5 and those observed on the NMR spectrum in that problem.

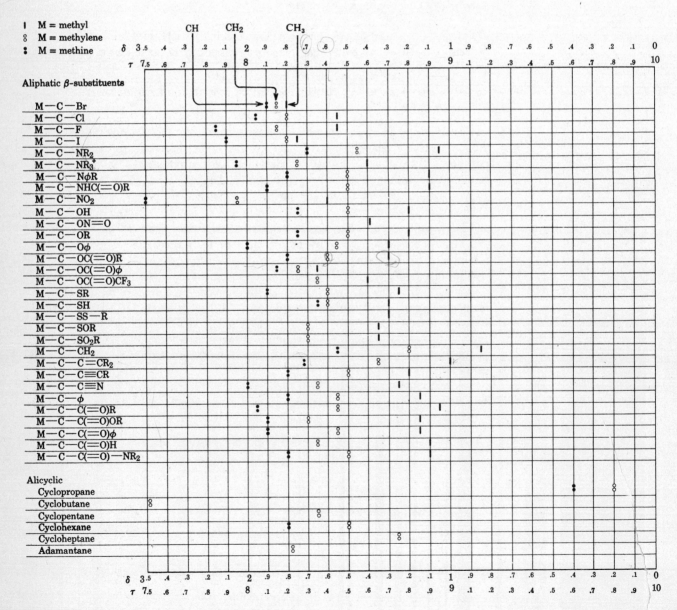

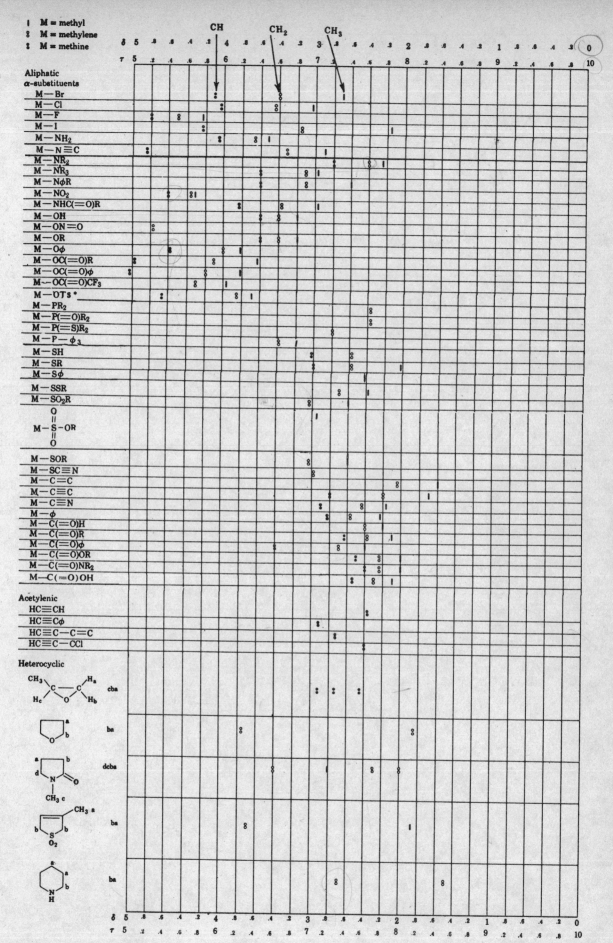

*OTs=p-toluenesulfonyloxy

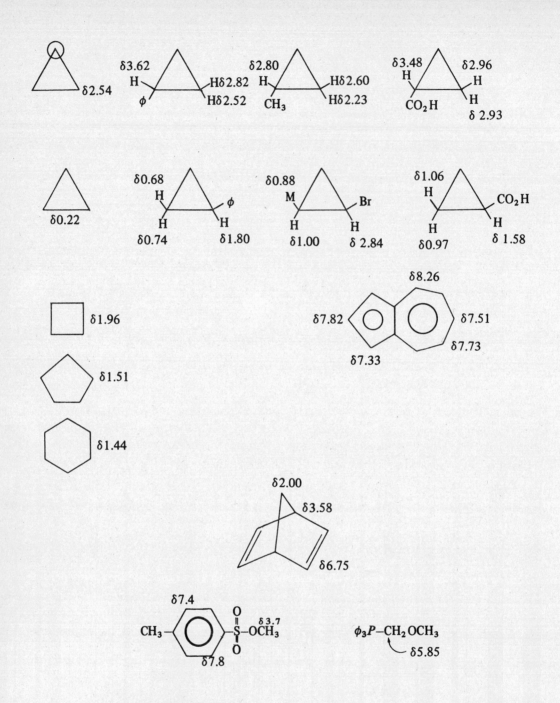

Substituent Effects on Chemical Shift

$$\underset{\beta}{C}-\underset{\alpha}{C}-H$$

Substituent	Type of Hydrogen[b]	Alpha Shift[c]	Beta Shift
$\overset{O}{\overset{\|}{-OC}}-R$, $\overset{O}{\overset{\|}{-OC}}-OR$,	CH$_3$	2.88 (12) ± 0.08	0.38 (16) ± 0.08
$-OAr$	CH$_2$	2.98 (13) ± 0.10	0.43 (2) ± 0.03
	CH	3.43 (2) ± 0.00 (esters only)	–
$-OH$	CH$_3$	2.50 (1)	0.33 (4) ± 0.05
	CH$_2$	2.30 (5) ± 0.03	0.13 (4) ± 0.03
	CH	2.20 (3) ± 0.10	–[d]
$-Cl$	CH$_3$	2.43 (1)	0.63 (5) ± 0.05
	CH$_2$	2.30 (7) ± 0.08	0.53 (4) ± 0.08
	CH	2.55 (3) ± 0.10	0.03 (1)
$-Br$	CH$_3$	1.80 (1)	0.83 (4) ± 0.03
	CH$_2$	2.18 (7) ± 0.08	0.60 (3) ± 0.08
	CH	2.68 (3) ± 0.10	0.25 (1)
$-I$	CH$_3$	1.28 (1)	1.23 (3) ± 0.03
	CH$_2$	1.95 (4) ± 0.05	0.58 (3) ± 0.05
	CH	2.75 (3) ± 0.00	0.00 (1)
$-OR$ (R is saturated)	CH$_3$	2.43 (5) ± 0.03	0.33 (2) ± 0.05
	CH$_2$	2.35 (5) ± 0.10	0.15 (2) ± 0.03
	CH	2.00 (1)	–[d]
Aryl	CH$_3$	1.40 (10) ± 0.03	0.35 (5) ± 0.03
	CH$_2$	1.45 (2) ± 0.03	0.53 (2) ± 0.03
	CH	1.33 (1)	–
$-NRR'$	CH$_3$	1.30 (5) ± 0.05	0.13 (3) ± 0.05
	CH$_2$	1.33 (5) ± 0.13	0.13 (3) ± 0.13
	CH	1.33 (1)	–
$\overset{O}{\overset{\|}{-CR}}$, where R is alkyl,	CH$_3$	1.23 (12) ± 0.08	0.18 (1)
aryl, OH, OR', H, CO,	CH$_2$	1.05 (12) ± 0.10	0.31 (2) ± 0.05
or N	CH	1.05 (1)	–
$-C=C-$	CH$_3$	0.78 (13)	–
	CH$_2$	0.75 (10)	0.10 (1)
	CH	–	–
$-C=C-\underset{\overset{\|}{X}}{C}-R$ (X=C or O)	CH$_3$	1.08 (7) ± 0.08	–

[a]Taken from the Ph.D. Dissertation of Thomas Curphey, Harvard University, by permission of T. J. Curphey.
[b]Standard positions are CH$_3$, $\delta 0.87 (\tau 9.13)$; CH$_2$, $\delta 1.20 (\tau 8.80)$; CH, $\delta 1.55 (\tau 8.45)$.
[c]Number in parentheses is the number of examples used in determining the shift.
[d]Assumedly negligible shift.

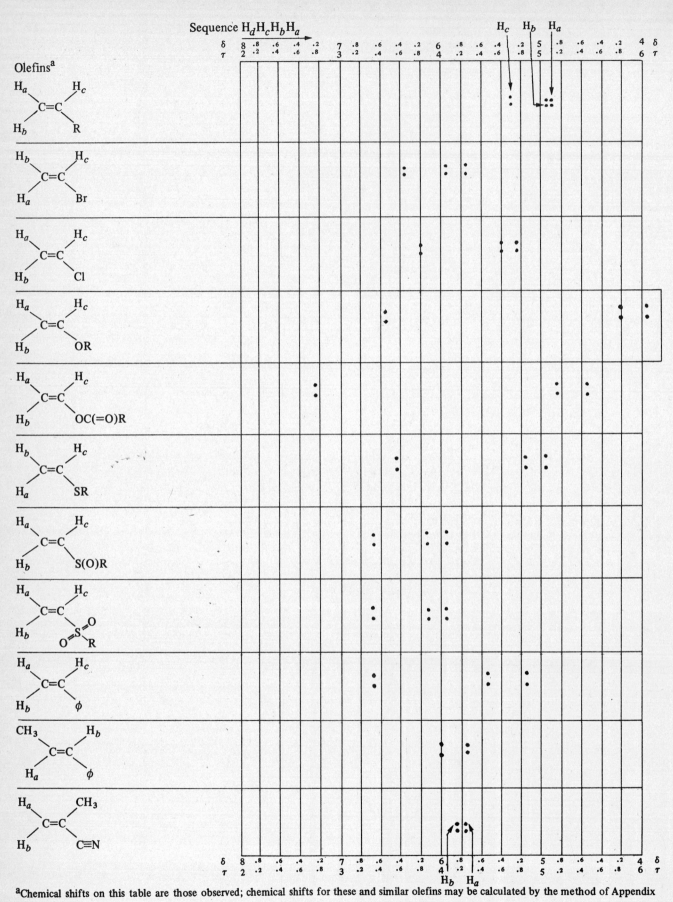

aChemical shifts on this table are those observed; chemical shifts for these and similar olefins may be calculated by the method of Appendix E for olefinic protons.

216   spectrometric identification of organic compounds

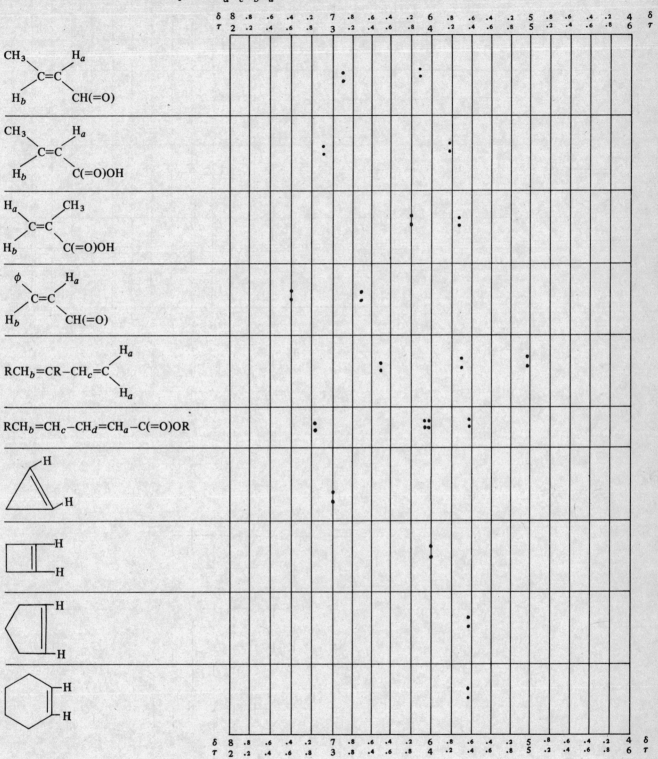

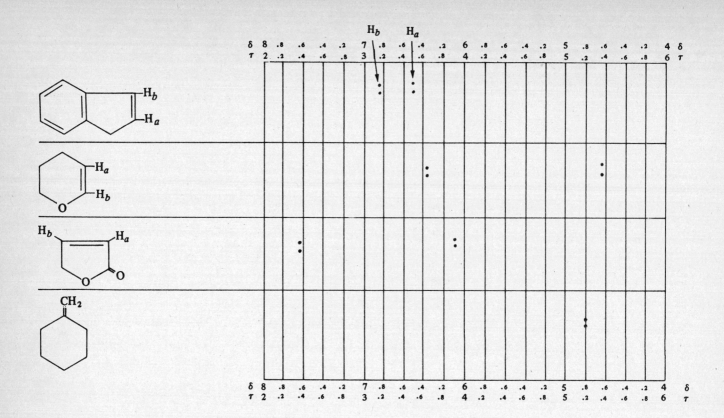

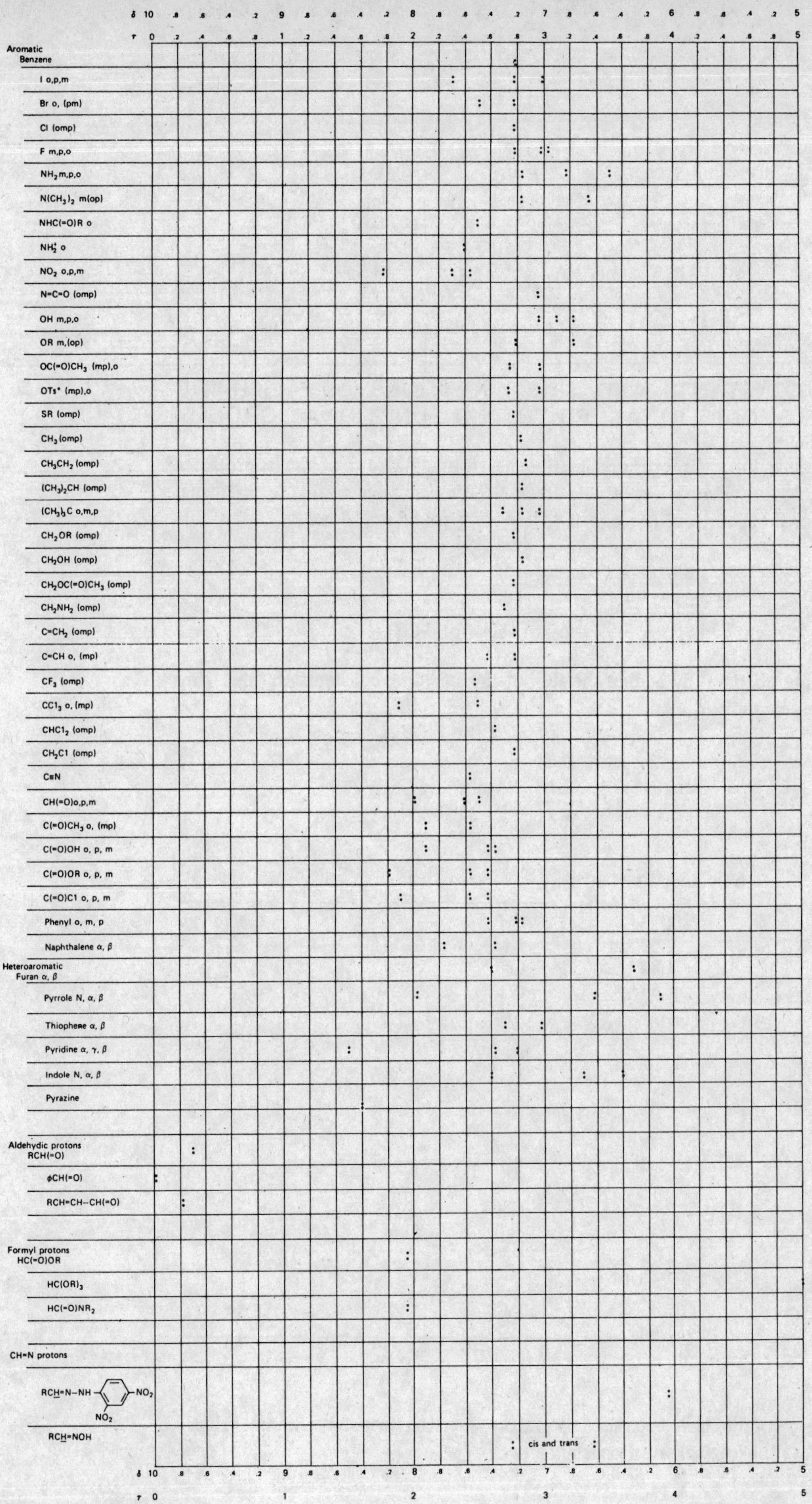
δ 10 .8 .6 .4 .2 9 .8 .6 .4 .2 8 .8 .6 .4 .2 7 .8 .6 .4 .2 6 .8 .6 .4 .2 5
τ 0 .2 .4 .6 .8 1 .2 .4 .6 .8 2 .2 .4 .6 .8 3 .2 .4 .6 .8 4 .2 .4 .6 .8 5
Aromatic
Benzene
I o,p,m
Br o, (pm)
Cl (omp)
F m,p,o
NH2m,p,o
N(CH3)2 m(op)
NHC(=O)R o
NH3+ o
NO2 o,p,m
N=C=O (omp)
OH m,p,o
OR m,(op)
OC(=O)CH3 (mp),o
OTs* (mp),o
SR (omp)
CH3 (omp)
CH3CH2 (omp)
(CH3)2CH (omp)
(CH3)3C o,m,p
CH2OR (omp)
CH2OH (omp)
CH2OC(=O)CH3 (omp)
CH2NH2 (omp)
C=CH2 (omp)
C=CH o, (mp)
CF3 (omp)
CCl3 o, (mp)
CHCl2 (omp)
CH2Cl (omp)
C≡N
CH(=O)o,p,m
C(=O)CH3 o, (mp)
C(=O)OH o, p, m
C(=O)OR o, p, m
C(=O)Cl o, p, m
Phenyl o, m, p
Naphthalene α, β
Heteroaromatic
Furan α, β
Pyrrole N, α, β
Thiophene α, β
Pyridine α, γ, β
Indole N, α, β
Pyrazine
Aldehydic protons
RCH(=O)
φCH(=O)
RCH=CH–CH(=O)
Formyl protons
HC(=O)OR
HC(OR)3
HC(=O)NR2
CH=N protons
RCH=N–NH
NO2
NO2
RCH=NOH
cis and trans
δ 10 .8 .6 .4 .2 9 .8 .6 .4 .2 8 .8 .6 .4 .2 7 .8 .6 .4 .2 6 .8 .6 .4 .2 5
τ 0 1 2 3 4 5

*OTs = toluenesulfonyloxy group

appendix c

SHIELDING CONSTANTS AND SHIFT POSITIONS OF ALIPHATIC METHYLENE GROUPS ATTACHED TO TWO FUNCTIONAL GROUPS.

To obtain a τ value for a disubstituted methylene group, the sum of the appropriate shielding constants for X and Y is subtracted from 9.77 which is the τ value for methane. Thus the value for the methylene protons of $\phi\text{-CH}_2-\text{Br}$ is found as follows:

$$
\begin{array}{ll}
\phi = 1.85 & 9.77 \\
\text{Br} = \underline{2.33} & \underline{-4.18} \\
\phantom{\text{Br} = }4.18 & 5.59 = \text{calc } \tau \text{ value for } -\text{CH}_2-.
\end{array}
$$

Delta values are obtained by adding the sum of the shielding constants to 0.23. Thus the δ value in this example is

$$
\begin{array}{r}
0.23 \\
+4.18 \\
\hline
4.41
\end{array}
$$

X or Y	Shielding Constants	X or Y	Shielding Constants
$-\text{Br}$	2.33	$-\text{CH}_3$	0.47
$-\text{Cl}$	2.53	$-\text{C}=\text{C}$	1.32
$-\text{I}$	1.82	$-\text{C}\equiv\text{C}$	1.44
$-\text{NR}_2$	1.57	$-\phi$	1.85
$-\text{NHCR}$	2.27	$-\text{CF}_2$	1.21
$\quad \overset{\parallel}{\underset{\text{O}}{}}$		$-\text{CF}_3$	1.14
		$-\text{C}\equiv\text{N}$	1.70
$-\text{N}_3$	1.97	O	
$-\text{OH}$	2.56	$\parallel$	
$-\text{OR}$	2.36	$-\text{CR}$	1.70
$-\text{O}\phi$	3.23	O	
$\quad \overset{\text{O}}{\underset{\parallel}{}}$		$\parallel$	
$-\text{OCR}$	3.13	$-\text{C}-\phi$	1.84
$\quad \overset{\text{O}}{\underset{\parallel}{}}$		OR	
$-\text{OS}-\text{R}$	3.13	$-\text{C}=\text{O}$	1.55
$\quad \underset{\text{O}}{\overset{\parallel}{}}$		NR_2	
$-\text{SR}$	1.64	$-\text{C}=\text{O}$	1.59

The shielding constants have been used to prepare the chart on p. 221. Several values have been added to the original set of constants.[18]

τ Values for Methylene Groups Attached to Two Functional Groups X—CH₂—Y

Note: The upper number in each box is an experimental value; the lower number is calculated from the shielding constants. Only τ values are given.

Functional Group	−Br	−Cl	−I	−NR₂	−N₃	−NHCR	O=Cφ	−OH	−OR	−Oφ	O=OCR	−SR	−CH₃	−C=C	−C≡C	−φ	−CF₂	−CF₃	−C≡N	O=CR	O=COR	O=CNR₂
−Br	5.06 / 5.11	4.84 / 4.91	5.62	5.87	5.47	5.17	5.60	4.88	5.08	3.64 / 4.21	4.31	5.80	6.57 / 6.97	6.07 / 6.12	6.10 / 6.00	5.65 / 5.59	6.24	6.30	5.74	5.74	6.30 / 5.89	6.08 / 5.85
−Cl		4.67 / 4.71	5.01 / 5.42	5.63	5.27	4.87	5.40	4.68	4.60 / 4.88	4.01	4.11	5.60	6.43 / 6.77	5.92	5.91 / 5.80	5.50 / 5.39	6.03	6.10	5.93 / 5.54	5.54	5.95 / 5.69	5.83 / 5.65
−I			6.10 / 6.13	6.38	5.98	5.68	6.11	5.39	5.59	4.84	4.94	6.31	6.80 / 7.48	6.13 / 6.63	6.51	6.10	6.74	6.44 / 6.81	6.35 / 6.25	6.25	6.40	6.35 / 6.36
−NR₂				6.90 / 6.63	6.23	5.93	6.36	5.65	5.85	4.97	5.07	6.56	7.37 / 7.73	6.70 / 6.88	6.50 / 6.76	6.52 / 6.35	6.99	7.06	6.50	6.50	6.83 / 6.65	6.61
−N₃					5.85	5.55	5.98	5.24	5.36	4.44	4.67	6.16	7.33	6.48	6.33	5.96	6.59	6.66	6.10	6.10	6.25	6.21
−NHCR						5.27	5.66	4.94	5.06	4.14	4.37	5.86	7.03	6.18	6.03	5.66	6.29	6.36	5.80	5.90 / 5.80	5.95	5.91
O=Cφ							6.09	5.37	5.57	4.70	4.80	6.29	7.46	6.61	6.49	6.08	6.72	6.79	6.23	5.58 / 4.90	6.38	6.34
−OH								5.45 / 4.65	4.85	3.98	4.08	5.57	6.30 / 6.74	5.87	5.72	5.42	5.99 / 5.99	6.07	5.51	5.51	5.66	5.62
−OR									5.45 / 5.05	4.18	4.28	5.77	6.60 / 6.94	6.05 / 6.09	5.97	5.30 / 5.56	6.20	5.27	5.80 / 5.71	5.71	5.87 / 5.78	5.82
−Oφ										3.31	3.41	4.90	6.07	5.22	5.10	5.10 / 4.69	5.33	5.40	4.84	4.84	4.91	5.55 / 4.95
O=OCR											3.54	5.00	5.75 / 6.17	5.32	5.29 / 5.20	4.92 / 4.79	5.43	5.46	4.90	4.90	5.09	5.05
−SR												6.49	7.47 / 7.66	6.92 / 6.81	6.69	6.32 / 6.28	6.92	6.99	6.43	6.43	6.58	6.54
−CH₃													8.83	8.10 / 7.98	7.86	7.45	8.09	8.16	7.60	7.53 / 7.60	7.75 / 7.75	7.77 / 7.71
−C=C														7.40 / 7.13	6.61 / 7.01	6.70 / 6.60	7.24	7.31	6.85 / 6.75	6.75	7.00 / 6.90	6.86
−C≡C															6.89	6.48	7.12	7.19	6.63	6.63	6.78	6.74
−φ																6.03 / 6.07	6.71	6.50 / 6.78	6.35 / 6.22	6.45 / 6.22	6.60 / 6.37	6.34
−CF₂																	7.37	7.42	6.88	6.88	7.01	6.97
−CF₃																		7.49	6.93	6.93	7.08	7.04
−C≡N																			6.37	6.37	6.52	6.48
O=CR																				6.40 / 6.37	6.63 / 6.52	6.48
O=COR																					6.65 / 6.67	6.63
O=CNR₂																						6.70 / 6.59

appendix d

PROTONS SUBJECT TO HYDROGEN-BONDING EFFECTS

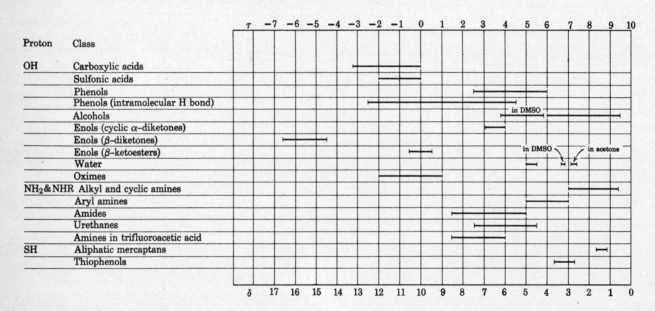

CALCULATION OF CHEMICAL SHIFTS FOR OLEFINIC PROTONS

Table A Formula and Substituent constants (Z_i) for Chemical Shift of Substituted Ethylenes in CCl_4

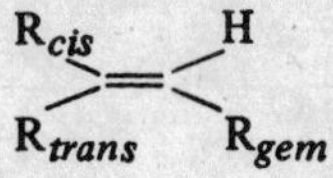

$$
\begin{array}{c}
R_{cis} \qquad H \\
\diagdown C = C \diagup \\
R_{trans} \qquad R_{gem}
\end{array}
$$

$$\delta = 5.28 + \sum_i Z_i$$

Substituent R	Z_i			Substituent R	Z_i		
	gem	cis	trans		gem	cis	trans
—H	0	0	0	$-\overset{\displaystyle H}{\underset{}{C}}=O$	1.03	0.97	1.21
—Alkyl	0.44	−0.26	−0.29				
—Alkyl-Ring	0.71	−0.33	−0.30	$-\overset{\displaystyle N}{\underset{}{C}}=O$	1.37	0.93	0.35
—CH_2O, —CH_2	0.67	−0.02	−0.07				
—CH_2S	0.53	−0.15	−0.15	$-\overset{\displaystyle Cl}{\underset{}{C}}=O$	1.10	1.41	0.99
—CH_2Cl, —CH_2Br	0.72	0.12	0.07	—OR, R: aliph.	1.18	−1.06	−1.28
—CH_2N	0.66	−0.05	−0.23	—OR, R[c]	1.14	−0.65	−1.05
—C≡C	0.50	0.35	0.10	—OCOR	2.09	−0.40	−0.67
—C≡N	0.23	0.78	0.58	—Aromatic	1.35	0.37	−0.10
—C=C[a]	0.98	−0.04	−0.21	—Cl	1.00	0.19	0.03
—C=C[b]	1.26	0.08	−0.01	—Br	1.04	0.40	0.55
—C=O[a]	1.10	1.13	0.81				
—C=O[b]	1.06	1.01	0.95	$-N\diagdown{\,}^{R}_{R}$ R: aliph.	0.69	−1.19	−1.31
—COOH[a]	1.00	1.35	0.74				
—COOH[b]	0.69	0.97	0.39				
—COOR[a]	0.84	1.15	0.56	$-N\diagdown{\,}^{R}_{R^{c}}$	2.30	−0.73	−0.81
—COOR[b]	0.68	1.02	0.33	—SR	1.00	−0.24	−0.04
				—SO_2	1.58	1.15	0.95

[a]For a single such group in conjugation with the C=C of interest.
[b]For such a group in conjugation with the double bond of interest and conjugated with a further substituent.
[c]R group contains unsaturation in conjugation with double bond of interest.

Examples:
trans-stilbene:

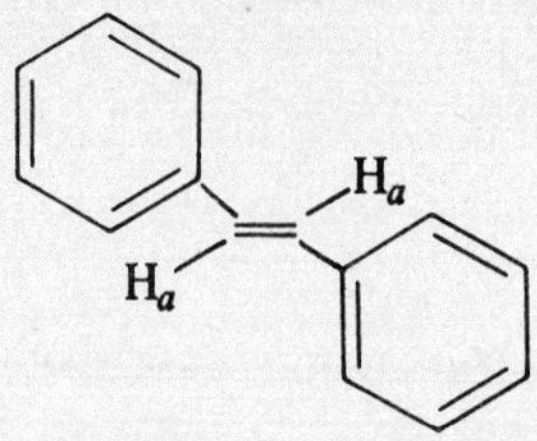

δ_{H_a} exp.: 6.99 ppm (CCl$_4$)

(for each H$_a$)
δ_{calc} = 5.28 + 1.35 + 0.37 = 7.00

cis-stilbene:

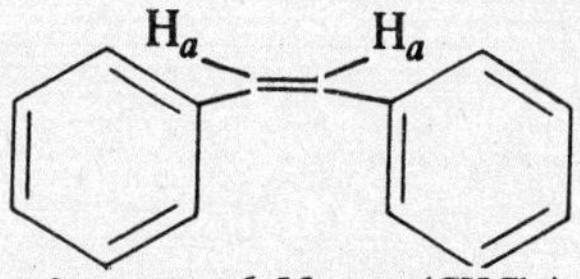

δ_{H_a} exp.: 6.55 ppm (CHCl$_3$)

δ_{calc} = 5.28 + 1.35 + (− 0.10) = 6.53.

Table B

Percent of 1070 Cases of Application of Above Table that Agree within Given Range ($\Delta\delta$) of Calculated vs. Observed Chemical Shift

$\Delta\delta$ (ppm)	%
0.10	51.0
0.15	73.3
0.20	86.3
0.30	96.3
0.45	99.6

Table C **Specific Systems for Which Calculated–Observed Ranges ($\Delta\delta$) are Greater Than 0.45 (Table A, p. 223, Applications)**

System	Solvent	δH_a exp.	δH_a calc
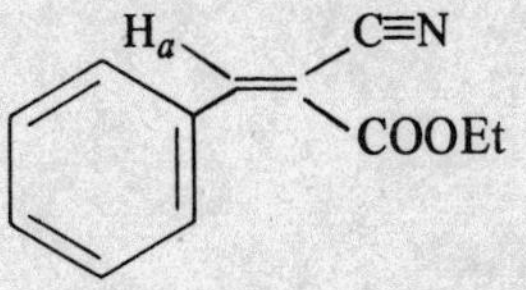	CDCl$_3$	8.22	7.74
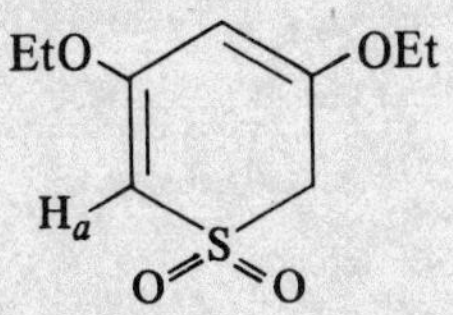	CDCl$_3$	5.20	5.79
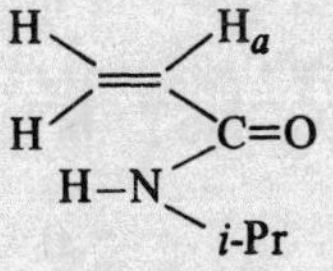	CDCl$_3$	6.14	6.65
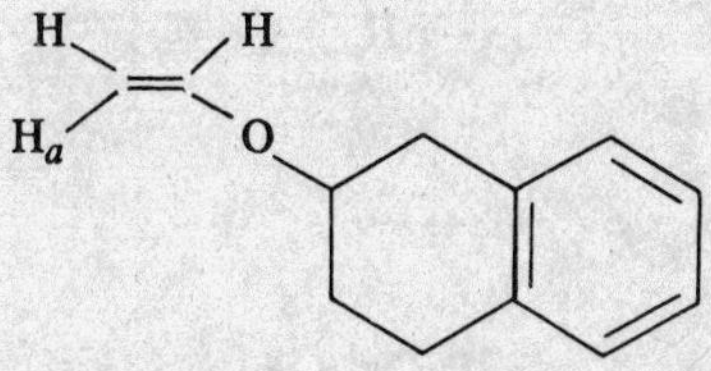	CCl$_4$	4.76	4.22

Reference: C. Pascual, J. Meier, and W. Simon, Helv. Chim. Acta., *49*, 164 (1966).

appendix f

PROTON SPIN-SPIN-COUPLING CONSTANTS* (Absolute Values)†

Type	J_{ab} (Hz)	J_{ab} Typical	Type	J_{ab} (Hz)	J_{ab} Typical
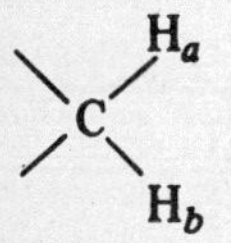 (C with H_a and H_b)	0 to 30	12 to 15	(cyclopropane ring with H_a, H_b) (cis or trans)	2-8	3-5
CH_a–CH_b (free rotation)	6-8	7	CH_a–OH_b (no exchange)	4-10	5
CH_a–C–CH_b	0-1	0	CH_a–$\overset{O}{\overset{\|}{C}}H_b$	1-3	2-3
(cyclohexane ring with H_a, H_b)			C=CH_a–$\overset{O}{\overset{\|}{C}}H_b$	5-8	6
ax.-ax.	6-14	8-10	(H_a / C=C / H_b, cis double bond)	12-18	17
ax.-eq.	0-5	2-3			
eq.-eq.	0-5	2-3	(C=C with H_a and H_b geminal)	0-3	0-2
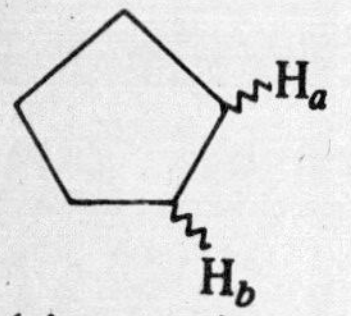 (cyclopentane ring with H_a, H_b) (cis or trans)	0-7	4-5	(H_a / C=C / H_b trans)	6-12	10
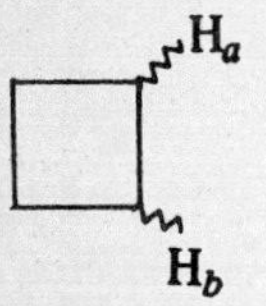 (cyclobutane ring with H_a, H_b) (cis or trans)	6-10	8	CH_a / C=C / CH_b	0-3	1-2

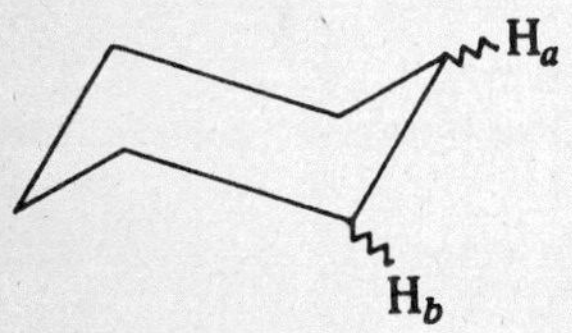

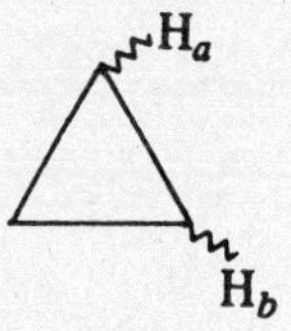

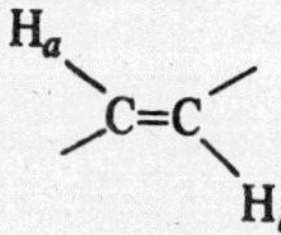

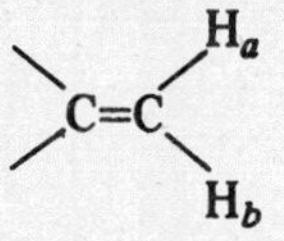

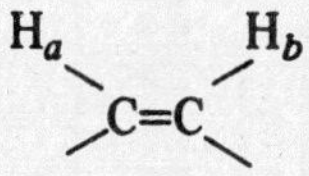

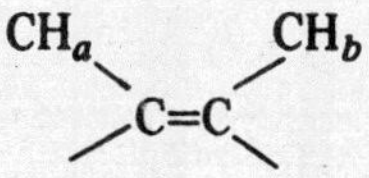

*Compiled by Varian Associates.

†Signs of typical coupling constants can be obtained from the Appendixes of Bovey's book.[1]

spectrometric identification of organic compounds

Type	J_{ab} (Hz)	J_{ab} Typical
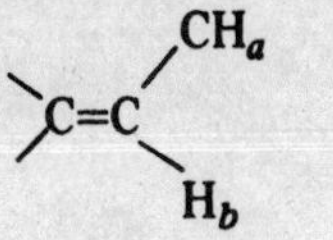 $C=C$ with CH_a / H_b	4-10	7
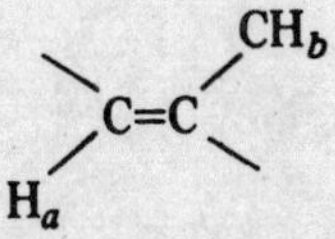 $C=C$ with CH_b / H_a	0-3	1.5
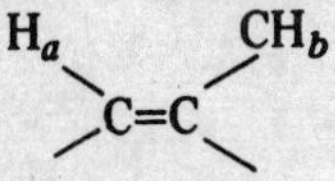 $C=C$ with H_a / CH_b	0-3	2
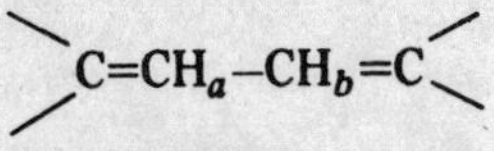 $C=CH_a-CH_b=C$	9-13	10
$CH_a-C\equiv CH_b$	2-3	
$-CH_a-C\equiv C-CH_b-$	2-3	
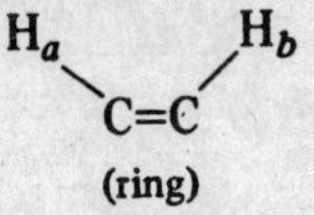 $C=C$ (ring) H_a / H_b — 3 mem.	0.5-2.0	
4 mem.	2.5-4.0	
5 mem.	5.1-7.0	
6 mem.	8.8-11.0	
7 mem.	9-13	
8 mem.	10-13	
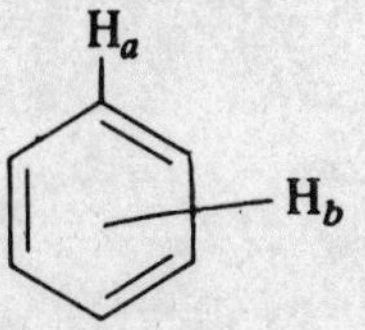 benzene ring H_a / H_b — J (ortho)	6-10	9
J (meta)	1-3	3
J (para)	0-1	~0
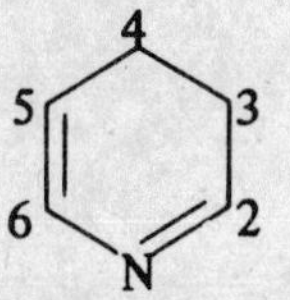 pyridine — J (2-3)	5-6	5
J (3–4)	7-9	8
J (2-4)	1-2	1.5
J (3-5)	1-2	1.5
J (2-5)	0-1	1
J (2-6)	0-1	~0
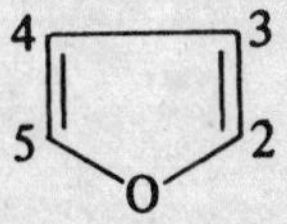 furan — J (2-3)	1.3-2.0	1.8
J (3-4)	3.1-3.8	3.6
J (2-4)	0-1	~0
J (2-5)	1-2	1.5
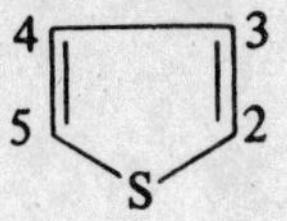 thiophene — J (2-3)	4.9-6.2	5.4
J (3-4)	3.4-5.0	4.0
J (2-4)	1.2-1.7	1.5
J (2-5)	3.2-3.7	3.4

Type	J_{ab} (Hz)	J_{ab} Typical
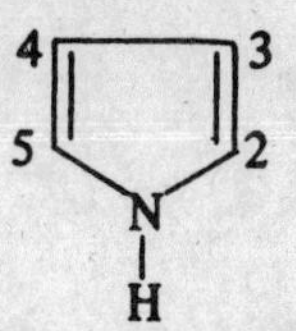 pyrrole — J (1-2)	2-3	
J (1-3)	2-3	
J (2-3)	2-3	
J (3-4)	3-4	
J (2-4)	1-2	
J (2-5)	1.5-2.5	
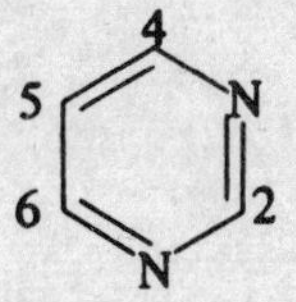 pyrimidine — J (4-5)	4-6	
J (2-5)	1-2	
J (2-4)	0-1	
J (4-6)	?	
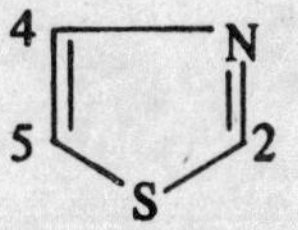 thiazole — J (4-5)	3-4	
J (2-5)	1-2	
J (2-4)	~0	

Proton-Fluorine

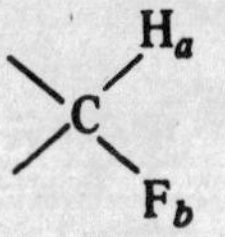

Type	J_{ab} (Hz)
C with H_a / F_b	44-81
CH_a-CF_b	3-25
CH_a-C-CF_b	0
$C=C$ with H_a / F_b (cis)	1-8
$C=C$ with H_a / F_b (trans)	12-40
fluorobenzene H_a	o 6-10
	m 5-6
	p 2

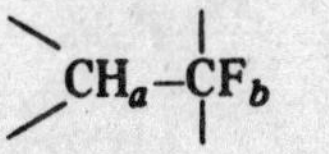

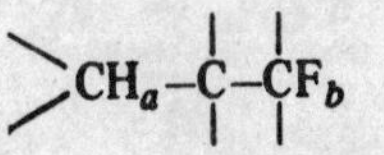

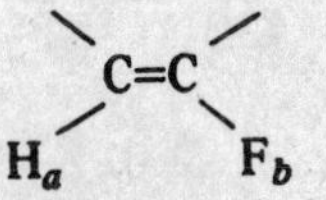

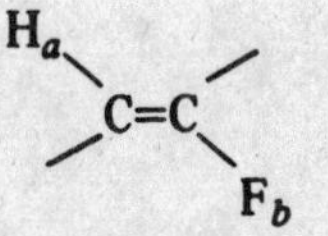

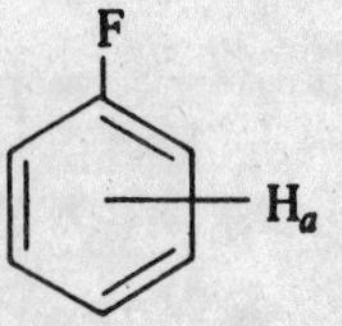

Proton-Phosphorous

Type	J_{ab} (Hz)
$(CH_3)_3PO$	13.4 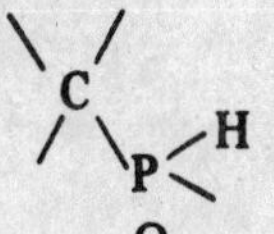
CH_3-C-P	10-18
$-C-CH_2-P$	0.5-13
$C-P(H)$	180-200
$P(O)(H)$	630-707
$(CH_3)_3P$	2.7

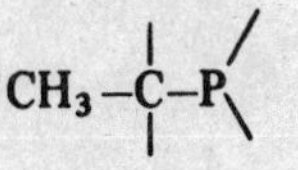

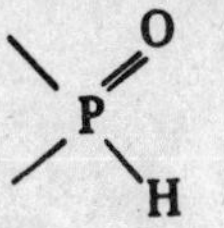

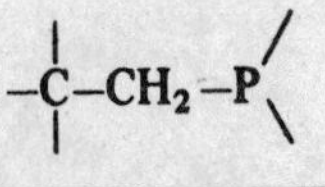

appendix g

PROPERTIES OF SEVERAL NUCLEI[a]

Isotope	NMR Frequency MHz for a 10 Kilogauss Field	Natural Abundance %	Relative Sensitivity at Constant Field	Magnetic Moment μ	Spin Number I	Electrical Quadrupole Moment $e \times 10^{-24}$ cm^2
^{1}H	42.576	99.9844	1.000	2.79268	1/2	—
^{2}H	6.5357	1.56×10^{-2}	9.64×10^{-2}	0.85739	1	2.77×10^{-3}
^{3}H	45.414	—	1.21	2.9788	1/2	—
^{10}B	4.575	18.83	1.99×10^{-2}	1.8005	3	7.4×10^{-2}
^{11}B	13.660	81.17	0.165	2.6880	3/2	3.55×10^{-2}
^{12}C	—	98.9	—	—	0	—
^{13}C	10.705	1.108	1.59×10^{-2}	0.70220	1/2	—
^{14}N	3.076	99.635	1.01×10^{-3}	0.40358	1	7.1×10^{-2}
^{15}N	4.315	0.365	1.04×10^{-3}	−0.28304	1/2	—
^{16}O	—	99.76	—	—	0	—
^{17}O	5.772	3.7×10^{-2}	2.91×10^{-2}	−1.8930	5/2	-4.0×10^{-3}
^{19}F	40.055	100	0.834	2.6273	1/2	—
^{28}Si	—	92.28	—	—	0	—
^{29}Si	8.458	4.70	7.85×10^{-2}	−0.55548	1/2	—
^{30}Si	—	3.02	—	—	0	—
^{31}P	17.236	100	6.64×10^{-2}	1.1305	1/2	—
^{32}S	—	95.06	—	—	0	—
^{33}S	3.266	0.74	2.26×10^{-3}	0.64274	3/2	−0.053
^{34}S	—	4.2	—	—	0	—
^{35}Cl	4.172	75.4	4.71×10^{-3}	0.82091	3/2	-7.9×10^{-2}
^{37}Cl	3.472	24.6	2.72×10^{-3}	0.68330	3/2	-6.21×10^{-2}
^{79}Br	10.667	50.57	7.86×10^{-2}	2.0991	3/2	0.34
^{81}Br	11.499	49.43	9.84×10^{-2}	2.2626	3/2	0.28
^{127}I	8.519	100	9.35×10^{-2}	2.7937	5/2	−0.75

[a]Varian Associates NMR Table, 4th ed., 1964.

MAGNETOGYRIC RATIOS (γ_N) OF A FEW NUCLEI

Nucleus	γ_N (radians sec^{-1} gauss^{-1})
^{1}H	26,753
^{13}C	6,728
^{15}N	−2,712
^{19}F	25,179
^{29}Si	−5,319
^{31}P	10,840
^{2}D	4,107
^{7}Li	10,398
^{14}N	1,934
^{17}O	−3,628
^{23}Na	7,081
^{33}S	2,054
^{35}Cl	2,624
^{37}Cl	2,184
^{39}K	1,250

ultraviolet spectrometry

I. INTRODUCTION

Molecular absorption in the ultraviolet and visible region of the spectrum is dependent on the electronic structure of the molecule. Absorption of energy is quantized and results in the elevation of electrons from orbitals in the ground state to higher-energy orbitals in an excited state. For many electronic structures, the absorption does not occur in the readily accessible portion of the ultraviolet region. In practice, ultraviolet spectrometry is for the most part limited to conjugated systems.

There is, however, an advantage to the selectivity of ultraviolet absorption: characteristic groups may be recognized in molecules of widely varying complexities. A large portion of a relatively complex molecule may be transparent in the ultraviolet so that we may obtain a spectrum similar to that of a much simpler molecule. Thus the spectrum of the male hormone testosterone closely resembles the spectrum of mesityl oxide. The absorption results from the conjugated enone structure of the two compounds.

A detailed mathematical treatment of the origin of ultraviolet or electronic spectra is beyond the scope of this chapter. Rather it is our objective to point out correlations between spectra and structure to be used by the organic chemist. However, enough theory to rationalize these correlations will be presented.

An ultraviolet spectrum is a plot of the wavelength or frequency of absorption versus the absorption intensity (transmittance or absorbance, see Figure 1). The data are frequently shown as a graphical plot or tabular presentation of wavelength versus molar absorptivity or log of the molar absorptivity, ϵ_{max} or log ϵ_{max} (see Fig. 1a). The use of molar absorptivity as the unit of absorption intensity has the advantage that all intensity values refer to the same number of absorbing species. A tabular presentation is used in Chapters 6 and 7 of this text.

231

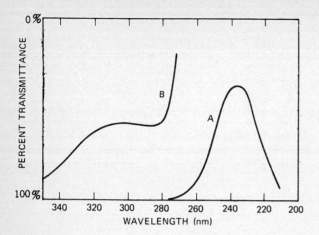

Fig. 1.

Ultraviolet spectrum of mesityl oxide in 95% ethanol: Trace A displays the π $\longrightarrow$ π* band, concentration of 6.25×10^{-5} mole per liter, and trace B displays the n $\longrightarrow$ π* band, concentration 6.29×10^{-3} mole per liter. (Recorded on a Bausch and Lomb Spectronic 505 ultraviolet and visible spectrophotometer.) (Source: Daniel J. Pasto and Carl R. Johnson, Organic Structure Determination, *copyright 1969, p. 107, Reprinted by permission of Prentice-Hall, Inc., Englewood Cliffs, N.J.)*

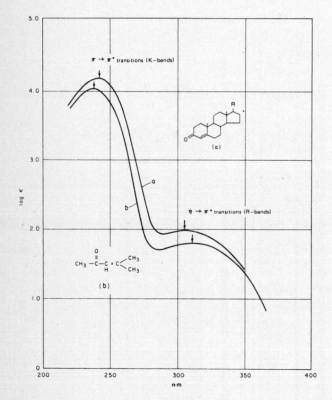

Fig. 1a.

Ultraviolet absorption of cholesta-4-ene-3-one (a) and mesityl oxide (b).

An abundance of reference material relating to the theory and interpretation of ultraviolet spectra is available.[1-11] Two of the most useful references for the organic chemist are the texts by Stern and Timmons[1] and by A. I. Scott.[2] The latter is particularly recommended to the natural-products chemist. The text by Jaffé and Orchin[3] is an excellent source of theory for both the spectroscopist and the organic chemist. Several compilations of ultraviolet spectra and absorption data are available.[12-22] The Sadtler spectra[12] and the volumes of "Electronic Spectral Data"[13-15] are particularly useful to the organic chemist.

II. THEORY

The ultraviolet portion of the electromagnetic spectrum is indicated in Figure 2. Wavelengths in the ultraviolet region are usually expressed in nanometers (1 nm = 10^{-7} cm) or angstroms, Å (1Å = 10^{-8} cm). Occasionally, absorption is reported in wavenumbers, ($\bar{\nu}$ = cm^{-1} = 10^7/nm = 10^8/Å). We are primarily interested in the near ultraviolet (quartz) region extending from 200 to 380 nm. The atmosphere is transparent in this region, which is readily accessible with quartz optics. Atmospheric absorption starts near 200 nm and extends into the shorter wavelength region which is accessible through vacuum ultraviolet spectrometry.

The total energy of a molecule is the sum of its binding or electronic energy, its vibrational energy and its rotational energy. The magnitude of these energies decreases in the following order: E_{elec}, E_{vib}, and E_{rot}. Energy absorbed in the ultraviolet region produces changes in the electronic energy of the molecule resulting from transitions of valence electrons in the molecule. These transitions consist of the excitation of an electron from a filled molecular orbital (usually a nonbonding p or bonding π-orbital) to the next higher energy orbital (an antibonding, π* or σ*, orbital). The antibonding orbital is designated by an asterisk. Thus, the transition of an electron from a π-bonding orbital to an antibonding (π*) orbital is indicated π → π*.

The concept of an antibonding orbital can be explained simply by consideration of the ultraviolet absorption of ethylene. The ethylenic double bond, in the ground state, consists of a pair of bonding σ-electrons and a pair of bonding π-electrons. On absorption of ultraviolet radiation near 165 nm, one of the bonding π-electrons is raised to the next higher energy orbital, an antibonding π-orbital. The orbitals occupied by the π-electron in the ground state and in the excited state are diagrammed in Figure 3.

The shaded volumes indicate regions of maximum electron density. It can be seen that the antibonding π-electron no longer contributes appreciably to the force constant of the C-to-C bond. In fact it negates the bonding power of the remaining unexcited π-electron; the olefinic

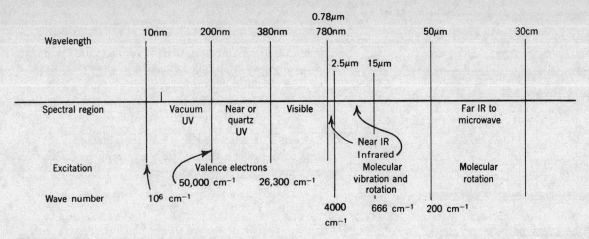

Fig. 2.
Electromagnetic spectrum.

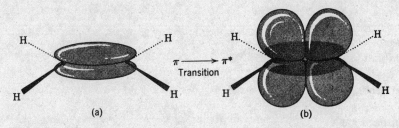

Fig. 3.
π and π* orbitals. (a) Bonding orbital, π. Both π electrons occupying bonding orbital. A bonding π orbital has a nodal plane in the plane of the molecule. (b) Antibonding orbital, π*. One π electron in bonding orbital, one in antibonding orbital. An (antibonding) π* orbital has an additional nodal plane perpendicular to the plane of the molecule and bisecting the carbon-carbon bond.

bond has considerable single-bond character in the excited state.

The relationship between the energy absorbed in an electronic transition and the frequency (ν), wavelength (λ), and wavenumber ($\bar{\nu}$) of radiation producing the transition is

$$\Delta E = h\nu = \frac{hc}{\lambda} = h\bar{\nu}c$$

where h is Planck's constant and c is the velocity of light. ΔE is the energy absorbed in an electronic transition in a molecule from a low-energy state (ground state) to a high-energy state (excited state). The energy absorbed is dependent on the energy difference between the ground state and the excited state; the smaller the difference in energy, the longer the wavelength of absorption. The excess energy in the excited state may result in disassociation or ionization of the molecule, or it may be reemitted as heat or light. The release of energy as light results in fluorescence or phosphorescence.

Since ultraviolet energy is quantized, the absorption spectrum arising from a single electronic transition should consist of a single, discrete line. A discrete line is not obtained since electronic absorption is superimposed upon rotational and vibrational sublevels. The spectra of simple molecules, in the gaseous state, consist of narrow absorption peaks, each representing a transition from a particular combination of vibrational and rotational levels in the electronic ground state to a corresponding combination in the excited state. This is shown schematically in Figure 4 in which the vibrational levels are designated ν_0, ν_1, ν_2, and so forth. At ordinary temperatures, most of the molecules in the electronic ground state will be in the zero vibrational level ($G\nu_0$); consequently, there are many electronic

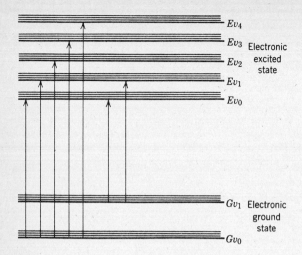

Fig. 4.
Energy-level diagram of a diatomic molecule.

$$\log_{10}(I_0/I) = kcb = A$$

where k = a constant characteristic of the solute,
c = concentration of solute,
b = path length through the sample,
A = absorbance (optical density in the older literature).

When c is expressed in moles per liter, and the path length (b) through the sample is expressed in centimeters, the preceding expression becomes

$$A = \epsilon cb$$

The term ϵ is known as the molar absorptivity, formerly called the molar extinction coefficient.[23]

If the concentration (c) of the solute is now defined as g/liter, the equation becomes:

$$A = abc$$

where a is the absorptivity and is thus related to the molar absorptivity by

$$\epsilon = aM$$

where M is the molecular weight of the solute.

The intensity of an absorption band in the ultraviolet spectrum is usually expressed as the molar absorptivity at maximum absorption, ϵ_{max} or $\log \epsilon_{max}$. When the constitution of an absorbing material is unknown, the intensity of absorption may be expressed as

$$E^{1\%}_{1\,cm} = \frac{A}{cb}$$

where c = concentration in grams per 100 ml,
b = path length through the sample in centimeters.

At this point it is necessary to define certain terms that are frequently used in the discussion of electronic spectra.

CHROMOPHORE. A covalently unsaturated group responsible for electronic absorption (for example, $C=C$, $C=O$, and NO_2).

AUXOCHROME. A saturated group which, when attached to a chromophore, alters both the wavelength and the intensity of the absorption maximum (for example, OH, NH_2, and Cl).

BATHOCHROMIC SHIFT. The shift of absorption to a longer wavelength due to substitution or solvent effect (a red shift).

HYPSOCHROMIC SHIFT. The shift of absorption to a shorter wavelength due to substitution or solvent effect (a blue shift).

HYPERCHROMIC EFFECT. An increase in absorption intensity.

transitions are from that level. In more complex molecules containing more atoms, the multiplicity of vibrational sublevels and the closeness of their spacing cause the discrete bands to coalesce, and broad absorption bands or "band envelopes" are obtained.

The principal characteristics of an absorption band are its position and intensity. The position of absorption corresponds to the wavelength of radiation whose energy is equal to that required for an electronic transition. The intensity of absorption is largely dependent on two factors: the probability of interaction between the radiation energy and the electronic system to raise the ground level to an excited state, and the polarity of the excited state. The probability of transition is proportional to the square of the transition moment. The transition moment, or dipole moment of transition, is proportional to the change in the electronic charge distribution occurring during excitation. Intense absorption occurs when a transition is accompanied by a large change in the transition moment. Absorption with $\epsilon_{max} > 10^4$ is high intensity absorption; low intensity absorption corresponds to ϵ_{max} values $< 10^3$. Transitions of low probability are "forbidden" transitions. The intensity of absorption may be expressed as transmittance (T), defined by

$$T = I/I_0$$

where I_0 is the intensity of the radiant energy striking the sample and I is the intensity of the radiation emerging from the sample. A more convenient expression of absorption intensity is that derived from the Lambert-Beer law which establishes a relationship between the transmittance, the sample thickness, and the concentration of the absorbing species. This relationship is expressed as

234 spectrometric identification of organic compounds

Table I Summary of Electronic Structure and Transitions

Electronic Structure	Example	Electronic Transition	$\lambda_{max}(nm)$	ϵ_{max}	Band[4,24] a
σ	Ethane	$\sigma \rightarrow \sigma^*$	135	–	–
n	Water	$n \rightarrow \sigma^*$	167	7,000	–
	Methanol	$n \rightarrow \sigma^*$	183	500	–
	1-Hexanethiol	$n \rightarrow \sigma^*$	224	126	–
	n-Butyl iodide	$n \rightarrow \sigma^*$	257	486	–
π	Ethylene	$\pi \rightarrow \pi^*$	165	10,000	–
	Acetylene	$\pi \rightarrow \pi^*$	173	6,000	–
π and n	Acetone	$\pi \rightarrow \pi^*$	about 150	–	–
		$n \rightarrow \sigma^*$	188	1,860	–
		$n \rightarrow \pi^*$	279	15	R
π-π	1,3-Butadiene	$\pi \rightarrow \pi^*$	217	21,000	K
	1,3,5-Hexatriene	$\pi \rightarrow \pi^*$	258	35,000	K
π-π and n	Acrolein	$\pi \rightarrow \pi^*$	210	11,500	K
		$n \rightarrow \pi^*$	315	14	R
Aromatic π	Benzene	Aromatic $\pi \rightarrow \pi^*$	about 180	60,000	E_1
		Aromatic $\pi \rightarrow \pi^*$	about 200	8,000	E_2
		Aromatic $\pi \rightarrow \pi^*$	255	215	B
Aromatic π-π	Styrene	Aromatic $\pi \rightarrow \pi^*$	244	12,000	K
		Aromatic $\pi \rightarrow \pi^*$	282	450	B
Aromatic π-σ (Hyperconjugated)	Toluene	Aromatic $\pi \rightarrow \pi^*$	208	2,460	E_2
		Aromatic $\pi \rightarrow \pi^*$	262	174	B
Aromatic π-π and n	Acetophenone	Aromatic $\pi \rightarrow \pi^*$	240	13,000	K
		Aromatic $\pi \rightarrow \pi^*$	278	1,110	B
		$n \rightarrow \pi^*$	319	50	R
Aromatic π-n (Auxochromic)	Phenol	Aromatic $\pi \rightarrow \pi^*$	210	6,200	E_2
		Aromatic $\pi \rightarrow \pi^*$	270	1,450	B

[a] R Band, German *radikalartig*; K-Band, German, *konjugierte*; B-Band, benzenoid; E-Band, ethylenic.[4,24]

HYPOCHROMIC EFFECT. A decrease in absorption intensity.

The absorption characteristics of organic molecules in the ultraviolet region depend on the electronic transitions that can occur and the effect of the atomic environment on the transitions. A summary of electronic structures and transitions that are involved in ultraviolet absorption is presented in Table I.

Representative UV spectra (plots of log ϵ vs. λ) are shown in Figure 1a. Note that the spectrum of the relatively simple model compound mesityl oxide very closely approximates the spectrum of the more complex steroid. Increased molecular structure complexity normally results in increased spectral complexity in NMR, infrared, and mass spectra; Figure 1a shows that this is not necessarily the case for UV spectra.

The relative ease with which the various transitions can occur is summarized in Figure 4a. Although the energy changes are not shown in scale, it is readily seen, for example, that an $n \rightarrow \pi^*$ transition requires less energy than a $\pi \rightarrow \pi^*$ or a $\sigma \rightarrow \sigma^*$ transition. Several notation systems are used to designate UV absorption bands (Appendix 1 of

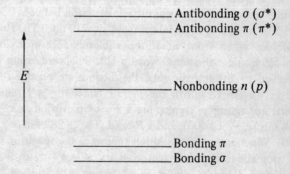

Fig. 4a.
Summary of electronic energy levels.

Reference 3). It seems simplest to use electronic transitions, or the letter designation assigned by Burawoy[24] as shown in Table I and described below.

The $n \rightarrow \pi^*$ transitions (also called R-bands)[4,24] of single chromophoric groups such as the carbonyl or nitro group

are forbidden and the corresponding bands are characterized by low molar absorptivities, ϵ_{max} generally less than 100. They are further characterized by the hypsochromic or blue shift observed with an increase in solvent polarity. They frequently remain in the spectrum when modifications in molecular structure introduce additional bands at shorter wavelengths. When additional bands make their appearance, the $n \rightarrow \pi^*$ transition is shifted to a longer wavelength but may be submerged by more intense bands.

Bands attributed to $\pi \rightarrow \pi^*$ transitions (K-Bands) appear in the spectra of molecules that have $\pi-\pi$ conjugated structures such as butadiene or mesityl oxide. Such absorptions also appear in the spectra of aromatic molecules possessing chromophoric substitution—styrene, benzaldehyde, or acetophenone. These $\pi \rightarrow \pi^*$ transitions are characterized by high molar absorptivity, $\epsilon_{max} > 10,000$.

The $\pi \rightarrow \pi^*$ transitions (K-bands) of conjugated di- or poly-ene systems can be distinguished from those of enone systems by observing the effect of changing solvent polarity. The $\pi \rightarrow \pi^*$ transitions of diene or polyene systems are essentially unresponsive to solvent polarity; the hydrocarbon double bonds are nonpolar. The corresponding absorptions of enones, however, undergo a bathochromic shift, frequently accompanied by increasing intensity, as the polarity of the solvent is increased. The red shift presumably results from a reduction in the energy level of the excited state accompanying dipole-dipole interaction and hydrogen bonding.

B-Bands (benzenoid bands) are characteristic of the spectra of aromatic or heteroaromatic molecules. Benzene shows a broad absorption band, containing multiple peaks or fine structure, in the near ultraviolet region between 230 and 270 nm. (ϵ of most intense peak ca. 230 nm.) The fine structure arises from vibrational sublevels accompanying the electronic transitions. When a chromophoric group is attached to an aromatic ring, the B-bands are observed at longer wavelengths than the more intense $\pi \rightarrow \pi^*$ transitions. For example, styrene has a $\pi \rightarrow \pi^*$ transition at λ_{max} 244 nm (ϵ_{max} 12,000), and a B-band at λ_{max} 282 nm (ϵ_{max} 450). When an $n \rightarrow \pi^*$ transition appears in the spectrum of an aromatic compound that contains $\pi \rightarrow \pi^*$ transitions (including B-bands), the $n \rightarrow \pi^*$ transition is shifted to longer wavelengths. The characteristic fine structure of the B-bands may be absent in spectra of substituted aromatics. The fine structure is often destroyed by the use of polar solvents.

E-Bands (ethylenic bands), like the B-bands, are characteristic of aromatic structures. The E_1- and E_2-bands of benzene are observed near 180 nm and 200 nm respectively. Auxochromic substitution brings the E_2 band into the near ultraviolet region, although in many cases it may not appear at wavelengths much over 210 nm. In auxochromic substitution, the heteroatom with the lone pair of

electrons shares these electrons with the π-electron system of the ring, facilitating the $\pi \rightarrow \pi^*$ transition and thus causing a red shift of the E-bands. The molar absorptivity of E-bands generally varies between 2000 and 14,000.

A bathochromically displaced E_2-band is probably responsible for the intense, fine-structured bands of polynuclear aromatics. With the appearance of the E-bands as a result of auxochromic substitution, the B-band shifts to longer wavelengths and frequently increases in intensity. Molecules such as benzylidene acetone, in which more complex conjugated chromophoric substitution occurs, produce spectra with both E- and K-bands; the B-bands are obscured by the displaced K-bands.

III. INSTRUMENTATION

A modern, recording photoelectric spectrophotometer consists of five sections or areas: (1) radiation source, (2) monochromator, (3) photometer, (4) sample area, and (5) detector area. The optical layout of a typical double-beam instrument is presented in Figure 5.

RADIATION SOURCE

The radiation source for the ultraviolet region of the spectrum is a hydrogen discharge tube. The hydrogen discharge tube can be replaced by a tungsten incandescent lamp when absorption in the visible region is to be determined. Mirror $M1$ is rotated manually to focus the light emitted from either source onto the entrance slit ($S1$) of the monochromator.

MONOCHROMATOR

The light from the source is dispersed into its separate wavelengths by the monochromator. The light that passes through entrance slit $S1$ is collimated into parallel rays by the spherical mirror $M3$ and is reflected to quartz prism $P1$. After the light has passed through prism $P1$, it is reflected back through the prism by a mirrored surface on the back of the prism. Dispersion takes place during both passes through the prism. The light that emerges from $P1$ is reflected to the intermediate slit $S2$ by mirrors $M3$ and $M4$. Since a dispersing prism inherently produces a curved image of a straight slit, entrance slit $S1$ is curved to compensate

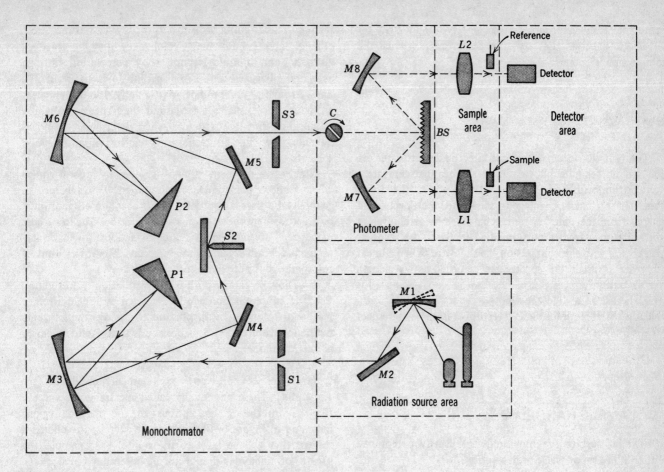

Fig. 5.
Optical layout of a double-beam ultraviolet spectrophotometer.

for the curvature produced in the first stage of the monochromator. Thus the image that strikes the intermediate slit $S2$ is essentially straight. Slit $S2$ consists of a single fixed jaw normal to a mirror. The second jaw of the slit is the image of the fixed jaw in the mirror. The mirror, at slit $S2$, reverses the field of the beam so that comatic aberration introduced by the first stage of the monochromator is removed by the second stage of the monochromator. The light path through the second stage ($M5$, $M6$, and $P2$) is the mirror image of that through the first stage.

The two prisms $P1$ and $P2$ are rotated simultaneously. The wavelength of radiation that appears at exit slit $S3$ is determined by the angular position of the prisms. The rotating mechanism for the prisms is coupled to the recording drum. The double monochromator has two advantages: it doubles dispersion and reduces stray light. Stray light consists of the unwanted wavelengths appearing at slit $S3$. Slit $S3$ is curved to compensate for curvature reintroduced by the second stage of the mono-chromometer.

PHOTOMETER

The monochromatic light that emerges from exit slit $S3$ is pulsed by chopper C and split into sample and reference beams by the beam splitter BS. The reference and sample beams, reflected from the beam splitter, are reflected by mirrors $M7$ and $M8$ through lenses $L1$ and $L2$ to the sample area. Lenses $L1$ and $L2$ serve to optimize the parallel nature of the rays that pass through the sample area. Optics for the transmission of ultraviolet radiation are made of quartz.

SAMPLE AREA

The beams entering the sample area become more concentrated as they pass through the area toward the detectors. Small cells are usually placed in the region nearest the detector area.

The photomultiplier tubes of the detector area are

protected from exposure to excessively bright light by automatic shutters that seal the detector area when the sample area is opened.

DETECTOR AREA

The radiation beams that pass into the detector area are focused on separate photomultiplier tubes that generate a voltage proportional to the energy that strikes the detectors. The off-balance voltage, resulting from absorption of energy from the sample beam, is balanced by an equivalent voltage tapped from a portion of a slidewire. The recorder pen travels with the contacts on the slidewire. When a linear slidewire is used, the spectra are recorded as wavelength versus transmittance. Since the absorbance (A) equals log $1/T$, the use of a slidewire whose resistance varies logarithmically with length results in a recording linear with respect to absorbance.

IV. SAMPLE HANDLING

Ultraviolet spectra of compounds are usually determined either in the vapor phase or in solution.

A variety of quartz cells is available for the determination of spectra in the gas phase. These cells are equipped with gas inlets and outlets and have path lengths from 0.1 mm to 100 mm. Cell jackets are available through which liquids may be circulated for temperature control.

Cells used for the determination of spectra in solution vary in path length from 1 cm to 10 cm. Quartz cells, 1 cm square, are commonly used. These require about 3 ml of solution. Filler plugs are available to reduce the volume and the path length of the 1 cm square cell. Small-volume cells with 1 cm path lengths are also available. Microcells may be used when only a small amount of solution is available: the use of a beam condenser, to minimize the loss of energy, is advisable when the microcells are used. Microcells with an internal width of 2 mm and a pathlength as short as 5 mm are available from Lightpath Cells, Inc., St. Louis, Mo.

In preparing a solution, a sample is accurately weighed and made up to volume in a volumetric flask. Aliquots are then removed and additional dilutions made until the desired concentration has been acquired. Clean cells are of utmost importance. The cells should be rinsed several times with solvent and checked for absorption between successive determinations. It may be necessary to clean the cells with a detergent or hot nitric acid to remove traces of previous samples.

Many solvents are available for use in the ultraviolet

region. Three common solvents are cyclohexane, 95% ethanol, and 1,4-dioxane. Cyclohexane may be freed of aromatic and olefinic impurities by percolation through activated silica gel, and is transparent down to about 210 nm. Aromatic compounds, particularly the polynuclear aromatics, are usually soluble and their spectra generally retain their fine-line structure when determined in cyclohexane. The fine structure is often lost in more polar solvents.

Ninety-five per cent ethanol is generally a good choice when a more polar solvent is required. This solvent can generally be used as purchased, but absolute ethanol must be freed of benzene used in its preparation. The last traces of benzene are removed by careful fractional distillation,[25] or by preparative gas chromatography. The lower limit of transparency for ethanol is near 210 nm.

1,4-Dioxane can be purified by distillation from sodium. Benzene contamination can be removed by the addition of methanol followed by distillation to remove the benzene-methanol azeotrope.[19] 1,4-Dioxane is transparent down to about 220 nm.

Many of the "spectral grade" solvents for ultraviolet spectral analysis are now commercially available (see Appendix). These are usually called spectral-grade and are free of interfering absorptions in the shaded regions indicated in Figure 6. Care should be exercised to choose a solvent that will be inert to the solute. For example, the spectra of aldehydes should not be determined in alcohols. Photochemical reactions may be detected by checking for changes in absorbance with time after exposure to the ultraviolet beam in the instrument.

V. CHARACTERISTIC ABSORPTION OF ORGANIC COMPOUNDS

In our discussion of the theory of electronic or ultraviolet spectra, it was shown that the ability of an organic compound to absorb ultraviolet radiation is dependent on its electronic structure. In the following sections we shall discuss the characteristic absorption of basic electronic structures, and the effects of molecular geometry and substitution on the absorption.

COMPOUNDS CONTAINING ONLY σ-ELECTRONS

Saturated hydrocarbons contain σ-electrons exclusively. Since the energy required to bring about a $\sigma \rightarrow \sigma^*$ transition is of the order of 185 kcal per mole and is available only in

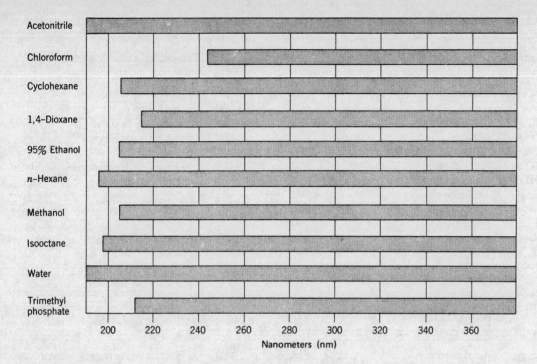

Fig. 6.
Useful transparency ranges of solvents in near ultraviolet region.

the far (vacuum) ultraviolet region, saturated hydrocarbons are transparent in the near ultraviolet region.

SATURATED COMPOUNDS CONTAINING *n*-ELEC-TRONS

Saturated compounds containing heteroatoms such as oxygen, nitrogen, sulfur, or the halogens, possess nonbonding electrons (*n*- or *p*-electrons) in addition to σ-electrons. The $n \rightarrow \sigma^*$ tfansition requires less energy than the $\sigma \rightarrow \sigma^*$ transition (see Figure 4*a*) but the majority of compounds in this class still show no absorption in the near ultraviolet.

Alcohols and ethers absorb at wavelengths shorter than 185 nm and therefore are commonly used as solvents for work in the near ultraviolet region. When these compounds are used as solvents, the intense absorption extends into the near ultraviolet, producing end-absorption or cut-off in the 200–220 nm region.

Sulfides, disulfides, thiols, amines, bromides, and iodides may show weak absorption in the near ultraviolet. Frequently the absorption appears only as a shoulder or an inflection so that its diagnostic value is questionable.

Absorption data for several saturated compounds bearing heteroatoms are presented in Table II.

Table II Absorption Characteristics of Saturated Compounds Containing Heteroatoms ($n \rightarrow \sigma^*$)

Compound	$\lambda_{max}(nm)$	ϵ_{max}	Solvent
Methanol	177	200	Hexane
Di-*n*-butyl sulfide	210	1200	Ethanol
	229 (*s*)		
Di-*n*-butyl disulfide	204	2089	Ethanol
	251	398	
1-Hexanethiol	224 (*s*)	126	Cyclohexane
Trimethylamine	199	3950	Hexane
N-Methylpiperidine	213	1600	Ether
Methyl chloride	173	200	Hexane
n-Propyl bromide	208	300	Hexane
Methyl iodide	259	400	Hexane

(*s*) shoulder or inflection.

COMPOUNDS CONTAINING π-ELECTRONS (CHROMOPHORES)

The absorption characteristics of a list of compounds containing single, isolated chromophoric groups are presented in Table III. All the compounds contain π-electrons, and many also contain nonbonding electron pairs. An examination of the absorption data shows that many of

Table III Absorption Data for Isolated Chromophores *

Chromophoric Group	System	Example	$\lambda_{max}(nm)$	ϵ_{max}	Transition	Solvent
Ethylenic	RCH=CHR	Ethylene	165	15,000	$\pi \to \pi^*$	Vapor
			193	10,000	$\pi \to \pi^*$	
Acetylenic	R–C≡C–R	Acetylene	173	6,000	$\pi \to \pi^*$	Vapor
Carbonyl	$RR_1C=O$	Acetone	188	900	$\pi \to \pi^*$	n-Hexane
			279	15	$n \to \pi^*$	
Carbonyl	RHC=O	Acetaldehyde	290	16	$n \to \pi^*$	Heptane
Carboxyl	RCOOH	Acetic acid	204	60	$n \to \pi^*$	Water
Amido	$RCONH_2$	Acetamide	<208	–	$n \to \pi^*$	–
Azomethine	>C=N–	Acetoxime	190	5,000	$\pi \to \pi^*$	Water
Nitrile	–C≡N	Acetonitrile	<160	–	$\pi \to \pi^*$	–
Azo	–N=N–	Azomethane	347	4.5	$n \to \pi^*$	Dioxane
Nitroso	–N=O	Nitrosobutane	300	100		Ether
			665	20		
Nitrate	$–ONO_2$	Ethyl nitrate	270	12	$n \to \pi^*$	Dioxane
Nitro	$–N\begin{smallmatrix}O\\\\O\end{smallmatrix}$	Nitromethane	271	18.6	$n \to \pi^*$	Alcohol
Nitrite	–ONO	Amyl nitrite	218.5	1,120	$\pi \to \pi^*$	Petroleum ether
			346.5[a]		$n \to \pi^*$	
Sulfoxide	>S=O	Cyclohexyl methyl sulfoxide	210	1,500		Alcohol
Sulfone	$>S\begin{smallmatrix}O\\\\O\end{smallmatrix}$	Dimethyl sulfone	<180	–		–

[a]Most intense peak of fine structure group.

* Taken from Gillam and Stern, *An Introduction to Electronic Absorption Spectroscopy in Organic Chemistry,* 2nd Ed., Edward Arnold, London, 1957.

these single chromophoric groups absorb strongly in the far ultraviolet region, with no absorption in the near ultraviolet. Those groups containing both π- and n-electrons can undergo three transitions: $n \to \sigma^*$, $\pi \to \pi^*$, and $n \to \pi^*$. Absorption of single chromophores in the near ultraviolet region results from the low-energy, forbidden $n \to \pi^*$ transition.

Ethylenic Chromophore

The isolated ethylenic chromophore is responsible for intense absorption that almost always occurs in the far ultraviolet region. Absorption is due to a $\pi \to \pi^*$ transition. Ethylene in the vapor phase absorbs at 165 nm (ϵ_{max} 10,000). A second band near 200 nm has been attributed to the elevation of two π electrons to π^* orbitals.[3] The intensity of olefinic absorption is essentially independent of solvent because of the nonpolar nature of the olefinic bond.

Alkyl substitution of the parent compound moves the absorption to longer wavelenghts. The bathochromic effect is progressive as the number of alkyl groups increases. A double bond exocyclic to two rings absorbs near 204 nm. The bathochromic shift, accompanying alkyl substitution, results from hyperconjugation, in which the σ-electrons of

the alkyl group are mobile enough to interact with the chromophoric group.

Attachment of a heteroatom, containing a nonbonding electron pair, to the ethylenic linkage brings about a bathochromic shift. Nitrogen and sulfur atoms are most effective, bringing the absorption well into the near ultraviolet region. Methyl vinyl sulfide, for example, absorbs at 228 nm (ϵ_{max} 8000).

The absorption characteristics of cyclic monoolefins resemble those of the acyclic compounds, the absorption bearing no apparent relationship to ring size.

When two or more ethylenic linkages appear in a single molecule, isolated from one another by at least one methylene group, the molecule absorbs at the same position as the single ethylenic chromophore. The intensity of

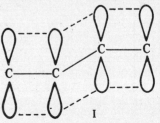

I

absorption is proportional to the number of isolated chromophoric groups in the molecule.

Allenes, which possess the C=C=C unit, show a strong absorption band in the far ultraviolet region near 170 nm, with a shoulder on the long wavelength side sometimes extending into the near ultraviolet region.

The single olefinic bond consists of two energy levels (two π orbitals) one bonding one antibonding. In conjugated diene molecules such as 1,3-butadiene, when coplanarity permits, there is an effective overlap of π orbitals resulting in a $\pi\text{--}\pi$ conjugated system (I).

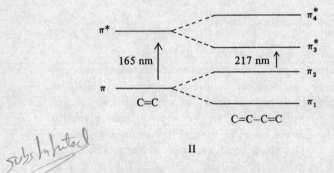

II

This overlap or interaction results in the creation of two new energy levels in butadiene (II). Thus the $\pi_2 \rightarrow \pi_3{}^*$ transition of butadiene is bathochromically shifted relative to the $\pi \rightarrow \pi^*$ transition of ethylene. There are other $\pi \rightarrow \pi^*$ transitions possible in the conjugated system. Their intensities depend on the "allowedness" of the transitions.[3]

Acyclic conjugated dienes show intense $\pi \rightarrow \pi^*$ transition bands (K-bands) in the 215–230 nm region. 1,3-Butadiene absorbs at 217 nm (ϵ_{max} 21,000). Further conjugation, in open-chain trienes and polyenes, results in additional bathochromic shifts accompanied by increases in absorption intensity. The spectra of polyenes are characterized by fine structure, particularly when the spectra are determined in the vapor phase or in nonpolar solvents. Absorption data for several conjugated olefins are presented in Table IV.

The bathochromic effect of alkyl substitution in 1,3-butadiene is apparent from the data for 2,3-dimethyl-1,3-butadiene.

In cases where *cis* and *trans* isomers are possible, the *trans* isomer absorbs at the longer wavelength with the greater intensity (see Table XIX). This difference becomes more pronounced as the length of the conjugated system increases. Coplanarity is required for the most effective overlap of the π-orbitals and increased ease of the $\pi \rightarrow \pi^*$ transition. Of the two isomers, the *cis* isomer is more likely to be forced into a nonplanar conformation by steric effects (see discussion of stilbene below). The greater absorption intensity of the *trans* isomer results from the greater overall length of the transition moment of the excited molecule.

Table IV   Absorption Data for Conjugated Olefins

Compound	$\pi \rightarrow \pi^*$ Transition (K-band)		Solvent
	$\lambda_{max}(nm)$	ϵ_{max}	
1,3-Butadiene	217	21,000	Hexane
2,3-Dimethyl-1,3-butadiene	226	21,400	Cyclohexane
1,3,5-Hexatriene	253	~50,000	Isooctane
	263	52,500	
	274	~50,000	
1,3-Cyclohexadiene	256	8,000	Hexane
1,3-Cyclopentadiene	239	3,400	Hexane

WOODWARD'S RULES

An empirical method for predicting the bathochromic effect of alkyl substitution in 1,3-butadiene has been formulated by Woodward.[26,27] These rules can be summarized as follows:

(1) Each alkyl group, or ring residue, attached to the parent diene (1,3-butadiene), shifts the absorption 5 nm toward the long wavelength region; and (2) the creation of an exocyclic double bond causes an additional bathochromic shift of 5 nm, the shift being 10 nm if the double bond is exocyclic to two rings. For example, 217 + (2 × 5) is the predicted λ_{max} value for 2,3-dimethyl-1,3-butadiene; the observed λ_{max} is 226.

Examination of the data in Table IV shows a marked bathochromic shift and a decrease in absorption intensity for the conjugated, monocyclic diene system compared with 1,3-butadiene. Butadiene exists in the preferred *s-trans* (transoid) conformation, whereas the cyclic dienes are forced into an *s-cis* (cisoid) conformation. The reason for the bathochromic shift in the *s-cis* structure brought about by cyclization, is not clear. The decrease in intensity is more easily explained, since the transition moment of the cyclic or homoannular system is less than that of the acyclic or the heteroannular systems.

Homoannular   Heteroannular

Heteroannular and acyclic dienes usually display molar absorptivities in the 8000–20,000 range, whereas homoannular dienes usually display molar absorptivities in the 5000–8000 range.

Rules for predicting the position of absorption of homo- and heteroannular systems are due largely to the work of Fieser.[29] These rules are summarized in Table V.

Poor correlations are obtained when the data of Table V are applied to such cross-conjugated polyene systems as

The value of these rules in structural studies of natural products will be obvious from two examples:

Cholesta-3,5-diene

1.

Calc. λ_{max}

214 (base)
+ 15 (3-ring residues, 1, 2, 3)
+ 5 (1-exocyclic C=C)
234

Obs. λ_{max} = 235 (ϵ_{max} 19,000)

Cholesta-2,4-diene

2.

Calc. λ_{max}

253 (base)
+ 15 (3-ring residues, 1, 2, 3)
+ 5 (1-exocyclic C=C)
273

Obs. λ_{max} = 275 (ϵ_{max} 10,000)

The examples above illustrate the typically much higher molar absorptivity of the heteroannular (*transoid*) diene compared to the homoannular (*cisoid*) diene.

Table V   Rules of Diene Absorption[2]

Base value for heteroannular diene	214
Base value for homoannular diene	253
Increments for	
Double bond extending conjugation	+30
Alkyl substituent or ring residue	+5
Exocyclic double bond	+5
Polar groupings: OAc	+0
OAlk	+6
SAlk	+30
Cl, Br	+5
N(Alk)$_2$	+60
Solvent correction	+0
	λ_{calc} = Total

Acetylenic Chromophore

The absorption characteristics of the acetylenic chromophore are more complex than those of the ethylenic chromophore. Acetylene shows a relatively weak band at 173 nm resulting from a $\pi \to \pi^*$ transition. Conjugated polyynes show two principal bands in the near ultraviolet which are characterized by fine structure. The short wavelength band is extremely intense and arises from $\pi \to \pi^*$ transition. Typical absorption data for conjugated polyynes are shown in Table VI.

Carbonyl Chromophore

The carbonyl group contains, in addition to a pair of σ-electrons, a pair of π-electrons and two pairs of nonbonding (n or p) electrons. Saturated ketones and aldehydes display three absorption bands, two of which are observed in the far ultraviolet region. A $\pi \to \pi^*$ transition absorbs strongly near 150 nm; an $n \to \sigma^*$ transition absorbs near 190 nm. The third band (*R*-band) appears in the near ultraviolet in the 270–300 nm region. The *R*-band is weak ($\epsilon_{max} < 30$) and results from the forbidden transition of a loosely held *n*-electron to the π^* orbital, the lowest unoccupied orbital of the carbonyl group. *R*-bands undergo a blue shift as the polarity of the solvent is increased. Acetone absorbs at 279 nm in *n*-hexane; in water the λ_{max} is 264.5. The blue shift results from hydrogen bonding which lowers the energy of the *n* orbital. The blue shift can be used as a measure of the strength of the hydrogen bond.

SATURATED KETONES AND ALDEHYDES. Absorption data for several saturated ketones and aldehydes are presented in Table VII.

The bathochromic effect accompanying the introduction of larger and more highly branched alkyl groups in the aliphatic ketones can be seen from the data in Table VII.

Since the $n \to \pi^*$ absorption of ketones and aldehydes is weak, spectra of derivatives such as the semicarbazones or 2,4-dinitrophenylhyrazones are often used for identification work.

The introduction of an α-halogen atom in an aliphatic ketone has little effect upon the $n \to \pi^*$ transition. However, α-substitution of halogen atoms in saturated cyclic ketones has a marked effect upon the absorption

Table VI   Absorption Data for Conjugated Polyynes

	λ_{max},	ϵ_{max}	λ_{max},	ϵ_{max}
2,4-hexadiyne	–	–	227	360
2,4,6-octatriyne	207	135,000	268	200
2,4,6,8-decatetrayne	234	281,000	306	180

Table VII Absorption Data for Saturated Aldehydes and Ketones

Compound	$n \rightarrow \pi^*$ Transition (R-band)		
	$\lambda_{max}(nm)$	ϵ_{max}	Solvent
Acetone	279	13	Isooctane
Methyl ethyl ketone	279	16	Isooctane
Diisobutyl ketone	288	24	Isooctane
Hexamethylacetone	295	20	Alcohol
Cyclopentanone	299	20	Hexane
Cyclohexanone	285	14	Hexane
Acetaldehyde	290	17	Isooctane
Propionaldehyde	292	21	Isooctane
Isobutyraldehyde	290	16	Hexane

characteristics.[28b] The λ_{max} of the parent compound is reduced by 5–10 nm when the substituent is equatorial and a bathochromic shift of 10–30 nm occurs when the substituent is axial. The bathochromic shift is usually accompanied by a strong hyperchromic effect. These effects are valuable in the structure determination of halogenated steroids and terpenes.[2]

The attachment of groups containing lone electron pairs to carbonyl groups has a marked effect upon the $n \rightarrow \pi^*$ transition. The R-band is shifted to shorter wavelengths with little effect upon intensity. The shift in absorption results from a combination of inductive and resonance effects. Substitution may change the energy levels of both the ground state and the excited state, but the important factor is the relative energies of the two levels. Absorption values for the $n \rightarrow \pi^*$ transitions of several simple carbonyl compounds are presented in Table VIII.

α-DIKETONES AND α-KETOALDEHYDES. Acyclic α-diketones, such as biacetyl, exist in the *s-trans* conformation. The spectrum of biacetyl shows the normal weak R-band at 275 nm and a weak band near 450 nm resulting from interaction between the carbonyl groups. The position of the long-wavelength band of α-diketones incapable of

Table VIII $n \rightarrow \pi^*$ Transitions (R-Bands) of Simple Carbonyl-Containing Compounds

Compound	$\lambda_{max}(nm)$	ϵ_{max}	Solvent
Acetaldehyde	293	11.8	Hexane
Acetic acid	204	41	Ethanol
Ethyl acetate	207	69	Pet. ether
Acetamide	220 (s)		Water
Acetyl chloride	235	53	Hexane
Acetic anhydride	225	47	Isooctane
Acetone	279	15	Hexane

enolization reflects the effect of coplanarity upon resonance, and hence depends on the dihedral angle ϕ between the carbonyl groups (I, II, III):

I Camphorquinone
$\phi = 0\text{-}10°$ λ_{max} 466 nm

II Benzil
$\phi = 90°$ λ_{max} 370 nm

III Isoduril
$\phi = 180°$ λ_{max} 490 nm

Cyclic α-diketones with α-hydrogen atoms exist in the enolic form, for example, diosphenol. The absorption characteristics of diosphenol appear later in this chapter.

Diosphenol

β-DIKETONES. The ultraviolet spectra of β-diketones depend on the degree of enolization. The enolic form is stabilized when steric considerations permit intramolecular hydrogen bonding. Acetylacetone is a classic example. The enolic species exists to the extent of about 15% in aqueous solution and 91–92% in the vapor phase or in solution in nonpolar solvents. The absorption is dependent on the concentration of the enol tautomer.

$\lambda_{max}^{H_2O}$ 274, ϵ_{max} 2050

$\lambda_{max}^{Isooctane}$ 272, ϵ_{max} 12,000

Table IX Absorption Data for Conjugated Ketones and Aldehydes

Compound	K-band		R-band		Solvent
	$\lambda_{max}(nm)$	$\log \epsilon_{max}$	$\lambda_{max}(nm)$	$\log \epsilon_{max}$	
Methyl vinyl ketone	212.5	3.85	320	1.32	Ethanol
Methyl isopropenyl ketone	218	3.90	315	1.4	Ethanol
Acrolein	210	4.06	315	1.41	Water
Crotonaldehyde	220	4.17	322	1.45	Ethanol
Crotonaldehyde	214	4.20	329	1.39	Isooctane
			341	1.38	
			352 (s)	1.25	

(s) shoulder.

Cyclic β-diketones, such as 1,3-cyclohexadione, exist almost exclusively in the enolic form even in polar solvents. The enolic structures show strong absorption in the 230–260 nm region due to the $\pi \rightarrow \pi^*$ transition in the s-trans enone system. 1,3-Cyclohexadione, in ethanol, absorbs at 253 nm (ϵ_{max} 22,000). The formation of the enolate ion, in alkaline solution, shifts the strong absorption band into the 270 to 300 nm region.

α,β-UNSATURATED KETONES AND ALDEHYDES. Compounds containing a carbonyl group in conjugation with an ethylenic group are called *enones*. Spectra of enones are characterized by an intense absorption band (K-band) in the 215–250 nm region (ϵ_{max} usually 10,000–20,000), and a weak $n \rightarrow \pi^*$ band (R-band) at 310–330 nm. The weak R-band is frequently poorly defined. Absorption data for several conjugated ketones and aldehydes are presented in Table IX.

Since carbonyl compounds are polar, the positions of the K- and R-bands of enones are both dependent on the solvent. The hypsochromic effect on the R-band with increasing solvent polarity has already been discussed above.

The K-bands of enones undergo a bathochromic shift with increasing solvent polarity. The solvent effect on the spectrum of mesityl oxide is summarized in Table X.

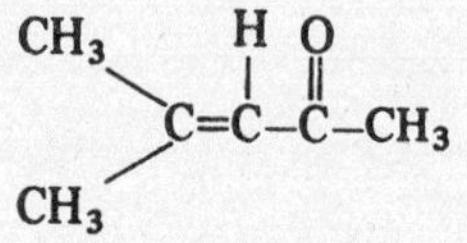

The intensity of the K-band may be reduced to less than 10^4 in cases where steric hindrance prevents coplanarity. This frequently occurs in cyclic systems such as IV.[28a]

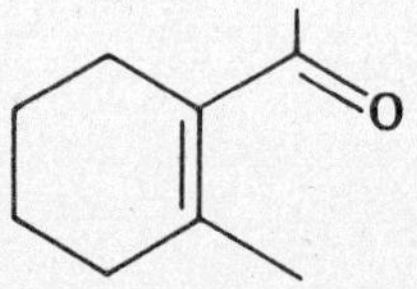

λ_{max} 243 nm, ϵ_{max} 1400

Woodward has derived empirical generalizations for the effect of substitution on the position of the $\pi \rightarrow \pi^*$ transition (K-band) in the spectra of α,β-unsaturated ketones.[26,27] The positions of the K-bands that result from substitution in the basic formula

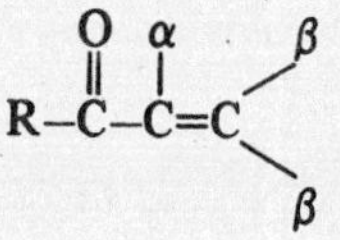

are summarized as follows:

Table X Effect of Solvent Polarity on the Spectrum of Mesityl Oxide

Solvent	Transition	
	$\pi \rightarrow \pi^*$ (λ_{max} nm)	$n \rightarrow \pi^*$ (λ_{max} nm)
Isooctane	230.6	321
Chloroform	237.6	314
Water	242.6	Submerged under K-band

Substitution		Probable λ_{max} (nm)
Unsubstituted		215
α or β	No exocyclic C=C	225
$\alpha\beta$ or $\beta\beta$	No exocyclic C=C	235
$\alpha\beta$ or $\beta\beta$	One exocyclic C=C	240
$\alpha\beta\beta$	No exocyclic C=C	247
$\alpha\beta\beta$	One exocyclic C=C	252

The spectra of α,β-unsaturated aldehydes are similar to those of the α,β-unsaturated ketones. The R-bands occur in the 350–370 nm region and exhibit some fine structure when the spectra are determined in nonpolar solvents.

Similar rules apply to the cyclopentenone system[28]

Parent system   214 nm[a]
α or β substituent   224 nm
α, β substituents   236 nm

[a]This value can be calculated from data in Table XI.
202 (Base value to form 5-membered cyclic enone)
<u> 12</u> (β-substituent, bond to form ring)
214

More extensive rules of enone absorption have been summarized by Fieser.[29] These rules appear as Table XI.

A few examples will serve to illustrate the usefulness of these correlations.

1-Acetylcyclohexene

1.

Calc. λ_{max}^{EtOH}

215 (base)
10 (α-subst)
<u>12</u> (β-subst)
237

Obs. λ_{max}^{EtOH} = 232

Cholesta-1,4-diene-3-one

2.

Calc. λ_{max}^{EtOH}

215 (base, $\Delta^{4,5}$ system)
24 (2-β-subst)
<u>5</u> (1-exo C=C)
244

Obs. λ_{max}^{EtOH} = 245

Cross conjugation has little effect on the λ_{max} of cholesta-1,4-diene-3-one. The calculations used were for an

Table XI   Rules of Enone and Dienone Absorption (α, β-Unsaturated Carbonyls)

Enone   Dienone

Base values:

Acyclic α,β-unsaturated ketones	215
Six membered cyclic α,β-unsaturated ketones	215
Five membered cyclic α,β-unsaturated ketones	202
α,β-Unsaturated aldehydes	210
α,β-Unsaturated carboxylic acids and esters	195

Increments for

Double bond extending conjugation		+30
Alkyl group, ring residue	α	+10
	β	+12
	γ and higher	+18
Polar groupings: —OH	α	+35
	β	+30
	δ	+50
—OAc	α,β,δ	+6
—OMe	α	+35
	β	+30
	γ	+17
	δ	+31
—SAlk	β	+85
—Cl	α	+15
	β	+12
—Br	α	+25
	β	+30
—NR$_2$	β	+95
Exo double bond		+5
Homodiene component		+39
Solvent correction (see table below)		variable

λ_{calc} = Total[a]

[a]The calculated values usually fall within ± 3 nm of the observed values. The molar absorptivities of *cisoid* enones are usually less than 10,000, whereas the molar absorptivities of *transoid* enones are greater than 10,000.

Solvent Corrections *

Solvent	Correction (nm)
Ethanol	0
Methanol	0
Dioxane	+5
Chloroform	+1
Ether	+7
Water	−8
Hexane	+11
Cyclohexane	+11

* E.g., if the absorption was determined in cyclohexane, 11 nm would be added to the calc. $\lambda_{max}^{cyclohexane}$ to change it to a calculated λ_{max}^{EtOH} I.e., ethanol causes a bathochromic shift of this band.

enone (β,β-disubstituted) with no correction required for the 1,2 double bond nor for the β' group. The corresponding calculation using the $\Delta^{1,2}$ system yields 227 nm as the predicted λ_{max}; this is an indication that, when such a choice is presented, the more reliable prediction arises from the system that is more highly substituted (longer wavelength $\pi \rightarrow \pi^*$ transition).

Enol of 1,2-cyclopentadione

3.

Calc. λ_{max}^{EtOH}	202 (base)
	12 (β-subst)
	35 (α-OH)
	249

Obs. λ_{max}^{EtOH} = 247

Diosphenol

4.

Calc. λ_{max}^{EtOH}	215 (base)
	24 (2 β-subst)
	35 (α-OH)
	274

Obs. λ_{max}^{EtOH} = 270

The spectrum of p-benzoquinone is similar to

that of a typical α,β-unsaturated ketone, the strong K-band appearing at 245 nm with a weak R-band near 435 nm.

There are cases where the C=O and the C=C are nonconjugated in the classical sense but where interaction does occur to produce an absorption band. In structures where this occurs, the C=O and C=C groups must be oriented so that there can be effective overlap of the π orbitals. For example, the structure $CH_2 = \langle\rangle = O$ shows a moderately strong band near 214 nm with a normal weak R-band at 284 nm. Similar effects are observed when there can be effective overlap of the π orbital of the C=O group and the p (n-) orbitals of a heteroatom. For example,

λ 238 nm, ϵ_{max} 2535

The interaction in these apparently nonconjugated systems is known as transannular conjugation.[30]

α,β-Unsaturated aldehydes present spectra similar to the α,β-unsaturated ketones. Calculations are based upon a base value of 207 nm for acrolein.

CARBOXYLIC ACIDS. Saturated carboxylic acids show a weak absorption band near 200 nm resulting from the forbidden $n \rightarrow \pi^*$ transition. The position of the band undergoes a small bathochromic shift with an increase in chain length. This band is of little diagnostic value.

α,β-Unsaturated acids display a strong K-band characteristic of the conjugated system. In the first member of the series, acrylic acid, the absorption occurs near 200 nm, ϵ_{max} 10,000. Alkyl substitution in this basic structure results in a bathochromic shift of the K-band in much the same manner as observed for α,β-unsaturated ketones. The extension of conjugation produces further bathochromic shifts accompanied by an increase in the band intensity and the appearance of fine structure (Tables XI and XII). The attachment of an electronegative group on the α-carbon also produces a bathochromic shift. One can calculate expected maxima for the compounds described in Table XII utilizing the data of Table XI.

The absorption characteristics of several α,β-unsaturated acids are summarized in Table XIII.

ESTERS AND LACTONES. Esters and sodium salts of carboxylic acids show absorption at wavelengths and intensities comparable to the parent acid. Conjugated, unsaturated lactones display spectra similar to unsaturated esters. The spectra of simple unsaturated lactones show end absorption in the 200–240 nm region. Extended conjugation produces a bathochromic shift of the $\pi \rightarrow \pi^*$ transition (K-band). Table XI may be used to predict the maxima of such compounds.

Table XII Absorption Maxima of Unsaturated Carboxylic Acids and Esters[2,31]

	λ_{max}^{EtOH} (± 5 nm)
α or β-monosubstituted	208
α,β- or β,β-disubstituted	217
α,β,β-trisubstituted	225

Table XIII   Absorption of α,β-Unsaturated Acids

Compound	λ_{max}^{EtOH}	ϵ_{max}
$CH_2=CH-COOH$	200	10,000
$CH_3CH=CH-COOH$ (trans)	205	14,000
$CH_3CH=CH-COOH$ (cis)	205.5	13,500
$CH_2=\overset{\overset{\displaystyle CH_3}{\vert}}{C}-COOH$	210	–
S $\overset{\overset{\displaystyle H}{}}{=}C-COOH$	220	14,000
S $\overset{\overset{\displaystyle CN}{}}{=}C-COOH$	235	12,500
$CH_3-(CH=CH)_2-COOH$	254	25,000
$CH_3-(CH=CH)_3-COOH$	294	37,000
$CH_3-(CH=CH)_4-COOH$	332	49,000

AMIDES AND LACTAMS. α,β-Unsaturated amides and lactams show absorption in the near ultraviolet at λ_{max} 200–220, $\epsilon_{max} < 10,000$. α,β-Unsaturated lactams show a second band near 250 nm ($\epsilon_{max} \sim 1000$).

Azomethines (Imines) and Oximes

These structures show no absorption in the near ultraviolet unless the $\diagdown C=N-$ group is involved in conjugation. In the spectra of conjugated azomethines and oximes the $\pi \rightarrow \pi^*$ transition (K-band) appears in the 220–230 nm region, $\epsilon_{max} > 10,000$. Acidification of the azomethines producing a positive charge on the nitrogen, shifts the absorption to the 270–290 nm region.

Nitriles and Azo Compounds

α,β-Unsaturated nitriles absorb just inside the near ultraviolet region, near 213 nm, $\epsilon_{max} \sim 10,000$.

The azo group is analogous to the ethylenic linkage with two σ bonds being replaced by two lone pairs of electrons ($-\ddot{N}=\ddot{N}-$). The $\pi \rightarrow \pi^*$ transition occurs in the far (vacuum) ultraviolet. The $n \rightarrow \pi^*$ band in aliphatic azo compounds appears near 350 nm with the expected low intensity, $\epsilon_{max} < 30$. trans-Azobenzene absorbs at 320 nm (ϵ_{max} 21,000). Comparable absorption for trans-stilbene occurs at 295 nm (ϵ_{max} 28,000).

Compounds With N to O Bonds

Four groups contain multiple nitrogen to oxygen linkages: nitro, nitroso, nitrates, and nitrites. All of these structures show weak absorption in the near ultraviolet region resulting from an $n \rightarrow \pi^*$ transition.

The absorption of several typical compounds containing nitrogen to oxygen linkages are presented in Table XIV.

The effect of conjugation upon the absorption characteristics of the nitro group is apparent from the data for 1-nitro-1-propene. The strong $\pi \rightarrow \pi^*$ transition (K-band) submerges the weak $n \rightarrow \pi^*$ transition (R-band).

Multiple Bonded Sulfur Groups

Aliphatic sulfones are transparent in the near ultraviolet region. The sulfur atom in sulfones has no lone-pair electrons, and the lone pairs of electrons associated with the oxygen atoms, appear to be tightly bound. In an α,β-unsaturated sulfone, such as ethyl vinyl sulfone, a band appears in the 210 nm region resulting from resonance between the S to O linkage and the ethylenic linkage.

Saturated sulfoxides absorb near 220 nm with intensities of the order of 1500. This absorption involves an $n \rightarrow \pi^*$ transition in the S=O group and thus undergoes a hypsochromic shift as solvent polarity is increased. Aromatic sulfoxides show an intense K-band in addition to the displaced B-band.

In compounds containing the $-\overset{\overset{\displaystyle O}{\parallel}}{S}-X$ grouping, the

Table XIV   Absorption of Compounds Containing Nitrogen-Oxygen Linkages

Compound	$n \rightarrow \pi^*$ Transition (R-band)		Solvent
	λ_{max}	ϵ_{max}	
Nitromethane	275	15	Heptane
2-Methyl-2-nitropropane	280.5	23	Heptane
1-Nitro-1-propene	229[a]	9400	Ethanol
	235[a]	9800	
Nitrosobutane	300	100	Ether
	665	20	
Octyl nitrate	270[b]	15	Pentane
Cyclohexyl nitrate	270[b]	22	–
n-Butyl nitrite	218	1050	Ethanol
	313-384[c]	17-45	

[a]These are $\pi \rightarrow \pi^*$ transitions.
[b]This is typically a point of inflection in the spectra of nitrates.
[c]This region is one of the fine structure with bands roughly 10 nm apart. The band with maximum absorption occurs at 357 nm.

position of the $n \rightarrow \pi^*$ band depends on the electronegativity of X; the greater the electronegativity, the shorter will be the wavelength of absorption.

Simple thioketones of the dialkyl- or alkylaryl types are unstable and generally exist as trimers. Thiobenzophenone is monomeric as are compounds in which the thione group is attached to an electron-donating group such as $-NR_2$. The $n \rightarrow \pi^*$ transition of the C=S group in thioketones occurs at a longer wavelength than the analogous C=O transition because the energy of the nonbonding electron pair of the sulfur atom lies at a higher level than the corresponding electrons of the oxygen atom. Compounds containing the C=S group also display intense bands in the 250–320 nm region which presumably arise from $\pi \rightarrow \pi^*$ and $n \rightarrow \sigma^*$ transitions in the C=S group.

The $n \rightarrow \pi^*$ bands of several thiocarbonyl compounds are summarized in Table XV.

Benzene Chromophore

Benzene displays three absorption bands: 184 nm (ϵ_{max} 60,000), 204 nm (ϵ_{max} 7900), and 256 nm (ϵ_{max} 200). These bands originate from $\pi \rightarrow \pi^*$ transitions. The intense band near 180 nm results from an allowed transition, whereas the weaker bands near 200 and 260 nm result from forbidden transitions in the highly symmetrical benzene molecule. Different notations have been used to designate the absorption bands of benzene; these are summarized in Table XVI. We shall use the designation of E and B bands employed by Braude.[4]

The B-band of benzene and many of its homologs is characterized by considerable fine structure. This is particularly true of spectra determined in the vapor phase or in nonpolar solvents. The fine structure originates from sub-levels of vibrational absorption upon which the electronic absorption is superimposed (Figure 7, p. 253). In polar solvents, interactions between solute and solvent tend to reduce the fine structure.

Substitution of alkyl groups on the benzene ring produces a bathochromic shift of the B-band, but the effect of alkyl substitution upon the E-bands is not clearly

Table XV $n \rightarrow \pi^*$ **Transitions in Thiocarbonyl Compounds**

Compound	$n \rightarrow \pi^*$ Transition (R-band)	
	$\lambda_{max}(nm)$	$\log \epsilon_{max}$
Thiobenzophenone	599	2.81
Thioacetamide	358	1.25
Thiourea	291 (s)	1.85

(s) shoulder.

Table XVI **Benzene Bands**

184 nm	204 nm	256 nm	Ref.
"180"	"200"	"260"	8
E_1	E_2	B	4
—	K	B	24
$A_{1g} \rightarrow E_{1u}$	$A_{1g} \rightarrow B_{1u}$	$A_{1g} \rightarrow B_{2u}$	3, 6
Second primary	First primary	Secondary	32, 33

defined. The absorption characteristics of the B-bands of several alkylbenzenes are presented in Table XVII.

The bathochromic shift is attributed to hyperconjugation in which the σ-electrons of an alkyl C–H bond participate in resonance with the ring. The methyl group is more effective in hyperconjugation than other alkyl groups.

The addition of a second alkyl group into the molecule is most effective in producing a red shift if it is in the *para* position. The *para* isomer absorbs at the longest wavelength with the largest ϵ_{max}. The *ortho* isomer generally absorbs at the shortest wavelength with reduced ϵ_{max}. This effect is attributed to steric interactions between the *ortho* substituents which effectively reduces hyperconjugation.

Substitution on the benzene ring of auxochromic groups (OH, NH_2, etc.) shifts the E- and B-bands to longer wavelengths, frequently with intensification of the B-band and loss of its fine structure, because of n–π conjugation (Table XVIII.)

Conversion of a phenol to the corresponding anion results in a bathochromic shift of the E_2- and B-bands and an increase in ϵ_{max} because an additional pair of nonbonding electrons in the anion is available for interaction with the π-electron system of the ring. When aniline is converted to the anilinium cation, the pair of nonbonding electrons of aniline is no longer available for interaction with the

Table XVII **Absorption Data for Alkylbenzenes (B-Bands)**

Compound	$\lambda_{max}(nm)$*	ϵ_{max}
Benzene	256	200
Toluene	261	300
m-Xylene	262.5	300
1,3,5-Trimethylbenzene	266	305
Hexamethylbenzene	272	300

*λ_{max} of most intense peak in band with fine structure.

Table XVIII Effect of Auxochromic Substitution on the Spectrum of Benzene

Compound	E_2-band		B-band		Solvent
	λ_{max}(nm)	ϵ_{max}	λ_{max}(nm)	ϵ_{max}	
Benzene	204	7,900	256	200	Hexane
Chlorobenzene	210	7,600	265	240	Ethanol
Thiophenol	236	10,000	269	700	Hexane
Anisole	217	6,400	269	1,480	2% Methanol
Phenol	210.5	6,200	270	1,450	Water
Phenolate anion	235	9,400	287	2,600	Aq. alkali
o-Catechol	214	6,300	276	2,300	Water (pH 3)
o-Catecholate anion	236.5	6,800	292	3,500	Water (ph 11)
Aniline	230	8,600	280	1,430	Water
Anilinium cation	203	7,500	254	160	Aq. acid
Diphenyl ether	255	11,000	272	2,000	Cyclohexane
			278	1,800	

π-electrons of the ring, and a spectrum almost identical to that of benzene results.

Confirmation of a suspected phenolic structure may be obtained by comparison of the ultraviolet spectra obtained for the compound in neutral and in alkaline solution (pH 13). Similar confirmatory information for a suspected aniline derivative may be obtained by a comparison of spectra determined in neutral and acid solution (pH 1).

When the spectra of pure acid and of pure conjugate base cross at some point, the spectra of all solutions containing various ratios of these two species (and no other absorbing species) must also go through this point (called the isosbestic point), provided that the sum of the concentrations of both species is constant, and that the spectral characteristics (i.e., the absorption coefficients) of both species are insensitive to the effect of pH and of the buffer used. An isosbestic point's presence can be used to verify that one is dealing with a simple acid-base reaction that is not complicated by further equilibria or other phenomena.[3]

Interaction between the nonbonding electron pair(s) of a heteroatom attached to the ring and the π-electrons of the ring is most effective when the p orbital of the nonbonding electrons is parallel to the π orbitals of the ring. Thus, bulky substitution in the ortho position of molecules such as N,N-dimethylaniline causes a hypsochromic shift in the E_2-band, accompanied by a marked reduction in ϵ_{max}.

N,N-Dimethylaniline	λ_{max} 251	ϵ_{max} 15,500
2-Methyl-N,N-dimethylaniline	λ_{max} 248	ϵ_{max} 6360

Direct attachment of an unsaturated group (chromophore) to the benzene ring produces a strong bathochromic shift of the B-band, and the appearance of a K-band (ϵ_{max} > 10,000) in the 200 to 250 nm region. (Table XIX). The overlap of absorption positions of the K-band, and the displaced E-bands of auxochromically substituted benzenes, may lead to confusion in the interpretation of ultraviolet spectra. Generally E-bands are less intense. The B-bands are sometimes buried under the K-bands.

The data in Table XIX show that in certain structures, such as acetophenone and benzaldehyde, displaced B- and R-bands can still be recognized.

The data of Table XIXa allow one to calculate an expected maximum for aromatic aldehydes, ketones, carboxylic acids and esters.[2] Application of Table XIXa to structure confirmation is illustrated with the following examples:

Example 1. 6-Methoxytetralone

Calc: λ_{max}^{EtOH} = 246 (parent chromophore) + 3(o-ring residue) + 25 (p-OMe)

= 274 nm

Obs: λ_{max}^{EtOH} = 276 nm

Example 2. 3-Carbethoxy-4-methyl-5-chloro-8-hydroxy-tetralone

Table XIX Absorption Characteristics of Chromophoric Substituted Benzenes

Compound	$\pi \to \pi^*$ Transition (K-band)		B-band		$n \to \pi^*$ Transition (R-band)		Solvent
	λ_{max}(nm)	ϵ_{max}	λ_{max}(nm)	ϵ_{max}	λ_{max}(nm)	ϵ_{max}	
Benzene	–	–	255	215	–	–	Alcohol
Styrene	244	12,000	282	450	–	–	Alcohol
Phenylacetylene	236	12,500	278	650	–	–	Hexane
Benzaldehyde	244	15,000	280	1,500	328	20	Alcohol
Acetophenone	240	13,000	278	1,100	319	50	Alcohol
Nitrobenzene	252	10,000	280	1,000	330	125	Hexane
Benzoic acid	230	10,000	270	800	–	–	Water
Phenyl cyanide	224	13,000	271	1,000	–	–	Water
Diphenyl sulfoxide	232	14,000	262	2,400	–	–	Alcohol
Phenyl methyl sulfone	217	6,700	264	977	–	–	–
Benzophenone	252	20,000	–	–	325	180	Alcohol
Biphenyl	246	20,000	submerged		–	–	Alcohol
Stilbene(*cis*)	283	12,300*	submerged		–	–	Alcohol
Stilbene(*trans*)	295†	25,000*	submerged		–	–	Alcohol
1-phenyl-1,3-butadiene							
cis-	268	18,500	–	–	–	–	Isooctane
trans-	280	27,000	–	–	–	–	Isooctane
1,3-pentadiene							
cis-	223	22,600	–	–	–	–	Alcohol
trans-	223.5	23,000	–	–	–	–	Alcohol

*Intense bands also occur in the 200–230 nm region.
†Most intense band of fine structure.

Calc.: λ_{max}^{EtOH} = 246 + 3 (*o*-ring residue) + 7 (*o*-OH)
= 256 nm
Obs.: λ_{max}^{EtOH} = 257 nm

Example 3. 3,4-Dimethoxy-10-oxo-octahydrophenanthrene

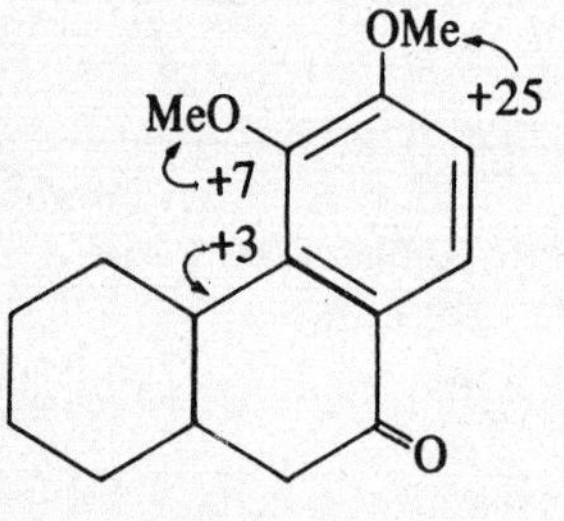

Calc.: λ_{max}^{EtOH} = 246 + 25 + 7 + 3 = 281 nm
Obs.: λ_{max}^{EtOH} = 278 nm

Application of Table XIX*a* does not give accurate predicted maxima in the 2,6-disubstituted phenyl carbonyl compounds; that this is due to disruption of the planarity necessary for conjugation of the carbonyl and phenyl groups is supported by the large decrease in molar absorptivity accompanying such substitution.

When auxochromic groups appear on the same ring as the chromophore, both groups influence the absorption. The influence is most pronounced when an electron donating group and electron attracting group are para to one another (complementary substitution) (Table XX).

The red shift and increase in intensity of the K-band are related to contributions of the following polar resonance forms:

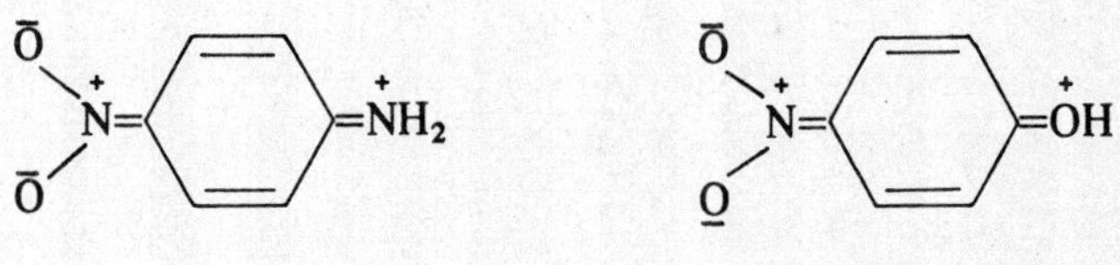

Biphenyl is the parent molecule of a series of compounds in which two aromatic rings are in conjugation. Resonance energy is at a maximum when the rings are coplanar and essentially zero when the rings are at 90° to one another.

Table XIXa Calculation of the Principal Band ($\pi \to \pi^*$ Transition) of Substituted Benzene Derivatives, Ar–COG (in EtOH)[2,*]

$ArCOR/ArCHO/ArCO_2 H/ArCO_2 R$		λ_{max}^{EtOH} (nm)
Parent chromophore: Ar = ϕ		
G = Alkyl or ring residue, (e.g. ArCOR)		246
G = H, (ArCHO)		250
G = OH, OAlk, (ArCO$_2$H, ArCO$_2$R)		230
Increment for each substituent on Ar:		
—Alkyl or ring residue	o-, m-	+3
	p-	+10
—OH, —OMe, —OAlk	o-, m-	+7
	p-	+25
—O⁻ (oxyanion)	o-	+11
	m-	+20
	p-	+78[a]
—Cl	o-, m-	+0
	p-	+10
—Br	o-, m-	+2
	p-	+15
—NH$_2$	o-, m-	+13
	p-	+58
—NHAc	o-, m-	+20
	p-	+45
—NHMe	p-	+73
—NMe$_2$	o-, m-	+20
	p-	+85

[a]This value may be decreased markedly by steric hindrance to coplanarity.

*Use of Table XIXa is illustrated by the examples on pp. 249–250 and in Ref. 2.

K-Band, λ_{max} 252, ϵ_{max} 19,000
Biphenyl

B-Band, λ_{max} 270, ϵ_{max} 800
2, 2'-Dimethylbiphenyl

The effect of forcing the rings out of coplanarity is readily seen from a comparison of the absorption characteristics of biphenyl and its 2,2'-dimethyl homolog whose absorption characteristics are similar to those of o-xylene.

Introduction of a methylene group between two chromophores is generally considered capable of destroying conjugation. Compare the data of diphenylmethane

λ_{max} = 262 nm

ϵ_{max} = 5000

Table XX Absorption Characteristics of Disubstituted Benzenes

Compound	$\pi \to \pi^*$ Transition (K-band)		B-band	
	λ_{max}	ϵ_{max}	λ_{max}	ϵ_{max}
o-NO$_2$ Phenol	279	6,600	351	3,200
m-NO$_2$ Phenol	274	6,000	333	1,960
p-NO$_2$ Phenol	318	10,000	submerged	
o-NO$_2$ Aniline	283	5,400	412	4,500
m-NO$_2$ Aniline	280	4,800	358	1,450
p-NO$_2$ Aniline	381	13,500	submerged	

with the ϵ_{max} for biphenyl above and the substituted diphenylmethanes below. In some substituted diphenylmethanes, there is an effective overlap of π orbitals of the two rings resulting in homoconjugation. The ϵ_{max} of 4-nitro-4'-methoxydiphenylmethane is not merely the sum of the ϵ_{max} of p-nitrotoluene and p-methoxytoluene:

λ_{max}	ϵ_{max}
274	9,490

λ_{max}	ϵ_{max}
277	2,190
285.5	1,786

λ_{max}	ϵ_{max}
280	24,400
287	16,800

The first homolog in the diphenyl polyene series (ϕ-(C=C)$_n$-ϕ) is stilbene. Stilbene offers an interesting example of steric effects in electronic spectra.

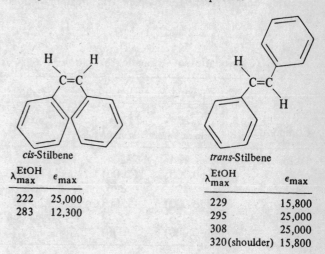

cis-Stilbene

λ_{max}^{EtOH}	ϵ_{max}
222	25,000
283	12,300

trans-Stilbene

λ_{max}^{EtOH}	ϵ_{max}
229	15,800
295	25,000
308	25,000
320 (shoulder)	15,800

The destruction of coplanarity by steric interference, in the cis-structure, is reflected by the lower intensity of the 283 nm band compared with the corresponding band (295 nm) in the trans isomer. The B-band appears to be swamped by this intense absorption.

The absorption bands of the parent stilbene molecule move to longer wavelengths and increase in intensity as n, in $\phi\text{-(CH=CH)}_n\text{-}\phi$, increases. When n equals 7, the molecule absorbs in the 400–465 region with ϵ_{max} 135,000.

Two common series of aromatic compounds are the linear series such as anthracene and the angular series such as phenanthrene. Although the polynuclear aromatics might well be treated as individual chromophores, a correlation between the bands of benzene and the acenes, such as naphthalene, can be made.[3] These correlations appear in Table XXI.

As the number of condensed rings increases in the acene series the absorption moves to progressively longer wavelengths until it occurs in the visible region.

Naphthacene
(Yellow)

Pentacene
(Blue)

The angular polycyclic compounds, the phenenes, also show a bathochromic shift of the three band system with an increase in the number of rings. However, the increase in λ_{max}, per ring added, is less than for the acenes. The three band system is still distinct for phenanthrene but in the spectrum of anthracene the E_2-band has already swamped the B-band.

The spectra of polynuclear aromatics are characterized by vibrational fine-structure as observed in the spectrum of benzene. Spectra of some polynuclear aromatics are shown in Figure 7.

Heteroaromatic Compounds

Saturated five- and six-membered heterocyclic compounds are transparent at wavelengths longer than 200 nm. Only the unsaturated heterocyclic compounds (heteroaromatics) show absorption in the near ultraviolet region.

FINE MEMBERED RINGS. The theoretical interpretation of the spectra of five-membered-ring heteroaromatic compounds is not simple. The absorption of these compounds has been compared to that of cyclopentadiene, the cis-diene analog, which shows strong diene absorption near 200 nm and moderately intense absorption near 238 nm. The aromatic properties increase in the order cyclopentadiene, furan, pyrrole, and thiophene. The absorption of some five-membered heteroaromatics is compared with cyclopentadiene in Table XXII. No attempt has been made to classify the bands, although the band near 200 nm has been likened to the E_2-band of benzene, and the long wavelength band frequently has fine structure analogous to the B-band of benzene.

Auxochromic or chromophoric substitution of the five-membered unsaturated heterocyclics causes a bathochromic shift and an increase in the intensity of the bands of the parent molecule (Table XXIII).

SIX MEMBERED RINGS. The spectrum of pyridine is similar to that of benzene. The B-band of pyridine, however, is somewhat more intense and has less distinct fine structure than that of benzene (Figure 8). This transition is allowed for pyridine, but forbidden for the symmetrical benzene molecule. The weak R-band expected for an $n \rightarrow \pi^*$ transition in pyridine has been observed in vapor phase spectra. This band is generally swamped by the more intense B-band when the spectrum is determined in solution.

Table XXI   Correlation of Aromatic Absorption

Compound	E_1-band λ_{max} (ϵ_{max})	E_2-band λ_{max} (ϵ_{max})	B-band λ_{max} (ϵ_{max})	λ_{max} (ϵ_{max})
Benzene	184 (60,000)	204 (7900)	256 (200)	
Naphthalene	221 (133,000)	286 (9300)	312 (289)	
Anthracene	256 (180,000)	375 (9000)	Submerged	221 (14,500)

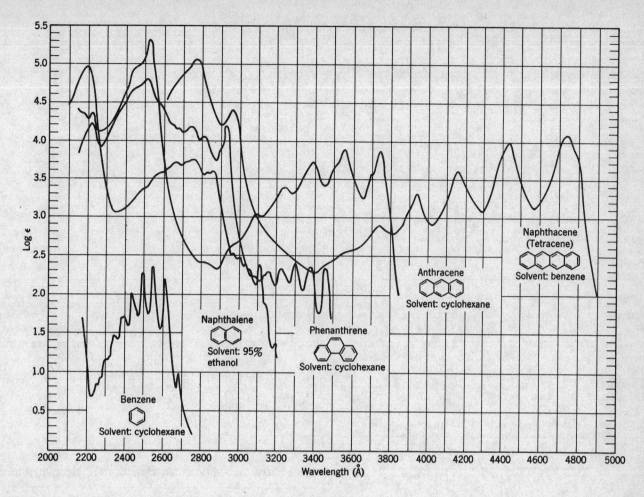

Fig. 7.

Electronic absorption spectra of benzene, naphthalene, phenanthrene, anthracene, and naphthacene.

An increase in solvent polarity has little or no effect on the position or intensity of the *B*-band of benzene, but produces a marked hyperchromic effect on the *B*-band of pyridine and its homologs. The hyperchromic effect undoubtedly results from hydrogen bonding through the lone pair of electrons of the nitrogen atom. The extreme case is the absorption of a pyridinium salt. The absorption characteristics of 2-methylpyridine (α-picoline), in several solvents, are shown in Table XXIV.

The effect of substitution on the 257 nm band (*B*-band)

Table XXII Absorption Data for Some Five-Membered Heteroaromatics

| Compound | Band I | | Band II | | |
	λ_{max}	ϵ_{max}	λ_{max}	ϵ_{max}	Solvent
Cyclopentadiene	200	10,000	238.5	3,400	Hexane
Furan	200	10,000	252	1[a]	Cyclohexane
Pyrrole[b]	211	15,000	240	300[a]	Hexane
Thiophene	231	7,100	269.5	1.5[a]	Hexane
Pyrazole	214	3,160	—	—	Ethanol

[a]These weak bands may be due to impurities rather than a forbidden transition ($n \rightarrow \pi^*$) of a heteroaromatic molecule.[34]

[b]See Table XXIII for different assignment of bands in the spectrum of pyrrole.

Table XXIII Absorption Characteristics of Five-Membered Heteroaromatics[3]

Parent	Substituent	Band I		Band II	
		λ_{max}(nm)	ϵ_{max}	λ_{max}(nm)	ϵ_{max}
Furan		200	10,000	252	1
Furan	2-CHO	227	2,200	272	13,000
Furan	2-C(=O)−CH$_3$	225	2,300	270	12,900
Furan	2-COOH	214	3,800	243	10,700
Furan	2-NO$_2$	225	3,400	315	8,100
Furan	2-Br, 5-NO$_2$	–	–	315	9,600
Pyrrole		183	–	211	15,000
Pyrrole	2-CHO	252	5,000	290	16,600
Pyrrole	2-C(=O)−CH$_3$	250	4,400	287	16,000
Pyrrole	2-COOH	228	4,500	258	12,600
Pyrrole	1-C(=O)−CH$_3$	234	10,800	288	760
Thiophene		231	7,100	–	–
Thiophene	2-CHO	265	10,500	279	6,500
Thiophene	2-C(=O)−CH$_3$	252	10,500	273	7,200
Thiophene	2-COOH	249	11,500	269	8,200
Thiophene	2-NO$_2$	268-272	6,300	294-298	6,000
Thiophene	2-Br	236	9,100	–	–

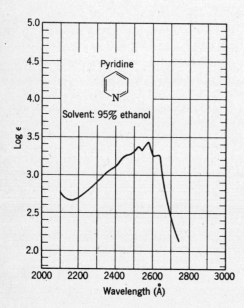

Fig. 8.
Ultraviolet spectrum of pyridine.

Table XXIV Absorption Characteristics of 2-Methylpyridine

Solvent	λ_{max}(nm)	ϵ_{max}
Hexane	260	2000
Chloroform	263	4500
Ethanol	260	4000
Water	260	4000
Ethanol-HCl(1:1)	262	5200

attributed to the pyridone structures. Tautomerism in hydroxy- and aminopyridines is discussed thoroughly in Reference 2.

of pyridine is illustrated in the data presented in Table XXV.

The absorption of the 2-OH and 4-OH pyridines is

Table XXV Absorption Characteristics of Pyridine Derivatives

Derivative	$\lambda_{max}^{(pH > 7)}$	ϵ_{max}
Pyridine	257	2,750
	270	450
2-CH_3	262	3,560
3-CH_3	263	3,110
4-CH_3	255	2,100
2-F	257	3,350
2-Cl	263	3,650
2-Br	265	3,750
2-I	272	400
2-OH	230	10,000
	295	6,300
4-OH	239	14,100
3-OH	260	2,200

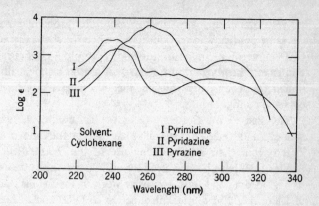

Fig. 9.

The spectra of heteroaromatics appear to be related to their isocyclic analogs (Table XXVI).

The spectra of the diazines are similar to those of pyridine (Figure 9). In addition to the enhanced B-band, still retaining some fine structure, the enhanced $n \rightarrow \pi^*$ bands are quite prominent.

The absorption of the diazines, for example the pyrazines, respond to solvent polarity in a manner similar to the pyridines. The B-band undergoes a hyperchromic shift with an increase in solvent polarity with little or no effect upon λ_{max}. The R-band, of course, disappears in acid solution since the nonbonding electrons of the free base are now involved in salt formation.

OPTICAL ROTATORY DISPERSION[35-38]

The specific rotation, $[\alpha]$, of a compound varies with the wavelength of light at which the measurement is made.

For many simple compounds, the rotation simply increases in absolute value as we proceed from visible light farther and farther into the ultraviolet region. The function which describes the variation of the rotation of a compound with wavelength is called the *optical rotatory dispersion (ORD) curve.* The type of curve described above is referred to as a *plain curve.*

ORD facilitates organic structure determination most when a compound yields the characteristically shaped optical rotatory dispersion curve referred to as a *Cotton effect curve:* Near and through regions of absorption where electronic transitions occur (UV spectrum) optical rotatory dispersions often behave quite irregularly. They often characteristically pass through one or more maxima ("peaks") and minima ("troughs"). Such irregularities are called the *Cotton effect.* Such curves are called positive if, when going from long toward short wavelength the curve first increases in the positive direction, and then becomes more negative. A Cotton effect curve is called negative if the signs of the segments are reversed. Enantiomers always give mirror-image ORD curves. If the ORD curve of an optically active compound of known configuration and that of a chemically similar molecule of unknown configuration

Table XXVI Absorption Characteristics of Some Heteroaromatics Containing Nitrogen and Their Isocyclic Analogs

Compound	E_1-band[a] λ_{max}(nm)	ϵ_{max}	E_2-band[a] λ_{max}(nm)	ϵ_{max}	B-band[a] λ_{max}(nm)	ϵ_{max}	Solvent
Benzene	184	60,000	204	7,900	256	200	—
Naphthalene	221	100,000	286	9,300	312	280	—
Quinoline	228	40,000	270	3,162	315	2,500	Cyclohexane
iso-Quinoline	218	63,000	265	4,170	313	1,800	Cyclohexane
Anthracene	256	180,000	375	9,000	—	—	
Acridine	250	200,000	358	10,000	—	—	Ethanol

[a]All of the bands contain fine structure.

are very similar in shape, the two molecules probably have the same configurations. Conversely, when the shape of the ORD curve of a compound of known absolute configuration is very nearly the mirror image of that of a chemically similar, optically active species, the two most likely have opposite configurations.

The sign of the Cotton effect of substituted cyclohexanones (and thus keto-steroids) can successfully be predicted from their structures by use of the *Octant Rule*. Similar rules have been evolved for other compound classes.

The leading and most extensive work in the field is that of Djerassi.[35] For a brief introduction to ORD and a clear explanation of the Octant Rule, consult Eliel.[36]

REFERENCES

1. Stern, E. S., and T. C. J. Timmons, *Electronic Absorption Spectroscopy in Organic Chemistry*, St. Martin's Press, New York, 1971. (effectively the third edition of Gillam and Stern).

2. Scott, A. I., *Interpretation of the Ultraviolet Spectra of Natural Products*, Pergamon Press (The Macmillan Co.), New York, 1964.

3. Jaffé, H. H., and Milton Orchin, *Theory and Application of Ultraviolet Spectroscopy*, Wiley, New York, 1962.

4. Braude, E. A., "Ultra-Violet Light Absorption and the Structure of Organic Compounds," *Ann. Repts. on Progress Chem.*, Chemical Society of London, Vol. XLII (1945), pp. 105–30.

5. Braude, E. A., "Ultraviolet and Visible Light Absorption," Chap. 4, pp. 131–94, in *Determination of Organic Structures by Physical Methods*, Academic Press, New York, 1955.

6. Duncan, A. B. F., and F. A. Matsen, "Electronic Spectra in the Visible and Ultraviolet," Vol. IX, pp. 581–706, in *Technique of Organic Chemistry*, A. Weissberger, Ed., Interscience, New York, 1956.

7. Bauman, R. P., *Absorption Spectroscopy*, Wiley, New York, 1962.

8. Rao, C. N. R., *Ultra-Violet and Visible Spectroscopy*, 2nd Ed., Plenum, New York, 1967.

9. Brand, J. C. D., and G. Eglinton, *Applications of Spectroscopy to Organic Chemistry*, Oldbourne, London, 1965.

10. Phillips, J. P., *Spectra Structure Correlation*, Academic Press, New York, 1964.

11. Forbes, W. F., Chap. I in *Interpretive Spectroscopy*, S. K. Freeman, Ed., Reinhold, New York, N.Y. 1965.

12a. *Ultraviolet Reference Spectra*, Sadtler Research Laboratories, 3314–20 Spring Garden St., Philadelphia, Pa. *Sadtler–Reference Spectra–Commonly Abused Drugs UV Spectra*, 1972, 300 UV Spectra Sadtler Research Laboratories, Inc. *Sadtler Standard Ultraviolet Spectra*. 1972, 32,000 spectra in 74 volumes, Sadtler Research Laboratories, Inc.

12b. Bolshakow, G. F., V. S. Vatago and F. B. Agrest, *Ultraviolet Spectra of Hetero-Organic Compounds*, Khimia Publ. House, Leningrad, U.S.S.R., 1969.

12c. Hirayama, K., *Handbook of Ultraviolet and Visible Absorption Spectra of Organic Compounds*, Plenum Press Data Div., New York, 1967.

13. *Electronic Spectral Data*, Vol. I, M. J. Kamlet, Ed., Interscience, New York, 1960. Covers literature 1946–1952.

14. *Organic Electronic Spectral Data*, Vol. II, H. G. Ungnade, Ed., Interscience, New York, 1960. Covers literature 1953–1955.

15. *Organic Electronic Spectral Data*, Vol. IV, J. P. Phillips and F. C. Nachod, Ed., Interscience, New York, 1963. Covers literature 1958–1959.

16. Hershenson, H. M., *Ultraviolet Absorption Spectra*, Index for 1954–1957, Academic Press, New York, 1959.

17. Lang, L., *Absorption Spectra in the Ultraviolet and Visible Region*, Academic Press, New York, 1961–1965, 5 volumes.

18. *Ultraviolet Spectral Data*. Manufacturing Chemists Association Research Project, Carnegie Institute of Technology, Pittsburgh, Pa.

19. Friedel, R. A., and Milton Orchin, *Ultraviolet Spectra of Aromatic Compounds*, Wiley, New York, 1951. Revised ed., 1958.

20. American Petroleum Institute Research Project 44. *Selected Ultraviolet Spectral Data*. (looseleaf data sheets) 1945–1970 Vol. I–III 1147 pp. Thermodynamics Research Center, Texas A & M.

21. ASTM index to Ultraviolet and Visible Spectra. ASTM Technical Publication No. 357, American Society for Testing Materials, 1916 Race St., Philadelphia, Pa. (1963).

22. UV-Atlas of Organic Compounds, Photoelectric Spectrometry Group, London, and Institut für Spektrochemie und Angewandte Spektroskopie, Dortmund, Butterworths, London.

23. "Spectrometry Nomenclature," *Anal. Chem.*, December Issue, Annually.

24. Burawoy, A., *Ber.*, *63*, 3155 (1930); *J. Chem. Soc.*, 1177 (1939).

25. Riddick, J. A., and E. E. Toops, Jr., "Organic Solvents," Vol. VII, p. 339, in *Technique of Organic Chemistry*, A. Weissberger, Ed., Interscience, New York, 1955.

26. Woodward, R. B., *J. Am. Chem. Soc.*, *63*, 1123 (1941).

27. Woodward, R. B., *ibid.*, *64*, 72, 76 (1942).

28a. Frank, R. L., R. Armstrong, J. Kwiatek, and H. A. Price, *J. Am. Chem. Soc.*, *70*, 1379 (1948).

28b. Turner, R. B., and D. M. Voitle, *J. Am. Chem. Soc.*, *73*, 1403 (1951).

29a. Fieser, L. M., and Fieser, M., *Natural Products Related to Phenanthrene*, Reinhold, New York, 1949, p. 184 ff.

29b. Fieser, L. M. and M. Fieser, *Steroids*, Reinhold, New York, 1959.

30. Ferguson, N. F., and J. C. Nnadi, *J. Chem. Ed.*, *42*, 529 (1965).

31. Nielson, A. T., *J. Org. Chem.*, *22*, 1539 (1957).

32. Doub, L., and J. M. Vandenbelt, *J. Am. Chem. Soc.*, *69*, 2714 (1947).

33. Doub, L., and J. M. Vandenbelt, *J. Am. Chem. Soc.*, *71*, 2414 (1949).

34. Katritzky, A. R., Ed., Vol. II, *Physical Methods in Heterocyclic Chemistry*, Academic Press, New York and London, 1963.

35. Djerassi, C., *Optical Rotatory Dispersion*, McGraw-Hill, New York, 1960.

36. Eliel, E. L., *Stereochemistry of Carbon Compounds*, McGraw-Hill, New York, 1962, Chapter 14.

37. Klyne, W., "Optical Rotatory Dispersion" in R. A. Raphael, E. C. Taylor, and H. Wynberg, Eds., *Advances in Organic Chemistry, Methods and Results*, Vol. I, Interscience, New York, 1960.

38. Crabbe, P., *Optical Rotatory Dispersion and Circular Dichroism in Organic Chemistry*, Holden-Day, San Francisco, 1965.

appendix

"BAKER ANALYZED" REAGENT, SPECTROPHOTOMETRIC GRADE

Acetone	Cyclohexane	Hexane	2-Propanol
Benzene	1,2-Dichloroethane	Methanol	Toluene
Carbon Tetrachloride	Dimethylformamide	Methylene Chloride	2,2,4-Trimethylpentane
Chloroform	Ethyl Acetate	2-Methyl-1-propanol	

For the absorption free regions of some of the above solvents, consult Figure 6 of this chapter.

sets of spectra translated into compounds

Compound Number	Name, Page	Compound Class
6-1	Benzyl acetate, 261	aromatic ester
6-2	Diisopropyl ether, 264	aliphatic ether
6-3	Diisoamyl disulfide, 266	aliphatic sulfide (thio-ether)
6-4	Methyl 2-furoate, 269	heteroaromatic ester
6-5	β, β'-Dibromodiethyl ether, 272	halo-ether
6-6	2-Mercaptoethanol, 275	alcohol-mercaptan (thiol)
6-7	Isoeugenol, 278	aromatic, olefinic, ether, phenol
6-8	Divinyl β, β'-thiodipropionate, 281	olefinic, ester, sulfide (thio-ether)
6-9	Levulinic acid, 273	keto-carboxylic acid
6-10	Heptaldehyde, 286	aliphatic aldehyde
6-11	γ-Valerolactone, 288	lactone (cyclic ester)
6-12	Diphenyl sulfoxide, 290	aromatic sulfoxide
6-13	2,3-Dihydropyran, 293	heterocycle
6-14	Methyl n-butyrate, 296	aliphatic ester
6-15	β, β'-thiodipropionitrile, 298	nitrile (cyanide), sulfide (thio ether)
6-16	2-Dimethylaminoethyl acetate, 300	amine, ester
6-17	p-Fluorostyrene, 302	halo, olefinic, aromatic
6-18	4-Chloroindole, 305	halo-heterocycle, polynuclear
6-19	ϵ-Caprolactam, 307	lactam, cyclic amide
6-20	Dibutyldifluoromethylphosphine oxide, 309	halo-aliphatic, phosphine oxide
6-21	Ethyl sorbate, 312	unsaturated ester

In practice, identification of an organic compound begins with a history, and large areas are quickly excluded from further consideration. Since, with a few exceptions, we dispense with a history, the limits will be set as follows: The compounds are quite pure; they may contain carbon, hydrogen, oxygen, nitrogen, sulfur, and the halogens in any combination; since the table of isotope contributions runs only to a molecular weight of 250, the examples will be limited to this range. Within these restrictions, we can still cover a vast expanse of organic chemistry.

Additional information, usually a history, would probably be necessary to identify a compound that contained boron, silicon, or phosphorus in addition to the above elements. But it is hardly likely that a chemist would encounter such compounds without prior knowledge that these elements are present.

The interpreter and correlator of data has always lived on the edge of uncertainty. How pure is his compound, how reliable his data, how relevant his reference material? There are few unequivocal data, either spectrometric or chemical; thus there is no substitute for experience and a broad knowledge of chemistry. Nor is there a prescribed procedure. In general, we attempt as a first step to establish a molecular formula from the parent peak and the isotope contributions. In a number of cases, the parent peak is so small that the isotope contributions cannot be accurately measured. We settle for the molecular weight. In some cases, the parent peak may be missing. We then try to establish the molecular weight from other evidence. In many cases this can be done from the fragmentation pattern and from the other spectra at hand. In other cases, we may resort to preparation of appropriate derivatives, to other methods of obtaining molecular weights, or to other methods of determining elemental composition. Recent improvements in combustion analysis permit C, H, and N determinations on less than a milligram of sample.

We use the obvious features of one spectrum to bring out the more subtle aspects of another. The power of this methodology lies in the complementary features of the four spectra. When enough information is accumulated, a structure is postulated and spectral features predicted on the basis of the postulated structure are compared with the spectra actually observed. Appropriate structural modifications are made to accomodate the discrepancies. Confirmation is obtained by comparison with the spectra of an authentic sample. Let us work through the twenty one samples comprising this chapter.

We may not always come up with an unequivocal structure; nor do we in any system of organic analysis. However, we should at least be able to narrow down the possibilities to several structures (often isomers) and to indicate the steps required to complete the identification.

Mass spectra were obtained on the Consolidated Electrodynamics Corporation Model 21-103C. The lower limit of the fragmentation patterns is mass 20. Peaks of less than 3% relative intensity are not reported except for the parent and isotope peaks. Infrared spectra were run on the Perkin-Elmer Corporation Model 221 and 137. NMR spectra were run on the Varian Associates Model A-60, HR-60, and HA-100; samples were dissolved in carbon tetrachloride or deuterated chloroform, and 1% tetramethylsilane in the solution was used as a reference. Ultraviolet spectra were obtained from the literature, or were run on Applied Physics Corporation's Cary Model 14M at pH 7, pH 1, and pH 13. Only the pH 7 spectrum is recorded unless a change occurred at pH 1 or pH 13.

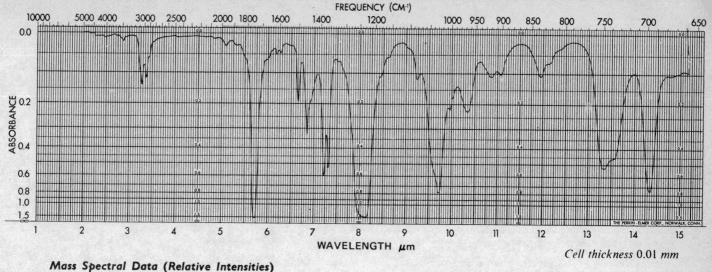

FREQUENCY (CM⁻¹)

WAVELENGTH μm

Cell thickness 0.01 mm

THE PERKIN-ELMER CORP., NORWALK, CONN.

Mass Spectral Data (Relative Intensities)

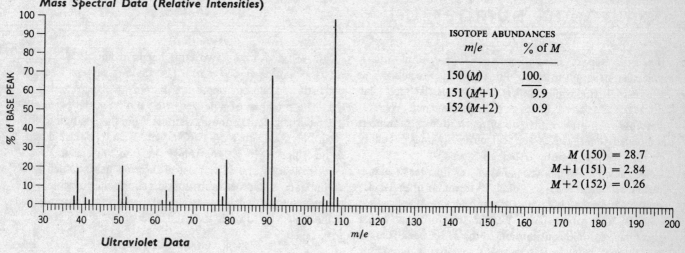

m/e

ISOTOPE ABUNDANCES	
m/e	% of *M*
150 (*M*)	100.
151 (*M*+1)	9.9
152 (*M*+2)	0.9

$M (150) = 28.7$
$M+1 (151) = 2.84$
$M+2 (152) = 0.26$

Ultraviolet Data

λ_{max}^{EtOH}	ϵ_{max}		
		252	153
268	101	248 (s)	109
264	158	243 (s)	78
262	147		
257	194	(s) = shoulder	

NMR Spectrum (*Solvent CCl₄*)

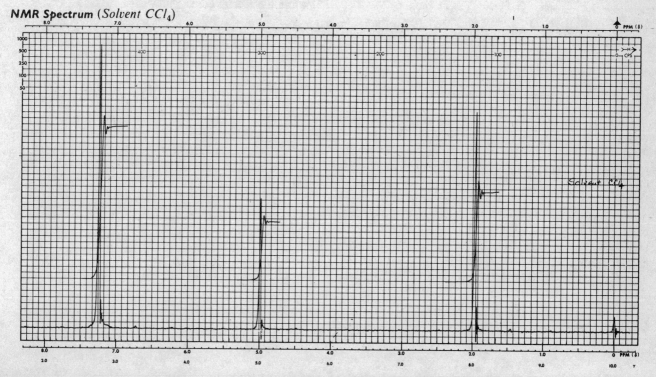

Solvent CCl₄

compound number 6-1

The first step in translating these four spectra into a molecular structure is to establish a molecular formula. The parent peak (molecular ion) is 150; this is the molecular weight. The parent peak is an even number. We are, therefore, permitted either no nitrogen atoms, or an even number of them. The $M + 2$ peak obviously does not allow for the presence of sulfur or halogen atoms.

We now look in Appendix A of Chapter 2 under molecular weight 150, and find 29 formulas of molecular weight 150 containing only CHN and O. Our $M + 1$ peak is 9.9% of the parent peak. We list the formulas whose calculated isotopic contribution to the $M + 1$ peak falls—to be arbitrary—between 9.0 and 11.0; we also list their $M + 2$ values:

Formula	$M + 1$	$M + 2$
$C_7H_{10}N_4$	9.25	0.38
$C_8H_8NO_2$	9.23	0.78
$C_8H_{10}N_2O$	9.61	0.61
$C_8H_{12}N_3$	9.98	0.45
$C_9H_{10}O_2$	9.96	0.84
$C_9H_{12}NO$	10.34	0.68
$C_9H_{14}N_2$	10.71	0.52

Three of these formulas can be eliminated because they contain an odd number of nitrogen atoms. The $M + 2$ peak is 0.9% of the parent; this best fits $C_9H_{10}O_2$, which we shall tentatively designate as our molecular formula. We make a mental note that both the intensity of the parent peak, and the C-to-H ratio of the formula indicate aromaticity.

The infrared spectrum shows a C=O band at about 1745 cm^{-1} (5.73 μm). This, together with the presence of two O atoms in the formula, suggests an ester. We look for confirmation in the C–O–C stretching region and note the large broad band at about 1225 cm^{-1} (8.15 μm) character-istic of an acetate. Two large bands at about 749 cm^{-1} (13.35 μm) and 697 cm^{-1} (14.35 μm) suggest a singly substituted benzene ring.

The presence of a benzene ring and an acetate group is established. Furthermore, we note from the position of the carbonyl band that the C=O moiety is not conjugated with the ring. This is confirmed by the wavelengths and intensities of the ultraviolet absorption peaks which also eliminate a ketone from consideration. Subtraction of a singly substituted benzene ring and an acetate group from the molecular formula gives the following:

$$\text{Molecular formula} \quad C_9H_{10}O_2$$

$$C_6H_5 + CH_3\overset{\displaystyle O}{\overset{\displaystyle \|}{C}}{-}O \quad C_8H_8O_2$$
$$\text{Remaining} \qquad\qquad CH_2$$

It takes no great imagination to insert the CH_2 between the ring and the acetate group, and write

The NMR spectrum provides almost conclusive confirmation for the above structure. We see three sharp unsplit peaks in the following positions and with the following integrated intensities

τ	δ	Intensity
2.78	7.22	5
5.00	5.00	2
8.04	1.96	3

The five protons at δ 7.22, τ 2.78, of course, are the five benzene-ring protons. The singlet of two protons at δ 5.00, τ 5.00 represents the methylene group substituted by a phenyl and an ester group. And, of course, the singlet of three protons at δ 1.96, τ 8.04 represents the methyl group.

We can obtain additional confirmation by returning to the mass spectrum and considering the fragmentation pattern (see Appendix B, Chapter 2) in view of the information at hand. The base peak at 108 is a rearrangement peak representing cleavage of an acetyl group (43) and rearrangement of a single hydrogen atom (Chapter 2, p. 23). The large peak at mass 91 is the benzyl (or tropylium) ion formed by cleavage beta to the ring. And the large peak at mass 43, of course, represents the acetyl fragment. The peaks at 77, 78, and 79 are additional evidence for the benzene ring.

We can state with a high degree of confidence that the compound represented by these spectra is:

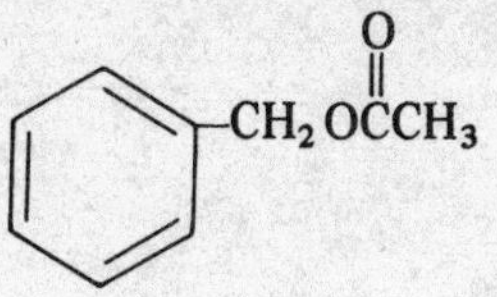

Benzyl acetate

There are, of course, a number of other sequences to the identity of this compound. Having established the molecular formula, we could note at once the characteristic benzene ring peak at δ 7.22, τ 2.78, in the NMR spectrum. We could confirm this by the typical "benzenoid" fine structure absorption in the ultraviolet spectrum. The base peak in the mass spectrum is a common rearrangement peak, and the mass 91 peak immediately calls to mind the benzyl (or tropylium) structure. The large mass 43 peak strongly suggests the CH_3CO group in view of the C=O peak in the infrared. Subtraction of a benzyl and an acetyl group from the formula leaves a mass of 16; consideration of the infrared spectrum leaves very little question as to how to handle this oxygen atom.

The student will find it instructive to write the possible isomeric structures and to eliminate them on spectrometric grounds.

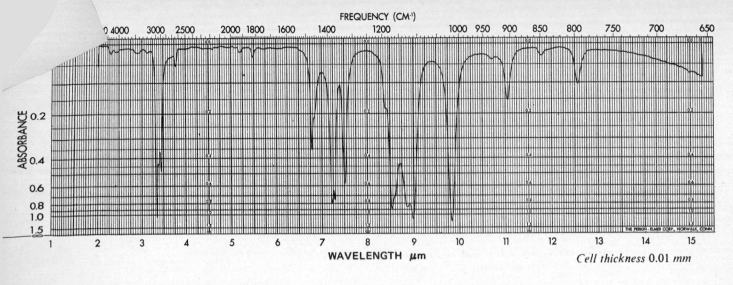

FREQUENCY (CM⁻¹)

WAVELENGTH μm

Cell thickness 0.01 mm

Mass Spectral Data (Relative Intensities)

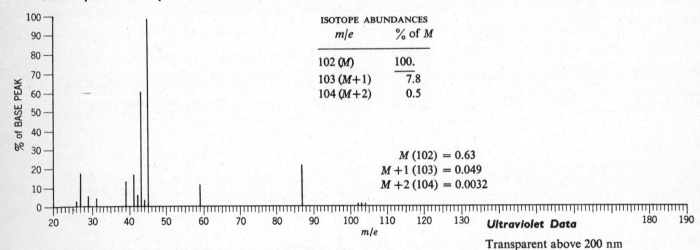

ISOTOPE ABUNDANCES

m/e	% of M
102 (M)	100.
103 (M+1)	7.8
104 (M+2)	0.5

$M (102) = 0.63$
$M+1 (103) = 0.049$
$M+2 (104) = 0.0032$

m/e

Ultraviolet Data

Transparent above 200 nm

NMR Spectrum (Solvent CDCl₃)

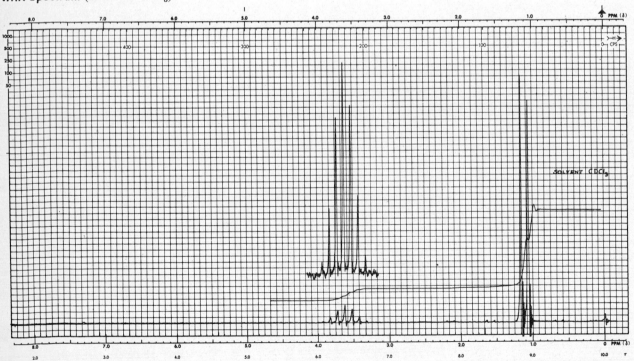

SOLVENT CDCl₃

compound number 6-2

In accordance with our usual procedure, we write the possible molecular formulas (together with the $M + 1$ and $M + 2$ peaks) under the molecular weight, in this case, 102:

Formula	$M + 1$	$M + 2$
$C_5H_{14}N_2$	6.39	0.17
$C_6H_2N_2$	7.28	0.23
$C_6H_{14}O$	6.75	0.39
C_7H_2O	7.64	0.45
C_8H_6	8.74	0.34

The best fit to our $M + 1$ (7.8) and $M + 2$ (0.5) values is C_7H_2O. But this is a trivial formula as is $C_6H_2N_2$. The next best fit is $C_6H_{14}O$. We shall proceed on this basis. Sulfur or halogen is not permitted.

The compound is aliphatic on the basis of the C-to-H ratio, the general appearance of the IR spectrum (though the C–H stretch is at rather a high frequency), and lack of absorption in the ultraviolet spectrum, and the absence of peaks at low field in the NMR spectrum.

There is no carbonyl or hydroxyl absorption in the infrared spectrum. Since the formula requires an oxygen atom, we consider some type of ether, and look for C–O absorption which we find (split) at about 1130 to 1110 cm^{-1} (8.7 to 9.0 μm).

The NMR spectrum is quite definitive. The doublet and the symmetrical heptet in the integrated ratio of 6:1 spell out an isopropyl group. Obviously the molecule is symmetrical about the oxygen atom. We write.

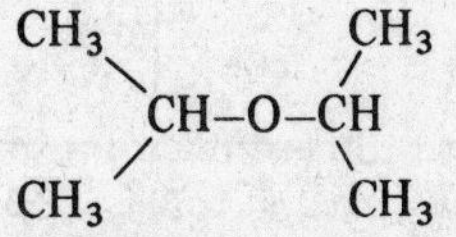

$$CH_3 \qquad\qquad CH_3$$
$$CH-O-CH$$
$$CH_3 \qquad\qquad CH_3$$

Diisopropyl ether

The infrared doublet at 1380 cm^{-1} (7.23 μm) and 1370 cm^{-1} (7.28 μm) confirms the isopropyl group. The rather high-frequency C–H stretching peak is explained by the presence of the oxygen atom. The strong band at 1170 cm^{-1} (8.55 μm) is a C–C stretch intensified by branching.

The base peak (45) in the mass spectrum results from double cleavage with rearrangement of a hydrogen atom. This is prominent in α-substituted ethers. Removal of the methyl group accounts for the mass 87 peak. C–O cleavage with retention of the charge on the alkyl portion results in the large mass 43 peak. (See Chapter 2, page 24.)

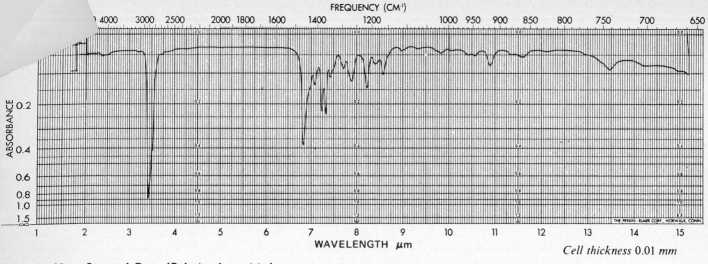

FREQUENCY (CM⁻¹)

ABSORBANCE

WAVELENGTH μm

Cell thickness 0.01 mm

Mass Spectral Data (Relative Intensities)

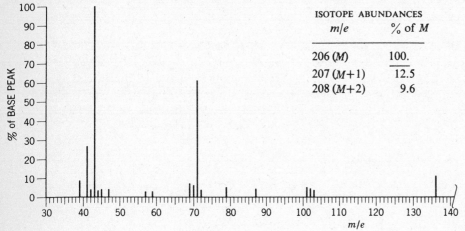

% of BASE PEAK

m/e

ISOTOPE ABUNDANCES

m/e	% of M
206 (M)	100.
207 (M+1)	12.5
208 (M+2)	9.6

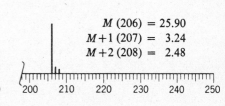

$M (206) = 25.90$
$M+1 (207) = 3.24$
$M+2 (208) = 2.48$

Ultraviolet Data

λ_{max}^{EtOH}	$\log \epsilon_{max}$
248	2.55

NMR Spectrum (*Solvent CDCl₃*)

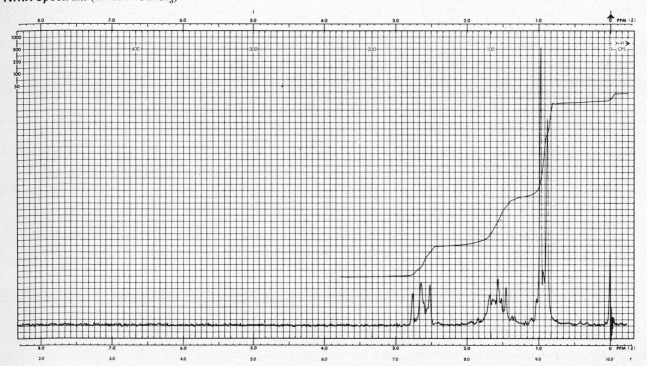

PPM (δ)

compound number 6-3

Following our usual practice, we strike immediately for the molecular ion (parent peak, 206) and start to list possibilities under that molecular weight. But consideration of the $M+2$ peak brings us up short. Obviously, we are no longer dealing with compounds containing only C, H, O, and N. The $M+2$ peak is too small for a chlorine or a bromine atom (see Table II, Chapter 2), and too large for a sulfur atom. But it will accommodate two sulfur atoms very nicely.

We subtract the mass of two sulfur atoms from 206 and get 142, which is the weight of the rest of the molecule. We now compile the list of possibilities from the table under 142, using the $M+1$ peak (and, of course, the fact that 142 is an even number) to narrow the possible molecular formulas. The $M+1$ peak becomes 12.5 minus 2×0.78 which gives 10.9; this removes the contribution of the two ^{33}S atoms.

Formula	$M+1$
$C_{10}H_6O$	10.94
$C_{10}H_{22}$	11.16
$C_{11}H_{10}$	12.05

The infrared and the NMR spectra convey a strong impression that we are dealing with an aliphatic compound. The ultraviolet spectrum is not especially informative. The infrared gives no evidence for the presence of an oxygen atom; in fact, it is rather featureless save for the strong aliphatic C–H stretching bands at 2915 to 2841 cm^{-1} (3.43 to 3.52 μm) the CH_2 and CH_3 bending vibration at 1464 cm^{-1} (6.83 μm) and the twin peaks at 1381 and 1364 cm^{-1} (7.24 and 7.33 μm) which we may often associate with chain branching.

Although $C_{11}H_{10}$ is a possible formula, we can find no support for unsaturated character. We find no evidence for the presence of oxygen, so we write $C_{10}H_{22}S_2$.

The base peak, mass 43, in the mass spectrum allows us to write $CH_3CH_2CH_2$ or CH_3CHCH_3. We choose the latter for several reasons. A base peak is more likely to result from cleavage at a branch. The large doublet in the NMR spectrum is familiar as the methyl protons of an isopropyl group. We also associated a pair of peaks in the infrared spectrum with chain branching.

We note a slightly distorted triplet in the NMR spectrum centered on δ 2.65, τ 7.35. This represents two protons (possibly CH_2) if we assign six protons to the large methyl doublet. A sulfur atom adjacent to the methylene group would account for its downfield shift.

The integrator shows that the multiplet centered at about δ 1.55, τ 8.45 contains three protons. It cannot be a methyl group because that would have produced a quartet rather than a triplet at the downfield position. It must then be another methylene and contain the CH group whose proton is responsible for producing the large doublet upfield. The extraneous peaks in and around the triplet at δ 1.55, τ 8.45 must then belong to the CH proton. We now have enough information to write

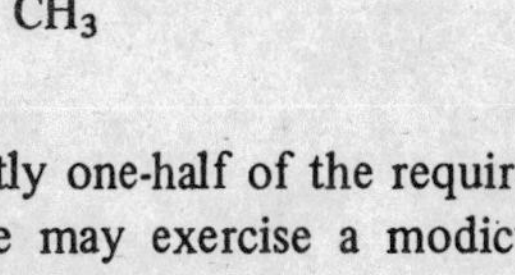

$$\begin{array}{c} CH_3 \\ \\ CH_3 \end{array} \!\!\! > \!\! CH-CH_2-CH_2-$$

Since this is exactly one-half of the required weight of the alkyl portion, we may exercise a modicum of chemical sense and write the full structure

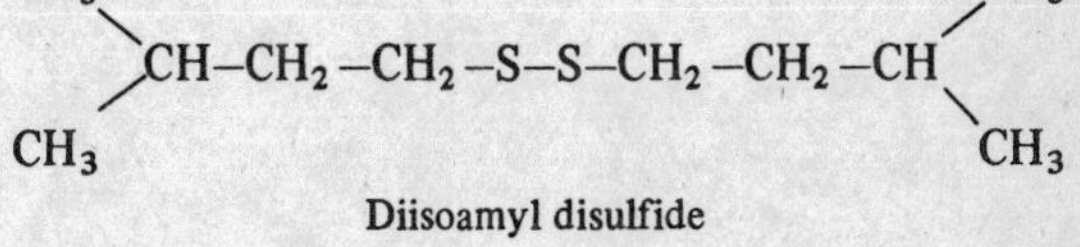

$$\begin{array}{ccc} CH_3 & & CH_3 \\ & CH-CH_2-CH_2-S-S-CH_2-CH_2-CH & \\ CH_3 & & CH_3 \end{array}$$

Diisoamyl disulfide

mentation pattern, although complex, bears this ...ture out. The large peak at 71 represents cleavage next to the sulfur with retention of the charge on the alkyl fragment. The peak at 136 results from the same cleavage with shift of a hydrogen atom to the sulfur-containing fragment which retains the charge.

We have not rigorously proved that the two sulfur atoms are contiguous, although the ultraviolet spectrum supports the disulfide structure. But it would be difficult to write another structure to fit the spectra. We could carry out reductive cleavage and obtain conclusive spectral data on the resulting mercaptan.

Note the typical distortions in the NMR spectrum due to virtual coupling. We have taken liberties in treating a fairly strongly coupled spin system as though it consisted of several first order spin systems.

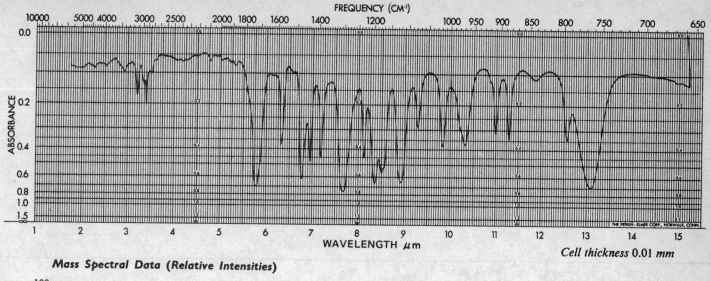

FREQUENCY (CM⁻¹)

ABSORBANCE

WAVELENGTH μm

Cell thickness 0.01 mm

Mass Spectral Data (Relative Intensities)

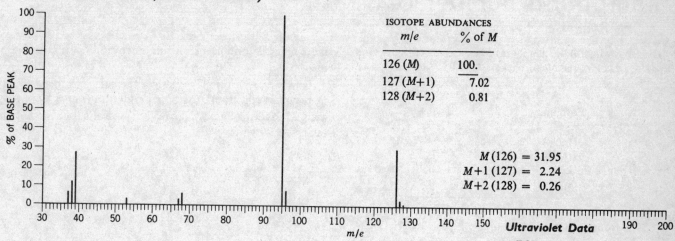

% of BASE PEAK

m/e

ISOTOPE ABUNDANCES	
m/e	% of M
126 (M)	100.
127 (M+1)	7.02
128 (M+2)	0.81

M (126) = 31.95
$M+1$ (127) = 2.24
$M+2$ (128) = 0.26

Ultraviolet Data

λ_{max}^{EtOH}	log ϵ_{max}
220.0 (s)	3.47
250.5	4.13

(s) = shoulder

NMR Spectrum (Solvent CCl₄)

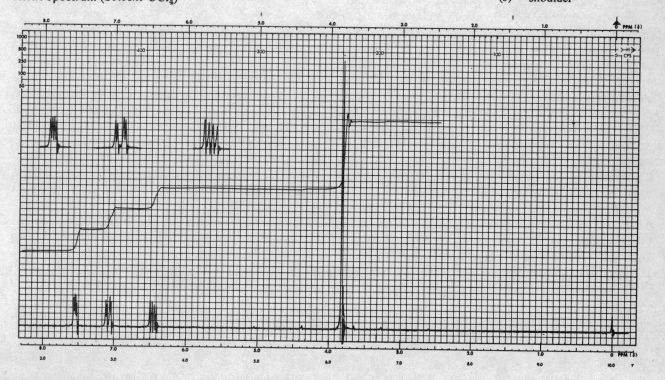

PPM (δ)

compound number 6-4

The following molecular formulas fit our data for the parent peak and the $M + 1$ peak:

Formula	$M + 1$	$M + 2$
$C_5H_6N_2O_2$	6.34	0.57
$C_5H_{10}N_4$	7.09	0.22
$C_6H_6O_3$	6.70	0.79
$C_6H_{10}N_2O$	7.45	0.44

The formula that best fits our $M + 2$ peak is $C_6H_6O_3$. We should bear in mind, however, that the $M + 2$ peak may be higher than the calculated figure, and we accept $C_6H_6O_3$ only as a tentative formula.

The formula and the general appearance of the infrared spectrum suggest aromaticity. We note strong peaks beyond 800 cm^{-1} (12.5 μm), two strong peaks at 1587 cm^{-1} (6.30 μm) and 1479 cm^{-1} (6.76 μm), and a medium peak at 3106 cm^{-1} (3.22 μm). The intense band in the ultraviolet spectrum at 250 nm is suggestive of a chromophore conjugated with an aromatic ring. The striking pattern at the low-field end of the NMR spectrum demands an aromatic ring of some sort.

A conspicuous feature of the infrared spectrum is the C=O peak at 1730 cm^{-1} (5.78 μm). Bearing in mind that we are probably dealing with a conjugated chromophore, we can make a choice between a ketone and an ester. We lean toward an ester because the conjugated ketone C=O bands are usually at lower frequency. We also look in vain for the long wavelength R-band of conjugated ketones in the ultraviolet spectrum. (We should recall, however, that heteroaromatic ketones do not show a detectable R-band.) There are a number of strong bands between 1420 and 1110 cm^{-1} (7.05 and 9.0 μm) some of which may be associated with an ester C—O absorption. And, of course, the empirical formula permits an ester group plus another oxygen.

Now we can profitably consider the base peak, mass 95, in the mass spectrum. The base peak arises from a loss of mass 31, and this loss is practically diagnostic for a methyl ester. We can write

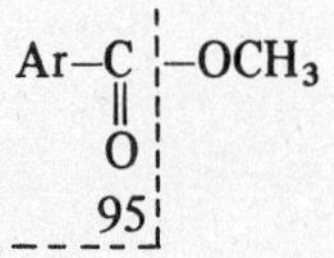

$$\text{Ar}\!-\!\underset{\underset{95}{\overset{\|}{O}}}{C} \overset{\vdots}{}\!-\!\text{OCH}_3$$

The mass 95 component then must be $C_4H_3O\!-\!C\!=\!O$, and we write the structural formula

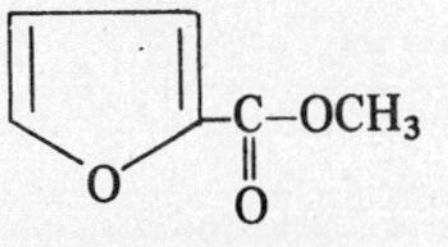

Methyl 2-furoate

The NMR spectrum is nicely in accord with this structure. We see three separate ring-protons at low field, and the three protons of the methyl group as a sharp singlet at δ 3.81, τ 6.19. From low to high field the three low-field peaks (multiplets) represent the five-proton, the three-proton, and the four-proton, respectively. Each is shifted downfield, by the carboxylate substituent, from its position in an unsubstituted furan ring, the three-proton being most strongly affected.

These multiplets afford a tidy demonstration of spin-spin coupling. The system is *AMX* with three coupling constants. The five-proton is coupled with the four-proton (J_{54} = 2 Hz) and with the three-proton (J_{53} = 1 Hz).

The five-proton, therefore, shows two pairs of peaks.

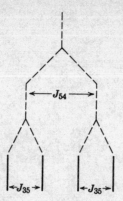

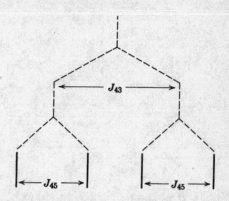

The four-proton is coupled with the three-proton (J_{43} = 3.5 Hz) and with the five-proton (J_{45} = 2 Hz).

Again we see a doublet of doublets.

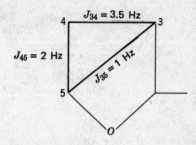

The three-proton is coupled with the four-proton (J_{34} = 3.5 Hz) and with the five-proton (J_{35} = 1 Hz). The three-proton, thus, shows a doublet of doublets.

The couplings can be summed up as follows:

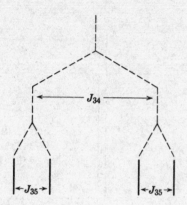

Note that methyl 3-furoate would give an entirely different NMR pattern.

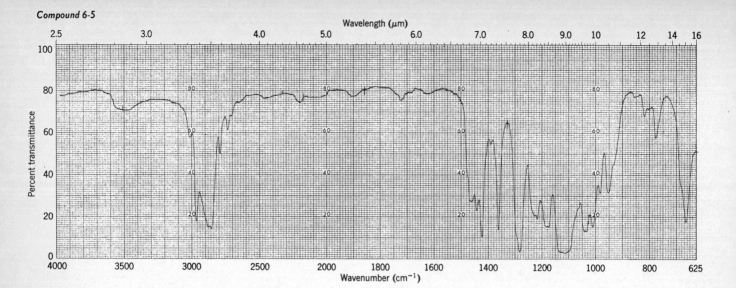

Wavelength (μm)

Percent transmittance

Wavenumber (cm⁻¹)

Mass Spectral Data (Relative Intensities)

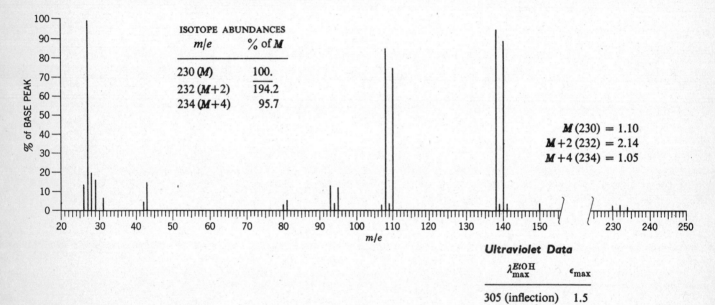

% of BASE PEAK

ISOTOPE ABUNDANCES	
m/e	% of M
230 (M)	100.
232 (M+2)	194.2
234 (M+4)	95.7

M (230) = 1.10
$M+2$ (232) = 2.14
$M+4$ (234) = 1.05

m/e

Ultraviolet Data

λ_{max}^{EtOH}	ϵ_{max}
305 (inflection)	1.5

NMR Spectrum (Solvent CCl₄)

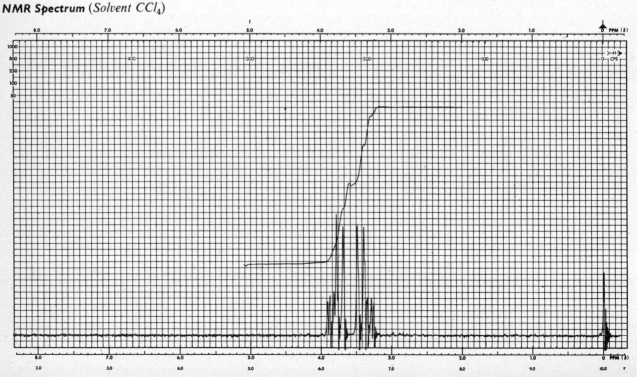

compound number 6-5

The characteristic $M+2$ and $M+4$ pattern indicates the presence of two bromine atoms. Subtracting 2×79 from the molecular weight leaves a mass of 72 for the rest of the molecule.

The strongest peak in the infrared spectrum at 1117 cm^{-1} (8.95 μm) strongly suggests an aliphatic ether. Since no nitrogen- or other oxygen-containing group appears to be present, we can assume, tentatively, that the remaining fragment is C_4H_8O. The weak inflection in the ultraviolet spectrum may simply result from the accumulation of atoms with nonbonding electrons.

The striking symmetry of the NMR spectrum suggests a symmetrical molecule. If we dispose our fragments, Br_2, $-O-$, and C_4H_8, in a symmetrical pattern, there are two possibilities

$$CH_3\underset{\underset{Br}{|}}{C}HOC\underset{\underset{Br}{|}}{H}CH_3$$

or

$$BrCH_2CH_2OCH_2CH_2Br$$

The first compound would give an A_3X NMR spectrum consisting of a one-proton quartet quite far downfield ($\delta \sim 5.0$, τ 5.0) and a doublet rather upfield ($\delta \sim 2.0$, τ 8.0). The second compound would give an $AA'BB'$ pattern in accord with that observed.

The fragmentation pattern, though complex, affords ample confirmation. Characteristic bromine-containing pairs are discernible at m/e 138 and 140 (loss of CH_2Br with H transfer), 108 and 110 (loss of OCH_2CH_2Br with H transfer), and 93 and 95 (CH_2Br^+).

The compound is

$$BrCH_2CH_2OCH_2CH_2Br$$
$\beta\beta'$-dibromodiethyl ether

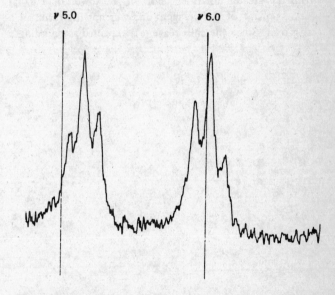

Sample 4 (see table, next page) of Shift Reagent study; 600 Hz Sweep Width, CCl_4 Solvent.

The strong peak in the infrared spectrum at 1279 cm^{-1} (7.82 μm) is attributed to a CH_2Br wag.

The CH_2 adjacent to O is deshielded by O by about 2.15 units (the normal CH_2 absorption is δ 1.25, τ 8.75, and the position for CH_2O is δ 3.40, τ 6.60). This methylene group is also deshielded by the Br atom by about 0.48 units. Its calculated position, then, is about δ 3.88, τ 6.12, in fair agreement with the downfield absorption. Similarly, the position of the methylene adjacent to Br can be calculated to be about δ 3.40, τ 6.60, again in fair agreement with the upfield absorption.

These NMR assignments are in accord with a shift reagent [Eu(fod)$_3$] study. The following table describes a

Sample	Mole Ratio Eu(fod)$_3$/ether	Chemical Shifts (δ)		Change in Chemical Shifts	
		Upfield Signal	Downfield Signal	Upfield Signal	Downfield Signal
1	0.00	3.35	3.75	—	—
2	0.12	3.60	4.16	+0.25	+0.41
3	0.25	3.80	4.53	+0.45	+0.78
4	0.38	3.98	4.87	+0.63	+1.12

[a]In all experiments, there was 30 mmoles of the bromoether in about 0.5 ml of CCl$_4$.

downfield progression of both NMR signals as a function of increased amounts of shift reagent. Such results are consistent with the chemical shift changes being inversely dependent upon distance. The NMR spectrum of Sample 4 is shown. The Table shows that the separation has resulted from the greater downfield shift of the lower field signal per increment of shift reagent and was not the result of a cross-over. Since the shift reagent is expected to coordinate more strongly to the (more nucleophilic) oxygen atom (rather than bromine), the signal at lower field (δ 3.75) in the original spectrum is due to the methylene protons alpha to oxygen. The shift reagents has begun to convert a higher order system (sample 1, $\Delta\nu/J$ = ca. 24/6 = ca. 4) to a first order system (sample 4, $\Delta\nu/J$ = ca. 53/6 = ca. 9). The benefits clearly outweight the slight line broadening effect of the shift reagent.

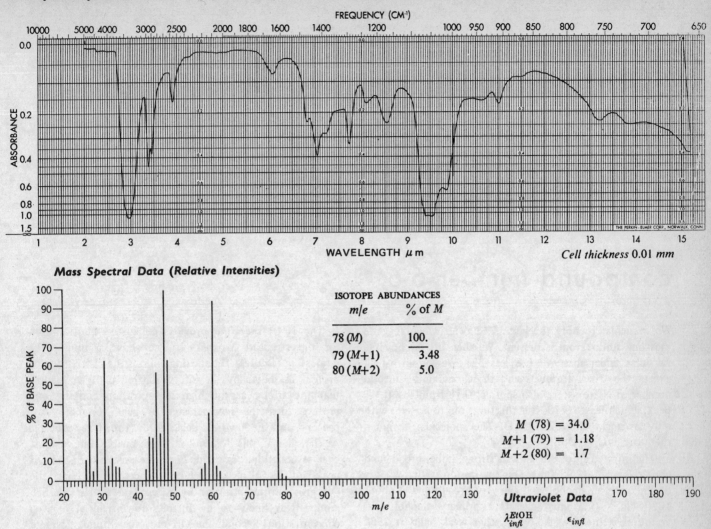

FREQUENCY (CM⁻¹)

ABSORBANCE

WAVELENGTH μm

Cell thickness 0.01 mm

Mass Spectral Data (Relative Intensities)

% of BASE PEAK

m/e

ISOTOPE ABUNDANCES

m/e	% of M
78 (M)	100.
79 (M+1)	3.48
80 (M+2)	5.0

M (78) = 34.0
M+1 (79) = 1.18
M+2 (80) = 1.7

Ultraviolet Data

λ_{infl}^{EtOH}	ϵ_{infl}
232	136

infl = inflection

NMR Spectrum (Solvent CDCl₃)

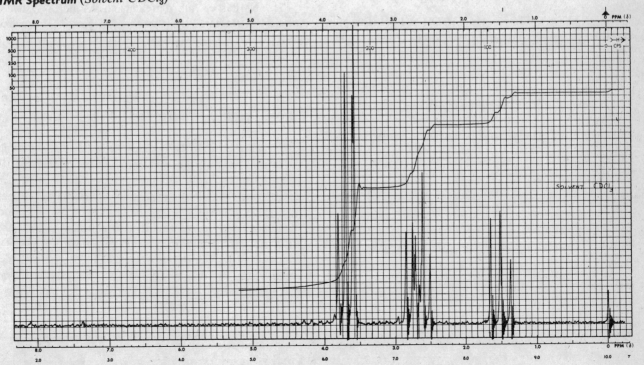

SOLVENT CDCl₃

compound number 6-6

We immediately note the large $M+2$ peak which suggests that one sulfur atom is present. We then list the possible formulas under mass 46 (78 less 32). We should list all except the trivial formulas and those containing an odd number of N atoms; we also subtract 0.78 from the $M + 1$ peak, which leaves 2.70. But this turns out to leave us with only a single choice, C_2H_6O. The molecular formula, therefore, is C_2H_6OS.

The infrared spectrum shows a strong, rather broad band at 3367 cm^{-1} (2.97 μm). Our impression is that we are dealing with an alcohol, and the very broad band at about 1050 cm^{-1} (9.5 μm) suggests a primary alcohol. Our attention is then caught by a rather weak band at 2558 cm^{-1} (3.91 μm), which practically spells out a mercaptan group. In this case, the infrared spectrometer is at some disadvantage with respect to the nose. Had this been a thin film spectrum, we might have missed the S–H stretching band.

We now have the fragments: CH_2OH and SH. This only leaves a CH_2 group to fit in, and we write

$$HOCH_2CH_2SH$$
2-Mercaptoethanol

Some of the major fragmentation peaks can be assigned as follows:

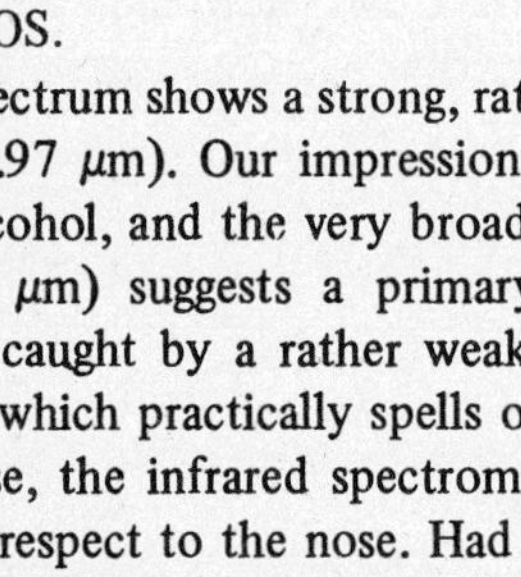

$$HO-CH_2-CH_2-SH$$
$$60$$

$$HO-CH_2-CH_2-SH$$
$$31 \quad 47$$
$$45$$

The NMR spectrum provides exhaustive confirmation for the structure written. It also shows a number of interesting features. The starting point is the distribution of protons as shown by the integration curve. If we assume that the triplet at the high-field position contains one proton, then the next cluster of peaks contains two protons, and the low-field peaks account for three protons. At first glance, this does not seem reasonable. But we must bear several things in mind. First, the positions of the OH peak and the SH peak depend on concentration. Since the OH-group forms stronger hydrogen bonds than the SH-group, it is likely to be further down-field at a given concentration. Second, the OH proton will undergo rapid exchange under normal condition and will usually appear as a single peak; the SH proton, under the same conditions will not exchange rapidly (at least in a nonaqueous solvent), and the peak will be split by the adjacent methylene group.

The NMR spectrum represents an $AA'MM'X$ system (if we neglect the rapidly exchanging OH proton), but we can treat it as A_2M_2X and examine the A_2M_2 and M_2X couplings:

$$HO-CH_2-CH_2-SH$$
$$AA'MM'X = \sim A_2M_2X$$

We see then that the upfield triplet (with slight second order splitting of the middle peak) represents the SH proton coupled with (split by) the adjacent methylene group; the coupling constant is 8 Hz. The adjacent methylene group is split into a doublet by the SH proton (coupling constant, of course, is 8 Hz), and again into a triplet by the other methylene group with a coupling constant of 6 Hz. This is a somewhat distorted A_2M_2X system with two coupling constants. An idealized diagram is as follows:

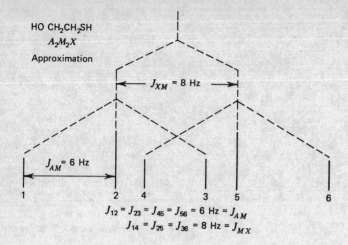

HO CH$_2$CH$_2$SH

A_2M_2X

Approximation

$J_{XM} = 8$ Hz

$J_{AM} = 6$ Hz

1   2   4   3   5   6

$J_{12} = J_{23} = J_{45} = J_{56} = 6$ Hz $= J_{AM}$

$J_{14} = J_{25} = J_{36} = 8$ Hz $= J_{MX}$

The low-field peaks must then consist of a triplet with a coupling constant of 6 Hz and it must also contain the OH-proton. We do indeed see a triplet with the upfield peak distorted by the peak of the OH proton with which it almost coincides. The hydroxyl proton peak can be shifted by change of concentration, solvent, or temperature. Both the OH and SH peaks can be removed by shaking with deuterium oxide; this, of course, would collapse the CH$_2$S multiplet to a symmetrical "apparent triplet" signal.

There are distortions in the spectrum because of marginal $\Delta \nu/J$ values.

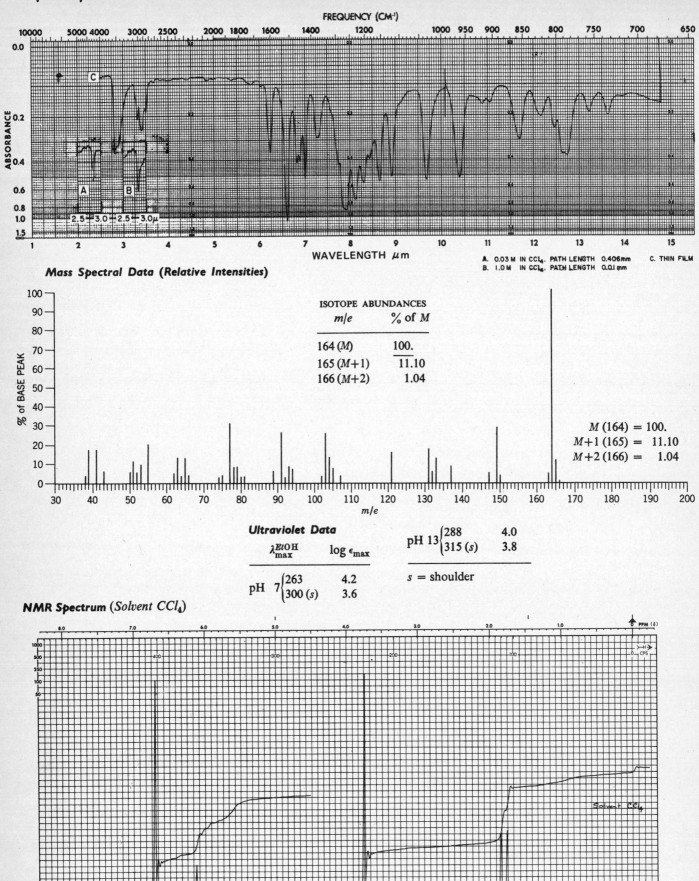

Infrared Spectrum

FREQUENCY (CM⁻¹)

A. 0.03 M IN CCl₄. PATH LENGTH 0.406mm   C. THIN FILM
B. 1.0 M IN CCl₄. PATH LENGTH 0.01 mm

Mass Spectral Data (Relative Intensities)

ISOTOPE ABUNDANCES	
m/e	% of M
164 (M)	100.
165 (M+1)	11.10
166 (M+2)	1.04

M (164) = 100.
$M+1$ (165) = 11.10
$M+2$ (166) = 1.04

Ultraviolet Data

λ^{EtOH}_{max}	$\log \epsilon_{max}$
pH 7 $\begin{cases} 263 \\ 300 \ (s) \end{cases}$	4.2 \\ 3.6

| pH 13 $\begin{cases} 288 \\ 315 \ (s) \end{cases}$ | 4.0 \\ 3.8 |

s = shoulder

NMR Spectrum (*Solvent CCl₄*)

Solvent CCl₄

compound number 6-7

The following molecular formulas fit the data for the molecular ion (parent peak) and the $M+1$ peak. We have eliminated trivial formulas and those containing an odd number of nitrogen atoms.

Formula	$M+1$	$M+2$
$C_8H_8N_2O_2$	9.61	0.81
$C_8H_{12}N_4$	10.36	0.49
$C_9H_8O_3$	9.97	1.05
$C_9H_{12}N_2O$	10.72	0.72
$C_{10}H_{12}O_2$	11.08	0.96
$C_{10}H_{16}N_2$	11.83	0.64

On the basis of the $M+2$ peak, we can tentatively eliminate $C_8H_{12}N_4$ and $C_{10}H_{16}N_2$. The presence of an aromatic ring is indicated by the NMR peak at δ 6.70, τ 3.30 (distinctly upfield from the position of an unsubstituted benzene ring) and the general appearance of the infrared spectrum. We note a small peak at 3030 cm^{-1} (3.30 μm), strong peaks between 1600 and 1430 cm^{-1} (6.2 and 7.0 μm), and several moderately strong peaks in the low-frequency (long wavelength) region. As additional confirmation of aromaticity, we note that the parent peak in the mass spectrum is also the base peak.

A conspicuous feature of the infrared spectrum is the rather strong absorption at 3510 cm^{-1} (2.85 μm). This is either an OH or an NH stretching band. We look to the region between 1230 and 1010 cm^{-1} (8.13 and 9.90 μm) for confirmation of OH absorption, but we are frustrated by the large number of bands in this region.

The ultraviolet spectrum gives us a good deal of information. The shift in wavelength at pH 13 is diagnostic for a phenol. Furthermore, the intense K-band at 263 nm is indicative of a chromophore conjugated with the ring. In the absence of a carbonyl band in the infrared spectrum, we would suspect a C=C group. We quickly confirm this possibility by the olefinic proton absorption in the NMR spectrum at about δ 6.0, τ 4.0. A strong peak in the infrared spectrum at 965 cm^{-1} (10.36 μm) is evidence for *trans* olefinic hydrogens. The peak at 1605 cm^{-1} (6.25 μm) can partially be ascribed to the conjugated C=C stretching vibration.

At this point we have the following information:

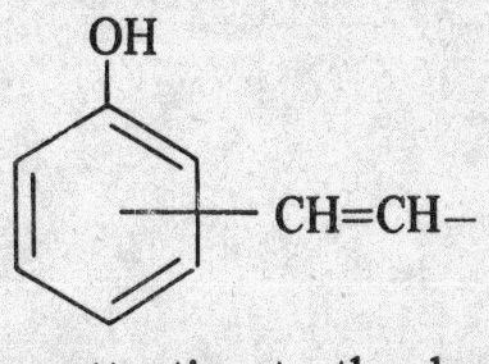

We now turn our attention to the doublet centered at δ 1.81, τ 8.19. The shift position is right for an allylic methyl group, and its existence as a doublet can be justified. We write

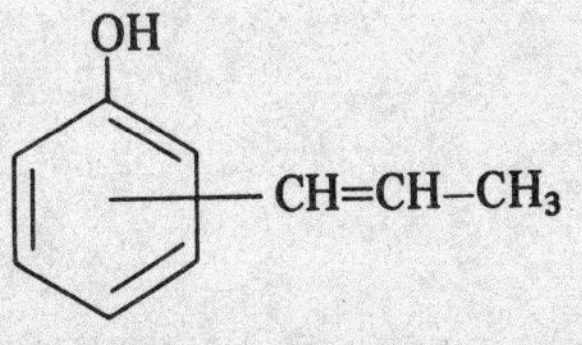

We consider the singlet in the NMR spectrum at δ 3.75, τ 6.25. The integrator indicates that this peak also contains three protons; this methyl group could be on a ring, an oxygen, a nitrogen, or a quarternary carbon. Its shift position is best satisfied by putting it on an oxygen and putting the oxygen on the ring. We have the following:

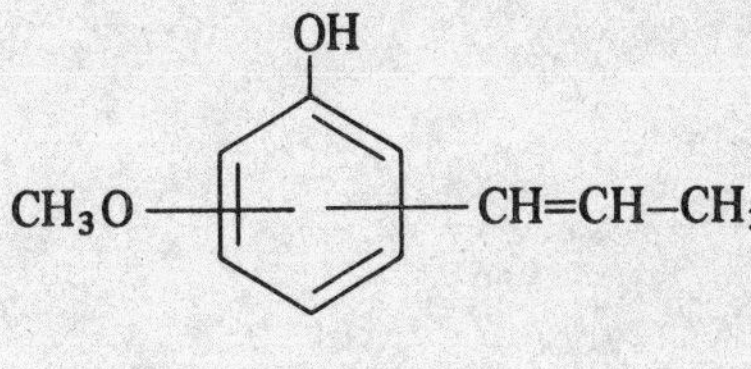

This adds up to 164 and to the formula $C_{10}H_{12}O_2$. The benzene peak at δ 6.70, τ 3.30 contains three protons; the complex olefinic multiplet also accounts for three protons. Obviously, the phenolic proton is hidden under the olefinic multiplet; it could either be moved by change of temperature, solvent, or concentration, or eliminated by shaking with D_2O. It is probably the broadened peak at $\sim \delta$ 5.60, τ 4.40. The olefinic proton near the ring is shown by the badly distorted doublet ($J \sim 16$) whose center of gravity is $\delta \sim 6.2$, τ 3.8.

Assignment of the positions of the substituents is not easy. We cannot rely upon the C—H out-of-plane deformations in the long-wavelength region of the infrared spectrum because of the polar nature of the substituents. The NMR, which is usually very helpful for distinguishing among isomers, is of little use because the ring protons are all at the same shift value. We might note that the phenolic OH band in the infrared spectrum is at an unusually high frequency (short wavelength) for a neat sample, and is not shifted on dilution; this suggests intramolecular hydrogen bonding to the ether oxygen. The methoxy group being

ortho to the hydroxy group satisfies the geometric requirements for such intramolecular hydrogen bonding. The fragmentation spectrum is complex.

However, an organic chemist would recognize isoeugenol as a likely choice, and he would indeed find that the compound is

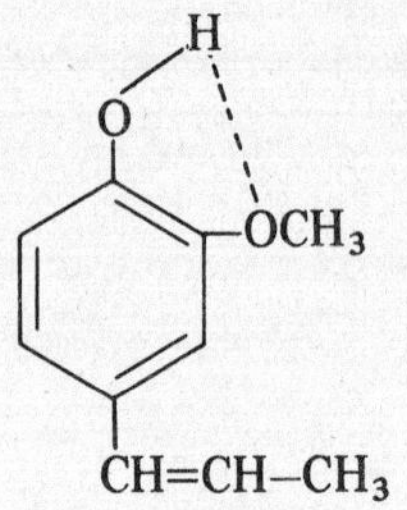

Isoeugenol

If isoeugenol did not come to mind, the organic chemist has enough information to decide on a degradation procedure. He might well ozonize the double bond and proceed to identify vanillin.

Infrared Spectrum

FREQUENCY (CM⁻¹)

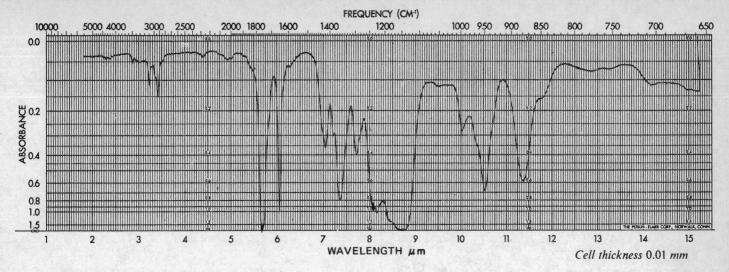

WAVELENGTH μm

Cell thickness 0.01 mm

Mass Spectral Data (Relative Intensities)

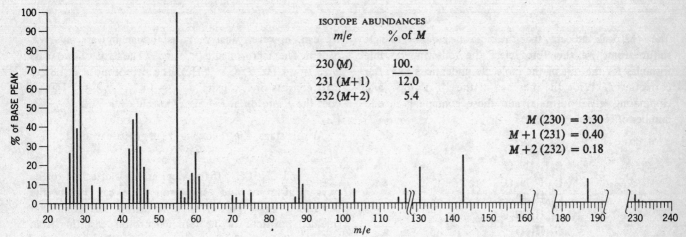

ISOTOPE ABUNDANCES	
m/e	% of M
230 (M)	100.
231 (M+1)	12.0
232 (M+2)	5.4

$M (230) = 3.30$
$M + 1 (231) = 0.40$
$M + 2 (232) = 0.18$

m/e

Ultraviolet Data

Featureless above 210 nm

NMR Spectrum (Solvent CCl₄)

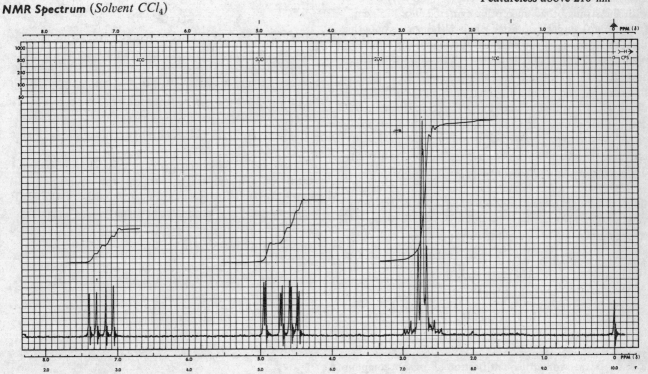

compound number 6-8

The $M+2$ peak indicates that the molecule contains a single sulfur atom. We therefore select the following possible formulas for the rest of the molecule under mass 198 after correcting $M+1$ for the presence of the ^{33}S isotope, and eliminating trivial formulas and those containing an odd number of N atoms.

$$C_8H_{14}N_2O_2$$
$$C_9H_{14}N_2O$$
$$C_9H_{18}N_4O$$
$$C_{10}H_{14}O_4$$
$$C_{10}H_{18}N_2O_2$$
$$C_{11}H_{18}O_3$$

The infrared spectrum affords a wealth of obvious information. The band at 1760 cm^{-1} (5.68 μm) together with the strong broad absorption centered at ~ 1150 cm^{-1} (~ 8.70 μm) suggests an ester group, although we note that the C=O band is at rather short wavelength for esters. An olefinic bond is certainly indicated by the sharp absorptions at 3090 cm^{-1} (3.24 μm) and at 1650 cm^{-1} (6.06 μm). In this context, and noting the unusual intensity of the latter peak, we are justified in considering the strong bands at 949 cm^{-1} (10.54 μm) and 878 cm^{-1} (11.39 μm) to be displaced vinyl bands. There is no evidence for aromaticity, nitrogen- or other oxygen-containing groups, or a sulfhydryl group. We are probably dealing with a sulfide.

The NMR pattern is nicely in accord with our assumptions. We should expect an ABX or an ABC pattern from a vinyl group. The pattern is strikingly ABX because of strong deshielding of one of the vinyl protons.

We can, however, analyze it as though it were an AMX system. The X proton at δ 7.21, τ 2.79 consists of two pairs (J_{AX} = 14 Hz, J_{BX} = 7 Hz). The A protons at δ 4.81, τ 5.19 consists of two pairs (J_{AX} = 14 Hz, J_{AB} = 2 Hz), as does the B proton at δ 4.52, τ 5.48 (J_{BX} = 7 Hz, J_{AB} = 2 Hz).

At this stage, the following fragments are evident: $-S-$,

$$-\overset{\overset{\text{O}}{\|}}{C}O-$$ and CH=CH$_2$. If the vinyl group is attached to the oxygen atom of the ester group, we can justify the deshielded vinyl proton and, incidently, the somewhat shielded positions of the terminal protons resulting from the canonical contribution:

$$-\overset{\overset{\text{O}}{\|}}{C}-\overset{+}{O}=CH-\overset{-}{C}H_2$$

The C=O band in the infrared spectrum fits a vinyl ester.

The proton ratio, from low to high field, in the NMR spectrum is 1:2:4. The four-proton multiplet centered at about δ ~ 2.7 τ 7.3 can be recognized as a very close A_2B_2 pattern resulting from two adjacent, similarly shielded methylene groups, $-CH_2CH_2-$ (see Chapter 4, Figure 38d).

The sum of these fragments is equal to one-half the weight of the sulfur-free part of the molecule. This observation and the symmetrical appearance of the NMR spectrum lead us to assemble the fragments as a symmetrical sulfide, which is in accord with one of the possible molecular formulas $C_{10}H_{14}O_4(S)$.

$$CH_2=CHO\overset{\overset{\text{O}}{\|}}{C}CH_2CH_2SCH_2CH_2\overset{\overset{\text{O}}{\|}}{C}OCH=CH_2$$

Divinyl β, β'-thiodipropionate

The fragmentation pattern is complex, but confirmational information can be gleaned. The peak at m/e 187 represents loss of $OCH=CH_2$, that at m/e 131 represents cleavage at one C–S bond with charge retention on the sulfur. The base peak at m/e 55 probably results from successive cleavage of a C–S and C–O bond with transfer of H to give the $CH_2=CHC=O$ moiety. The large peak at m/e 27 is the $CH_2=CH$ fragment. Absence of an M minus H_2S fragment is additional evidence in favor of a sulfide over a mercaptan.

By utilizing the α and β shift positions, given in the NMR Chapter Appendix, we can calculate the position for the $CH_2C=O$ absorption to be δ 2.70, τ 7.30, and that for the CH_2S absorption, δ 2.75, τ 7.25. Agreement with the observed positions is good.

Infrared Spectrum

FREQUENCY (CM⁻¹)

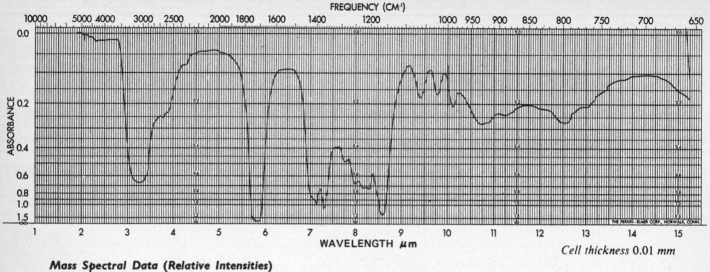

WAVELENGTH μm

Cell thickness 0.01 mm

Mass Spectral Data (Relative Intensities)

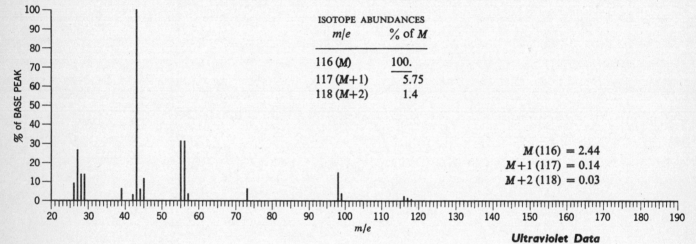

ISOTOPE ABUNDANCES

m/e	% of M
116 (M)	100.
117 (M+1)	5.75
118 (M+2)	1.4

M (116) = 2.44
$M+1$ (117) = 0.14
$M+2$ (118) = 0.03

m/e

Ultraviolet Data

λ_{max}^{EtOH}	log ϵ_{max}
262	1.5

NMR Spectrum (Solvent CCl₄)

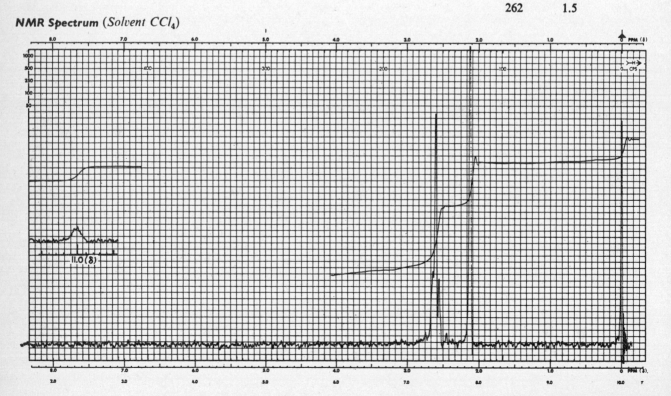

11.0 (δ)

compound number 6-9

We list the molecular formulas under mass 116 (molecular ion) that give a reasonable fit for the $M+1$ value of 5.75%.

Formula	$M+1$	$M+2$
$C_3H_8N_4O$	4.94	0.30
$C_4H_4O_4$	4.54	0.88
$C_4H_8N_2O_2$	5.29	0.52
$C_4H_{12}N_4$	6.04	0.16
$C_5H_8O_3$	5.65	0.73
$C_5H_{12}N_2O$	6.40	0.37
$C_6H_{12}O_2$	6.75	0.59

The formulas that best fit our $M+1$ value and our $M+2$ value (1.4%) are $C_4H_4O_4$, $C_4H_8N_2O_2$, $C_5H_8O_3$, and $C_6H_{12}O_2$.

The infrared spectrum points to an aliphatic carboxylic acid. We note the very broad, bonded OH-stretching absorption extending between about 3333 and 2300 cm^{-1} (3.0 and 4.3 μm) with its characteristic pattern on its low frequency (long wavelength) side. The strong C=O band at 1715 cm^{-1} (5.83 μm) satisfies the requirement for an aliphatic carboxylic acid. We find ready confirmation by noting the carboxylic acid proton at δ 11.0, τ 1.0 in the NMR spectrum.

But the ultraviolet spectrum interposes a caveat. The low-intensity absorption at 262 nm, suggests an aliphatic unconjugated ketone. There is only one C=O band in the infrared spectrum, but it is rather broad; it must accommodate both the ketone and the acid C=O band. The fragmentation pattern does not resemble that of an ordinary aliphatic carboxylic acid; e.g., there is no pronounced peak at mass 60. Since we are dealing with a ketone group in the molecule, we assume that the base peak of 43 in the mass spectrum arises from $CH_3C=O$. We now have the fragments $CH_3C=O$ and COOH which add up to mass 88. These leave us with an unknown mass of 28 for which we can write either C=O or CH_2-CH_2 or N_2.

The NMR spectrum allows us to make an unequivocal choice. The large singlet at δ 2.12, τ 7.88 must be the CH_3 group on the ketone carbonyl. If this peak contains three protons, then the absorption at δ 2.60, τ 7.40 contains four protons. These must result from adjacent methylene groups which have almost the same chemical shift ($AA'BB'$ system). We now write the structure

$$CH_3\overset{\|}{\underset{O}{C}}CH_2CH_2COOH$$

Levulinic acid

Infrared Spectrum

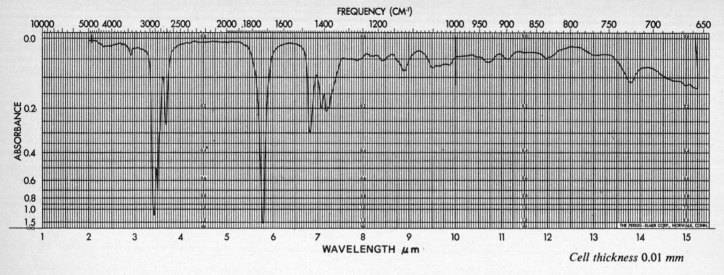

FREQUENCY (CM⁻¹)

ABSORBANCE

WAVELENGTH μm

Cell thickness 0.01 mm

Mass Spectral Data (Relative Intensities)

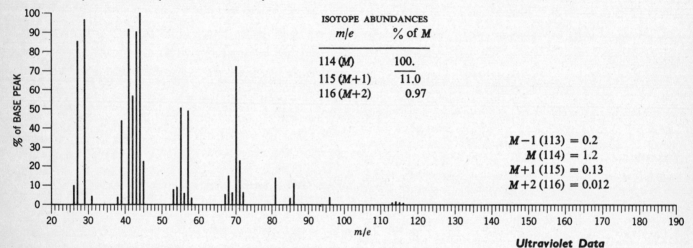

% of BASE PEAK

m/e

ISOTOPE ABUNDANCES	
m/e	% of M
114 (M)	100.
115 (M+1)	11.0
116 (M+2)	0.97

$$M-1\ (113) = 0.2$$
$$M\ (114) = 1.2$$
$$M+1\ (115) = 0.13$$
$$M+2\ (116) = 0.012$$

Ultraviolet Data

$\lambda_{max}^{Cyclohexane}$	ϵ_{max}
292	23.2

NMR Spectrum (*Solvent* CDCl₃)

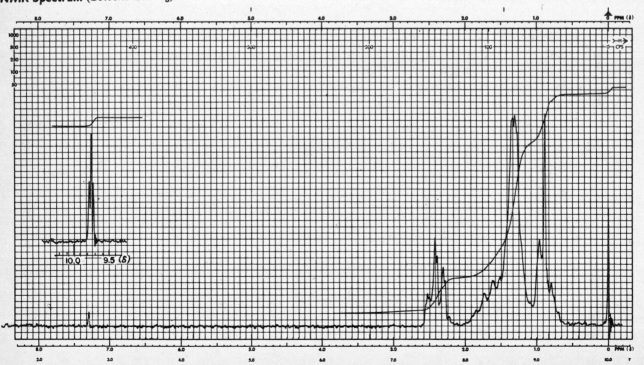

10.0   9.5 (δ)

compound number 6-10

In accordance with our usual procedure, we look at the molecular formulas under the parent mass (114) and try to select formulas that match our $M+1$ and $M+2$ peaks. But in this case, we find that our $M+1$ peak is too high to match any of the formulas listed. The high value of the $M+1$ peak can mean impurities. A more likely explanation is that this spectrum was obtained at a rather high inlet pressure in order to see the weak parent peak; we see a contribution to the $M+1$ peak from a bimolecular addition of hydrogen to the parent ion. This is a common occurrence with molecules containing a hetero atom, which, as we shall see, is present. No sulfur or halogen is present.

The C=O band in the infrared spectrum at 1730 cm^{-1} (5.78 μm), together with the CH stretching band at 2703 cm^{-1} (3.70 μm), spells out an aldehyde group. This is readily confirmed by the characteristic ultraviolet spectrum, by the triplet in the NMR spectrum at δ 9.75, τ 0.25, and by the typical aldehyde base peak at mass 44. Furthermore, we know that we are dealing with an aliphatic primary aldehyde by the triplet structure of the aldehydic proton in the NMR (split by CH$_2$), by the weak parent peak and the mass 44 base peak in the mass spectrum, and by the lack of aromatic structure in either the NMR or the infrared spectra.

The absorption at δ 2.42, τ 7.58 in the NMR must represent the CH$_2$ next to the aldehyde group. It is split into a triplet by an adjacent CH$_2$ and again into a doublet by the aldehydic proton (small coupling constant). We now have

$$-CH_2 CH_2 CHO$$
mass 57

and we need a moiety of mass 57 to account for a molecular weight of 114. We could write

$$C_4 H_9$$
$$C_3 H_5 O$$
$$C_2 H_5 N_2$$

The NMR integrator permits an unequivocal choice. If we assign one proton to the aldehydic absorption, and two protons to the adjacent methylene, there are eleven protons under the high-field absorption of which two belong to the β-methylene group. That leaves nine protons for the missing moiety which is obviously C$_4$H$_9$. Furthermore, we can make out a distorted triplet centered on δ 0.89, π 9.11 which represents a CH$_3$ group. The ratio of the area of this triplet to the area of the absorption of the methylene groups between δ 2.0, π 8.0, and δ 1.1, π 8.9 is 3:8. Therefore, the chain is not branched. The compound is

$$CH_3 CH_2 CH_2 CH_2 CH_2 CH_2 CHO$$
Heptaldehyde

The peaks in the mass spectrum at mass 96 ($M-H_2O$), mass 86 ($M-CO$), and at mass 70 ($M-CH_2CHOH$) confirm the choice of mass 114 as the parent peak despite its low intensity. The aliphatic chain is apparent in the sequence m/e 29, 43, 57, and 71.

FREQUENCY (CM⁻¹)

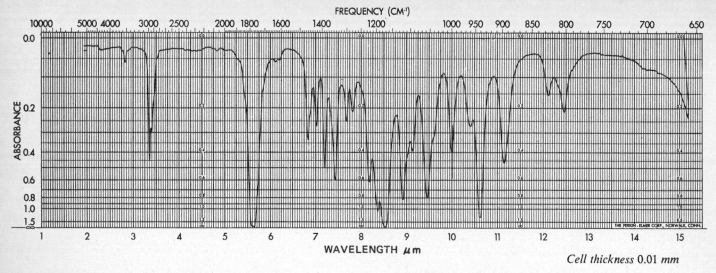

WAVELENGTH μm

Cell thickness 0.01 mm

Mass Spectral Data (Relative Intensities)

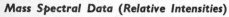

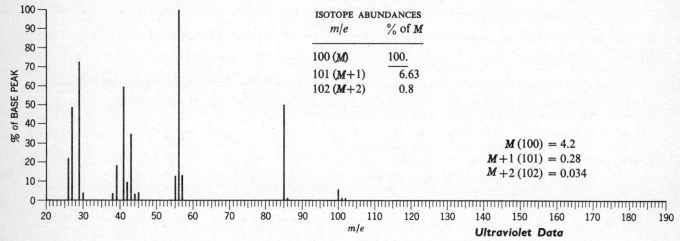

ISOTOPE ABUNDANCES	
m/e	% of *M*
100 (*M*)	100.
101 (*M*+1)	6.63
102 (*M*+2)	0.8

$M(100) = 4.2$
$M+1 (101) = 0.28$
$M+2 (102) = 0.034$

m/e

Ultraviolet Data

Transparent above 200 nm

NMR Spectrum (*Solvent CCl₄*)

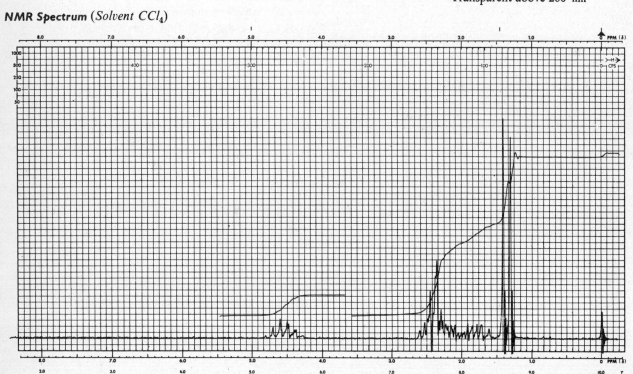

compound number 6-11

The likely molecular formulas under mass 100 on the basis of our $M+1$ peak (6.63%) are

Formula	$M + 1$	$M + 2$
$C_4H_8N_2O$	5.25	0.31
$C_5H_8O_2$	5.61	0.53
$C_5H_{12}N_2$	6.36	0.17
$C_6H_{12}O$	6.72	0.39

Our $M+2$ peak (0.8%) would tend to favor $C_5H_8O_2$, but $C_6H_{12}O$ is a possibility because of its close agreement with the $M + 1$ peak.

The strong C=O band at 1780 cm^{-1} (5.62 μm) in the infrared, together with the strong broad absorption band at 1170^{-1} (8.55 μm) suggests an ester, but the C=O band is at a rather high frequency (short wavelength) for an ordinary ester. Halogen substitutuents would account for this, but, obviously, chlorine or bromine are absent. Even more obviously, iodine is not present, but fluorine must still be kept in mind.

There is moderate absorption in the infrared spectrum in the low-frequency (long wavelength) end of the spectrum, but none between 1667 and 1471 cm^{-1} (6.0 and 6.8 μm); nor is there anything to suggest aromatic CH stretching vibration. On balance, we do not seem to be dealing with an aromatic compound. This impression is reinforced by the ultraviolet and the NMR spectra.

Since two oxygen atoms are probably present, we adopt $C_5H_8O_2$ as the molecular formula, (forgetting about fluorine for the moment). The formula suggests unsaturation, but none of the spectra bear out this possibility. We therefore consider a ring and, recalling the position of the C=O band in the infrared spectrum, we are led to a five-membered ring lactone. The peak at m/e 85 ($M - 15$) and the base peak at m/e 56 (elimination of $CH_3CH{=}O$) spell out

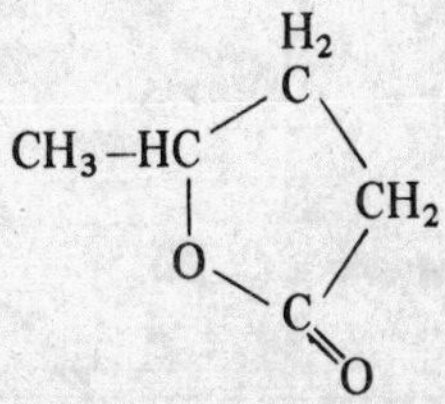

γ-Valerolactone

The upfield, 3-proton doublet in the NMR spectrum represents the C$\underline{H}_3$—CH absorption. The strongly de-shielded position of the CH absorption (a slightly distorted sextet at δ 4.55, τ 5.45), places the CH on the oxygen atom. The integrator shows four protons between about δ 2.6, τ 7.4, and δ 1.5, τ 8.5.

Compound 6-12

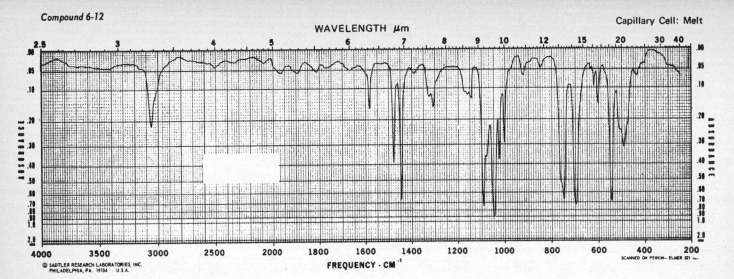

WAVELENGTH μm

Capillary Cell: Melt

FREQUENCY · CM⁻¹

© SADTLER RESEARCH LABORATORIES, INC.
PHILADELPHIA, PA. 19104   U.S.A.

SCANNED ON PERKIN-ELMER 521 ∞

Source: Aldrich Chemical Company, Milwaukee, Wis.

Mass Spectral Data (Relative Intensities)

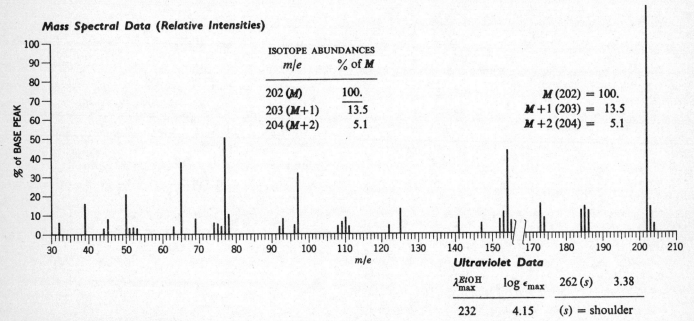

ISOTOPE ABUNDANCES	
m/e	% of M
202 (M)	100.
203 ($M+1$)	13.5
204 ($M+2$)	5.1

M (202) = 100.
$M+1$ (203) = 13.5
$M+2$ (204) = 5.1

m/e

Ultraviolet Data

λ_{max}^{EtOH}	log ϵ_{max}	262 (s)	3.38
232	4.15	(s) = shoulder	

NMR Spectrum (Solvent CCl₄)

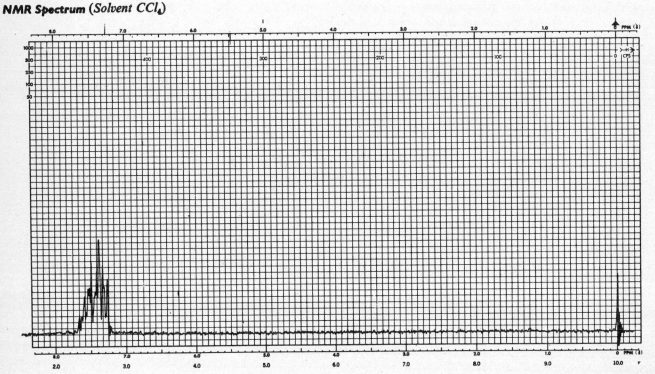

compound number 6-12

The $M+2$ peak indicates the presence of one sulfur atom. We select the following possible formulas for the rest of the molecules under mass 170 after correcting the $M+1$ peak for the ^{33}S isotope:

$$C_{10}H_6N_2O$$
$$C_{10}H_{22}N_2$$
$$C_{11}H_6O_2$$
$$C_{11}H_{10}N_2$$
$$C_{11}H_{22}O$$
$$C_{12}H_{10}O$$
$$C_{12}H_{26}$$

The intensity of the parent peak and general appearance of the IR spectrum strongly suggest aromaticity as does the UV spectrum, which also points to a chromophore conjugated with the aromatic system. The NMR spectrum is equally definitive; in fact, all the protons are aromatic. For want of evidence to the contrary, we can tentatively assume that no nitrogen atoms are present, and we are left with only two likely molecular formulas.

$$C_{11}H_6O_2S$$
$$\text{and}$$
$$C_{12}H_{10}OS$$

The long wavelength bands in the IR are accepted, with the usual reservations, as evidence for a singly substituted benzene ring (one of the bands is split). A conspicuous peak at m/e 77 (and a smaller one at m/e 78) can be adduced as additional evidence for a benzene ring. Given this evidence and the similarity of the shift positions of all the protons, we are justified in choosing $C_{12}H_{10}OS$ as the molecular formula and assuming that two benzene rings are present. If, in fact, the rings are singly substituted, we are forced to write

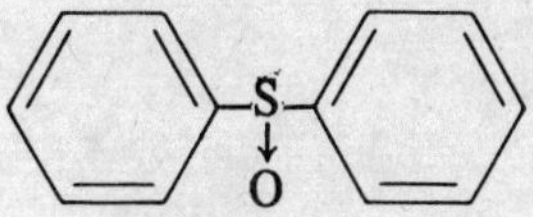

Diphenyl sulfoxide

This might also have been deduced from the strong $S \rightarrow O$ band in the infrared system at 1048 cm^{-1} (9.54 μm). The strong K-band in the UV spectrum is in accord with this structure as are the main fragmentation peaks. Expulsion of SO and formation of a C–C bond explain the prominent peak at m/e 154.

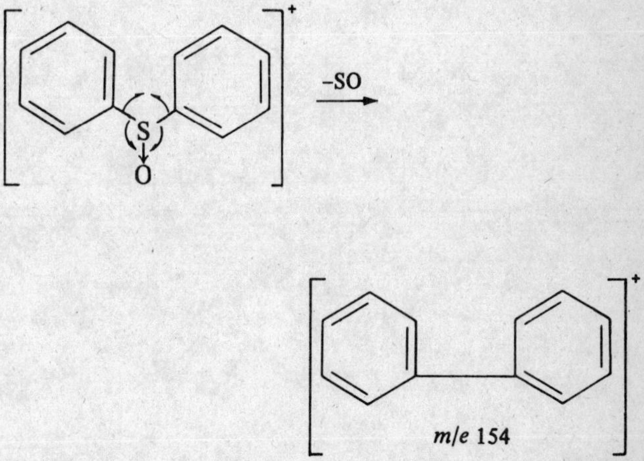

m/e 154

The peak at m/e 125 results from cleavage of one S–C bond to give

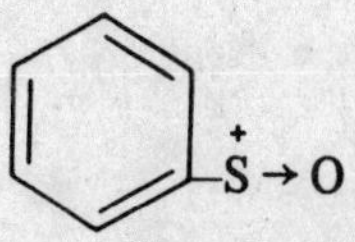

m/e 125

The peak at *m/e* 97 is more difficult to rationalize. Presumably, a complex rearrangement yields the ion

$$[C_5H_5S]^+$$

The strong IR band at 1090 cm^{-1} (9.18 μm) has been described as a phenyl-S stretch, presumably with some contribution from the canonical form

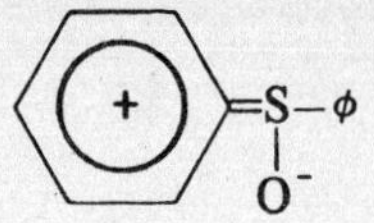

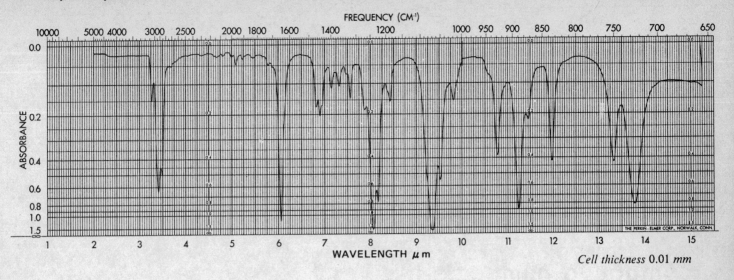

FREQUENCY (CM⁻¹)

WAVELENGTH μm

Cell thickness 0.01 mm

Mass Spectral Data (Relative Intensities)

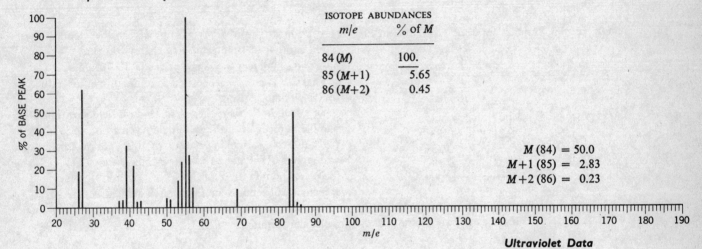

ISOTOPE ABUNDANCES	
m/e	% of M
84 (M)	100.
85 ($M+1$)	5.65
86 ($M+2$)	0.45

M (84) = 50.0
$M+1$ (85) = 2.83
$M+2$ (86) = 0.23

m/e

Ultraviolet Data

Transparent in
near ultraviolet.
(200—380 nm)

NMR Spectrum (Solvent CCl₄)

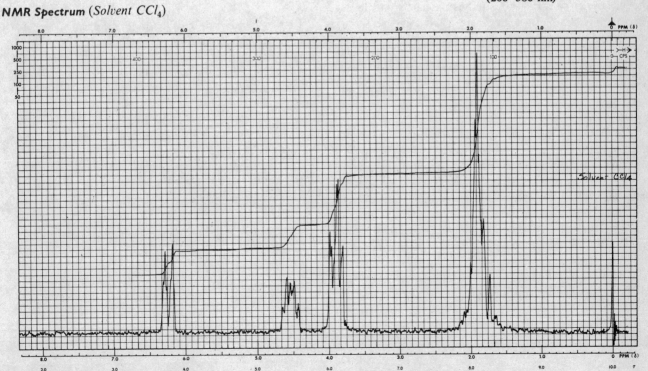

compound number 6-13

The possible molecular formulas for a molecular weight of 84 are

Formulas	$M + 1$	$M + 2$
$C_4H_4O_2$	4.47	0.48
$C_4H_8N_2$	5.21	0.11
C_5H_8O	5.57	0.33
C_6H_{12}	6.67	0.19

The best match for the $M+1$ peak (5.65%) and for the $M+2$ peak (0.45%) is C_5H_8O. This means two "unsaturated sites."

The general impression given by the infrared spectrum is that we are dealing with an aromatic carbonyl compound. But discrepancies rapidly become apparent. For one thing, the ultraviolet spectrum effectively rules out aromatic or heteroaromatic ring systems; there is no evidence for a ketone or aldehyde group. A closer look at the infrared spectrum reveals a disturbing ratio of the intensities of the aliphatic CH stretching absorption at 2933 cm^{-1} (3.41 μm) and the "carbonyl" band at 1650 cm^{-1} (6.06 μm). The molecular formulas do not permit a heavily alkylated ring system.

The NMR spectrum shows a doublet (with additional splitting) at δ 6.21, τ 3.79. This is at the high-field end for aromatic and heteroaromatic protons, and at the low-field end for olefinic protons. If we assume that we are dealing with an olefin, the spectra all make more sense. The "carbonyl" band in the infrared now becomes an intensified C=C stretching band. The small peak at 3058 cm^{-1} (3.27 μm) is obviously the olefinic CH stretching band, and the long wavelength absorption, 13.80 μm, 725 cm^{-1}, must be a strong cis CH=CH out-of-plane bending band.

If we assign one olefinic proton to the downfield absorption in the NMR spectrum, we can then tentatively assign another olefinic proton to the multiplet at δ 4.55, τ 5.45. We can detect the same spacing (J = 7 Hz) in both sets of peaks; this coupling constant is of the proper magnitude for cis olefinic protons.

In order to explain the extreme downfield position of one olefinic CH, we will place the oxygen atom, allowed by the empirical formula, adjacent to this CH. We can now write $-O-CH=CH-$. The intense bands at 1241 cm^{-1} (8.06 μm) and 1070 cm^{-1} (9.35 μm) in the infrared spectrum will support an unsaturated ether structure.

We now subtract the unsaturated ether moiety from the molecular formula

$$\begin{array}{c} C_5H_8O \\ C_2H_2O \\ \hline C_3H_6 \end{array}$$

Further consideration of the NMR spectrum allows us to make a rational distribution of the C and H atoms. The six protons are distributed under two peaks in the ratio of 2:4. The smaller peak, an apparent triplet with additional splitting, is moved strongly downfield; we are justified in putting a methylene on the oxygen atom

$$-H_2C-O-CH=CH-$$

The upfield absorption is caused by CH_2-CH_2—whose shift positions are similar. We can now close the gap with $-CH_2-CH_2-$ and write

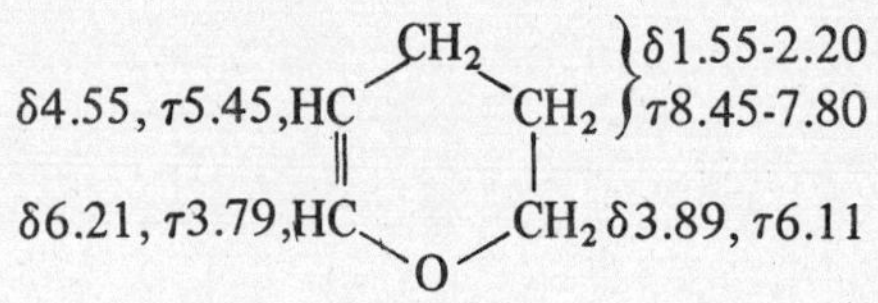

Dihydropyran

Note the very small allylic coupling evident in the downfield absorption in the NMR spectrum. The distortion in the δ 3.89, τ 6.11 signal (OCH_2) is due to a rather small $\Delta\nu/J$ ratio and to virtual coupling. In addition, these methylene protons are chemical shift equivalent (rapid conformational flip interchange), but are magnetically nonequivalent. The relatively high field position for the olefinic proton β to the O atom can be rationalized by contributions from this resonance form:

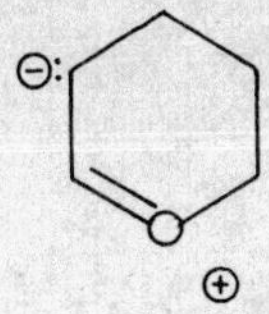

FREQUENCY (CM⁻¹)

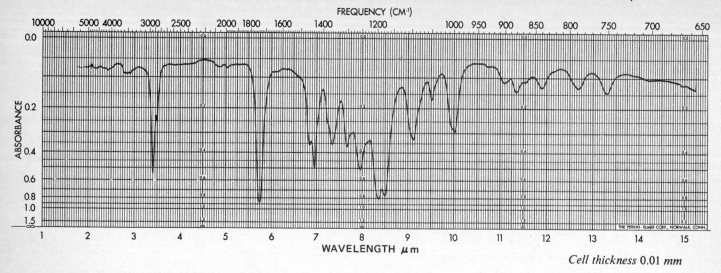

WAVELENGTH μm

Cell thickness 0.01 mm

Mass Spectral Data (Relative Intensities)

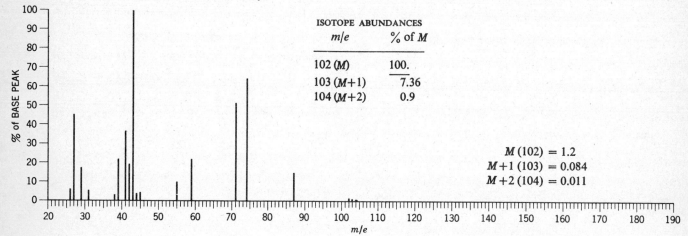

ISOTOPE ABUNDANCES	
m/e	% of *M*
102 (*M*)	100.
103 (*M+1*)	7.36
104 (*M+2*)	0.9

$M (102) = 1.2$
$M+1 (103) = 0.084$
$M+2 (104) = 0.011$

Ultraviolet Data

Transparent above 210 nm

NMR Spectrum (*Solvent CCl₄*)

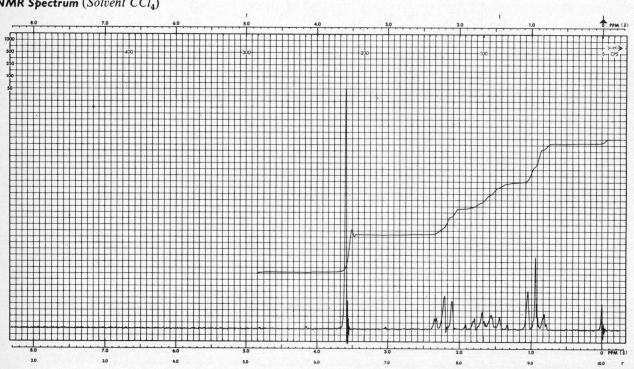

compound number 6-14

The following are likely molecular formulas.

$$C_5H_{14}N_2$$
$$C_6H_{14}O$$

However, the IR spectrum strongly indicates an ester group. Furthermore, an intense peak at M minus 31 in the mass spectrum and a singlet in the NMR spectrum at δ 3.60, τ 6.40 suggest a methyl ester. Apparently the $M+1$ peak is unreliable; in fact, it is too large because the mass spectrum was run at high inlet pressure in order to see the weak molecular ion.

All spectra show lack of aromaticity. If we subtract the mass of $\overset{O}{\overset{\|}{C}}OCH_3$ from the molecular weight, we have mass 43 for the rest of the molecule which must be either a propyl or an isopropyl fragment. The apparent triplet, sextet, triplet (all slightly distorted) in a 3:2:2 ratio in NMR spectrum certainly spell out a propyl fragment. The compound is

$$CH_3CH_2CH_2\overset{O}{\overset{\|}{C}}OCH_3 \qquad A_3BB'XX'$$

Methyl n-butyrate   NMR system

The fragmentation spectrum offers further confirmation. The peak at m/e 74 is the typical methyl ester rearrangement ion,* $\left[\begin{array}{c} CH_2=C-OCH_3 \\ | \\ OH \end{array}\right]^+$; that at m/e 59 is $\left[\begin{array}{c} CH_3O\overset{\|}{\underset{O}{C}} \end{array}\right]^+$, and that at m/e 43 is $[CH_3CH_2CH_2]^+$.

*McLafferty rearrangement

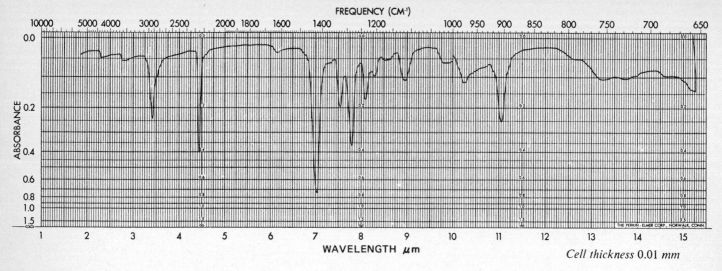

FREQUENCY (CM⁻¹)

ABSORBANCE

WAVELENGTH μm

Cell thickness 0.01 mm

THE PERKIN - ELMER CORP., NORWALK, CONN.

Mass Spectral Data (Relative Intensities)

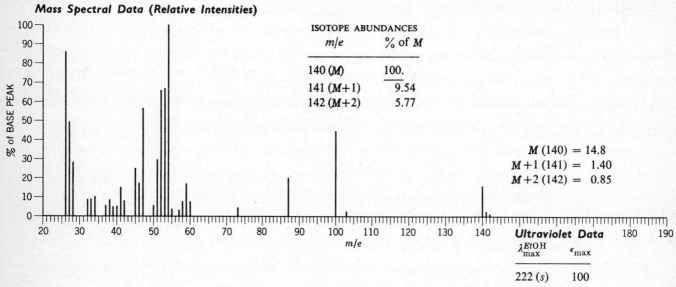

% of BASE PEAK

m/e

ISOTOPE ABUNDANCES	
m/e	% of M
140 (M)	100.
141 (M+1)	9.54
142 (M+2)	5.77

$M\,(140) = 14.8$
$M+1\,(141) = 1.40$
$M+2\,(142) = 0.85$

Ultraviolet Data

λ_{max}^{EtOH}	ϵ_{max}
222 (s)	100

(s) = shoulder

NMR Spectrum (*Solvent CDCl₃*)

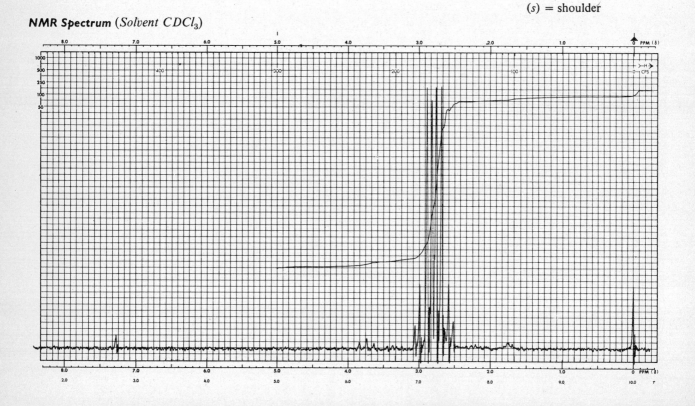

PPM (δ)

compound number 6-15

The molecular weight is 140, and the compound obviously contains one atom of sulfur. We look under mass 108, allowing for a 0.78 contribution to the $M+1$ peak. We write the partial formulas starting arbitrarily with $M+1 = 6.27$ and eliminating formulas containing an odd number of nitrogen atoms:

Formula	$M + 1$
$C_5H_4N_2O$	6.27
$C_6H_4O_2$	6.63
$C_6H_8N_2$	7.38
C_7H_8O	7.73
C_8H_{12}	8.84

The infrared spectrum conveys an aliphatic impression; nor is there any indication of unsaturation or aromaticity in the ultraviolet or NMR spectrum. The sharp band at 2247 cm^{-1} (4.45 μm) in the infrared stands out conspicuously. There are only a few possibilities: An isocyanate or a nitrile are the most probable candidates.

Thus far, we can account for a sulfur atom, presumably as a sulfide, and for a nitrogen atom. In order to account for the even mass, we need another nitrogen atom which could be another isocyanate or nitrile group or possibly a tertiary amine. The correct empirical formula must be either $C_5H_4N_2O$ or $C_6H_8N_2$. Neither empirical formula will accommodate two isocyanate groups.

Let us consider the fragmentation pattern. The base peak is mass 54 which is just half of the nonsulfur containing moiety. We seem to be dealing with a symmetrical molecule, and a glance at the NMR spectrum confirms this. We think in terms of a symmetrical dinitrile. Confirmational details now become apparent. The next largest peak in the fragmentation pattern is mass 26, obviously a C≡N group. The mass 41 peak (CH_2–C≡N + H), which we look for in aliphatic nitriles, is prominent. The mass 100 peak is large, and this may result from loss of a CH_2C≡N fragment. There appear to be a number of rearrangement peaks; this is reasonable in a molecule containing three heteroatoms.

The base peak of mass 54 must represent

$$CH_2-CH_2-C\equiv N$$

and the complete molecule must be

$$N\equiv C-CH_2-CH_2-S-CH_2-CH_2-C\equiv N$$
$$\beta, \beta'\text{-Thiodipropionitrile}$$

The NMR spectrum is an example of an $AA'BB'$ coupling with a small $\Delta \nu/J$ ratio. The large peak in the IR spectrum at 1420 cm^{-1} (7.04 μm) represents the methylene scissoring band shifted by the functional groups.

FREQUENCY (CM⁻¹)

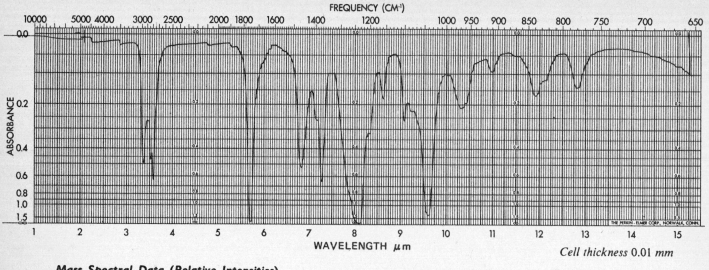

WAVELENGTH μm

Cell thickness 0.01 mm

Mass Spectral Data (*Relative Intensities*)

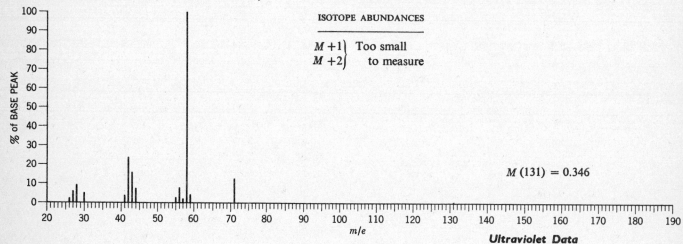

ISOTOPE ABUNDANCES

$M+1$ } Too small
$M+2$ } to measure

$M(131) = 0.346$

m/e

Ultraviolet Data

Transparent above 210 nm

NMR Spectrum (*Solvent CCl₄*)

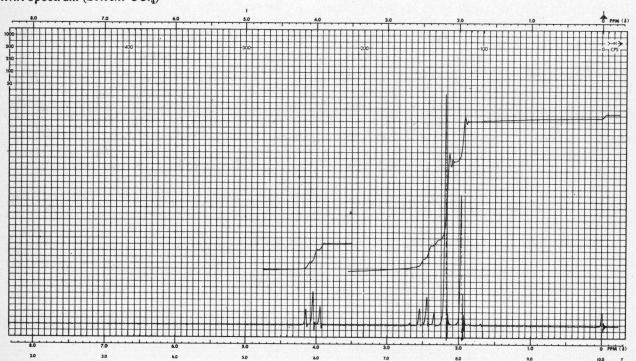

compound number 6-16

The molecular (parent) ion peak (mass 131) of this compound was very small. The $M+1$ and $M+2$ peaks were too small for accurate measurement of intensity, and we cannot arrive at a molecular formula.

The molecular weight calls for an odd number of nitrogen atoms; let us begin with a single nitrogen atom. In the absence of evidence for primary or secondary amines, nitriles, amides, nitro compounds, or heteroaromatic compounds, we shall assume we may be dealing with a tertiary amine. There is no evidence for unsaturation or aromaticity in any of the spectra.

The infrared spectrum shows a strong carbonyl band at 1748 cm^{-1} (5.72 μm) and a typical broad strong C–O band at about 1235 cm^{-1} (8.10 μm). This combination is evidence for the presence of an acetate group. As supporting evidence there is a prominent mass 43 (CH$_3$C=O) in the mass spectrum, and a singlet at δ 1.95, τ 8.05 in the NMR spectrum, which we may attribute to the CH$_3$ of the acetate group.

The NMR spectrum shows two triplets of equal areas with the same spacings. We are justified in writing –CH$_2$–CH$_2$– and in placing the more deshielded methylene group on the oxygen of the acetate group; thus,

$$-CH_2-CH_2-O-\overset{\overset{\displaystyle O}{\|}}{C}-CH_3$$

This is an $AA'XX'$ system that results in an NMR spectrum essentially identical to that of an A_2X_2 system.

We have postulated the presence of a tertiary amine group. The molecular weight allows for C_2H_6N. The singlet at δ 2.20, τ 7.80 in the NMR, with double the area of the acetate CH$_3$ group, permits us to write

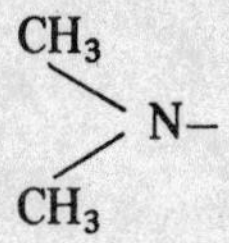

$$\begin{matrix} CH_3 \\ \diagdown \\ \qquad N- \\ \diagup \\ CH_3 \end{matrix}$$

We can now write the complete structure

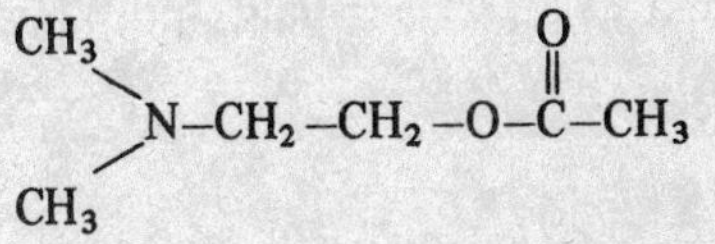

$$\begin{matrix} CH_3 \\ \diagdown \\ \qquad\qquad N-CH_2-CH_2-O-\overset{\overset{\displaystyle O}{\|}}{C}-CH_3 \\ \diagup \\ CH_3 \end{matrix}$$

2-Dimethylaminoethyl acetate

There are other possible lines of observation and reasoning we could have followed. The base peak, mass 58, is a characteristic amine fragmentation peak which results from cleavage of the C–C bond next to the nitrogen atom. This, together with consideration of the other spectra, would have lead us directly to the fragment

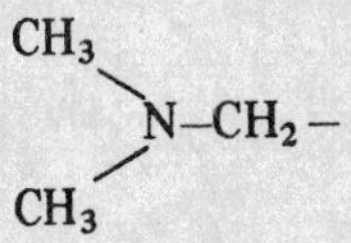

$$\begin{matrix} CH_3 \\ \diagdown \\ \qquad N-CH_2- \\ \diagup \\ CH_3 \end{matrix}$$

The bands in the IR spectra at 2850 cm^{-1} (3.51 μm) and 2790 cm^{-1} (3.58 μm) are the C–H stretchings of the methyl groups on nitrogen. The band at 1040 cm^{-1} (9.62 μm) is characteristic of acetates of primary alcohols.

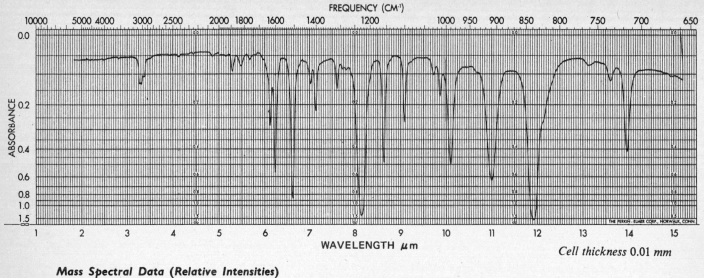

FREQUENCY (CM⁻¹)

WAVELENGTH μm

Cell thickness 0.01 *mm*

THE PERKIN-ELMER CORP., NORWALK, CONN.

Mass Spectral Data (Relative Intensities)

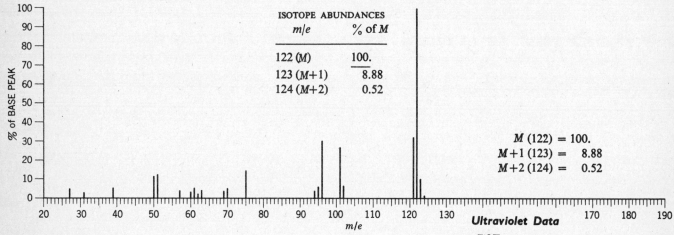

ISOTOPE ABUNDANCES	
m/e	% of M
122 (M)	100.
123 (M+1)	8.88
124 (M+2)	0.52

$M\ (122) = 100.$
$M+1\ (123) = 8.88$
$M+2\ (124) = 0.52$

m/e

Ultraviolet Data

λ_{max}^{EtOH}	log ϵ_{max}		
245	4.10	283	2.92
256 (s)	3.88	293	2.74
277	2.91		

(s) = shoulder

NMR Spectrum (*Solvent* CCl_4)

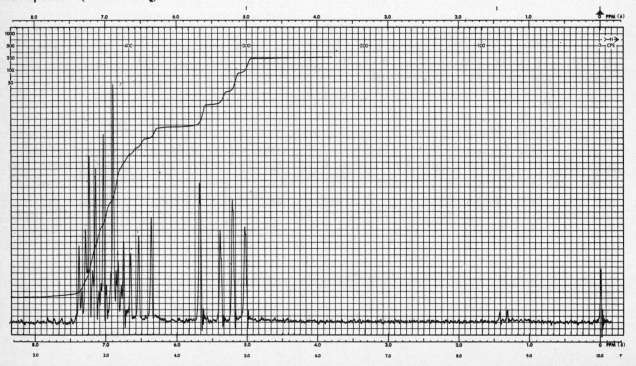

compound number 6-17

Possible molecular formulas are:

$$C_7H_6O_2$$
$$C_7H_{10}N_2$$
$$C_8H_{10}O$$
$$C_9H_{14}$$

molecular ion

The intensity of the molecular ion (parent peak), the IR, UV, and NMR spectra suggest aromaticity; in fact, the latter spectrum shows no saturated aliphatic protons.

The strongest band in the IR spectrum (838 cm^{-1}, 11.9 μm) suggests a p-disubstituted benzene. The next most intense band (1226 cm^{-1}, 8.15 μm) may be the aromatic C–O stretch of an aromatic ether. The prominent bands at 990 cm^{-1} (10.1 μm) and 909 cm^{-1} (11.0 μm) together with the sharp band at 1633 cm^{-1} (6.12 μm) are good evidence for a vinyl group. The intense K-band in the UV spectrum at 245 nm argues for conjugation of the vinyl group with the benzene ring. A vinyl ABX pattern can be recognized in the olefinic region of the NMR spectrum.

Given these fragments

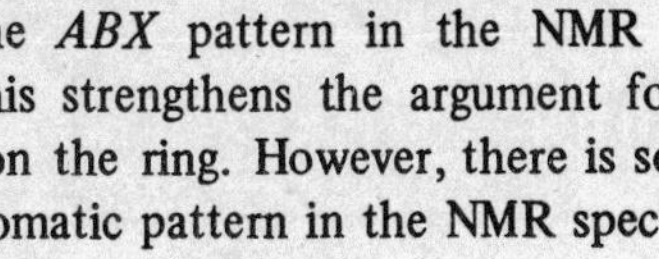

and a molecular weight of 122, we cannot write a p-vinyl aromatic ether. In fact, none of the listed molecular formulas seem to fit, and we are in trouble.

The evidence for a vinyl group seems sound. The X proton of the ABX pattern in the NMR is rather far downfield; this strengthens the argument for placing the vinyl group on the ring. However, there is something odd about the aromatic pattern in the NMR spectrum; it lacks the typical symmetry and apparent simplicity of the $AA'BB'$ pattern for a p-disubstituted ring. Thus far, we seem to be justified in writing

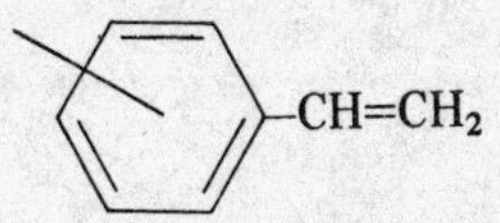

All difficulties vanish when we subtract the mass of this moiety from the molecular weight; the difference is mass 19 which can be accounted for by a fluorine atom. An aromatic C–F stretch explains the band in the IR spectrum at 1226 cm^{-1} (8.15 μm). If we place the fluorine atom para to the vinyl group to justify the IR band at 838 cm^{-1} (11.9 μm) we can rationalize the aromatic splitting.

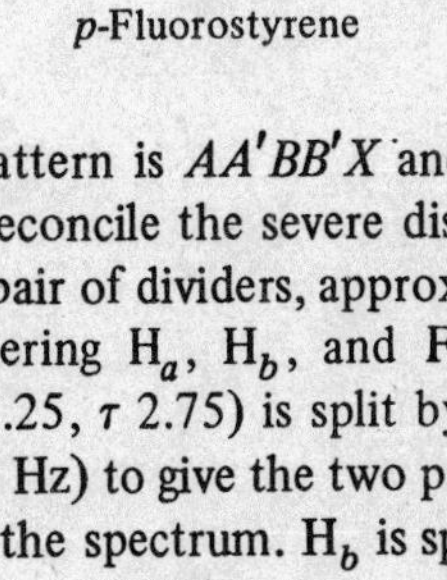

p-Fluorostyrene

Actually, the pattern is $AA'BB'X$ and is very complex. However, we can reconcile the severe distortions and extra peaks, and, with a pair of dividers, approximate a first-order analysis by considering H$_a$, H$_b$, and F to give an AMX pattern. H$_a$ ($\delta \sim 7.25$, τ 2.75) is split by H$_b$ ($J_{ab} \sim 8$ Hz) and by F ($J_{aF} \sim 6$ Hz) to give the two pairs of peaks at the downfield edge of the spectrum. H$_b$ is split by H$_a$ ($J_{ab} \sim 8$ Hz) and by F ($J_{bF} \sim 8$ Hz); we can pick out the triplet centered at δ 6.90, τ 3.10, one peak of which overlaps one peak of the X proton of the vinyl groups.

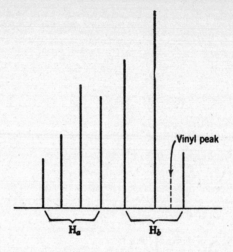

The mass spectral fragmentation pattern is not much help.

FREQUENCY (CM⁻¹)

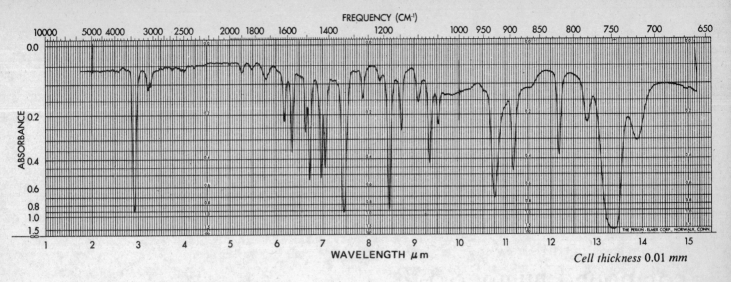

WAVELENGTH μm

ABSORBANCE

Cell thickness 0.01 *mm*

THE PERKIN - ELMER CORP., NORWALK, CONN.

Mass Spectral Data (*Relative Intensities*)

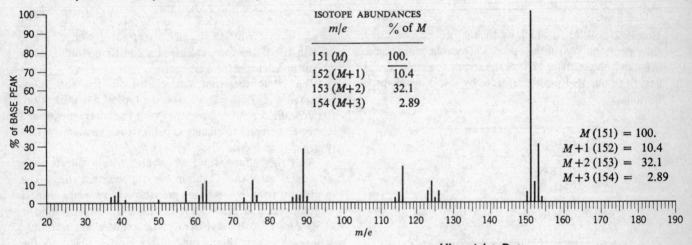

% of BASE PEAK

m/e

ISOTOPE ABUNDANCES

m/e	% of M
151 (M)	100.
152 (M+1)	10.4
153 (M+2)	32.1
154 (M+3)	2.89

M (151) = 100.
M+1 (152) = 10.4
M+2 (153) = 32.1
M+3 (154) = 2.89

Ultraviolet Data

λ_{max}^{EtOH}	log ϵ_{max}		
		272	3.88
		278	3.89
218	4.61	288	3.77

NMR Spectrum (*Solvent CCl₄*)

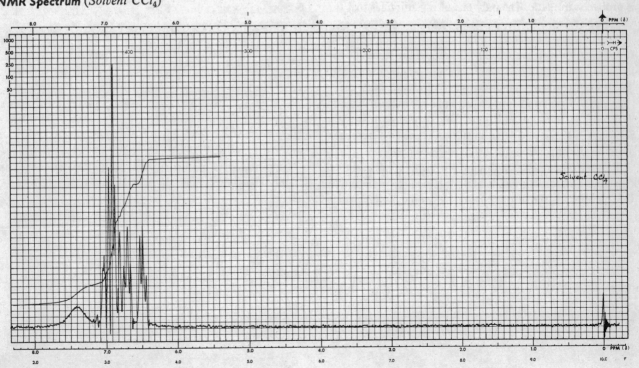

Solvent CCl₄

PPM (δ)

CPS

compound number 6-18

The $M+2$ peak (32.1%) allows for the presence of a single chlorine atom. We subtract mass 35 from the molecular ion mass and obtain mass 116 for the rest of the molecule. The best fit to our $M+1$ peak is afforded by the following partial formulas:

Formula	$M+1$
$C_7H_4N_2$	8.39
$C_7H_{16}O$	7.86
C_8H_4O	8.75
C_8H_6N	9.12
C_9H_8	9.85

Since the molecular ion (parent) mass is an odd number, the molecule contains an odd number of nitrogen atoms. Since the parent peak is also the base peak, the compound is probably aromatic. The only one of our formulas that fits is C_8H_6N.

Supporting evidence for a high degree of aromaticity is found in all of the spectra. The CH stretching region of the infrared, in fact, shows only aromatic C–H absorption. There are five strong bands between 1667 and 1429 cm^{-1} (6.0 and 7.0 μm) and four strong bands between 1000 and 715 cm^{-1} (10.0 and 14.0 μm). The strong sharp band at 3413 cm^{-1} (2.93 μm) is an invitation to place a hydrogen atom on the nitrogen atom we know to be present.

We are obviously not dealing with an aliphatic amine. Nor does an aromatic amine or a pyridine type molecule fit the picture; in the former case, we would have noted a hypsochromic shift in the ultraviolet absorption at pH 1, and in the latter case, enhanced absorption. No change in the ultraviolet spectrum is reported.

The NMR spectrum shows a broad, flat absorption centered at δ 7.40, τ 2.60. This is a typical NH absorption. Its downfield position, together with the failure mentioned above to respond to change in pH is strongly suggestive of a pyrrole or an indole NH.

With this information, we would write a chloroindole structure with the Cl atom on the benzene ring. The protons on the pyrrole ring are visible as apparent triplets centering at δ 6.71, τ 3.29 for the 2-H, and at δ 6.50, τ 3.50 for the 3-H. Apparently the protons are not only coupled to each other, but each is also coupled to the proton on nitrogen It is not possible from the available data to assign the position of Cl substitution because a Cl atom has very little effect on the position of ring protons. Reference spectra are needed to tell us that the compound is

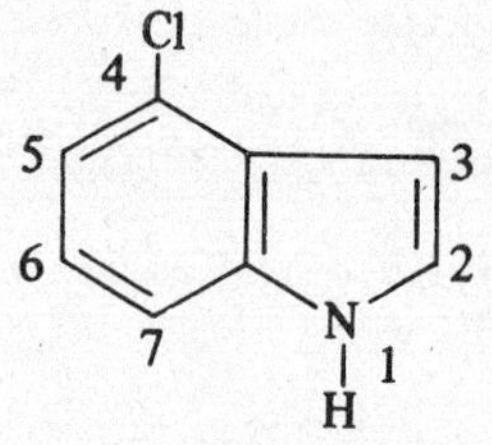

4-Chloroindole

The half-mass peaks at m/e $57\frac{1}{2}$, $75\frac{1}{2}$, and $76\frac{1}{2}$ are doubly charged particles of mass 115, 151, and 153 respectively.

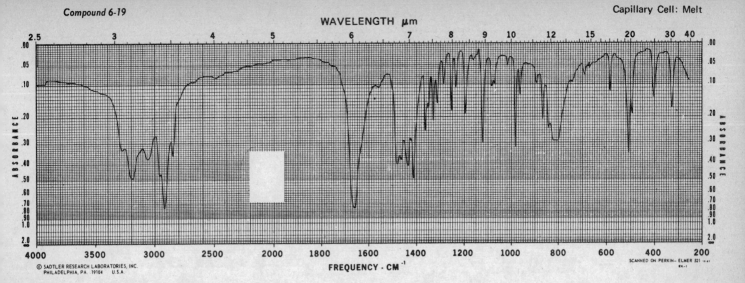

WAVELENGTH μm

ABSORBANCE

FREQUENCY - CM⁻¹

© SADTLER RESEARCH LABORATORIES, INC.
PHILADELPHIA, PA. 19104 U.S.A.

SCANNED ON PERKIN- ELMER 521

Source: Fluka AG, Buchs, Switzerland

Mass Spectral Data (Relative Intensities)

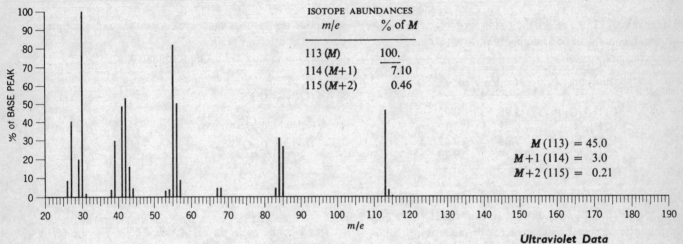

% of BASE PEAK

m/e

ISOTOPE ABUNDANCES

m/e	% of M
113 (M)	100.
114 (M+1)	7.10
115 (M+2)	0.46

M (113) = 45.0
M+1 (114) = 3.0
M+2 (115) = 0.21

Ultraviolet Data

EtOH–Featureless
above 210 nm

NMR Spectrum (Solvent CCl₄)

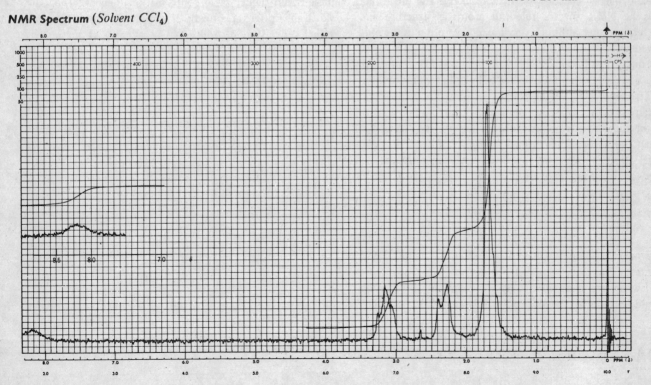

PPM (δ)

compound number 6-19

Under mass 113, we select the following molecular formulas which contain an odd number of nitrogen atoms:

Formula	$M + 1$	$M + 2$
$C_5H_7NO_2$	5.98	0.55
$C_5H_{11}N_3$	6.72	0.19
$C_6H_{11}NO$	7.08	0.42
$C_7H_{15}N$	8.19	0.29

The best fit is $C_6H_{11}NO$; there are two "unsaturated sites." The strong carbonyl band at 1669 cm^{-1} (5.99 μm) in the infrared spectrum, together with the series of bands in the region between 3448 and 3077 cm^{-1} (2.90 and 3.25 μm) suggests an amide group, but the absence of an amide II band makes a lactam seem more likely. The ultraviolet and the NMR spectra rule out aromatic structure and olefinic protons, although very broad absorption in the long wavelength region of the infrared spectrum may have raised the question of aromaticity. This long wavelength absorption is probably the out-of-plane NH bending vibration.

The molecular weight permits us to write a six-carbon lactam structure. All that remains is to determine the size of the ring and positions of substituents. We can ascribe the broad flat absorption centered at about δ 8.2, τ 1.8 in the NMR spectrum to a hydrogen on a lactam nitrogen. We have the following fragment:

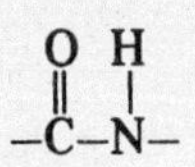

The broad apparent triplet at δ 3.2, τ 6.8 in the NMR spectrum must represent a CH$_2$ group on the N atom, and a somewhat less deshielded CH$_2$ group on the C=O group would account for the absorption at about δ 2.3, τ 7.7. The

integrator tells us that there are six protons under the peak centered at δ 1.70, τ 8.30.

The compound can now be formulated as

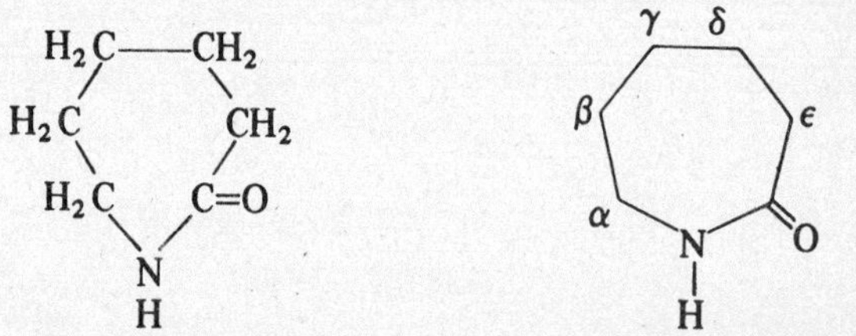

ε-Caprolactam

This lactam results in an $AA'BB'CC'DD'EE'MX$ ($M = N\underline{H}$, $X = N$) NMR system. The protons of each of the individual methylene groups are chemical shift equivalent if there is sufficiently rapid conformational change. The individual absorptions are broadly distorted since this is far from a first order system. Additional broadening of the δ 3.2, τ 6.8 signal is the result of coupling to the NH proton.

The base peak at m/e 30 of the mass spectrum may be rationalized as follows:

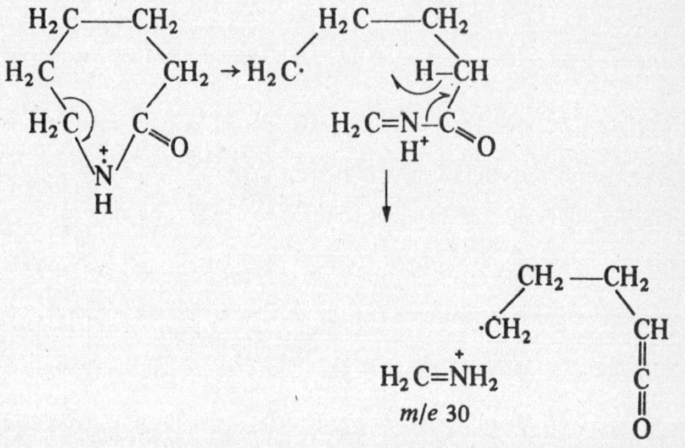

m/e 30

spectrometric identification of organic compounds

FREQUENCY (CM⁻¹)

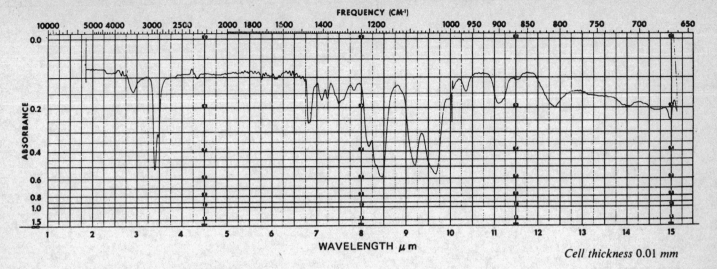

WAVELENGTH μm

Cell thickness 0.01 mm

Mass Spectral Data (Relative Intensities)

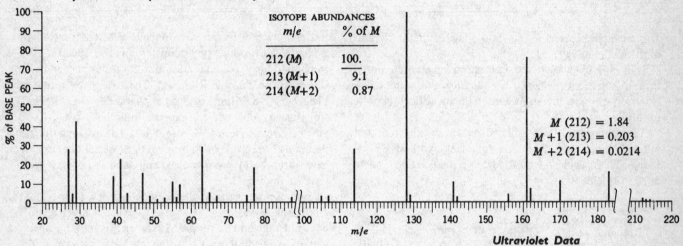

ISOTOPE ABUNDANCES	
m/e	% of M
212 (M)	100.
213 (M+1)	9.1
214 (M+2)	0.87

M (212) = 1.84
M +1 (213) = 0.203
M +2 (214) = 0.0214

Ultraviolet Data

Featureless beyond 210 nm

NMR Spectrum (Solvent CCl₄)

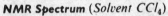

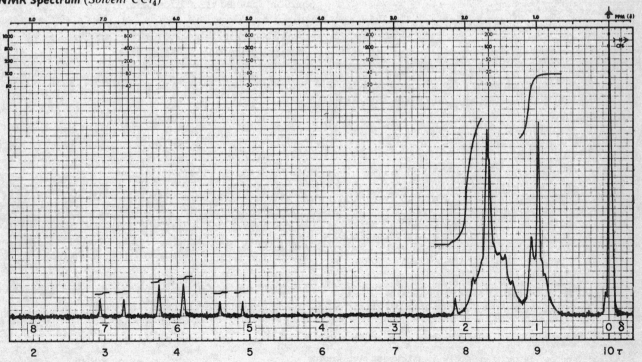

compound number 6-20

This example is presented to show some of the limitations of the methodology.

It is not likely that the four spectra presented would furnish sufficient evidence to identify this compound. However, given its history, we make the identification quite readily.

The compound was isolated as an unexpected, but rational, by-product from the following synthesis (S. A. Fuqua, W. G. Duncan, and R. M. Silverstein, *J. Org. Chem.*, **30**, 2543 (1965)):

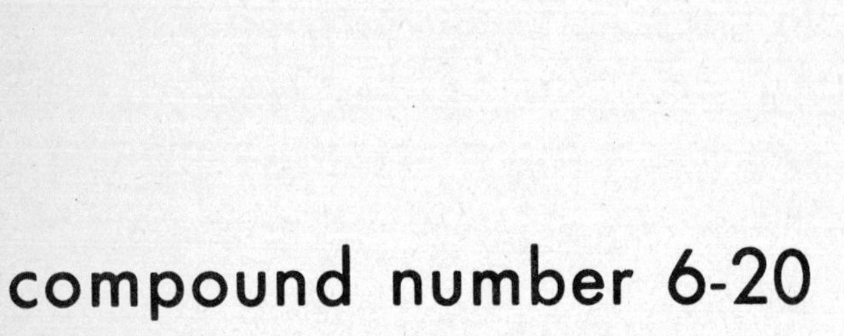

The reaction should proceed by formation of a Wittig reagent from the phosphine and the CF_2 carbene formed by pyrolysis of sodium chlorodifluoroacetate. The Wittig reagent should react with the ketone.

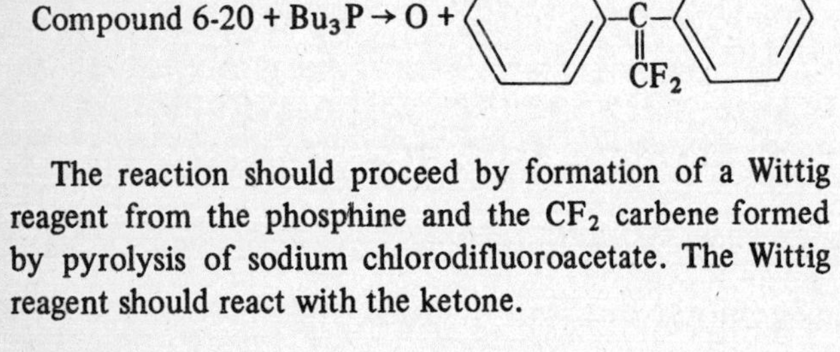

The molecular weight of the by-product is 212. It is obvious from the spectra (e.g., lack of aromatic character)

that the material is not a benzophenone derivative. Rather, aliphatic absorption predominates in the spectra. The compound may contain phosphorus or fluorine or both. Obviously no chlorine is present. The strong absorption at 1182 cm^{-1} (8.46 μm) may be aliphatic $P \rightarrow O$ absorption, and that at 1087 cm^{-1} (9.20 μm) and 1033 cm^{-1} (9.68 μm) may result from C—F stretching. The puzzling thing is that if we have the $P \rightarrow O$ group present, we cannot possibly have three butyl groups; the compound is not the usual product $Bu_3P \rightarrow O$.

Suppose we assume displacement of one of the butyl group; this leaves mass 51 short of the molecular weight. It does not take long to propose CHF_2 as a possible fragment, and to confirm it with the striking NMR pattern. If the upfield protons represent the 18 protons of the butyl groups, the entire downfield pattern represents a single proton, centered at δ 6.07, π 3.93. The proton is split into a triplet by the geminal F atoms (J_{HF} 49.4 Hz), and each peak is again split into a doublet by the P atom (J_{HP} 19.4 Hz). The compound is therefore

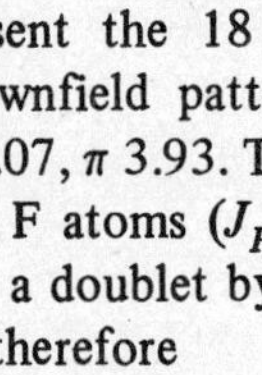

Dibutyldifluoromethylphosphine oxide

Some of the major fragmentation peaks can be accounted for: m/e 183 (M minus CH_3CH_2), m/e 161 (M minus CHF_2), m/e 47 ($P \rightarrow O$). The base peak, m/e 128 must result from complex rearrangements. The $M+1$ and $M+2$ peaks are in agreement with the calculated values for the proposed structure.

The rationale for this unexpected product depends on rearrangement of the initial Wittig reagent.

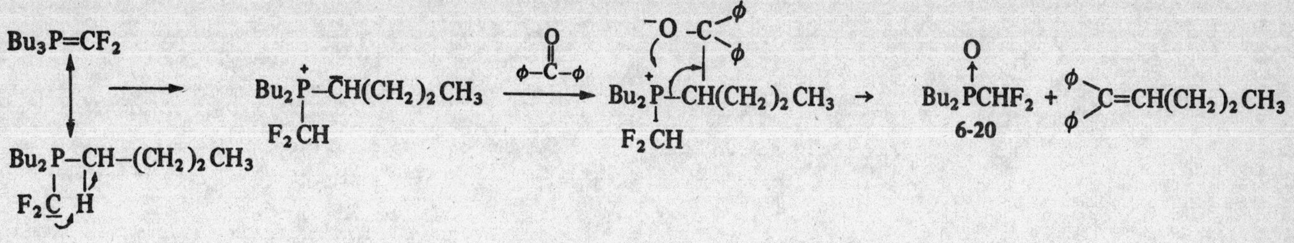

$$\text{Bu}_3\text{P}=\text{CF}_2$$

$$\updownarrow$$

$$\underset{\underset{\displaystyle \text{H}}{|}}{\underset{\displaystyle \text{F}_2\text{C}}{\text{Bu}_2\text{P}-\text{CH}-(\text{CH}_2)_2\text{CH}_3}} \longrightarrow \underset{\underset{\displaystyle \text{F}_2\text{CH}}{|}}{\overset{+}{\text{Bu}_2\text{P}}-\bar{\text{C}}\text{H}(\text{CH}_2)_2\text{CH}_3} \xrightarrow{\phi-\overset{\displaystyle \text{O}}{\overset{\|}{\text{C}}}-\phi} \underset{\underset{\displaystyle \text{F}_2\text{CH}}{|}}{\overset{+}{\text{Bu}_2\text{P}}-\text{CH}(\text{CH}_2)_2\text{CH}_3} \rightarrow \underset{\displaystyle \textbf{6-20}}{\overset{\displaystyle \overset{\displaystyle \text{O}}{\uparrow}}{\text{Bu}_2\text{PCHF}_2}} + \underset{\displaystyle \phi}{\overset{\displaystyle \phi}{\text{C}}}=\text{CH}(\text{CH}_2)_2\text{CH}_3$$

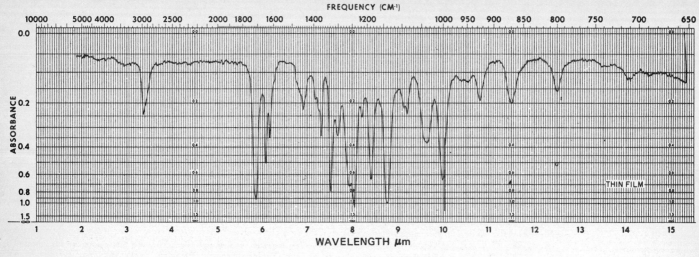

FREQUENCY (CM⁻¹)

ABSORBANCE

WAVELENGTH μm

THIN FILM

Mass Spectral Data (Relative Intensities)

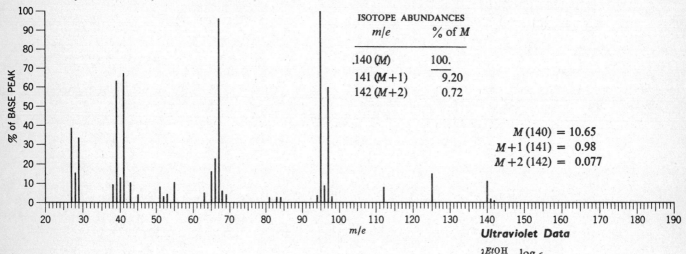

% of BASE PEAK

m/e

ISOTOPE ABUNDANCES	
m/e	% of M
.140 (M)	100.
141 (M+1)	9.20
142 (M+2)	0.72

$M (140) = 10.65$
$M+1 (141) = 0.98$
$M+2 (142) = 0.077$

Ultraviolet Data

λ_{max}^{EtOH}	$\log \epsilon_{max}$
259	4.4

NMR Spectrum

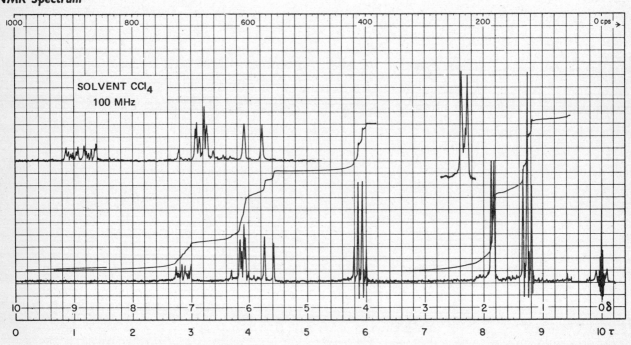

SOLVENT CCl₄
100 MHz

compound number 6-21

The molecular ion (m/e 140) and $M+1$ and $M+2$ data dictate, after elimination of formulas containing an odd number of N atoms, consideration of the following formulas:

Formula	$M + 1$	$M + 2$
$C_7H_{12}N_2O$	8.56	0.52
$C_8H_{12}O_2$	8.92	0.75
$C_8H_{16}N_2$	9.66	0.42

The formula containing two oxygen atoms is chosen on the basis of the infrared spectrum: the strong absorption at 1709 cm^{-1} (5.85 μm) and the multiple bands in the 1300–1160 cm^{-1} (7.69–8.62 μm) region are indicative of $\overset{\displaystyle O}{\overset{\|}{C}}=O$ and C–$\overset{\displaystyle O}{\overset{\|}{C}}$–$O$ stretching, respectively. We therefore anticipate an ester of molecular formula $C_8H_{12}O_2$. An ethyl ester is suggested by the triplet at δ 1.25 (τ 8.75, three protons) and the quartet at δ 4.10 (τ 5.90, two protons); the chemical shift of the latter signal is consistent with a methylene group of $R\overset{\displaystyle O}{\overset{\|}{C}}OCH_2CH_3$. The R group, by difference from the molecular formula, is C_5H_7– and contains two sites of unsaturation, at least one of which must be due to a double bond in view of the olefinic protons, δ 5.0–6.5 (τ 5.0–4.5), in the NMR spectrum. It is also clear that a double bond is in conjugation with the carbonyl group since the C=O stretch cited above is at a lower wavenumber (higher wavelength) than for normal, unconjugated, aliphatic esters (1730–1715 cm^{-1}, 5.71–5.76 μm). The UV spectrum also supports a highly conjugated structure. A C=C–C=C system is delineated by the coupled (respectively, symmetric and asymmetric) C=C

stretching vibrations at 1645 and 1634 cm^{-1} (6.08 and 6.22 μm). Since there is no other reason for low field protons in our structure (e.g., no indication of aromaticity), the signal at δ 7.10 (τ 2.90) is also an olefinic proton of unusually low field position. Integration shows a total of four olefinic protons. The remaining NMR absorption (δ 1.85, τ 8.15) is quite consistent with a methyl group (integration intensity) attached to a double bond (see chemical shift tables) bearing a single proton (hence, the doublet). These data can be satisfied only by placing the methyl group at the end of the conjugated sequence (remote to the ester grouping), thus:

$$CH_3-CH=CH-CH=CH-CO_2CH_2CH_3$$

Ethyl sorbate

This structure is supported by data from Tables XI and XII in the Chapter 5:

	nm
R–C=CCO_2R basic unit	208
One C=C extending conjugation	+30
One δ substituent	+18
(λ_{max}^{EtOH}) calc	$= 256 \pm 5$

The calculated value agrees very well with the observed 259 nm maximum.

The mass spectral fragmentation pattern is easily rationalized by the following cleavage patterns:

$$
\begin{array}{c}
m/e\ 67 \\
(M - CO_2CH_2CH_3) \\
CH_3-CH=CH-CH=CH \\
\end{array}
+
\begin{array}{c}
O \\
\| \\
C-OCH_2 \\
\end{array}
CH_3
$$

m/e 125
(M – CH_3)

Base peak, m/e 95

$[M^{\ddagger} - OCH_3]$

$$CH_3CH=CH-CH=CH-C\!\!\mid\!\!O\!\!\mid\!\!O\!\!\mid\!\!CH_2CH_3$$

$H\cdot$ transfer, m/e 112

$(M^{\ddagger} - C_2H_4)$

The infrared absorption at 1000 cm^{-1} (10.00 μm), C–H bend) is evidence that the configuration about the C=C double bonds is *trans-trans* (see footnote on Table III, Appendix C, Chapter 5).

The NMR spectrum possesses some unusual features that can be interpreted only after spin decoupling experiments:

$$\begin{array}{cccc}
\delta 1.85 & & \delta 6.1 & \\
\tau 8.15 & & \tau 3.9 & \delta 5.65 \\
\epsilon \ \ CH_3 & & H_\gamma & \tau 4.35 \\
\end{array}$$

Irradiation at δ 7.2 (τ 2.8, H_β) removes $J_{\beta\delta}$ and $J_{\beta\alpha}$, simplifying the signal for H_α at about δ 5.65 (τ 4.35). Note that the *trans* assignment for the α,β double bond is reinforced by the vicinal J value of about 18 Hz for the H_α absorption. The absorptions for H_δ and H_γ at about δ 6.1 (τ 3.9) are also partially collapsed. The unusually low field of H_β, compared to usual olefinic protons, can be rationalized by the resonance form that shows a decreased electron density at the β-carbon:

$$CH_3-CH=CH-\overset{\beta}{\underset{\oplus}{CH}}-CH=C-OCH_2CH_3 \ \ \overset{O^{\ominus}}{|}$$

Decoupling at δ 5.65 (τ 4.35, H_α) results in partial collapse of the signal at δ 7.2, τ 2.8, as does decoupling at δ 1.85 (τ 8.15, H_ϵ).

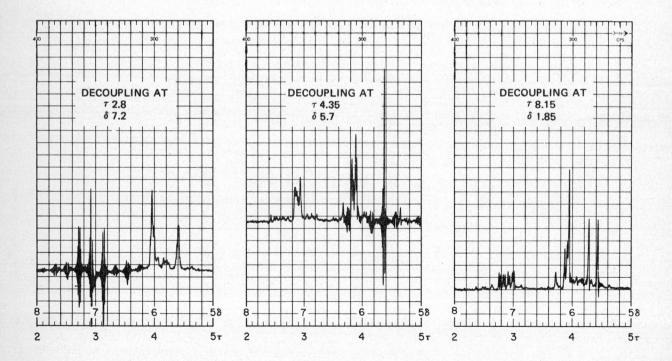

seven

sets of spectra with
beilstein references

This chapter consists of ten sets of spectra. Each compound represented by a set of spectra is identified by a reference to Beilstein. We again remind the reader that the compounds contain only C, H, O, N, S, and the halogens.

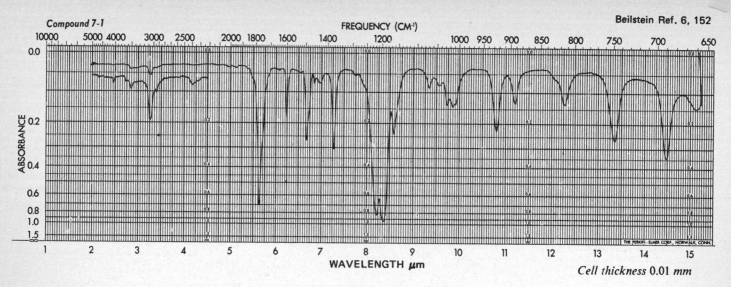

FREQUENCY (CM⁻¹)

ABSORBANCE

WAVELENGTH μm

Cell thickness 0.01 mm

Mass Spectral Data (Relative Intensities)

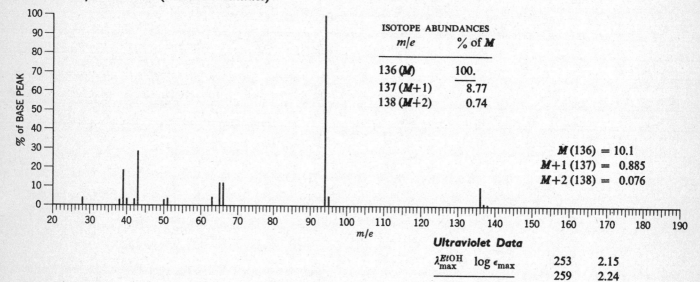

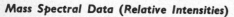

% of BASE PEAK

m/e

ISOTOPE ABUNDANCES	
m/e	% of M
136 (M)	100.
137 (M+1)	8.77
138 (M+2)	0.74

M (136) = 10.1
$M+1$ (137) = 0.885
$M+2$ (138) = 0.076

Ultraviolet Data

λ^{EtOH}_{max}	log ϵ_{max}		
		253	2.15
		259	2.24
248	2.02	265	2.12

NMR Spectrum (Solvent CCl₄)

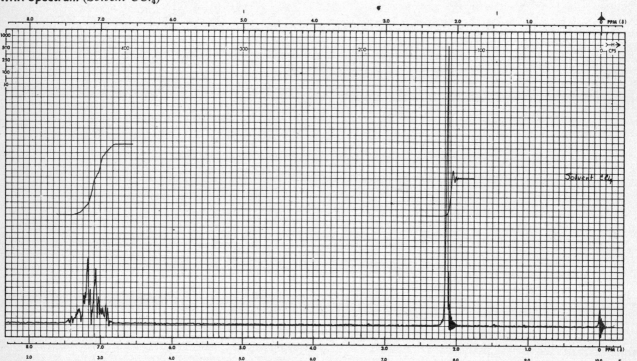

Solvent CCl₄

PPM (δ)

spectrometric identification of organic compounds

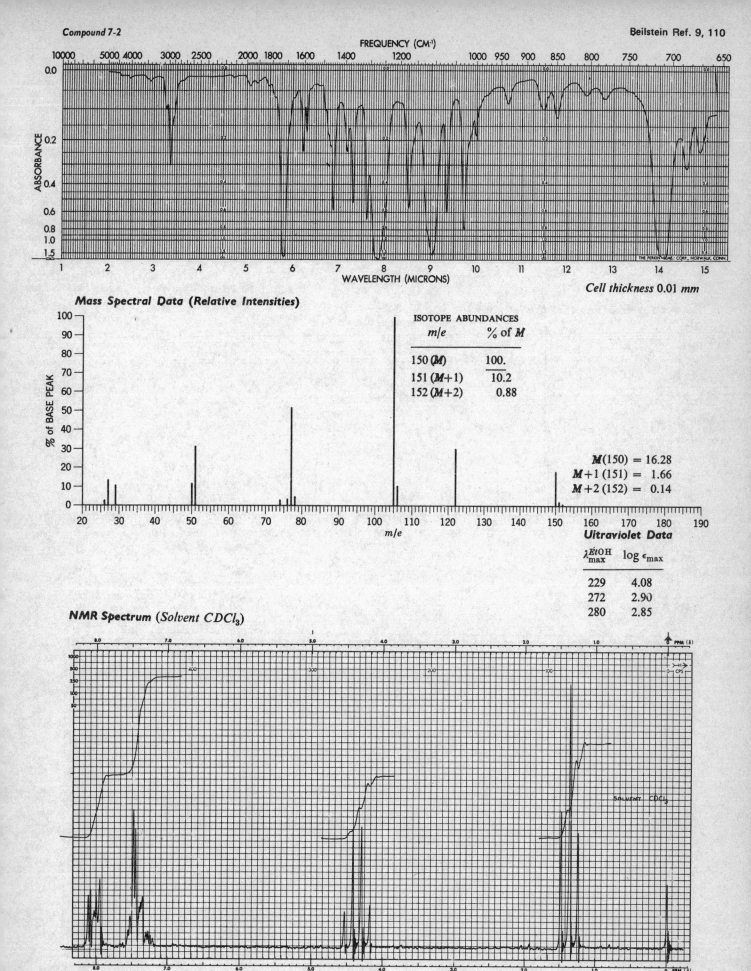

FREQUENCY (CM⁻¹)

ABSORBANCE

WAVELENGTH (MICRONS)

Cell thickness 0.01 mm

Mass Spectral Data (Relative Intensities)

% of BASE PEAK

m/e

ISOTOPE ABUNDANCES	
m/e	% of M
150 (M)	100.
151 (M+1)	10.2
152 (M+2)	0.88

M(150) = 16.28
M+1 (151) = 1.66
M+2 (152) = 0.14

Ultraviolet Data

λ_{max}^{EtOH}	log ϵ_{max}
229	4.08
272	2.90
280	2.85

NMR Spectrum (Solvent CDCl₃)

SOLVENT CDCl₃

Beilstein Ref. 1, 408

FREQUENCY (CM⁻¹)

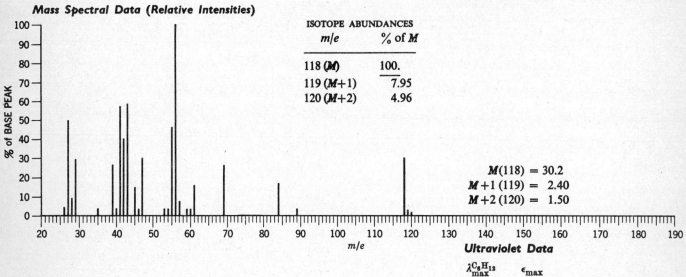

a. *Cell thickness 0.01 mm*   b. *Cell thickness 0.1 mm*

Mass Spectral Data (Relative Intensities)

ISOTOPE ABUNDANCES	
m/e	% of M
118 (M)	100.
119 (M+1)	7.95
120 (M+2)	4.96

$$M(118) = 30.2$$
$$M+1\,(119) = 2.40$$
$$M+2\,(120) = 1.50$$

Ultraviolet Data

$\lambda_{max}^{C_6H_{12}}$	ϵ_{max}	
225 (s)	163	(s) = shoulder

NMR Spectrum (*Solvent CDCl₃*)

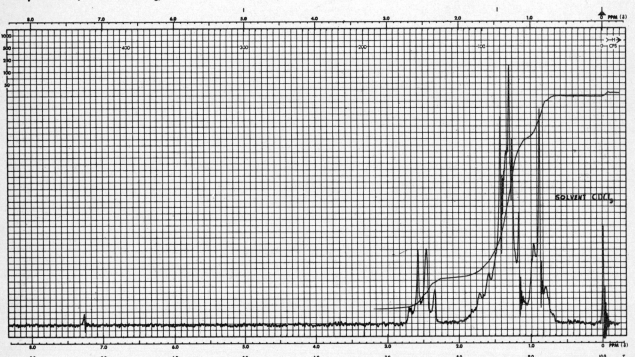

Compound 7-4

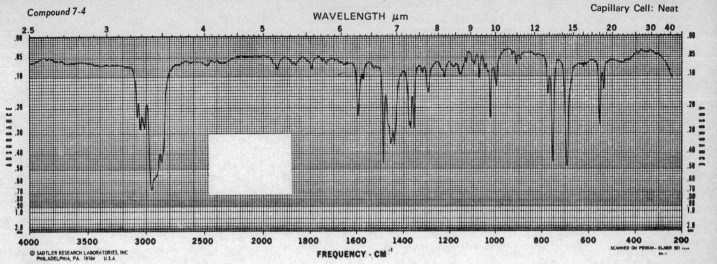

WAVELENGTH μm

FREQUENCY - CM⁻¹

© SADTLER RESEARCH LABORATORIES, INC
PHILADELPHIA, PA. 19104   U.S.A.

SCANNED ON PERKIN- ELMER 521

Source: Aldrich Chemical Company, Milwaukee, Wis.

Mass Spectral Data (Relative Intensities)

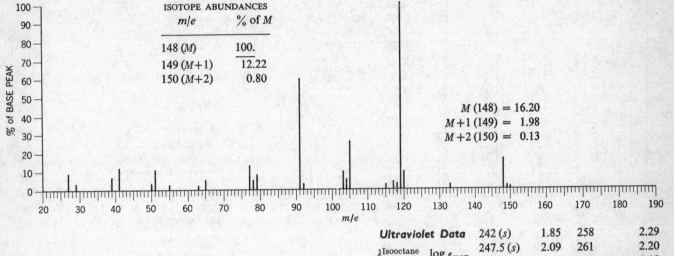

ISOTOPE ABUNDANCES

m/e	% of M
148 (M)	100.
149 (M+1)	12.22
150 (M+2)	0.80

M (148) = 16.20
$M+1$ (149) = 1.98
$M+2$ (150) = 0.13

Ultraviolet Data

$\lambda_{max}^{Isooctane}$	log ϵ_{max}		
217 (s)	3.60	242 (s)	1.85	258	2.29
236 (s)	1.57	247.5 (s)	2.09	261	2.20
		252.5	2.19	264	2.18

(s) = shoulder

NMR Spectrum (Solvent CCl₄)

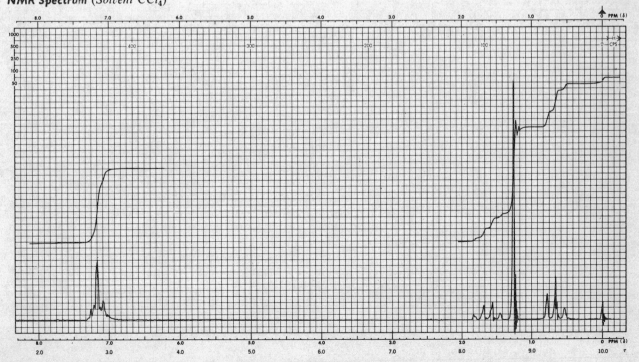

Beilstein Ref. 1, 106
Capillary Cell: Neat

WAVELENGTH μm

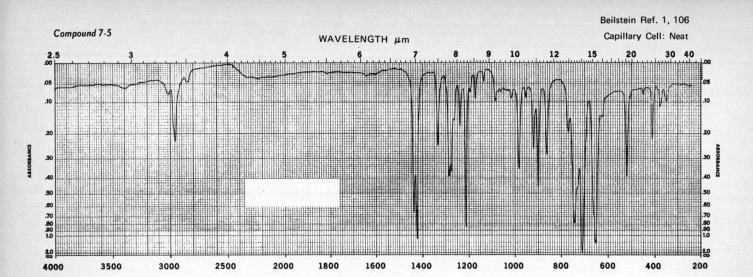

FREQUENCY - CM⁻¹

SCANNED ON PERKIN- ELMER 521

Source: Aldrich Chemical Company, Milwaukee, Wisconsin

Mass Spectral Data (Relative Intensities)

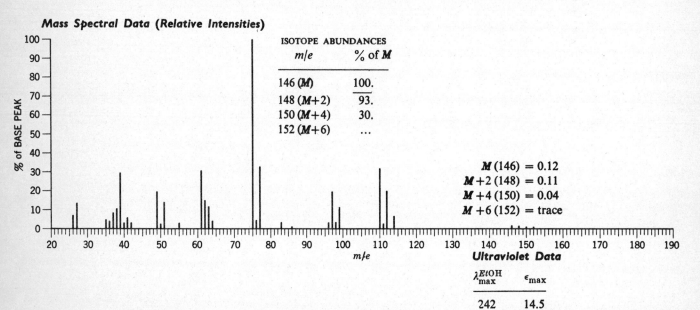

ISOTOPE ABUNDANCES	
m/e	% of M
146 (M)	100.
148 ($M+2$)	93.
150 ($M+4$)	30.
152 ($M+6$)	...

M (146) = 0.12
$M+2$ (148) = 0.11
$M+4$ (150) = 0.04
$M+6$ (152) = trace

Ultraviolet Data

λ_{max}^{EtOH}	ϵ_{max}
242	14.5

NMR Spectrum (Solvent CCl₄)

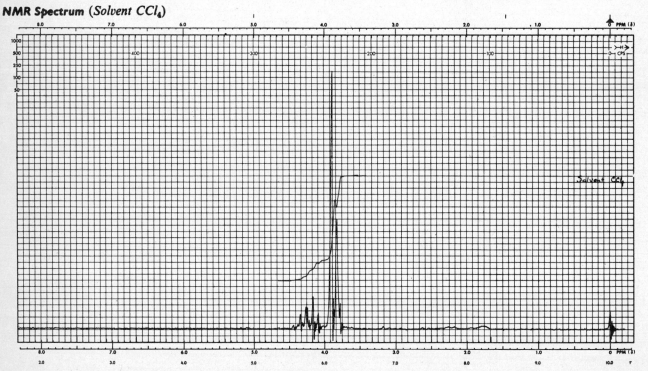

Solvent CCl₄

spectrometric identification of organic compounds

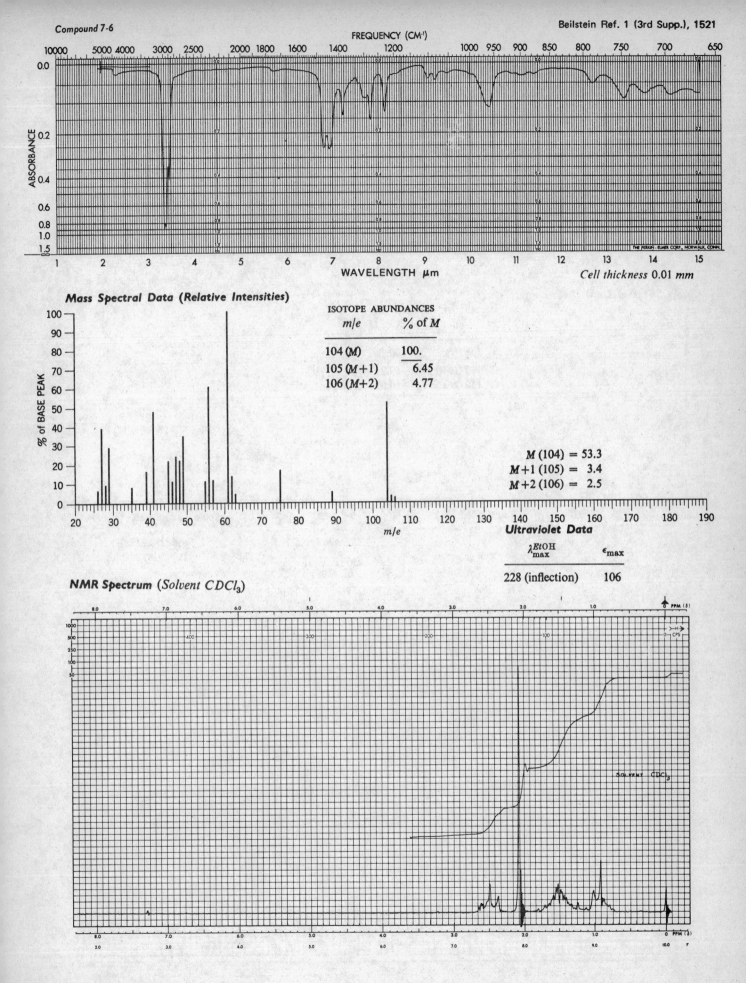

FREQUENCY (CM⁻¹)

WAVELENGTH μm

Cell thickness 0.01 mm

THE PERKIN-ELMER CORP., NORWALK, CONN.

Mass Spectral Data (Relative Intensities)

ISOTOPE ABUNDANCES	
m/e	% of M
104 (M)	100.
105 ($M+1$)	6.45
106 ($M+2$)	4.77

M (104) = 53.3
$M+1$ (105) = 3.4
$M+2$ (106) = 2.5

m/e

Ultraviolet Data

λ_{max}^{EtOH}	ϵ_{max}
228 (inflection)	106

NMR Spectrum (*Solvent CDCl₃*)

SOLVENT CDCl₃

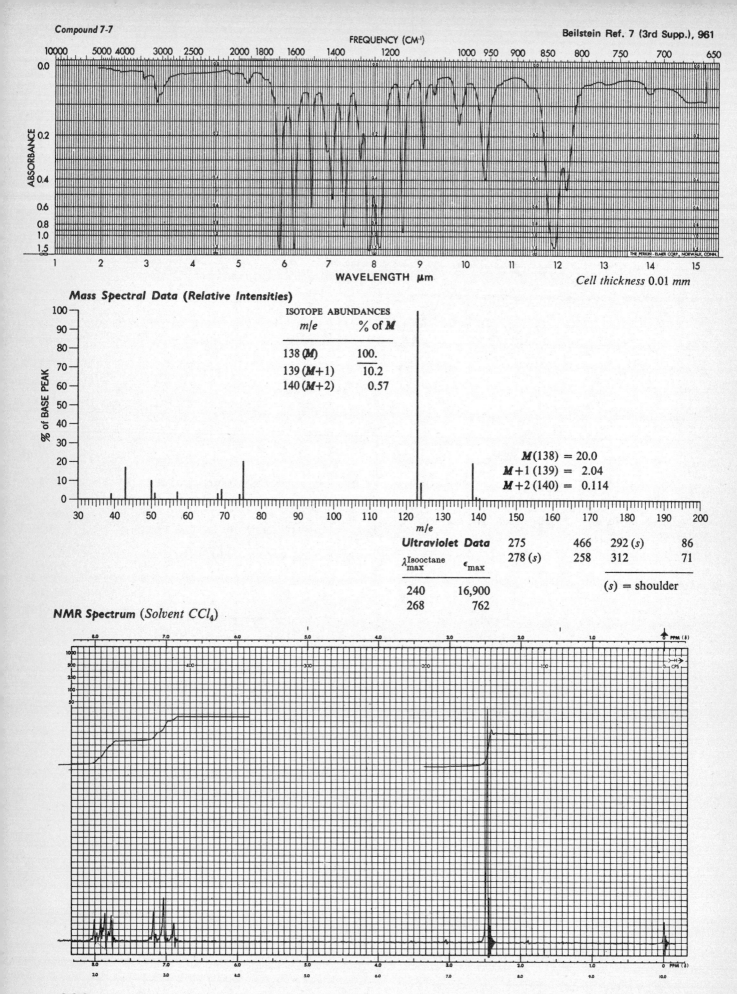

FREQUENCY (CM⁻¹)

ABSORBANCE

WAVELENGTH μm

THE PERKIN - ELMER CORP., NORWALK, CONN.

Cell thickness 0.01 *mm*

Mass Spectral Data (Relative Intensities)

% of BASE PEAK

ISOTOPE ABUNDANCES

m/e	% of M
138 (M)	100.
139 (M+1)	10.2
140 (M+2)	0.57

m/e

$M(138) = 20.0$
$M+1 (139) = 2.04$
$M+2 (140) = 0.114$

Ultraviolet Data

$\lambda_{max}^{Isooctane}$	ϵ_{max}
240	16,900
268	762

275	466	292 (s)	86
278 (s)	258	312	71

(s) = shoulder

NMR Spectrum (*Solvent CCl₄*)

PPM (δ)

CPS

PPM (δ)

FREQUENCY (CM⁻¹)

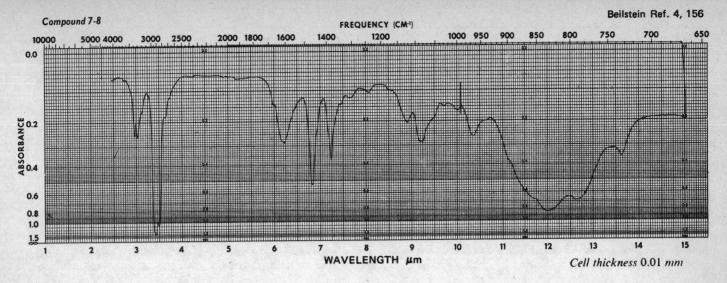

WAVELENGTH μm

Cell thickness 0.01 mm

Mass Spectral Data (Relative Intensities)

ISOTOPE ABUNDANCES

m/e	% of M
73 (M)	100.
74 (M+1)	5.5
75 (M+2)	0.2

$M\,(73) = 5.44$
$M+1\,(74) = 0.30$
$M+2\,(75) = 0.011$

m/e

Ultraviolet Data

λ_{max}^{EtOH}	ϵ_{max}
225 (s)	56

(s) = shoulder

NMR Spectrum (Solvent CCl₄)

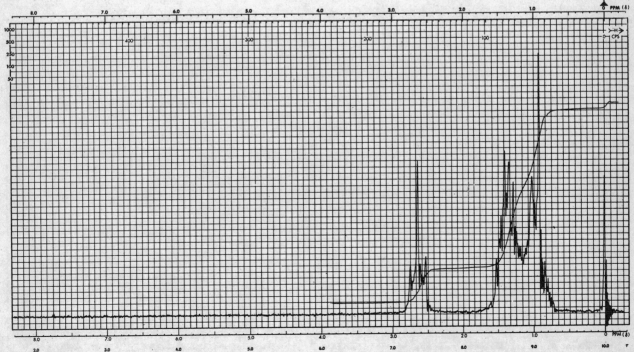

WAVELENGTH μm

Beilstein Ref. 6, 140
Capillary Cell: Neat

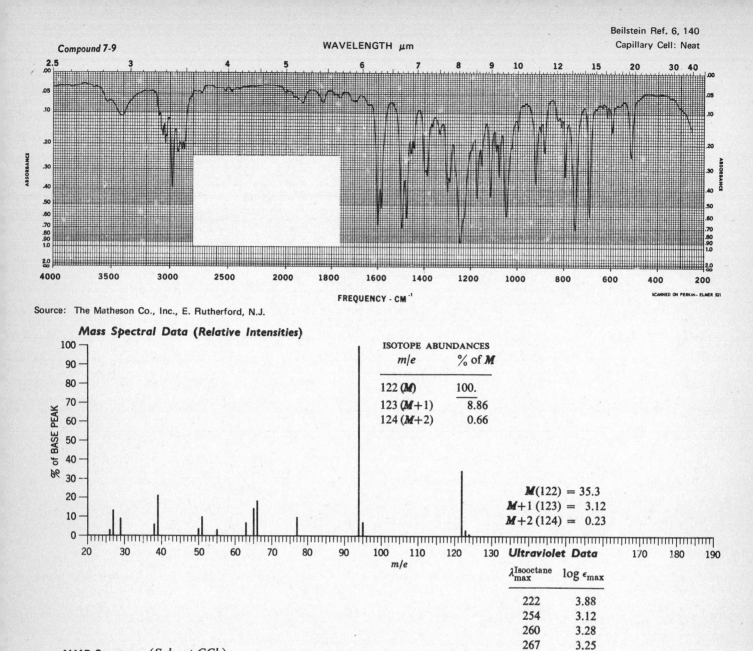

FREQUENCY - CM⁻¹

SCANNED ON PERKIN-ELMER 521

Source: The Matheson Co., Inc., E. Rutherford, N.J.

Mass Spectral Data (Relative Intensities)

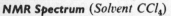

% of BASE PEAK

m/e

ISOTOPE ABUNDANCES

m/e	% of M
122 (M)	100.
123 (M+1)	8.86
124 (M+2)	0.66

$M(122) = 35.3$
$M+1 (123) = 3.12$
$M+2 (124) = 0.23$

Ultraviolet Data

$\lambda_{max}^{Isooctane}$	$\log \epsilon_{max}$
222	3.88
254	3.12
260	3.28
267	3.25

NMR Spectrum (Solvent CCl₄)

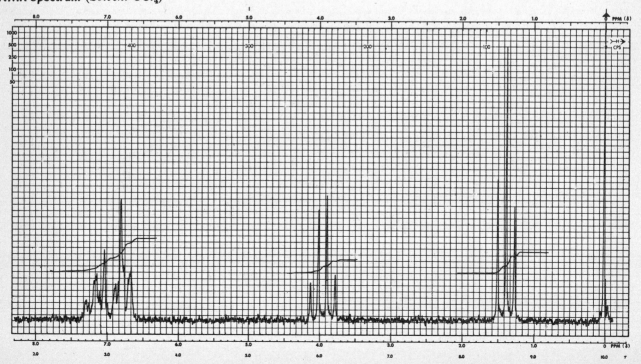

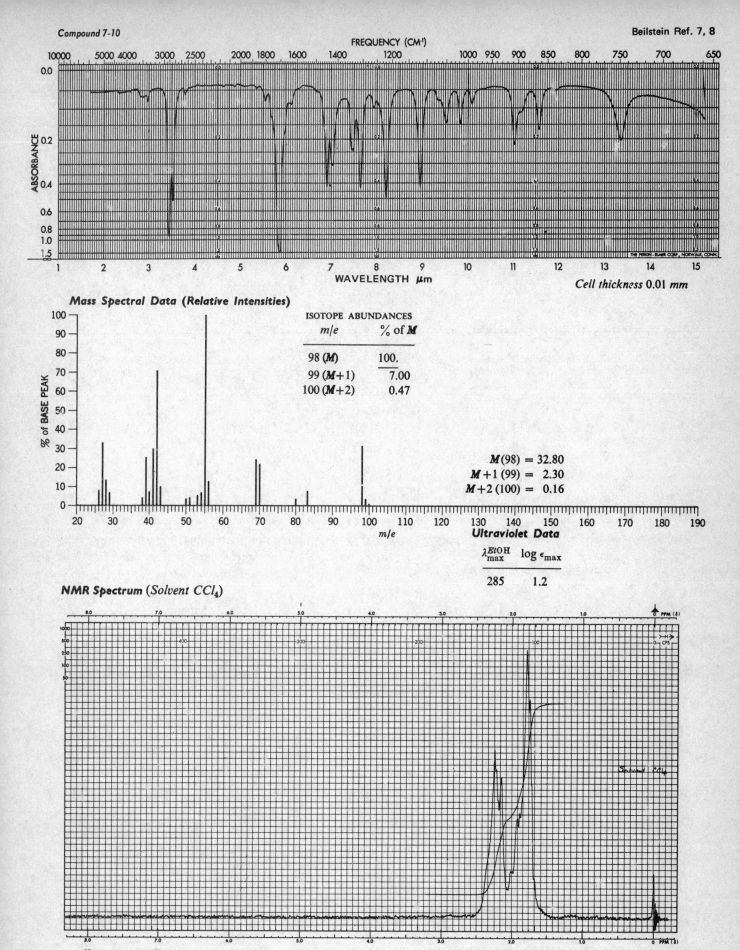

FREQUENCY (CM⁻¹)

ABSORBANCE

WAVELENGTH μm

Cell thickness 0.01 mm

Mass Spectral Data (Relative Intensities)

% of BASE PEAK

ISOTOPE ABUNDANCES

m/e	% of M
98 (M)	100.
99 (M+1)	7.00
100 (M+2)	0.47

$$M(98) = 32.80$$
$$M+1\ (99) = 2.30$$
$$M+2\ (100) = 0.16$$

m/e

Ultraviolet Data

λ_{max}^{EtOH}	$\log \epsilon_{max}$
285	1.2

NMR Spectrum (Solvent CCl₄)

WAVELENGTH μm

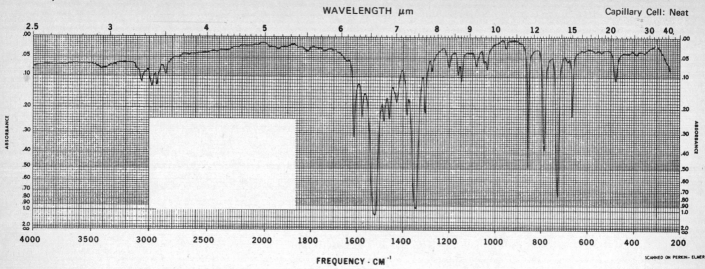

FREQUENCY - CM⁻¹

SCANNED ON PERKIN- ELMER 521

Source: E.I. duPont de Nemours & Co., Wilmington, Del.

Mass Spectral Data (Relative Intensities)

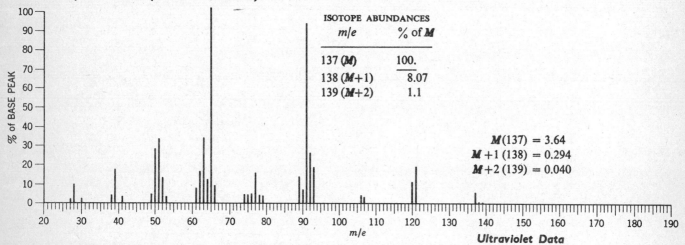

ISOTOPE ABUNDANCES

m/e	% of M
137 (M)	100.
138 (M+1)	8.07
139 (M+2)	1.1

M(137) = 3.64
M+1 (138) = 0.294
M+2 (139) = 0.040

m/e

Ultraviolet Data

$\lambda_{max}^{Isooctane}$	log ϵ_{max}
251	3.78
285 (s)	3.21
330	2.44

(s) = shoulder

NMR Spectrum

SOLVENT CCl₄
60MHz

PPM (δ)

CPS

PPM (δ)

τ

spectrometric identification of organic compounds

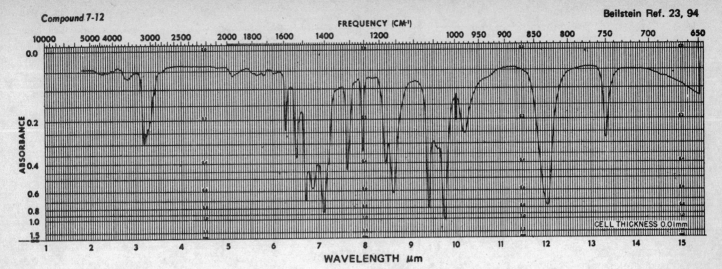

FREQUENCY (CM⁻¹)

WAVELENGTH μm

CELL THICKNESS 0.01mm

Mass Spectral Data (Relative Intensities)

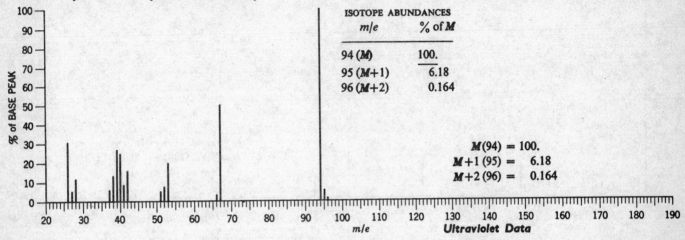

ISOTOPE ABUNDANCES	
m/e	% of M
94 (M)	100.
95 (M+1)	6.18
96 (M+2)	0.164

$M(94) = 100.$
$M+1 (95) = 6.18$
$M+2 (96) = 0.164$

m/e

Ultraviolet Data

	λ_{max}^{EtOH}	ϵ_{max}			
pH 7	266	6,600	pH 1	266	10,200
	272.5	6,000		273	9,700
	305	900			

NMR Spectrum

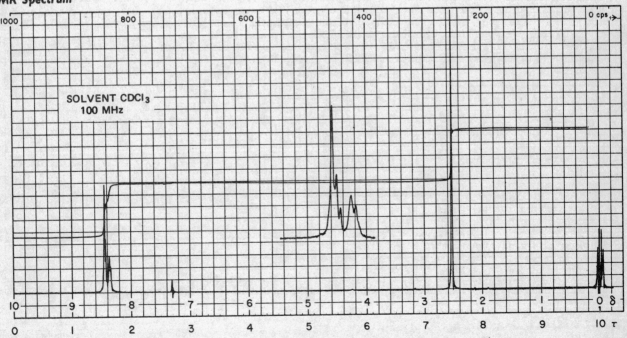

SOLVENT CDCl₃
100 MHz

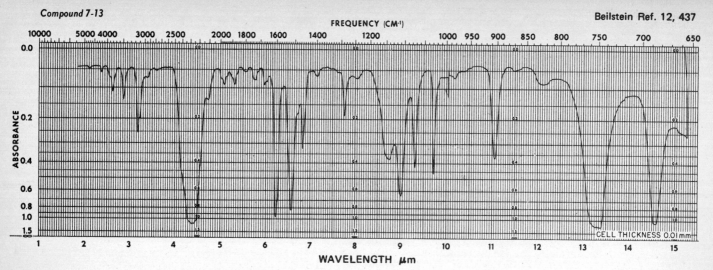

FREQUENCY (CM⁻¹)

WAVELENGTH μm

CELL THICKNESS 0.01mm

Mass Spectral Data (Relative Intensities)

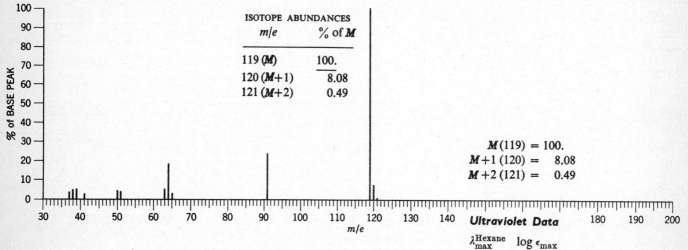

ISOTOPE ABUNDANCES	
m/e	% of M
119 (M)	100.
120 (M+1)	8.08
121 (M+2)	0.49

$M\,(119) = 100.$
$M+1\,(120) = 8.08$
$M+2\,(121) = 0.49$

m/e

Ultraviolet Data

λ_{max}^{Hexane}	$\log \epsilon_{max}$		
226	4.04		
256	2.59	270	2.76
263	2.66	277	2.67

NMR Spectrum

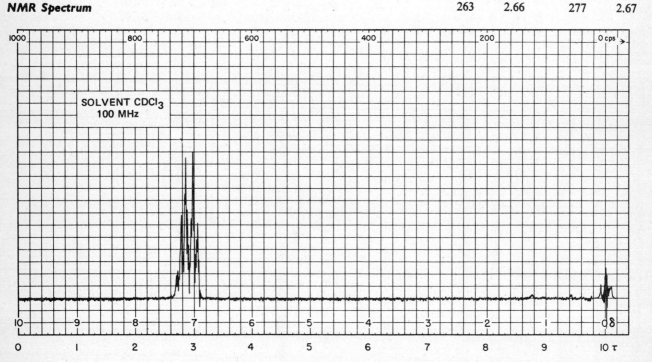

SOLVENT CDCl₃
100 MHz

Beilstein Ref. 1, (Supp. 3), 727

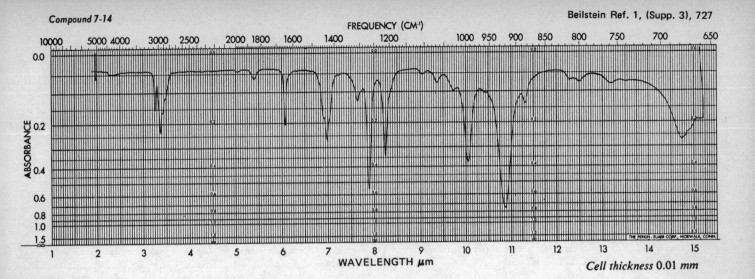

FREQUENCY (CM⁻¹)

ABSORBANCE

WAVELENGTH μm

THE PERKIN-ELMER CORP., NORWALK, CONN.

Cell thickness 0.01 mm

Mass Spectral Data (Relative Intensities)

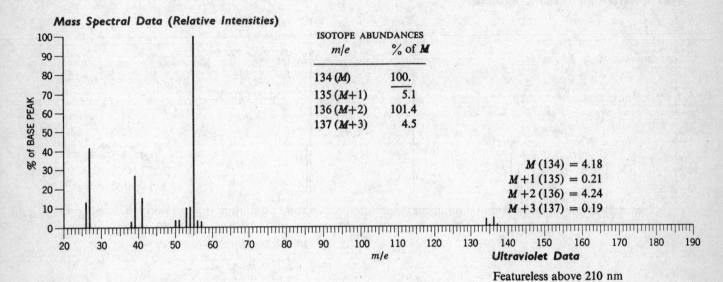

% of BASE PEAK

m/e

ISOTOPE ABUNDANCES

m/e	% of M
134 (M)	100.
135 (M+1)	5.1
136 (M+2)	101.4
137 (M+3)	4.5

$M (134) = 4.18$
$M+1 (135) = 0.21$
$M+2 (136) = 4.24$
$M+3 (137) = 0.19$

Ultraviolet Data

Featureless above 210 nm

NMR Spectrum (Solvent CCl₄)

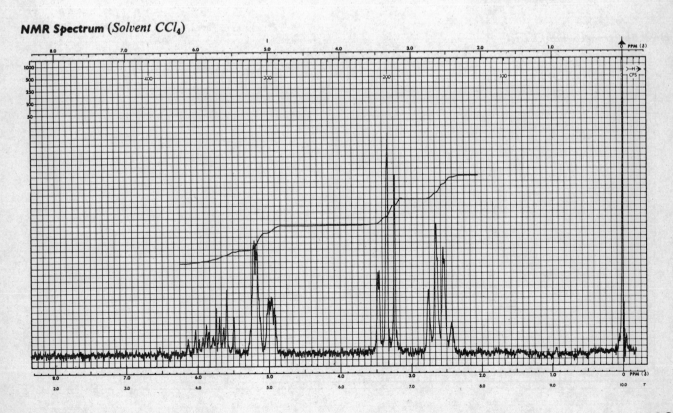

PPM (δ)

CPS

PPM (δ)

τ

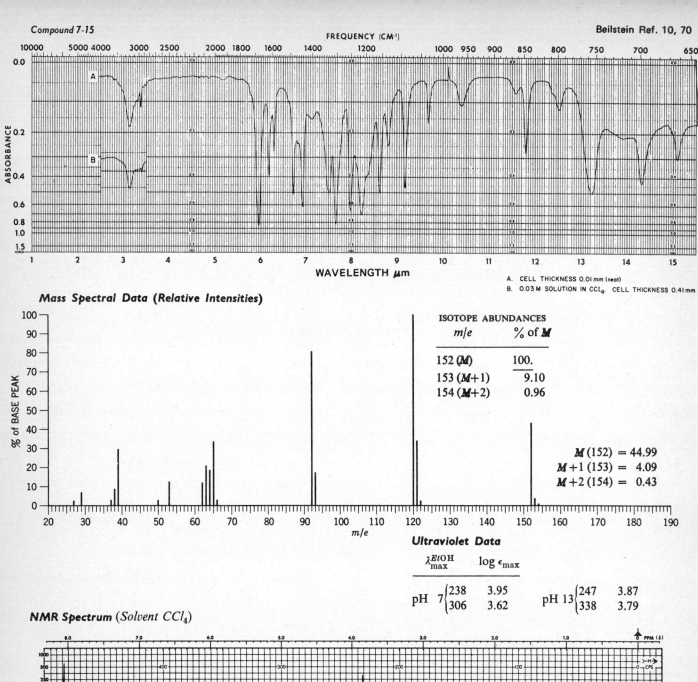

A. CELL THICKNESS 0.01mm (neat)
B. 0.03M SOLUTION IN CCl₄. CELL THICKNESS 0.41mm

Mass Spectral Data (Relative Intensities)

ISOTOPE ABUNDANCES

m/e	% of M
152 (M)	100.
153 (M+1)	9.10
154 (M+2)	0.96

M (152) = 44.99
M+1 (153) = 4.09
M+2 (154) = 0.43

Ultraviolet Data

λ_{max}^{EtOH}	$\log \epsilon_{max}$		λ_{max}^{EtOH}	$\log \epsilon_{max}$
pH 7 {238	3.95	pH 13 {247	3.87	
306	3.62	338	3.79	

NMR Spectrum (*Solvent CCl₄*)

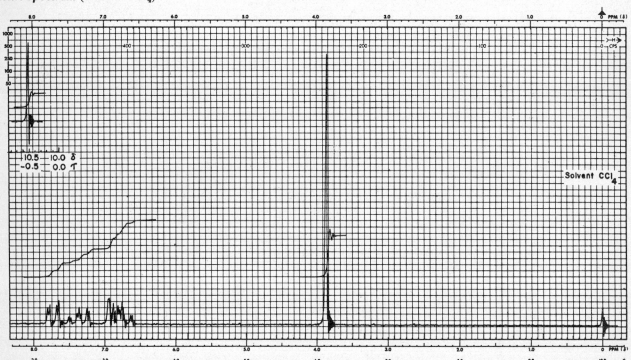

Solvent CCl₄

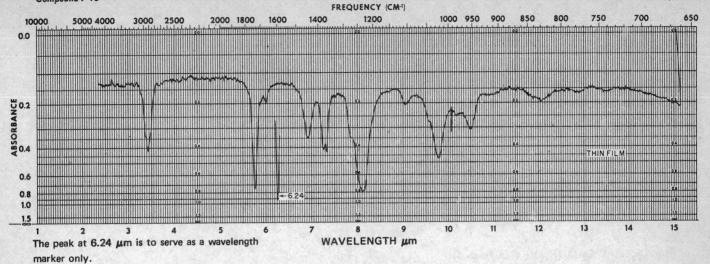

FREQUENCY (CM⁻¹)

WAVELENGTH μm

The peak at 6.24 μm is to serve as a wavelength marker only.

Mass Spectral Data (Relative Intensities)

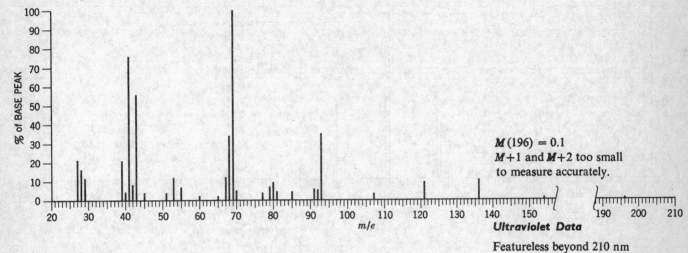

$M(196) = 0.1$
$M+1$ and $M+2$ too small
to measure accurately.

Ultraviolet Data

Featureless beyond 210 nm

NMR Spectrum

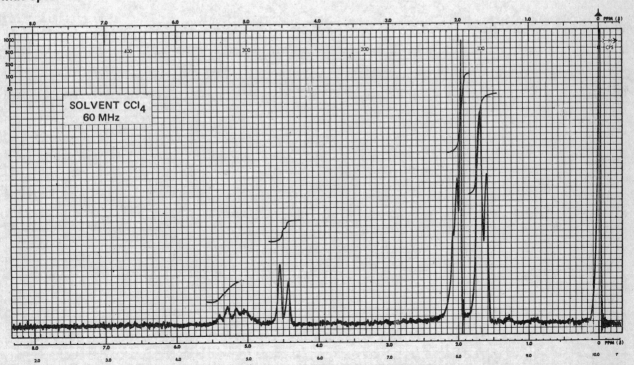

SOLVENT CCl₄
60 MHz

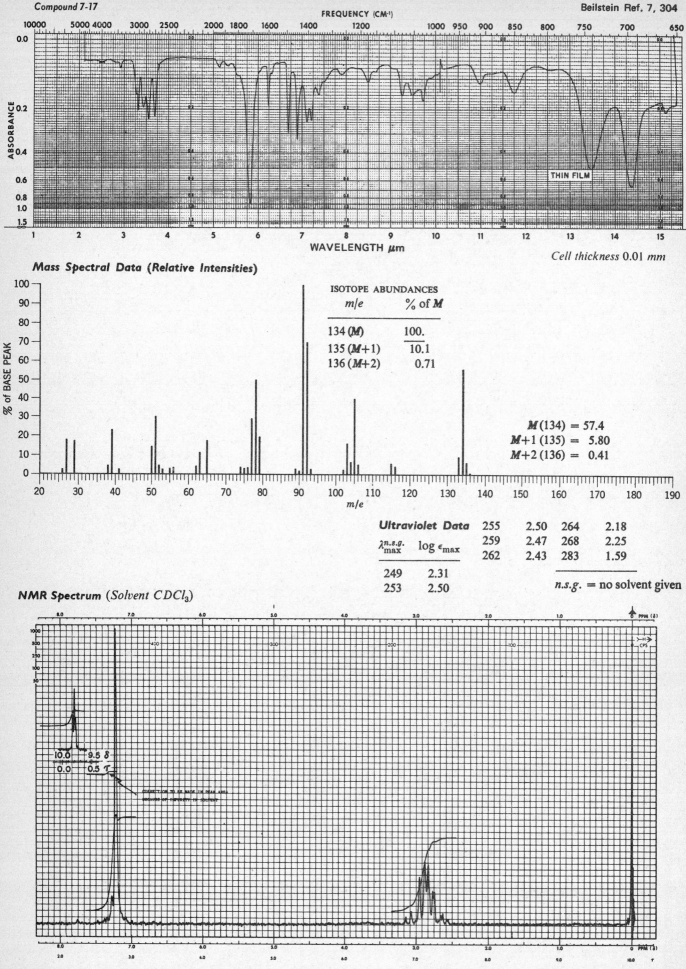

FREQUENCY (CM⁻¹)

ABSORBANCE

THIN FILM

WAVELENGTH μm

Cell thickness 0.01 mm

Mass Spectral Data (Relative Intensities)

% of BASE PEAK

m/e

ISOTOPE ABUNDANCES

m/e	% of M
134 (M)	100.
135 (M+1)	10.1
136 (M+2)	0.71

M (134) = 57.4
$M+1$ (135) = 5.80
$M+2$ (136) = 0.41

Ultraviolet Data

$\lambda^{n.s.g.}_{max}$	log ϵ_{max}
249	2.31
253	2.50
255	2.50
259	2.47
262	2.43
264	2.18
268	2.25
283	1.59

n.s.g. = no solvent given

NMR Spectrum (Solvent CDCl₃)

10.0   9.5 δ
0.0   0.5 τ

CORRECTION TO BE MADE IN PEAK AREA
BECAUSE OF IMPURITY IN SOLVENT

Compound 7-18

WAVELENGTH μm

Capillar Cell: Neat

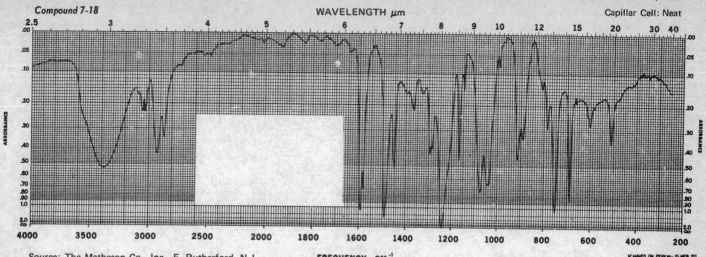

2.5 3 4 5 6 7 8 9 10 12 15 20 30 40

ABSORBANCE

4000 3500 3000 2500 2000 1800 1600 1400 1200 1000 800 600 400 200

FREQUENCY - CM⁻¹

Source: The Matheson Co., Inc., E. Rutherford, N.J.

SCANNED ON PERKIN-ELMER 521

Mass Spectral Data (Relative Intensities)

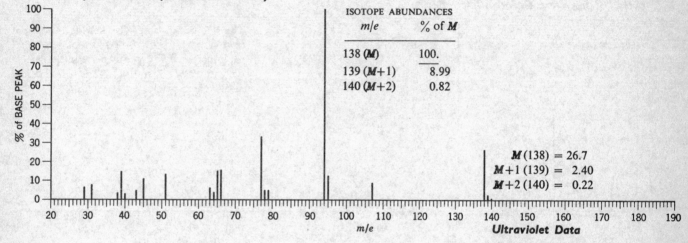

ISOTOPE ABUNDANCES

m/e	% of M
138 (M)	100.
139 ($M+1$)	8.99
140 ($M+2$)	0.82

M (138) = 26.7
$M+1$ (139) = 2.40
$M+2$ (140) = 0.22

% of BASE PEAK

20 30 40 50 60 70 80 90 100 110 120 130 140 150 160 170 180 190

m/e

Ultraviolet Data

$\lambda_{max}^{Isooctane}$	log ϵ_{max}
219	3.96
253	3.22
260	3.36
267	3.35

NMR Spectrum (Solvent CCl₄)

8.0 7.0 6.0 5.0 4.0 3.0 2.0 1.0 PPM (δ)

Solvent CCl₄

Beilstein Ref. 1 (3rd Supp.), 1994

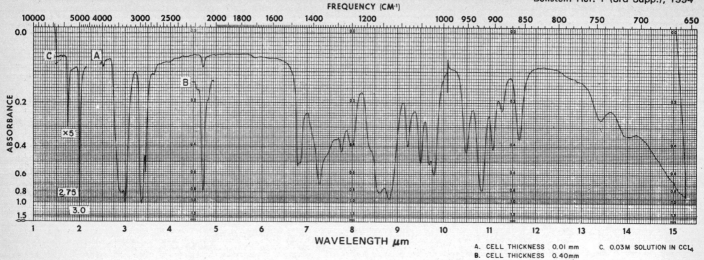

FREQUENCY (CM⁻¹)

WAVELENGTH μm

A. CELL THICKNESS 0.01 mm   C. 0.03M SOLUTION IN CCl₄
B. CELL THICKNESS 0.40mm

Mass Spectral Data (Relative Intensities)

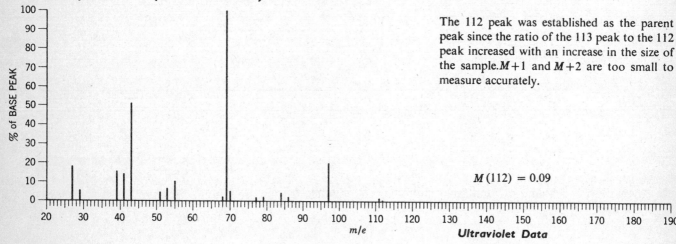

The 112 peak was established as the parent peak since the ratio of the 113 peak to the 112 peak increased with an increase in the size of the sample. $M+1$ and $M+2$ are too small to measure accurately.

$$M(112) = 0.09$$

m/e

Ultraviolet Data

Transparent beyond 200 nm

NMR Spectrum

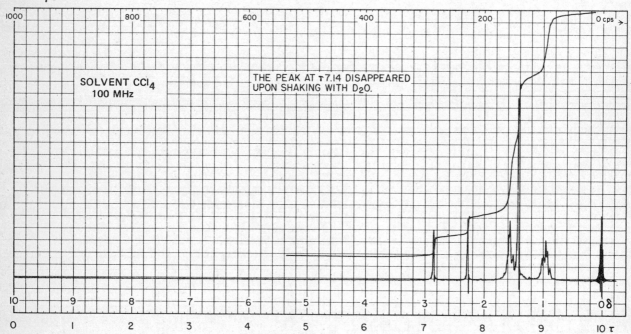

SOLVENT CCl₄
100 MHz

THE PEAK AT τ 7.14 DISAPPEARED
UPON SHAKING WITH D₂O.

SUBJECT INDEX

Key: **MS** = Mass Spectrometry (Chapter 2)
    IR = Infrared Spectrometry (Chapter 3)
    NMR = Nuclear Magnetic Resonance Spectrometry (Chapter 4)
    UV = Ultraviolet Spectrometry (Chapter 5)

Specific compounds, for which complete spectra are referenced, are listed in the Compound Index.

AA'BB' system, **NMR**, 183, 184
AA'XX system, **NMR**, 183, 184
AA'BB' system, **NMR**, 183
AB system, **NMR**, 171, 172, 181
A_3B_2 system, **NMR**, 183
ABC system, **NMR**, 184, 186
Absorbance, A., 2
Absorptivity, a, 2
  molar, ϵ, 2
ABX system, **NMR**, 186, 187
Acetylenes, **IR**, 89, 135
Acid anhydrides, **IR**, 104, 137
Acid halides, **IR**, 104, 137
  UV, 243
Acids, carboxylic, **MS**, aliphatic, 27
    UV, 243
  unsaturated, **UV**, 246, 247
Alcohols, **IR**, 91–94
    hydrogen bonding of, 92, 135
  MS, 21
    aliphatic, 21
    aromatic, 22
  NMR, 174
    classification of, 175, 176
Aldehydes, **IR**, 99, 136
Aldehydes, **MS**, 27
  UV, 243
    conjugated, **UV**, 244
Alkanes, **IR**, 85, 135
  branched, 86
  cyclic, 87
Alkenes, **IR**, 87, 135
Alkynes, **IR**, 89, 135
Amides, **IR**, 104–107, 135
  aliphatic, **MS**, 32
Amine salts, **IR**, 109, 137
  NMR, 178–179
Amines, **IR**, 107–109, 137
  classification, **NMR**, 179

Amino acids, **IR**, 109
$A_3MM'XX'$ system, **NMR**, 187, 188
AMX system, **NMR**, 186
Angstrom, A, 3
Anisotropy, **NMR**, 166–169
Aralkyl hydrocarbons, **MS**, 21
Aromatics, **IR**, mononuclear, 89–90
  polynuclear, 91
Auxochrome, **UV**, 234
A_2X_2 system, **NMR**, 183
AX system, **NMR**, 170–172, 181
Axial protons, chemical shift, **NMR**, 169
Azo compounds, **IR**, 111

Band positions, calculated, **IR**, 76
Base peak, **MS**, 11
Bathochromic shift, **UV**, 234
B-Bands, **UV**, 236
Beer's law, 3
Benzenoids, **UV**, 248–252
Bimolecular collisions, **MS**, 15
Bromides, aliphatic, **MS**, 37

^{13}C, **NMR**, 197
Carboxylate anion, **IR**, 101–102, 136
Carboxylic acid, **IR**, 99–101, 136
Carboxylic acids, **MS**, 27
  NMR, 177
  UV, 246, 247
Cells, **IR**, 81
Chemical shift, **NMR**, 164
Chemical shift equivalence, **NMR**, 181
Chemical shift tables, 211–215
  α substituents, 213, 215, 220, 221
  aromatic systems, 219
  β substituents, 212, 215
  exchangeable protons, 222
  olefinic systems, 214, 216–218, 223–225
  ring systems, 214

Chlorides, aliphatic, **MS**, 36
p-Chlorostyrene, **NMR**, 187
Chromophore, **UV**, 234
Computer signal averaging, **NMR**, 162
Concentration, c, 3
Conjugation, **UV**, 240–246
Coupling constant, **NMR**, 226–227

Deceptively simple patterns, **NMR**, 188
Decoupling, **NMR**, 192–194
Definitions, **UV**, 234
Delta scale, **NMR**, 164
$\Delta v/J$, **NMR**, 169, 170, 172
Derivatives, **MS**, 18
Detector (thermocouple), **IR**, 80
Dienes, **UV**, 241, 242
Disulfides, **IR**, 113
    aliphatic, **MS**, 35
Double focusing, **MS**, *see* High resolution
DSS, **NMR**, 164

E-bands, **UV**, 236
Electromagnetic spectrum, Chart, 233
Electronegativity, **NMR**, 165
Electronic transitions, **UV**, 235
Electrostatic field, **MS**, 9, 10
Enols, **NMR**, 177
Enones, **UV**, 244–247
Epoxides, **IR**, 95–96, 136
Equatorial protons, chemical shift, **NMR**, 169
Esters, **IR**, 102–104, 136
    MS, aliphatic, 28
    MS, aromatic, 30
    UV, unsaturated, 246
Ethers, **IR**, 94, 136
    MS, aliphatic, 23
    MS, aromatic, 25

Fermi Resonance, **IR**, 77–78
Ferromagnetic impurities, **NMR**, 163–166
Field strength, **NMR**, 162, 165, 167
    effect upon J, 165
First order, **NMR**, 172
"Fish-hook," **MS**, 17
Fluorides, **MS**, 37
Fourier transform, (FT), **NMR**, 163
Fragmentation, **MS**, 16–18
    rules, 17
Frequency, 3
"Frequency," **IR**, 74

Gas chromatography - **MS** Interface, 11
Gem-dimethyl groups, **IR**, 86
Group frequencies, **IR**, 135–152

Halides, **IR**, 114, 115, 139, 144, 146, 148
Halogen compounds, **MS**, 35–38
    aliphatic bromides, 37
    aliphatic chlorides, 36
    aliphatic fluorides, 36
    aliphatic iodides, 36
    aromatics, 38
    benzyl halides, 38
Heteroaromatic compounds, **MS**, 38
Heteroaromatics, **IR**, 116, 117, 151, 152
    UV, 252–255

High resolution, **MS**, 6, 9–11
Hydrocarbons, **MS**, aralkyl, 21
    olefins, 19
    saturated, 19
Hydrogen bonding, **IR**, 77, 78
Hydrogen deficiency, index of, **MS**, 16
Hydrohalides, **NMR**, 179
Hydroxy compounds, **MS**, 21
Hydroxy multiplets, **NMR**, 175–176
Hyperchromic shift, **UV**, 234
Hypochromic shift, **UV**, 235
Hypsochromic shift, **UV**, 234

Index of hydrogen deficiency, **MS**, 16
Infrared, definition, 3
    far **IR**, 73
    introduction, 73
    near **IR**, 73
    theory, 74
    reference discussion, 74
    units, 74
Imines, **IR**, 111
Instrumentation, **IR**, 78–80
    sampling, 78
    MS, 6–11
    UV, 236–238
Interface, GC–MS, 11
Iodides, **MS**, 37
Ionization, chemical, **MS**, 15
    electron impact, **MS**, 6
Isocyamides, **IR**, 111, 138
Isothiocyanates, **IR**, 111, 138
Isotopes, **MS**, 12–14
    abundances, 13
    masses, 12

Karplus correlations, **NMR**, 190–191
K-Bands, **UV**, 236
Ketones, **IR**, 96–99, 136
    MS, 25–27
    aliphatic, 25
    aromatic, 26
    cyclic, 26
Ketones, **UV**, 243
    conjugated, 244

Lactams, **IR**, 107, 137
Lactones, **IR**, 102–104, 137
    MS, 30–31
Long range coupling, **NMR**, 191

M + 1 Peak, **MS**, 12–14
Magnet, **MS**, 6, 7
Magnetic equivalence, **NMR**, 181, 183–187
Magnetogyric constant, **NMR**, 160
Magnetogyric ratio, effect upon coupling, 179
Mass spectrum trace, **MS**, 6–7
Mercaptans, **IR**, 113, 138
    MS, 34, 35
Metastable Peak, **MS**, 8, 12
Micrometer (μm), 3, 74
Miscellaneous compound classes, references, **MS**, 38
Molecular formula, determination, **MS**, 12–14
Molecular (Parent) ion, definition, **MS**, 11
    determination, 14–16
Monochromator, **IR**, 80

n → π* transition, UV, 232–236
n → π* transition, UV, 239
Nanometer (nm), 3
Natural Products, reference tabulation, MS, 38
N-H Compounds, NMR, 177
Nitrates, aliphatic, MS, 34
  IR, 112, 138
Nitriles, aliphatic, MS, 32
  IR, 110, 138
Nitrites, aliphatic, MS, 33
  IR, 112, 138
Nitro compounds, IR, 111, 138
  MS, 33
Nitrogen compounds, UV, 247
Nitrogen rule, corollary, MS, 15
  MS, application, 14
    definition, 15
Nitroso compounds, IR, 112, 138
NMR, introduction, 159
N-O compounds, UV, 247
Nomenclature, spectrometry, 2
Norbornenyl chloride, NMR, 170–171, 188
Notation, NMR, 172, 181
Nuclei, properties, NMR, 228, 229
Nuclei, set, NMR, 181

O-H Compounds, NMR, 174–177
Olefins, IR, 87
  cyclic, 88, 135
Optical Rotatory Dispersion (ORD), UV, 255–256
ORD, UV, 255–256

Paraffins, IR, 85, 135
Parent Peak, MS, see Molecular Ion
Pascal's Triangle, NMR, 173
Path length (b), sample, 3
Pellets, IR, 81–82
Peroxides, IR, 95, 136
Phenols, IR, 91–94
  hydrogen bonding, 92, 135
Phenols, MS, 22
  NMR, 175
Phosphorous Compounds, IR, 115–116, 139, 140, 142, 151
Photometer, IR, 78
π → π* transitions, UV, 232–236
Polyynes, UV, 242
Precessional Angular Velocity, NMR, 160

Quadrupole, MS, 6, 9, 10

Radiation Source, IR, 78
R-Bands, UV, 235
Rearrangements, MS, 18
Recorder, MS, 7
Recording, MS, 8
References, general, 3
Relaxation Times, NMR, 162
Resolution, MS, 6
Ring-Current Effect, NMR, 168

Sample Handling, IR, 80

MS, 6
  UV, 238
Shielding, diamagnetic, NMR, 166
Shift reagents, NMR, 194–196
Silicon Compounds, IR, 115, 139, 150
Sodium 2,2-dimethyl-2-silapentane-5-sulfonate (DSS), NMR, 164
Solvent Corrections, UV, 245
Solvents, effect upon chemical shift, NMR, 170, 171
  IR, 120
  NMR, 163, 200, 201
  spectral grade, UV, 239, 258
Spectrometer, NMR, 162
  optical, 3
Spectrometry, definition, 3
Spectrophotometer, 3
Spin Numbers, NMR, 160
Spin-Spin Coupling, NMR, 169
Spin System, NMR, 181
Spinning Side Bands, NMR, 163–164
Sulfates, IR, 114
Sulfides, aliphatic, MS, 34
  IR, 113
Sulfonamides, R, 114, 138
Sulfonates, IR, 114, 138
Sulfonic Acids, IR, 114
Sulfonyl Chlorides, IR, 114, 138
Sulfoxides, IR, 114, 138
Sulfur Compounds, MS, 34, 35
  NMR, 179
  UV, 247
Symmetry Operations, NMR, 181–182

Tan Scale, NMR, 165
Tetramethylsilane (TMS), NMR, 164
Theory, UV, 232, 233
Thiocarbonyl Compounds, IR, 113, 138
Thioethers, MS, 34
Thiols, MS, 34
TMS, NMR, 164
Toluene, MS, 8–9
Trace, MS, 8
Transmittance (T), 3
Tropylium cation, MS, 17, 21
Tubes, NMR, 162

Ultraviolet, 3
Unit resulution, MS, 6
"Unsaturation," degree of, MS, 16

Vibrational Modes, IR, 75
"Virtual" Coupling, NMR, 187, 188
Visible, 3

Wave length–wave number conversion table, IR, 153–157
Wavenumber, 3
Wavenumbers, IR, 74
Woodward's Rules, UV, 241–246

XX'AA'XX', System, NMR, 184, 186

COMPOUND INDEX

For each of the compounds below, complete spectra (of the type denoted in boldface) are referenced. Entries marked UV data are exceptions; for these λ_{max} and ϵ_{max} values for the major bands are referenced.

Acetaldehyde, UV data, 240, 243
Acetamide, UV data, 240, 243
Acetic Acid, UV data, 240, 243
Acetoxime, UV data, 240
Acetic Anhydride, UV data, 243
Acetone, **IR**, 29
  NMR, 202
  UV data, 235, 240, 243
Acetyl Chloride, UV data, 243
Acetylene, UV data, 235
Acetonitrile, UV data, 240
Acetophenone, **IR**, 98
  UV data, 235, 250
Acridine, UV data, 255
Acrolein, UV data, 235, 244
Amyl Nitrite, UV data, 240
Aniline, UV data, 249
Anilinium cation, UV data, 249
Anisole, **IR**, 96
  UV data, 249
Anthracene, **UV**, 253
  UV data, 252, 255
Antifoam A, **NMR**, 202
Apiezon L Grease, **NMR**, 203
Azomethane, UV data, 240

Benzene, **IR**, 122
  NMR, 203
  UV, 253
  UV data, 235, 248, 249, 250, 252, 255
Benyaldehyde, UV data, 250
Benzoate, Ammonium Salt, **IR**, 102
Benzoic Acid, UV data, 250
Benzophenone, UV data, 250
Benzoyl Chloride, **IR**, 105
Benzyl Acetate, **IR, MS, NMR, UV**, 261
Benzyl Alcohol, **IR**, 93
Benzyl Mercaptan, **IR**, 33
Biphenyl, UV data, 250, 251
1,3-Butadiene, UV data, 235, 241
2-Butanone, **IR**, 129
n-Butyl Iodide, UV data, 235

n-Butyl Nitrite, UV data, 247

ε-Caprolactam, **IR, MS, NMR, UV**, 307
Carbon Disulfide, **IR**, 133
Carbon Tetrachloride, **IR**, 125
  MS, 37
o-Catechol, UV data, 249
  anion UV data, 249
Cholesta-4-ene-3-one, **UV**, 232
Chlorobenzene, UV data, 249
Chloroform, **IR**, 125
  NMR, 203
4-Chloroindole, **IR, MS, NMR, UV**, 305
Crotonaldehyde, UV data, 244
Cumene, **NMR**, 174
1,3-Cyclohexadiene, UV data, 241
Cyclohexane, **IR**, 122
  NMR, 204
Cyclohexanone, **IR**, 130
Cyclohexyl methyl sulfoxide, UV data, 240
Cyclohexyl nitrate, UV data, 247
1,3-Cyclopentadiene, UV data, 241, 253
Cyclopentanone, **IR**, 79
  UV data, 243

2,4,6,8-Decatetrayne, UV data, 242
1-Decene, **IR**, 88
Di-*n*-amyl sulfide, **MS**, 36
β,β'-Dibromodiethyl Ether, **IR, MS, NMR, UV**, 272
Di-*n*-butyl Di-Sulfide, UV data, 239
Di-*n*-butyl Sulfide, UV data, 239
Dibutyldifluoromethylphosphine Oxide, **IR, MS, NMR, UV**, 310
1,2-Dichloroethane, **IR**, 126
1,3-Dichloropropane, **NMR**, 186
Difluoroacetic Acid, **NMR**, 204
2,3-Dihydropyran, **IR, MS, NMR, UV**, 292
Di-isoamyl Sulfide, **IR, MS, NMR, UV**, 266
Di-isobutyl Ketone, UV data, 243
Di-isopropyl Ether, **IR, MS, NMR, UV**, 264
2-Dimethylaminoethyl acetate, **IR, MS, NMR, UV**, 300
1,3-Dimethyl-1,3-butadiene, UV data, 241
N,N-Dimethylformamide, (DMF), **IR**, 134

NMR, 204
2,4-Dimethylpentane, **IR**, 87
Dimethyl Sulfone, **UV** data, 240
Dimethylsulfoxide (DMSO), **IR**, 133
Dimethylsulfoxide-d$_6$ (DMSO-d$_6$), **NMR**, 205
p-Dioxane, **IR**, 132
  NMR, 205
Diphenyl Ether, **UV**, data, 249
Diphenyl Sulfoxide, **IR, MS, NMR, UV**, 290
  UV data, 250
Divinyl β,β'-Thiodipropionate, **IR, MS, NMR, UV**, 281
DMF, **NMR**, 204
DMSO, **IR**, 133
DMSO-d$_6$, **NMR**, 205
Dodecane, **IR**, 85
Dow-Corning (DC) Stopcock Grease, **NMR**, 205
DSS, **NMR**, 209

Ethane, **UV** data, 235
Ethanol, **IR**, 128
  NMR, 206
Ethyl Acetate, **NMR**, 206
  UV data, 243
Ethyl sec-Butyl Ether, **MS**, 24
Ethyl Chloride, **NMR**, 173
Ethyl Ether, **IR**, 131
Ethyl N-Methyl Carbamate, **NMR**, 178
Ethyl Nitrate, **UV** data, 240
Ethyl Sorbate, **IR, MS, NMR, UV**, 312, 314
Ethyl *p*-Toluenesulfonate, **IR**, 115
Ethylene, **UV** data, 235, 240
Ethylene Glycol, **NMR**, 206

Fluorolube ®, **IR**, 127
p-Fluorostyrene, **IR, MS, NMR, UV**, 302
Furan, **UV** data, 253
  substituted, **UV** data, 254

Heptaldehyde, **IR, MS, NMR, UV**, 286
Heptanoic Acid, **IR**, 101
n-Hexadecane, **MS**, 20
2,4-Hexadiyne, **UV** data, 242
Hexamethylacetone, **UV** data, 243
Hexane, **IR**, 121
Hexamethylbenzene, **UV** data, 248
1-Hexanethiol, **UV** data, 235, 239
1,3,5-Hexatriene, **UV** data, 235, 241
1-Hexyne, **IR**, 89
Indene, **IR**, 124
Isobutyraldehyde, **UV** data, 243
Isobutyramide, **IR**, 106, 137
Isoeugenol, **IR, MS, NMR, UV**, 277
Isopropyl Alcohol, **NMR**, 207
Isoquinoline, **UV** data, 255

(±) Leucine, **IR**, 110
Levulinic Acid, **IR, MS, NMR, UV**, 284
Lubriseal ®, **NMR**, 207

MEK, **NMR**, 208
2-Mercaptoethanol, **IR, MS, NMR, UV**, 275
Mesityl Oxide, **UV**, 232
  UV data, 244
Mesitylene, **IR**, 91
Methanol, **IR**, 128

NMR, (acid free), 208
NMR (plus acid), 207
  UV data, 235, 239
2-Methyl-2-butanol, **MS**, 23
Methyl *n*-Butyrate, **IR, MS, NMR, UV**, 296
Methyl Caprylate, **MS**, 29
Methyl Chloride, **UV** data, 239
Methyl Ethyl Ketone, **IR**, 129
  NMR, 208
  UV data, 243
Methyl 2-Furoate, **IR, MS, NMR, UV**, 269
Methyl Iodide, **UV** data, 239
Methyl Isopropenyl Ketone, **UV** data, 244
2-Methyl-2-Nitropropane, **UV** data, 247
5-Methylpentadecane, **MS**, 20
N-Methylpiperidine, **UV** data, 239
Methyl Sulfoxide (DMSO), 133
Methyl Vinyl Ketone, **UV** data, 244

Naphthacene, **UV**, 253
  UV data, 252
Naphthalene, **UV**, 253
  UV data, 251
Nitroanilines (o,m,p), **UV** data, 251
Nitrobenzene, **IR**, 112
  UV data, 250
Nitromethane, **UV** data, 240
Nitrosobutane, **UV** data, 247
Nitrophenols (o,m,p), **UV** data, 244
1-Nitropropane, **NMR**, 188
1-Nitro-1-propene, **UV** data, 247
Nitrosobutane, **UV** data, 247
Nonanal, **MS**, 28
Nujol ®, **IR**, 121
2,4,6-Octatriyne, **UV** data, 242
Octyl Amine, **IR**, 108

Pentacene, **UV** data, 252
1,3-pentadiene (*cis, trans*), **UV** data, 250
1-Pentanol, **MS**, 7
2-Pentanol, **IR**, 93
  MS, 7
2-Pentanone, **IR**, 97
t-Pentyl Alcohol, **MS**, 7
Petroleum Ether, **IR**, 132
Phenanthrene, **UV**, 253
Phenol, **IR**, 94
  NMR, 177
  UV data, 235, 249
Phenolate Anion, **UV** data, 249
Phenyl Acetate, **IR**, 103
Phenyl Acetylene, **UV** data, 250
1-Phenyl-1,3- butadiene (*cis, trans*), **UV** data, 250
Phenyl Cyanide, **UV** data, 250
Phenyl methyl sulfone, **UV** data, 250
2-Phenylpropionaldehyde, **IR**, 100
Phthalic Acid, Diethyl Ester, **IR**, 131
Phthalic Acid, Bis [2-ethylhexyl] ester, **NMR**, 208
Polystyrene, **IR**, 84, 124
Propionaldehyde, **UV** data, 243
Propionic Anhydride, **IR**, 105
n-Propyl Bromide, **UV** data, 239
Pyrazine, **UV**, 255
Pyrazole, **UV** data, 253
Pyridazine, **UV**, 255

Pyridine-d$_5$, NMR, 209 ,
Pyridine, IR, 116
Pyridine, UV data, 254
  substituted, UV data, 254
Pyrimidine, UV, 255
Pyrrole, UV data, 253
  substituted, UV data, 253

Quinoline, UV data, 255

Silicone Lubricant, IR, 134
Silicone Oil, NMR, 202
Soap Flakes, NMR, 210
Stilbene (cis, trans), UV data, 250, 251
Stop Cock Grease (DC), 205
Styrene, UV data, 235, 250

Tetrachloroethylene, IR, 127
$\beta,\beta,'$-Thiodipropionitrile, IR, MS, NMR, UV, 298
Thioacetamide, UV data, 248
Thiobenzophenone, UV data, 248
Thiophene, UV data, 253
  substituted, UV data, 254

Thiophenol, UV data, 249
Thiourea, UV data, 248
Toluene, IR, 123
  MS, 8–9
  NMR, 209
  UV data, 235, 248
o-Toluenethiol, IR, 113
o-Tolunitrile, IR, 111
Trichloroethylene, IR, 126
Trimethylamine, UV data, 239
1,3,5-Trimethylbenzene, UV data, 248
3-(Trimethylsilyl)-propanesulfonic acid, sodium salt (DSS), NMR, 209

γ-Valerolactone, IR, MS, NMR, UV, 288
Vertex Flakes, NMR, 210

Water, UV data, 235

m-Xylene, IR, 123
  UV data, 248
o-Xylene, IR, 90